Contents

Impact of Climate Change on Medicinal and Herbal Plant microRNA

Edited by
Kanchanlata Tungare, Parul Johri,
Sachidanand Singh, and Surojeet Das

CRC Press
Taylor & Francis Group
Boca Raton London New York

CRC Press is an imprint of the
Taylor & Francis Group, an **informa** business

First edition published 2026
by CRC Press
2385 NW Executive Center Drive, Suite 320, Boca Raton FL 33431

and by CRC Press
4 Park Square, Milton Park, Abingdon, Oxon, OX14 4RN

CRC Press is an imprint of Taylor & Francis Group, LLC

ISBN: 9781032792828 (hbk)
ISBN: 9781032792842 (pbk)
ISBN: 9781003491378 (ebk)

DOI: 10.1201/9781003491378

Typeset in Times
by codeMantra

Impact of Climate Change on Medicinal and Herbal Plant microRNA

Climate change poses unprecedented challenges to plant growth, biodiversity, and productivity, necessitating innovative strategies for sustainability. *Impact of Climate Change on Medicinal and Herbal Plant microRNA* delves into the intricate relationship between climate-induced stress and the molecular mechanisms underpinning plant adaptation, with a special focus on microRNAs (miRNAs). This book provides an in-depth exploration of miRNAs as pivotal regulators in plant biology, offering insights into their biogenesis, functional roles, and applications in stress management and crop improvement.

Highlighting the interdisciplinary approach to understanding plant resilience, this book examines critical topics, including the impact of abiotic stressors like heavy metals and elevated CO_2 levels, regulatory roles of miRNAs in photosynthesis and productivity, and the integration of bioinformatics and epigenetics in miRNA research. Through comprehensive chapters, readers gain knowledge about miRNA-mediated bioengineering, genome stability, and the emerging potential of omics technologies to combat the effects of climate change on agriculture.

Key Features:

- A thorough analysis of miRNA biogenesis, regulation, and degradation, along with their myriad functional roles in plant biology.
- Exploration of abiotic stress tolerance mechanisms in medicinal, cereal, legume, tuber, fruit, biofuel, and beverage crops.
- Insights into bioinformatics tools and databases for miRNA analysis and their implications for stress tolerance studies.
- Discussions on miRNA-mediated bioengineering for climate-resilient crops and recent advances in omics approaches.

Designed for researchers, students, and professionals in plant sciences, bioinformatics, and climate studies, this book bridges fundamental and applied research, making it an essential resource for addressing climate variability through molecular innovations.

Exploring Medicinal Plants

Series Editor:
Azamal Husen, Wolaita Sodo University, Ethiopia

Medicinal plants render a rich source of bioactive compounds used in drug formulation and development; they play a key role in traditional or indigenous health systems. As the demand for herbal medicines increases worldwide, supply is declining as most of the harvest is derived from naturally growing vegetation. Considering global interests and covering several important aspects associated with medicinal plants, the Exploring Medicinal Plants series comprises volumes valuable to academia, practitioners, and researchers interested in medicinal plants. Topics provide information on a range of subjects including diversity, conservation, propagation, cultivation, physiology, molecular biology, growth response under extreme environment, handling, storage, bioactive compounds, secondary metabolites, extraction, therapeutics, mode of action, and healthcare practices.

Led by Azamal Husen, PhD, this series is directed to a broad range of researchers and professionals consisting of topical books exploring information related to medicinal plants. It includes edited volumes, references, and textbooks available for individual print and electronic purchases.

Plants as Medicine and Aromatics: Uses of Botanicals
Edited by Mohd Kafeel Ahmad Ansari, Mushtaq Ahmad, and Gary Owens

Exploring Traditional Wild Edible Plants
Edited by Vibhor Agarwal, Sachidanand Singh, and Rhaul Datta

Propagation to Pharmacopeia: Modern Approaches in Medicinal Plants
Edited by Abha Manohar K, Gopal Shukla, Saikat Gantait, A P Das, and Sumit Chakravarty

Exploring Medicinal Orchids
Edited by Azamal Husen and Manu Pant

Symbiotic Association of Microorganisms with Medicinal and Herbal Plants
Edited by Allah Ditta, Sajid Mehmood, and Azamal Husen

Tissue Culture Techniques and Medicinal Plants: Enhancing Propagation and Production
Edited by Azamal Husen and Manu Pant

Genetic Improvement and Conservation Practices of Medicinal Plants
Edited by Manu Pant, Sayed Sartaj Sohrab, and Azamal Husen

Sustainable Use of Plants and Their Products in Neurodegenerative Disease Management
Edited by Johnson Olaleye Oladele, Ebenezer I. O. Ajayi, María Liliana Flores-López, and Oluwaseyi E. Okoro

Wild Edibles of Deserts: Food Security and Sustainability
Salman Majeed, Muhammad Zafar, Mushtaq Ahmad, Mohd Kafeel Ahmad Ansari, Shazia Sultana, and Andrea Pieroni

Impact of Climate Change on Medicinal and Herbal Plant microRNA
Kanchanlata Tungare, Parul Johri, Sachidanand Singh, and Surojeet Das

Exploring Medicinal Bamboos
Edited by Manu Pant and Azamal Husen

For more information about this series, please visit: www.routledge.com/Exploring-Medicinal-Plants/book-series/CRCEMP

About the Editors

Dr. Kanchanlata Tungare is a faculty member at the School of Biotechnology and Bioinformatics, D. Y. Patil Deemed to be University, Navi Mumbai. She is a multifaceted and accomplished researcher with over 17 years of extensive experience in the field of biotechnology. Dr. Tungare has expertise in multidisciplinary research, including cancer research, molecular biology, nanotechnology, and population genetics. She is the proud recipient of the prestigious Dr. Ramrao Adik Gold Medal for academic excellence and has been awarded several intramural research grants by D. Y. Patil University. Additionally, Dr. Tungare has received a project grant from Research Society for the Study of Diabetes in India for her research on diabetes. Her research portfolio includes 65 publications in peer-reviewed journals of national and international repute. Dr. Tungare has presented her research at 30 national and international conferences and holds two patents in her name. She is currently guiding three PhD students in their research. In addition to her research, Dr. Tungare has successfully coordinated workshops and has guided 50 undergraduate and postgraduate students in completing their dissertation projects. A sought-after speaker, she has been featured in numerous webinars and serves as a peer reviewer for several prestigious scientific journals. Dr. Tungare can be contacted at dr.kanchanlata19@gmail.com.

Dr. Parul Johri is currently working as a faculty member in the Department of Biotechnology at Dr. Ambedkar Institute of Technology for Divyangjan in Kanpur, Uttar Pradesh. Her education includes a B.Tech. in Biotechnology, an M.Tech. in Bioinformatics, and a Ph.D. in Biotechnology from Amity University Uttar Pradesh. She is a recipient of a Gold Medal from Padmashree Dr. D. Y. Patil University, Navi Mumbai, Maharashtra. She has more than 15 years of teaching experience at various universities. She has published over 30 research and review papers in various journals of national and international repute. Dr. Johri has edited three books and written more than 30 book chapters with reputable publication houses like Springer, Elsevier, Taylor and Francis, Wiley, and many more. She has six Indian patents published, of which three have been granted to her credit. She has received various funds, including DBT CTEP and NCW, for conducting sensitization programs, seminars, and workshops. Her current research areas include algorithm development for protein secondary structure prediction, metagenomic analysis, transcriptome analysis, microarray data analysis, and genomic and proteomic studies.

Dr. Sachidanand Singh is currently an Associate Professor and Head of Biotechnology at the School of Energy and Technology, Pandit Deendayal Energy University, Gujarat, India, where he has academic and administrative responsibilities. He has an engineering background, holding a B.Tech. in Biotechnology, an M.Tech. in Bioinformatics, and a Ph.D. in Bioinformatics. His postdoctoral experience was gained at Ohio State University, College of Pharmacy, USA. He has more than 13 years of experience in teaching, administration, and research, with work experience that includes Karunya University, Shri Ramswaroop Memorial University, and Vignan University. His research areas include Plant Biotechnology, Metagenomics, Systems Biology, Network Biology, microRNA-Based Studies, Expression Data Analytics, Next-Generation Sequencing, and

Statistical Genetics. His current focus also includes understanding the canonical and non-canonical pathways of miRNA biogenesis, as well as the dynamics and mechanisms of miRNA-mediated gene regulation. He teaches various subjects, especially Genomics and Proteomics, Microarray and Image Analysis, Drug Design, and Molecular Modeling for graduate and postgraduate students. He has published more than 60 research and review articles in peer-reviewed journals and holds nine patents, five of which have been granted. Of his research and review publications, 40% (including book chapters, edited books, review articles, and research articles) focus on plant-based studies related to soil microbiomes, herbal drug formulation, drug design, and microbe–microbe interactions and their expression-based studies. He has completed one SERB, a Government of India–funded project, and two university research grant projects. He has received several awards for his research and teaching strategies. He has organized various conferences and workshops, receiving government funding for the same. He has shared his expertise as an invited speaker at different colleges and universities during various conferences and workshops. He has guided 36 B.Tech. and 10 M.Tech. projects, awarded three Ph.D. degrees, and currently supervises six Ph.D. students working in the fields of herbal products, drug design, and network biology integration. He has edited three books published by Springer and Taylor & Francis. He collaborates with Mendel University in the Czech Republic, Lund University in Sweden, and Monash University in Australia.

Dr. Surojeet Das is a Biotechnology researcher specializing in Regenerative Medicine, currently serving as an Assistant Professor in the Department of Biotechnology, School of Energy and Technology at Pandit Deendayal Energy University, Gandhinagar, Gujarat, India. He earned his Ph.D. from the European Molecular Biology Laboratory (EMBL) Australia at the Australian Regenerative Medicine Institute (ARMI), Monash University, Melbourne, Australia, and holds an M.Tech. in Biotechnology from the Indian Institute of Technology (IIT) Guwahati. His research primarily focuses on immune modulation strategies and biomaterial-based therapeutics for tissue regeneration and wound healing, utilizing advanced *in vivo* models, cellular assays, and molecular biology techniques. Dr. Das has authored nine peer-reviewed publications, holds two Australian patents in therapeutic biomaterials and controlled drug-delivery systems, and has received numerous prestigious national and international scholarships. With four years of prior academic experience at Shri Ramswaroop Memorial University, Barabanki, Uttar Pradesh, India, he has taught specialized courses in Nanobiotechnology, Tissue Engineering, and Biomedical Instrumentation, supervising multiple innovative research projects. His research on silk-based hydrogels earned him the Gandhian Young Technological Innovation Award, and he was recognized as a top 16 finalist in the DuPont Power of Shunya innovation program. Dr. Das actively collaborates on interdisciplinary research initiatives, translating advanced biotechnology innovations into clinically relevant therapeutic applications.

Contributors

Ankita Acharya
Department of Pharmaceutical Technology
JIS University
Kolkata, India

Namisha Arora
Department of Biological Sciences
University of Alberta, Edmonton
Alberta, Canada

Priyamvada Amol Arte
Amity Institute of Biotechnology
Amity University
Maharashtra, Mumbai, India
and
Anatek Services Pvt Ltd
Mumbai, India

Nida Azam
School of Biotechnology and Bioinformatics
DY Patil Deemed to be University
Navi Mumbai, India

Srimoyee Banerjee
Department of Microbiology
Mount Carmel College, Autonomous
Bengaluru, India

Megha Barot
Research and Development Cell
Parul University
Vadodara, India
and
Department of Environmental Science
Parul Institute of Applied Science
Parul University
Vadodara, India

Raviraj Barot
Department of Biotechnology
School of Energy and Technology
Pandit Deendayal Energy University
Gandhinagar, Gujarat, India

Riya Bhiwandkar
School of Biotechnology and Bioinformatics
DY Patil Deemed to be University
Navi Mumbai, India

Mustansir Bhori
Inveniolife Technology Pvt Ltd
Navi Mumbai, India

Harsh Bokadia
Department of Biological Sciences
Sunandan Divatia School of Science
NMIMS Deemed to Be University
Mumbai, India

Prerona Boruah
School of Biotechnology and Bioinformatics
DY Patil Deemed to be University
Navi Mumbai, Maharashtra, India
and
Shanghai Veterinary Research Institute
Shanghai, China

Arundhuti Chakrabarty
Department of Genetics and Plant Breeding
Central Agricultural University
Imphal, India

Rituparna Chatterjee
Department of Pharmaceutical Technology
JIS University
Kolkata, India

Moulima Das
Department of Pharmaceutical Technology
JIS University
Kolkata, India

Reshmi Das
School of Biotechnology and Bioinformatics
DY Patil Deemed to be University
Navi Mumbai, India

Surojeet Das
Department of Biotechnology
School of Energy and Technology
Pandit Deendayal Energy University
Gandhinagar, Gujarat, India

Shine Devarajan
School of Biotechnology and Bioinformatics
DY Patil Deemed to be University
Navi Mumbai, India

Deepa Dhatwalia
Department of Biotechnology
Punjab University
Chandigarh, India

Samiksha Garse
School of Biotechnology and Bioinformatics
DY Patil Deemed to be University
Navi Mumbai, India

Soham Sunil Gumaste
School of Biotechnology and Bioinformatics
DY Patil Deemed to be University
Navi Mumbai, India

Pallavi Gurav
School of Biotechnology and Bioinformatics
DY Patil Deemed to be University
Navi Mumbai, India

Pratima Gurung
Division of Infectious Diseases
Boston Children's Hospital
Boston, Massachusetts

Anil A Hake
Department of Biotechnology
ICAR-Indian Institute of Rice Research
Hyderabad, India

Heeram Hashmi
Department of Biotechnology
Dr. Ambedkar Institute of Technology for Divyangjan
Kanpur, Uttar Pradesh, India

Samah Isaani
Department of Biological Sciences
Sunandan Divatia School of Science
NMIMS Deemed to Be University
Mumbai, India

Shaikhul Islam
Plant Pathology Division
Bangladesh Wheat and Maize Research Institute
Dinajpur, Bangladesh

Dilfuza Jabborova
Institute of Genetics and Plant Experimental Biology
Kibray, Uzbekistan

Pravin V Jadhav
Biotechnology Centre
Department of Agricultural Botany
Post Graduate Institute, Dr. Panjabrao Deshmukh Krishi Vidyapeeth [State Agriculture University]
Akola, India

Srishti Jagiasi
School of Biotechnology and Bioinformatics
DY Patil Deemed to be University
Navi Mumbai, India

Tanvi Jain
Department of Biotechnology
Shri Ramswaroop Memorial University
Barabanki, India

Pamela Jha
Department of Biological Sciences
Sunandan Divatia School of Science
NMIMS Deemed to Be University
Mumbai, India

Renitta Jobby
Amity Institute of Biotechnology
Amity University Maharashtra - Pune Expressway
Mumbai, Maharashtra 410206, India
and
Amity Centre of Excellence in Astrobiology
Amity University, Maharashtra-Pune Expressway
Mumbai, India

Parul Johri
Department of Biotechnology
Dr. Ambedkar Institute of Technology for Divyangjan
Kanpur, Uttar Pradesh, India

Aayushi Kadam
Department of Food and Human Nutritional Sciences
University of Manitoba
Winnipeg, Manitoba, Canada
and
Anatek Services Pvt Ltd
Mumbai, India

Noopur Khare
Department of Biotechnology
Shri Ramswaroop Memorial University
Barabanki, India
and
Department of Food Technology
Bhai Gurdas Institute of Engineering and Technology
Sangrur, India

Pragati Khare
Department of Pharmacy
BPS educational Institution of Pharmacy
Etah, U.P., India

Murali Krishna Kormutla
Lethbridge Research and Development Centre
Agriculture and Agri-Food Canada
Lethbridge, Canada

Priyashree Laishram
Department of Genetics and Plant Breeding
Central Agricultural University
Imphal, India

Muhammad Latif
Centre for Genetics and Inherited Diseases (CGID)
Taibah University
Madinah 42353, Kingdom of Saudi Arabia
and
Department of Basic Medical Sciences
College of Medicine, Taibah University
Madinah 42353, Kingdom of Saudi Arabia

R Shubhra Maithreyi
International Centre for Genetic Engineering and Biotechnology (ICGEB)
Aruna Asaf Ali Marg
New Delhi, India

Stuti Maurya
College of Medical Veterinary and Life Sciences
University of Glasgow
Glasgow, Scotland, UK

Pranavi Mishra
School of Biotechnology and Bioinformatics
DY Patil Deemed to be University
Navi Mumbai, Maharashtra, India

Arpan Modi
Axenic Agritech Private Limited
Ahmedabad, Gujarat, India

Gopika Mote
Biotechnology Centre
Department of Agricultural Botany
Post Graduate Institute
Dr. Panjabrao Deshmukh Krishi Vidyapeeth
[State Agriculture University]
Akola, India

Rishabh Mukherjee
Department of Science
College of Science
Northeastern University
Boston, Massachusetts

Sayali Mukherjee
Amity Institute of Biotechnology
Amity University Uttar Pradesh
Lucknow, India

Manisha Negi
Department of Plant Science
University of Manitoba (Fort Garry campus)
Winnipeg, Manitoba, Canada

Navami Pai
School of Biotechnology and Bioinformatics
DY Patil Deemed to be University
Navi Mumbai, India

Ramesh Palakurthi
Department of Plant and Soil Sciences
University of Kentucky
Lexington, Kentucky

Mala Parab
School of Biotechnology and Bioinformatics
DY Patil Deemed to be University
Navi Mumbai, India

Vidadala Rajendra
Department of Genetics and Plant Breeding
M.S. Swaminathan School of Agriculture
Centurion University
Bhubaneswar, India

Manish Singh Rajput
Department of Biotechnology
Dr. Ambedkar Institute of Technology for Divyangjan
Kanpur, Uttar Pradesh, India

Mehidi Rehimi
Department of Biotechnology
Institute of Science and High Technology and Environmental Sciences
Graduate University of Advanced Technology
Kerman, Iran

Mayuresh M Sahasrabuddhe
School of Biotechnology and Bioinformatics
DY Patil Deemed to be University
Navi Mumbai, India

Manisha Saini
Division of Seed Science and Technology
ICAR-Indian Agricultural Research Institute
New Delhi, India

Samina Shabbir
Department of Chemistry
The Women University Multan
Multan, Pakistan
and
Plant Protection Research Institute
Guangdong Academy of Agricultural Sciences
Guangzhou, China

Mukesh Sharma
Department of Genetics and Plant Breeding
Central Agricultural University
Imphal, Manipur, India

Pooja Sharma
Department of Bio-Sciences and Technology
Maharishi Markandeshwar Deemed To Be University
Ambala, India

Jiban Shrestha
Scientist, Nepal Agricultural Research Council
National Plant Breeding and Genetics Research Centre
Lalitpur, Nepal

Sachidanand Singh
Department of Biotechnology
School of Energy and Technology
Pandit Deendayal Energy University
Gandhinagar, Gujarat, India

Shweta Singh
School of Biotechnology and Bioinformatics
DY Patil Deemed to be University
Navi Mumbai, India

Pratyaksha Srivastava
Department of Biotechnology
Dr. Ambedkar Institute of Technology for Divyangjan, Kanpur
Uttar Pradesh, India

Pranita P Thakur
School of Biotechnology and Bioinformatics
DY Patil Deemed to be University
Navi Mumbai, India

Mala Trivedi
Amity Institute of Biotechnology
Amity University Uttar Pradesh
Lucknow, India

Kanchanlata Tungare
School of Biotechnology and Bioinformatics
DY Patil Deemed to Be University
Navi Mumbai, India

Khadija Turabi
Eppley Institute for Research in Cancer and Allied Diseases
Fred & Pamela Buffet Cancer Center
University of Nebraska Medical Center
Omaha, Nebraska

Sekar Vijayakumar
College of Material Science and Engineering, Huaqiao University
Engineering Research Center of Environment-Friendly Functional Materials, Ministry of Education
Xiamen, PR China

Preface

The impact of climate change on agriculture and biodiversity is an enduring challenge, bringing significant shifts in the growth, adaptation, and productivity of medicinal and herbal plants. One of the most promising areas of study in addressing these challenges lies in the intricate world of microRNAs (miRNAs). These small, non-coding RNA molecules play pivotal roles in regulating plant responses to various abiotic stresses, including drought, salinity, temperature extremes, and heavy metal toxicity. Their regulatory roles offer a glimpse into the potential for bioengineering climate-resilient crops and safeguarding global food and medicinal plant systems.

This book, *Impact of Climate Change on Medicinal and Herbal Plant microRNA*, is an effort to bridge knowledge gaps and provide a comprehensive understanding of miRNAs in the context of climate-induced abiotic stresses. As editors, we have curated diverse, cutting-edge research and insights to unravel the multifaceted roles of miRNAs in plant stress biology, adaptation, and resilience.

The chapters are structured to provide a holistic overview of the subject. The journey begins with an introduction to the broad effects of climate change–induced abiotic stress on plant growth, biodiversity, and adaptation strategies. This foundational discussion sets the stage for understanding the physiological and molecular repercussions of environmental changes on medicinal plants and their active components.

We then extend the intricacies of miRNA biogenesis, regulation, and degradation. A chapter exploring the myriad functions of miRNAs in plant biology highlights the dynamic roles they play across various biological processes, offering a deep dive into the regulatory networks influenced by these small molecules.

This book also emphasizes the significance of miRNAs in stress tolerance, particularly in crops of agronomic importance. Dedicated chapters focus on miRNAs' functional roles in cereals, legumes, tuber, fruit, biofuel, and beverage crops. These chapters illustrate how miRNAs orchestrate stress responses and contribute to the development of climate-resilient crop varieties.

Technological advances have revolutionized our understanding of miRNAs, and this book captures the latest bioinformatics approaches, databases, and omics tools for miRNA analysis. These resources are invaluable for deciphering miRNA-mediated responses and leveraging this knowledge for agricultural innovation.

Climate change not only demands a molecular understanding of stress responses but also calls for translational approaches to enhance crop resilience. Chapters on epigenetics, genome stability, and bioengineering shed light on the potential of miRNAs as tools for genetic and epigenetic modifications, opening new avenues for sustainable agriculture.

Finally, we explore recent advances in omics technologies and their integration into studying plant resilience, providing a futuristic perspective on how these tools can help tackle climate variability and its impacts on global agriculture.

We envision this book as a valuable resource for researchers, academicians, and policymakers striving to understand and mitigate the adverse effects of climate change on plant systems. It reveals the promise of miRNAs in developing strategies for sustainable agriculture and the preservation of medicinal plant biodiversity.

We extend our gratitude to the contributing authors for their scholarly inputs and dedication to this emerging field. We also thank the publishing team for their support in bringing this book to fruition. Together, we hope to inspire new research and innovative solutions in this vital domain.

The Editors

1 Introduction to Impact of Climate Change on Plant microRNA

Heeram Hashmi, Parul Johri, Manish Singh Rajput, Pratyaksha Srivastava, Mala Trivedi, Sachidanand Singh, Surojeet Das, and Kanchanlata Tungare*

1.1 IMPACT OF CLIMATE CHANGE ON PLANT LIFE

Climate change can harm plant growth and survival by affecting temperature, precipitation patterns, and seasonal cycles (Abbass et al., 2022). This can impact photosynthesis, reproduction, and habitat suitability. The effects of climate change on plants are profound and multifaceted (van der Wiel & Bintanja, 2021). Changes in precipitation patterns can result in droughts or excessive rainfall, affecting water availability and soil conditions critical for plant growth (Sharma et al., 2022).

Shifts in the timing of seasons, such as earlier springs or altered frost dates, can also influence plant life cycles. For instance, plants may experience mismatches between their blooming periods and the availability of pollinators, disrupting reproduction and affecting plant populations. Additionally, climate change can result in shifts in plant ranges, with some species moving toward higher elevations or latitudes while others may face increased competition or habitat loss.

Climate change significantly impacts plants by altering their growth, distribution, and overall health through several mechanisms. Rising temperatures can induce heat stress, reducing photosynthesis, biomass, and crop yields, while extreme heat events can damage plant tissues. Shifts in precipitation patterns can lead to droughts or excessive rainfall, stressing plants, impairing root function, and increasing disease susceptibility (Anil et al., 2023; Jabborova et al., 2023; Verma et al., 2021). Changes in seasonal timing can disrupt plant phenology, affecting blooming and reproduction, and altering interactions with pollinators. Elevated atmospheric CO_2 can enhance photosynthesis for some species but may reduce nutrient content and affect pest and disease dynamics. Many plants may shift their geographic ranges in response to changing conditions, potentially leading to habitat loss and ecosystem disruptions (NOAA, 2018). Increased temperatures and altered precipitation can also heighten pest and disease pressures. Addressing these challenges requires adaptive strategies such as developing resilient plant varieties, improving water management, and reducing greenhouse gas emissions through sustainable practices to preserve plant health and ecosystem stability (Neufeldt et al., 2011).

To address these challenges, a multifaceted approach is required. Reducing greenhouse gas emissions is crucial; this can be achieved by transitioning to renewable energy sources, improving energy efficiency, and adopting sustainable transportation practices. Reforestation and afforestation efforts can help absorb CO_2 from the atmosphere, while sustainable land management practices can mitigate the impacts of deforestation and soil degradation. Additionally, supporting research and implementation of climate-resilient agricultural practices can help farmers adapt to changing conditions. Public awareness and policy measures also play a vital role in addressing climate change. By advocating for and supporting policies that promote environmental protection and climate action, individuals and communities can contribute to a collective effort to mitigate climate impacts.

* *Corresponding Author*

DOI: 10.1201/9781003491378-1

In summary, addressing climate change and its effects on plants requires concerted efforts across various sectors to reduce emissions, protect natural ecosystems, and support adaptive strategies for a more resilient environment (Little, 1933).

1.2 EFFECT OF HEAVY METALS ON PLANTS: IMPLICATIONS AND FUNCTIONS

Heavy metals are naturally occurring elements with high atomic weights and densities that can have detrimental effects on living organisms. When present in excessive amounts in the environment, these metals pose significant risks to plant health, ecosystems, and agricultural productivity. Understanding the implications and functions of heavy metals in plants is crucial for developing strategies to mitigate their adverse effects and enhance plant resilience.

Heavy metals, such as lead (Pb), cadmium (Cd), mercury (Hg), and arsenic (As), can originate from various sources, including industrial discharges, agricultural runoff, mining activities, and atmospheric deposition. These metals can enter soil and water systems, eventually making their way into plant tissues (Ghani, 2010).

Heavy metals adversely affect the physiological functions of plants, resulting in poor growth and reduced crop productivity. They can accumulate in edible plant parts, posing health risks to humans and animals that consume contaminated produce. This is a significant concern for food safety and public health (Yu et al., 2019). The effects of heavy metals on plants are multifaceted, impacting growth, nutrient uptake, and overall health. Understanding these effects and the mechanisms by which plants respond to metal stress is vital for developing effective management strategies. By addressing heavy metal contamination through soil remediation, strategic plant selection, and pollution prevention, we can safeguard agricultural productivity, ecosystem health, and human well-being (Khathi et al., 2016).

Heavy metals present a grave danger to both the environment and human health, primarily due to their persistent nature and bio-accumulative effects. Once these metals enter soil and water systems, they can remain for extended periods without breaking down, leading to long-term contamination. A particularly alarming aspect is the tendency of heavy metals to accumulate in the food chain. Plants that absorb these metals can transfer them to herbivores, which in turn pass them up the food chain to predators, including humans. This biomagnification effect means that even low levels of heavy metal pollution can lead to significant concentrations in higher trophic levels, potentially causing severe health issues such as neurological disorders, cancer, and kidney damage. Additionally, heavy metals can disrupt soil microbial communities, which are crucial for nutrient cycling and soil health, thereby exacerbating ecological imbalances and further compromising agricultural productivity (Das & Jayalekshmy, 2015). The complexity and persistence of heavy metal contamination make it a pressing environmental challenge, necessitating urgent and effective intervention strategies to mitigate its impacts. Heavy metal contamination poses severe and multifaceted dangers to the environment and human health. One of the most alarming issues is the long-term soil degradation it causes, leading to reduced fertility and impaired soil structure (Gupta et al., 2019). Metals like cadmium and lead disrupt essential soil microbial activity, which is crucial for nutrient cycling and organic matter decomposition, ultimately diminishing agricultural productivity and ecosystem health. Furthermore, heavy metals can leach into groundwater or surface waters, contaminating drinking supplies and aquatic ecosystems. This contamination presents significant health risks, including gastrointestinal issues, liver and kidney damage, and developmental disorders. Plants that absorb these metals can transfer them up the food chain, posing risks to humans who consume contaminated produce. In addition to direct health impacts, heavy metal contamination can impair plant reproduction, reducing seed germination rates and fruit yields, which disrupts plant populations and biodiversity (Ezenwobodo & Samuel, 2022). While some plants develop resistance mechanisms to cope with heavy metal stress, these adaptations can sometimes lead to further complications, such as the spread of metal-tolerant species and reduced plant growth. The economic repercussions are also substantial, encompassing the costs of soil and water remediation, health care expenses, and diminished agricultural output. Heavy metals disrupt entire ecosystems by altering soil health and plant communities, affecting predator–prey relationships and ecosystem services. Remediation efforts are fraught with challenges, as conventional methods can be costly and environmentally disruptive, while newer approaches like phytoremediation require careful management to avoid unintended consequences. Overall, the persistence and widespread impact of heavy metal contamination necessitate comprehensive strategies to mitigate its effects and protect both environmental and human health (Rodríquez-Serrano et al., 2008).

1.3 TYPES OF STRESS IN PLANTS: CAUSES, EFFECTS, AND RESPONSES

Plants are exposed to various types of stress that can significantly impact their growth, development, and overall health. These stresses are broadly categorized into abiotic and biotic types. Abiotic stresses include environmental factors such as water scarcity or excess, extreme temperatures, and soil conditions like salinity and nutrient imbalances. Water stress, including both drought and flooding, can severely affect plant physiology by disrupting nutrient uptake and causing wilting or root damage (Abdussalam et al., 2015). Temperature extremes, whether excessive heat or frost, can damage cellular structures and impair metabolic processes. Soil conditions, including salinity and pH imbalances, can lead to reduced nutrient availability and toxicity, further stressing plants. Additionally, exposure to heavy metals from soil contamination can inhibit plant growth and accumulate in tissues, posing risks to both plant and human health (Gull et al., 2019).

Biotic stresses arise from interactions with other living organisms, including pests, diseases, and competition with weeds. Insects and nematodes can damage plant tissues directly or transmit diseases, weakening the plant and reducing its yield. Fungal, bacterial, and viral pathogens can cause a range of symptoms, from leaf spots to root decay, severely affecting plant productivity and health. Weeds compete with crops for resources such as light, water, and nutrients, worsening the effects of other stresses. To cope with these diverse stressors, plants have evolved a variety of adaptive responses, including physiological adjustments like altering water use efficiency and synthesizing protective compounds. Understanding these stress responses is crucial for developing effective strategies to enhance plant resilience and ensure sustainable agricultural practices. Plants face a complex array of stressors that can severely impact their survival, growth, and productivity. These stresses are typically classified into two main categories: abiotic and biotic. Abiotic stressors stem from non-living environmental factors (Cohen, 1988). Water stress, which can manifest as drought or waterlogging, disrupts essential processes like nutrient uptake and photosynthesis. Drought causes plants to close their stomata to reduce water loss, but this also limits carbon dioxide intake and impairs growth. Conversely, waterlogging suffocates roots and inhibits nutrient absorption. Temperature extremes also pose significant challenges; excessive heat can denature proteins and enzymes, while frost can cause ice crystal formation within plant tissues, leading to cellular damage. Additionally, soil conditions such as salinity and pH imbalances can hinder plant growth by affecting root function and nutrient availability. Heavy metal contamination further exacerbates these issues, as metals like cadmium and lead can accumulate in plant tissues, leading to toxicity and reduced plant health (Radhika et al., 2015).

Biotic stresses, on the other hand, arise from interactions with living organisms, including pests, pathogens, and competing plants. Insect pests, such as aphids and caterpillars, can directly damage plant tissues, while nematodes can disrupt root function and facilitate disease entry. Pathogens, including fungi, bacteria, and viruses, can cause a range of symptoms from leaf spots and blights to systemic infections that compromise plant vitality. Weeds, competing for water, light, and nutrients, can further stress crops, reducing their growth and yield. In response to these biotic stresses, plants have developed various defense mechanisms, such as producing secondary metabolites that deter herbivores and employing hypersensitive responses to limit pathogen spread. Additionally, plants can activate systemic acquired resistance and induce localized defense responses to manage and mitigate damage. Understanding these diverse stress responses and mechanisms is crucial for developing effective agricultural strategies, such as breeding stress-resistant varieties and implementing integrated pest management practices, to enhance crop resilience and ensure sustainable food production (Chahal et al., 2022).

The myriad stresses that plants encounter—whether abiotic, such as water scarcity, extreme temperatures, and soil imbalances, or biotic, such as pests, diseases, and competition—pose significant challenges to their health and productivity. The ability of plants to cope with these stresses through various physiological, biochemical, and morphological adaptations is crucial for their survival and growth. Understanding these stress responses not only enhances our knowledge of plant biology but also informs practical approaches to improving crop resilience. By integrating insights from stress physiology with innovative agricultural practices, such as developing stress-resistant plant varieties and employing effective pest management strategies, we can better safeguard agricultural systems and ensure food security in the face of a changing environment. Addressing plant stress comprehensively is essential for fostering sustainable agriculture and preserving ecosystem stability in an increasingly unpredictable world.

1.4 miRNAs' ROLE IN REGULATING PLANT GROWTH AND STRESS RESPONSES

microRNAs (miRNAs) are small, non-coding RNA molecules that play a crucial role in regulating gene expression in plants (Noopur et al., 2025; Pooja et al., 2025; Rahul et al., 2024; Subhan et al., 2025; Tanvi et al., 2025). These molecules, typically 21–24 nucleotides in length, fine-tune various physiological processes by targeting specific messenger RNAs for degradation or translational repression (Meenakshi et al., 2025). In plants, miRNAs are integral to maintaining cellular homeostasis and coordinating developmental and stress responses. They function through post-transcriptional regulation, influencing a wide array of genes involved in growth, development, and environmental adaptation.

During plant growth, miRNAs regulate key developmental processes such as cell differentiation, organ formation, and tissue development. For example, the miR156 family controls the transition from juvenile to adult phases by modulating the expression of genes involved in developmental timing (Yang et al., 2021). Similarly, miR172 plays a role in flower development by regulating the timing of floral organ differentiation. By controlling these critical stages, miRNAs ensure that plants develop correctly and efficiently, adapting their growth patterns to the surrounding environment.

In response to abiotic stresses like drought, salinity, and extreme temperatures, miRNAs help plants adapt by modulating stress-responsive pathways. For instance, miR398 targets copper/zinc superoxide dismutase genes, which are crucial for managing oxidative stress caused by environmental stressors (Bajczyk et al., 2023). By regulating these stress-responsive genes, miRNAs enable plants to adjust their physiological processes, such as stomatal closure and osmotic adjustment, thereby improving their tolerance to adverse conditions. Additionally, miRNAs involved in stress responses often interact with other signaling pathways, such as those mediated by plant hormones like abscisic acid, to enhance stress resilience.

Biotic stress responses, including those triggered by pathogen infections and pest infestations, are also regulated by miRNAs. For example, miR393 modulates the expression of genes involved in defense signaling pathways, such as those related to the jasmonic acid and salicylic acid pathways. These miRNAs help orchestrate the plant's immune response by balancing defense and growth, ensuring that resources are allocated efficiently between combating pathogens and maintaining overall health. This regulatory balance is crucial for sustaining plant growth while managing biotic threats (Abbass et al., 2022).

The intricate roles of miRNAs in regulating plant growth and stress responses highlight their potential as targets for biotechnological interventions. By manipulating miRNA pathways, researchers can enhance crop resilience to environmental stressors and improve yield and quality. Genetic engineering approaches, such as the introduction of miRNA mimics or inhibitors, offer promising avenues for developing stress-resistant crop varieties. As our understanding of miRNA functions deepens, the potential for leveraging these small RNAs to advance agricultural practices and ensure food security amid a changing climate becomes increasingly significant.

1.5 FUNCTIONALITY OF miRNAs IN DIVERSE CONDITIONS

Biotic stresses, including pathogen infections and pest infestations, heavily rely on miRNA-mediated regulation. For instance, miR393 plays a crucial role in the plant immune response by targeting genes involved in the jasmonic acid signaling pathway, which is critical for defending against pests and pathogens. Additionally, miR159 regulates genes related to pathogen resistance and plant defense mechanisms, ensuring that plants can mount an effective response to biotic threats. The ability of miRNAs to fine-tune these defense responses helps plants balance their growth with the need for protection against harmful biotic interactions (van der Wiel & Bintanja, 2021).

As the challenges facing agriculture evolve, the demand for harnessing miRNA technology is rapidly increasing. Advances in genetic engineering and synthetic biology are making it possible to manipulate miRNA pathways to enhance crop resilience and productivity. For example, developing genetically modified crops with overexpressed or silenced miRNAs can lead to improved stress tolerance and better yield in challenging environmental conditions. This approach holds promise for addressing issues such as climate change, resource scarcity, and the need for sustainable agricultural practices. Looking ahead, the future of miRNA research is poised to play a transformative role in agriculture and plant science. With ongoing advancements in high-throughput sequencing (HTS) and functional genomics, researchers are uncovering new miRNAs and their specific roles in diverse environmental contexts. These discoveries are paving the way for innovative applications in crop improvement, including the development of stress-resistant varieties and enhanced management strategies. As the understanding of miRNA functions expands, their integration into agricultural practices will be crucial for advancing food security and adapting to an increasingly unpredictable climate.

As we look to the future, several challenges arise in harnessing the full potential of miRNAs for plant improvement and agriculture. One major challenge is the complex nature of miRNA interactions within the plant's regulatory networks. miRNAs do not act in isolation; they are part of intricate networks involving multiple miRNAs, target genes, and signaling pathways. This complexity makes it difficult to predict the outcomes of manipulating specific miRNAs, as changes in one part of the network can have cascading effects on other processes. Understanding these interactions in detail is crucial for designing effective genetic modifications and ensuring that interventions yield the desired effects without unintended consequences.

Another significant challenge is the accurate and efficient delivery of miRNA-based technologies to target plants. While advancements in genetic engineering techniques have made it possible to introduce or alter miRNA expression in plants, applying these methods in diverse agricultural settings remains complex. Ensuring that miRNA modifications are stable, heritable, and expressed at appropriate levels in the field is critical for successful outcomes. Additionally, developing delivery systems that can efficiently target specific plant tissues or organs without causing off-target effects is essential for maximizing the benefits of miRNA-based interventions. The potential for off-target effects and unintended interactions poses another challenge. miRNAs can have multiple target genes, and altering their levels can inadvertently affect genes unrelated to the intended target. This raises concerns about the safety and environmental impact of miRNA-based modifications, particularly when introducing genetically engineered crops into natural ecosystems. Rigorous testing and comprehensive risk assessments are necessary to evaluate the long-term effects of miRNA modifications on plant health, biodiversity, and ecosystem stability (Yang et al., 2021). Additionally, there is a need for robust methods to monitor and evaluate the effectiveness of miRNA-based interventions under real-world conditions. Laboratory and controlled-environment experiments provide valuable insights, but the performance of miRNA-modified plants in diverse and dynamic field conditions can differ significantly. Developing reliable field-testing protocols and performance metrics will be crucial for validating the practical utility of miRNA technologies and ensuring they deliver consistent benefits across various environments. Finally, addressing the regulatory and societal acceptance of miRNA-based technologies presents an ongoing challenge. The introduction of genetically modified organisms, including those involving miRNA modifications, often faces regulatory scrutiny. Engaging with stakeholders, including farmers, consumers, and policymakers, to communicate the benefits and safety of miRNA-based innovations is essential for gaining acceptance and facilitating the adoption of these technologies. Transparent research practices, clear risk communication, and demonstrating tangible benefits will be key to overcoming these hurdles and advancing the use of miRNAs in agriculture.

1.6 LEVERAGING BIOINFORMATICS FOR ANALYZING PLANT miRNAs

miRNAs regulate gene expression in plants, affecting key physiological processes like development, stress responses, and disease resistance. Bioinformatics has become essential for analyzing plant miRNAs, starting with HTS technologies that provide extensive small RNA profiles from plant tissues. Computational tools process this raw sequencing data to identify both known and novel miRNAs (Amrhein et al., 2013). Techniques include sequence alignment with databases such as BLAST and Bowtie, as well as de novo prediction using algorithms like miRDeep and RNAfold to detect characteristic hairpin structures. Once miRNAs are identified, they are annotated based on their genomic positions and functions using resources like the Plant miRNA Database.

To understand miRNA functions, bioinformatics tools predict their target genes with software like TargetFinder and psRNATarget, and further analyze these targets using Gene Ontology and KEGG pathway analysis to determine their roles in various biological processes. Expression profiling, using tools such as DESeq2 and EdgeR, identifies miRNAs that are upregulated or downregulated under different conditions. Experimental validation techniques like qRT-PCR and northern blotting confirm the accuracy of these bioinformatics predictions. Comparative genomics tools, such as ClustalW and MAFFT, study miRNA conservation and evolution across species, providing insights into their evolutionary history (Johri et al., 2022). Systems biology integrates miRNA data with transcriptomic, proteomic, and metabolomic data to build comprehensive regulatory networks and understand complex biological systems. Despite these advances, challenges like managing large-scale data and validating predictions remain. Future efforts in bioinformatics will aim to refine prediction algorithms, improve functional annotation tools, and integrate multi-omics data to enhance our understanding of miRNA functions and advance crop management and stress resilience (Bajczyk et al., 2023). Bioinformatics software is crucial for the analysis and understanding of plant miRNAs, playing an indispensable role at various stages of miRNA research. The process starts with sequence alignment tools such as BLAST (Basic Local Alignment Search Tool) and Bowtie. BLAST is used to align small RNA sequences with known miRNA databases, aiding in the identification of conserved miRNAs by detecting local similarities between the input sequences and those in the database (Verma et al., 2023).

In contrast, Bowtie is a high-speed aligner that maps small RNA sequences to reference genomes or transcriptomes, efficiently managing large datasets and providing the precise alignments necessary for detecting both known and novel miRNAs. miRDeep identifies novel miRNAs by analyzing HTS data to detect characteristic hairpin structures in precursor sequences (Amrhein et al., 2013). It refines its predictions using probabilistic models and compares these structures to established miRNA motifs. RNAfold contributes by predicting the secondary structures of RNA sequences, including miRNA precursors, using thermodynamic models, which aids in identifying potential miRNA candidates based on hairpin formations (Ogul & Akkaya, 2011).

For analyzing miRNA expression, tools like **DESeq2** and **EdgeR** are employed. DESeq2 performs differential expression analysis by normalizing read counts from sequencing data and applying statistical models to identify miRNAs with significant upregulation or downregulation under different conditions (Kurniawati, 2017). EdgeR also facilitates differential expression analysis, specifically suited for small RNA sequencing data, utilizing empirical Bayes methods to estimate variances and detect significant expression changes. These tools highlight how miRNAs impact specific pathways and cellular processes.

Collectively, these bioinformatics tools, with their specialized mechanisms, enable a comprehensive analysis of miRNAs, from their identification and characterization to their functional and evolutionary insights, thereby advancing our understanding of plant biology and enhancing agricultural practices.

1.7 HARNESSING miRNAs: INNOVATIONS IN HUMAN AND PLANT BIOTECHNOLOGY

miRNAs are proving to be transformative in both human medicine and plant biotechnology, offering groundbreaking solutions for disease management and agricultural advancement. In human medicine, miRNAs are increasingly valued for their potential as disease biomarkers. They are being utilized to create diagnostic tools that enable the early detection of conditions such as cancer, heart disease, and neurodegenerative disorders. By analyzing specific miRNA expression patterns, researchers can identify disease-related signatures, leading to more precise and timely diagnoses (Allmer, 2014). On the therapeutic front, miRNAs are opening new doors for treating various human health issues. Innovative treatments are emerging that use miRNA mimics or inhibitors to correct abnormal miRNA levels associated with diseases (Trivedi et al., 2017). For instance, therapies targeting cancer might involve restoring normal miRNA function or inhibiting harmful miRNAs to slow or stop tumor growth. This approach enables more personalized medicine, with treatments tailored to individual genetic profiles and specific disease mechanisms (Gangwar et al., 2023).

In plant biotechnology, miRNAs are crucial for managing plant growth, development, and responses to environmental stress. Researchers use miRNAs to enhance crop resilience against factors such as drought, salinity, and pests (Ogul & Akkaya, 2011). By modifying miRNA pathways, scientists can produce crops that better withstand adverse conditions, potentially leading to increased yields and more reliable food sources. This application is particularly important in the context of climate change and the need for greater food security (F. et al., 2011).

Additionally, miRNAs are used to improve the nutritional quality of plants. Through genetic modification and breeding techniques, miRNA-based methods can enhance the concentration of essential nutrients in crops. For example, adjusting specific miRNA levels can increase the vitamin content and other beneficial compounds in fruits and vegetables. This not only benefits human health but also helps address global nutritional deficiencies.

Overall, the use of miRNAs represents a significant advancement in both medicine and agriculture. In medicine, miRNAs offer innovative diagnostic and therapeutic options that could revolutionize disease treatment (Liu et al., 2014). In agriculture, miRNAs enable the creation of more resilient and nutritionally enriched crops, addressing important issues in food security and sustainability. Ongoing research and application of miRNAs promise substantial improvements across these diverse areas.

1.8 CONCLUSION

In summary, the use of miRNAs is proving to be revolutionary in both human medicine and plant biotechnology, demonstrating their substantial impact across various domains (Radhika et al., 2015). In human medicine, miRNAs are opening new avenues for early diagnosis and targeted treatment, allowing for more personalized and effective healthcare solutions. They act as crucial biomarkers for precise disease detection, while innovative therapies involving miRNA mimics and inhibitors offer new ways to tackle complex medical conditions at the genetic level. In the realm of plant biotechnology, miRNAs are transforming agricultural practices by improving crop resilience and enhancing nutritional quality. Their ability to influence key processes in plant growth and stress response helps develop crops that can better withstand environmental challenges and meet the increasing global food demand. Additionally, optimizing miRNA pathways to elevate the nutritional content of crops contributes to better global health by enriching the food supply with essential nutrients.

Overall, these advancements highlight the significant potential of miRNAs to drive progress in both medical and agricultural fields (Kuang et al., 2023). Continued research and development in miRNA technology are expected to further enhance our capabilities in disease management and crop improvement. As these innovations evolve, miRNAs are set to play a pivotal role in addressing critical health and food security issues, promising substantial benefits for the future.

REFERENCES

Abbass, K., Qasim, M. Z., Song, H., Murshed, M., Mahmood, H., & Younis, I. (2022). A review of the global climate change impacts, adaptation, and sustainable mitigation measures. *Environmental Science and Pollution Research*, *29*(28), 42539–42559. https://doi.org/10.1007/s11356-022-19718-6

Abdussalam, A. K., Ravindran, C. P., Ratheesh Chandra, P., Azeez, K., & Nabeesa, S. (2015). Physiological effects of heavy metal toxicity and associated histological changes in *Boerhavia diffusa* L. *Journal of Global Biosciences*, *4*(1), 1221–1234.

Allmer, J. (2014). Machine learning methods for microRNA gene prediction. *miRNomics: microRNA Biology and Computational Analysis*, 1107, 3333. https://doi.org/10.1007/978-1-62703-748-8

Amrhein, N., Apel, K., Baginsky, S., Buchmann, N., Geisler, M., Keller, F., Körner, C., Martinoia, E., Merbold, L., Müller, C., Schmid, B., & Paschke, M. (2013). *Plant Response to Stress*. Zurich-Basel Plant Science Center, Zürich, Switzerland. https://doi.org/10.3929/ethz-a-009779047

Anil, P., Dharmendra, P., Sachidanand, S., Yuriy, E., Snezhan, B., Dilfuza, J., & Chinmayi, J. (2023). Beneficial effects of biochar application on mitigating the drought and salinity stress implications on plants. *Plant Science Today*, *10*(2), 188–193. http://doi.org/10.14719/pst.2591

Bajczyk, M., Jarmolowski, A., Jozwiak, M., Pacak, A., Pietrykowska, H., Sierocka, I., Swida-Barteczka, A., Szewc, L., & Szweykowska-Kulinska, Z. (2023). Recent insights into plant miRNA biogenesis: Multiple layers of miRNA level regulation. *Plants*, *12*(2), 342. https://doi.org/10.3390/plants12020342

Chahal, G. K., Ghai, M., & Kaur, J. (2022). Stress mechanisms and adaptations in plants. *Journal of Pharmacognosy and Phytochemistry*, *11*(2), 109–114. https://doi.org/10.22271/phyto.2022.v11.i2b.14365

Ciarmiello, L. F., Woodrow, P., Fuggi, A., Pontecorvo, G., & Carillo, P. (2011). Plant genes for abiotic stress. In Shanker, A. (Ed), *Abiotic Stress in Plants - Mechanisms and Adaptations*. IntechOpen, London. https://doi.org/10.5772/22465

Cohen, R. L. (1988). Environmental stresses. *House Officer: Becoming a Medical Specialist*, *1988*, 203–214. https://doi.org/10.1007/978-1-4615-9525-0_19

Das, R. & Jayalekshmy, V. G. (2015). Mechanism of heavy metal tolerance and improvement of tolerance in crop plants. *Journal of Global Biosciences*, *4*(7), 2678–2698.

Ezenwobodo, V. G. & Samuel, S. (2022). Design of flexible pavement by CBR method. *International Journal of Research Publication and Reviews*, *4*(1), 1806–1812. https://doi.org/10.55248/gengpi.2023.4149

Gangwar, P., Johri, P., Dutta, R., Singh, S., & Trivedi, M. (2023). Overview of phytopharmaceuticals and biotechnology of herbal plants. *Phytopharmaceuticals and Biotechnology of Herbal Plants*, *2023*, 1–17.

Ghani, A. (2010). Toxic effects of heavy metals on plant growth and metal accumulation in maize (*Zea mays* L.). *Iranian Journal of Toxicology*, *3*(3), 325–334.

Gull, A., Ahmad Lone, A., & Ul Islam Wani, N. (2019). Biotic and abiotic stresses in plants. In de Oliveira, A. B. (Ed), *Abiotic and Biotic Stress in Plants*, pp. 1–8. IntechOpen, London. https://doi.org/10.5772/intechopen.85832

Gupta, V. , Lal, K. , & Uttreja, M. (2019). *Contamination of Heavy Metals in India - Health Effects and Remediation Measures*, p. 24. The Energy and Resources Institute, New Delhi.

Jabborova, D., Nigora, Z., Yuriy, E., Abdujalil, N., Anand, D., Khurshid, S., Zafarjon, J., Sachidanand, S., & Rahul, D. (2023). Growth of spinach as influenced by biochar and bacillus endophyticus IGPEB 33 in drought condition. *Pakistan Journal of Botany*, *55*(SI), 1–7. http://dx.doi.org/10.30848/PJB2023-SI(6)

Khathi, M. T., Farhood, A. T., & Issmer, A. H. (2016). Determination of some heavy metals in extraction of plants by using AAS technique. *Journal of Natural Sciences Research*, *6*(8), 130–137. https://www.iiste.org/Journals/index.php/JNSR/article/view/30202

Kuang, Z., Zhao, Y., & Yang, X. (2023). Machine learning approaches for plant miRNA prediction: Challenges, advancements, and future directions. *Agriculture Communications*, *1*(2), 100014. https://doi.org/10.1016/j.agrcom.2023.100014

Kurniawati, P. (2017). No Titleالابتزاز الإلكتروني.. جرائم تتغذى على طفرة «التواصل ال. In *Universitas Nusantara PGRI Kediri* (Vol. 01).

Little, C. C. (1933). Variability and individuality. *Science*, 77, 195–197. https://doi.org/10.1126/science.77.1990.195-a

Liu, B., Li, J., & Cairns, M. J. (2014). Identifying miRNAs, targets and functions. *Briefings in Bioinformatics*, *15*(1), 1–19. https://doi.org/10.1093/bib/bbs075

Meenakshi, S., Rahul, D., Sachidanand, S., Anita, G. M., & Rajpal, S. (2025). Plant microRNAs: An overview. In Dutta, R., Kapadia, C., Danish, S., & Singh, S. (Eds), *microRNA Advances and Application in Plant Biology Plant Biology, Sustainability and Climate Change*. Academic Press, New York.

Neufeldt, H., Sand, I. van de, & Dietz, J. (2011). Climate Change, Climate Variability and Adaptation Options. *To Climate Change:* https://www.worldagroforestry.org/downloads/Publications/PDFS/BC11312.pdf

NOAA. (2018). Climate Variability vs. Climate Change. *National Oceanic and Atmospheric Administration*, https://www.weather.gov/media/climateservices/VariabilityAndChange.pdf

Noopur, K., Pragati, K., Adarsh, K. S., Sachidanand, S., Tanvi, J., & Megha, B. (2025). Role of non coding RNA in effector trigger immunity. In Dutta, R., Kapadia, C., Danish, S., & Singh, S. (Eds), *Non-Coding RNA in Plants Modulation and Stress Responses Plant Biology, Sustainability and Climate Change*. Academic Press, New York.

Ogul, H. & Akkaya, M. S. (2011). Data integration in functional analysis of microRNAs. *Current Bioinformatics*, *6*(4), 462–472. https://doi.org/10.2174/157489311798072945

Pooja, G., Megha, B., Arpan, M., Chintan, K., Rahul, D., Subhan, D., & Sachidanand, S. (2025). Introduction to long noncoding RNA. In Dutta, R., Kapadia, C., Danish, S., & Singh, S. (Eds), *Non-Coding RNA in Plants Modulation and Stress Responses Plant Biology, Sustainability and Climate Change*, Vol. 1, pp. 1–12. Academic Press, New York.

Radhika, V. K., Rashmi, R., & Aswath, C. (2015). A Guide to in silico identification of miRNAs and their targets. *Journal of Horticultural Sciences*, *10*(1), 90–93. https://doi.org/10.24154/jhs.v10i1.166

Rahul, D., Chintan, K., Subhan, D., & Sachidanand, S. (2024). *Non-Coding RNA in Plants: Modulation and Stress Responses*. Academic Press, New York.

Rodríquez-Serrano, M., la Casa, N. M. De, Romero-Puertas, M. C., del Río, L. A., & Sandalio, L. M. (2008). Cadmium toxicity in plants. *Ecosistemas*, *17*(3), 139–146. https://doi.org/10.1016/b978-0-12-815794-7.00008-4

Sharma, G., Sharma, A., Sinha, N. K., Sharma, O. P., Singh, A., Pandey, A. K., Kumar, A., Trivedi, S. K., Sao, B., & Sahu, M. K. (2022). Assessment of long-term climate variability and its impact on the decadal growth of horticultural crops in central India. *Ecological Processes*, *11*(1), 61. https://doi.org/10.1186/s13717-022-00405-w

Subhan, D., Uzma, Y., Niaz, A., Khurram, S., Rahul, D., & Sachidanand, S. (2025). Fertilizer-responsive non-coding RNA and their significance. In Dutta, R., Kapadia, C., Danish, S., & Singh, S. (Eds), *Non-Coding RNA in Plants Modulation and Stress Responses Plant Biology, Sustainability and Climate Change*. Academic Press, New York.

Tanvi, J., Arpita, S., Praneta, Y., Palak, S., & Sachidanand, S. (2025). Epigenetic modulation of plant developmental pathways by non-coding RNA. In Dutta, R., Kapadia, C., Danish, S., & Singh, S. (Eds), *Non-coding RNA in Plants Modulation and Stress Responses Plant Biology, Sustainability and Climate Change*. Academic Press, New York.

Trivedi, M., Bhalla, R., & Johri, P. (2017). Hairy root technologies and beyond. *Biotechnology Trends and Applications*, *18*, 80–105

U.S. Global Change Research Program. (2018). Climate science special report: Fourth national climate assessment, volume I. *U.S. Global Change Research Program*, *1*, 470. https://doi.org/10.7930/J0J964J6

Van der Wiel, K. & Bintanja, R. (2021). Contribution of climatic changes in mean and variability to monthly temperature and precipitation extremes. *Communications Earth and Environment*, *2*(1), 1–11. https://doi.org/10.1038/s43247-020-00077-4

Verma, N., Sao, P., Srivastava, A., & Singh, S. (2021). Physiological and molecular responses to drought, submergence and excessive watering in plants. In: Husen, A. (Ed), *Harsh Environment and Plant Resilience*, pp. 305–321. Springer International Publishing, Cham.

Yang, X., Zhang, L., Yang, Y., Schmid, M., & Wang, Y. (2021). Mirna mediated regulation and interaction between plants and pathogens. *International Journal of Molecular Sciences*, *22*(6), 1–13. https://doi.org/10.3390/ijms22062913

Yu, G., Ma, J., Jiang, P., Li, J., Gao, J., Qiao, S., & Zhao, Z. (2019). The mechanism of plant resistance to heavy metal. *IOP Conference Series: Earth and Environmental Science*, *310*(5), 19–24. https://doi.org/10.1088/1755-1315/310/5/052004

2 Impact of Climate Change-Induced Abiotic Stress on Plant Growth, Biodiversity, and Adaptations

Ankita Acharya, Rituparna Chatterjee, Moulima Das, and Muhammad Latif*

2.1 INTRODUCTION

Over the past 65 years, climate change has been perceived as the most critical global issue, representing both current impacts and future predictions. This complex challenge involves governments worldwide and affects various realms, including the natural environment, ecology, politics, and economies (Feliciano et al., 2022; Leal Filho et al., 2021; Neil Adger et al., 2005). Immediate attention and early action may increase the likelihood of overcoming the damaging effects (Battisti & Naylor, 2009; Weisheimer & Palmer, 2005). However, forecasting the specific impacts of climate change on each sector remains challenging (Leppänen et al., 2014). This challenge is reflected in the increasing acknowledgement and incorporation of climate uncertainty into both local and national policies (Izaguirre et al., 2021; Jurgilevich et al., 2017).

Actually, climate change involves several changes in atmospheric conditions, which often encompass temperature, precipitation, pressure, and humidity. Among these consequences of climate change, the most dramatic is the alteration of weather patterns and the melting of ice layers across the globe, which raises sea levels (Lawlor & Mitchell, 1991; Lipczynska-Kochany, 2018; Murshed, 2020). These changes include carbon dioxide, methane (CH_4), nitrous oxide (N_2O), and water vapour (H_2O), whose origins can be traced to natural events such as volcanic eruptions, wildfires, and tectonic movements preceding the Industrial Revolution (Hussain et al., 2021; Murshed, 2020; Usman & Balsalobre-Lorente, 2022). On December 12, 2015, at the COP-21 conference in Paris, the United Nations Framework Convention on Climate Change (UNFCCC) achieved a landmark agreement to address climate change while fostering and strengthening the necessary actions and investments required for a sustainable low-carbon future (Chien et al., 2021; Sharif et al., 2020; Sharma et al., 2020).

The Paris Agreement builds on the Convention by uniting all nations for the first time around a common purpose to undertake ambitious actions to prevent and mitigate climate change, with increased financial support for developing countries (Anwar et al., 2021; Shahbaz et al., 2020; Usman & Balsalobre-Lorente, 2022). As an international landmark for climate change, the Paris Agreement seeks to enhance efforts against climate change by ensuring the strict limitation of this century's rise in global warming to at or below 2°C above pre-industrial levels, with a specific goal of limiting the rise to 1.5°C (Balsalobre-Lorente et al., 2020). Achieving these ambitious goals would require adequate financial investments and the development of a new, innovative technological framework along with capacity building. This approach would enable developing and at-risk countries to realize their national goals (Murshed & Dao, 2022).

* *Corresponding Author*

DOI: 10.1201/9781003491378-2

The agreement created a clear framework for action and support. All parties must put forth their best efforts through "NDCs" and maintain this commitment in the coming years (Murshed et al., 2022). Reporting on emissions will be required, and its implementation will be regular, while with a global stocktake every 5 years to assess collective progress toward the goals of the Paris Agreement, guiding further actions by participating countries (Huang et al., 2016). In fact, the Paris Agreement was officially adopted by the UN headquarters in New York on Earth Day, April 22, 2016. It took effect 30 days after this date, on November 4, 2016, when 55 countries ratified it, representing at least 55% of the world's emissions. By early 2017, the total number of ratifying parties had reached 125 (Abbass et al., 2022; Jannat et al., 2022; Usman & Balsalobre-Lorente, 2022; Velu et al., 2018).

With work programmes adopted, mechanisms, procedures, and recommendations can be identified to address the widespread concerns and facilitate the implementation of the Paris Agreement. Since 2016, parties have worked jointly in subgroups (APA/SBSTA/SBI) and entities (SBI). In November 2016, the Conference of the Parties (CMA) met in Marrakech concurrently with COP22 and adopted two resolutions. Completion of the work plan was anticipated in 2018. This sets a global aim: to hold the increase in the global average temperature to well below 2°C and to pursue efforts to limit it to 1.5°C. This goal is crucial for reducing the risks and impacts of climate change.

- There is a need to reach the peak of global emissions soon, considering that developing countries will have to cut their emissions more rapidly following the Paris Agreement.
- Establishing a balance between emissions reductions in Paris and removals after Paris, considering the best available science.

The strategies of adaptation aim to determine how the resilience of societies might be developed against climate change while, at the same time, sustaining and enhancing international support for the adaptation efforts of developing nations. Currently, humans are recognized as the leading contributors to carbon dioxide emissions. The Industrial Revolution is a significant factor in the increase in CO_2 emissions. Additionally, various human activities, including excessive fuel consumption, extensive agricultural practices, large-scale waste burning, deforestation, and reliance on fossil fuels for both domestic and national transportation, exemplify anthropogenic influences. These actions result in climate-related disasters, damage infrastructure locally and globally, harm human health, and reduce overall productivity. The majority of developing countries depend on fossil fuels for energy production. Energy consumption has raised greenhouse gas (GHG) levels, which are linked to increasing temperatures.

The impact of climate change is measurable and urgent. Mitigation refers to efforts aimed at decreasing GHG emissions or their concentrations in the atmosphere, whereas adaptation involves adjusting to the changing climate and its consequences. Mitigation has historically been viewed as a substitute for adaptation. Conversely, adaptation has been regarded as an alternative to mitigation, particularly concerning local advantages, while mitigation offers global benefits. It is now well recognized that both mitigation and adaptation strategies go hand in hand for addressing climate change issues. However, adaptive measures, at times, have unintended consequences on the environment that may undermine their success in achieving goals.

The effects of climate change on humans, natural systems, and their adaptation have been widely studied (IPCC, 2014) and across various sectors (Agriculture, Pastoral Systems, Transport Infrastructure; Winter Tourism (Abegg et al., 2008); Human Health (Abbass et al., 2022)). Guidelines on climate change and Environmental Assessment (EA), both SEA and EIA, can be found in the USA (CeQ, 1997; CEQ, 2010), Canada (CeAA, 2003; IEMA, 2010a, 2010b), and the UK (EU, 2013; EIA Directive, EU, 2014), where mitigation and adaptation measures are explicitly required. Following these guidelines, an EA must evaluate the impact of climate change and its effects within plans, programmes, and projects. However, a significant number of research activities must be undertaken to establish the impacts of adaptation and mitigation approaches on environmental outcomes. Adaptation efforts frequently remain overlooked in policy formulation processes, EIA guidelines, theoretical and conceptual discourses, and decision-making procedures.

2.2 CAUSES OF CLIMATE CHANGE

2.2.1 Natural Causes of Climate Change

One of the concerns regarding this change in climate is GHGs. This is the jargon used by scientists to describe how certain gases in the Earth's atmosphere retain heat that would otherwise be lost to space from the Earth's surface. This is important for the protection of life on our planet; otherwise, the Earth would have become uninhabitable because of extreme cold. Since the mid- to late 1800s, human activities have significantly amplified the greenhouse effect to levels never seen before. Consequently, Earth is now experiencing higher temperatures than at any point in human history, along with ongoing warming trends. Rising sea levels, extended droughts, increased flooding, severe storms, and extreme heat are all results of the significant disruption of natural cycles and weather patterns caused by global warming. Climate change refers to all of these detrimental and often fatal consequences. Tackling human-induced climate change isn't about shaming or blaming anyone; it's about acknowledging the problem to create effective solutions. Addressing the roots of climate change is crucial, even if it's uncomfortable. Over the past century, humanity has achieved remarkable advancements in productivity, and some of these developments have contributed to the current state of our planet. There is a need to harness this innovative spirit to promote shared responsibility for cooling the Earth, tackling climate change, and creating a more just and sustainable future. Figure 2.1 summarizes the factors driving climate change.

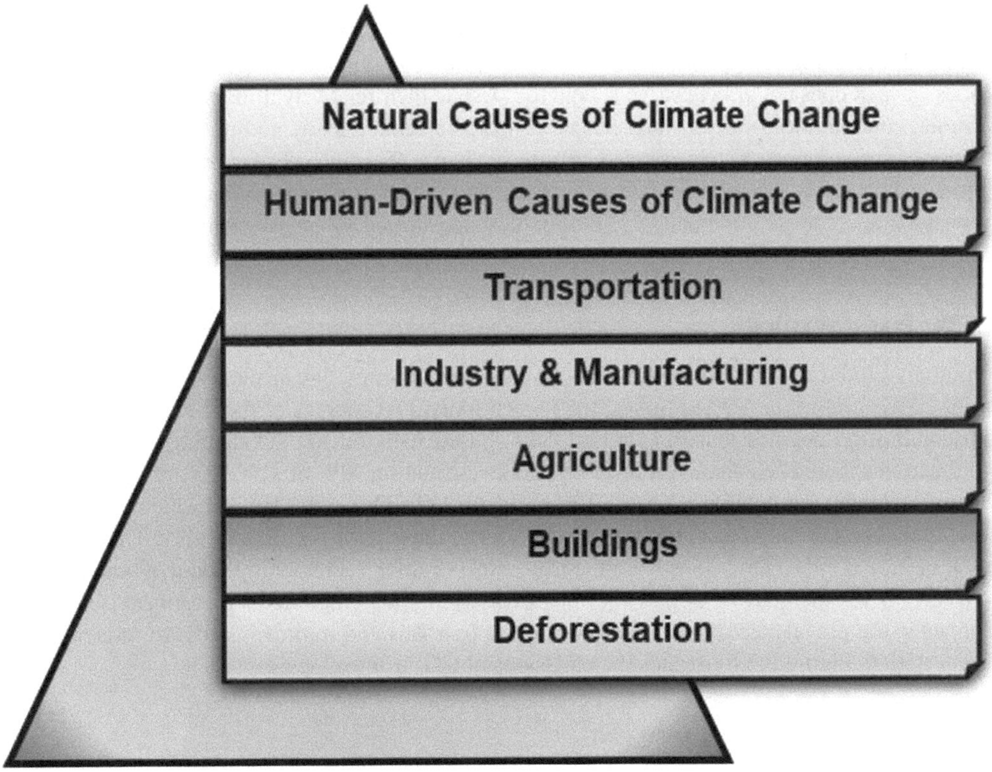

FIGURE 2.1 Factors driving climate change.

TABLE 2.1
Air Quality Index (AQI) Breakpoints

AQI Category	Index Values	Previous Breakpoints (1999 AQI) (µg/m³, 24-hour Average)	Revised Breakpoints (µg/m³, 24-hour Average)
Good	0–50	0.0–15.0	0.0–12.0
Moderate	51–100	>15.0–40	12.1–35.4
Unhealthy for sensitive groups	101–150	>40–65	35.5–55.4
Unhealthy	151–200	>65–150	55.5–150.4
Very unhealthy	201–300	>150–250	150.5–250.4
Hazardous	301–400	>250–350	250.5–350.4
	401–500	>350–500	350.5–500

2.2.2 Human-Driven Causes of Climate Change

The prevailing agreement among scientists is that human activities are the principal contributors to current global phenomena. This type of climate change is commonly termed "anthropogenic," indicating that it is caused by human actions. In the past century, burning fossil fuels has greatly increased GHG levels in the atmosphere, especially carbon dioxide. Additionally, deforestation and urban growth have significantly depleted forests, wetlands, and other natural carbon sinks that absorb carbon dioxide and prevent it from entering the atmosphere. Currently, GHG concentrations are at their highest levels in at least 800,000 years. Certain gases, like HFCs, aren't naturally occurring. Humans have increased the Earth's average temperature by about 1.5°F through constant emissions of these gases into the atmosphere during the 20th century. This increase has resulted in a period characterized by severe and increasingly frequent weather extremes. Although climate change impacts individuals in various ways, it does not affect everyone equally. Despite their relatively minor contribution to global warming, communities of colour and people residing in economically and politically marginalized situations are disproportionately affected by climate change. This includes factors such as how energy is produced for electricity, heating, and transportation; built environments and industries; interactions with nature; and consumption patterns. While the sources of GHG emissions may vary, they remain relatively constant overall. Air Quality Index (AQI) breakpoints are depicted in Table 2.1.

2.2.3 Transportation

Commonly used transportation modes include cars, trucks, ships, and airplanes. These global emitters of GHGs are, in the United States, the largest individual sources of those emissions. Vehicles with combustion engines that run on petroleum-based fuels release substantial amounts of CO_2 into the atmosphere each year. Automobiles alone account for 41% of all GHG emissions, with an average passenger automobile releasing almost 4.6 tons of CO_2 yearly. Trucks, which primarily use diesel fuel and operate almost continuously, are particularly harmful to air quality. They comprise a total of 23% of GHG emissions from the U.S. transportation sector, compared to a minuscule 4% from cars. While there are strategies to mitigate these emissions, significant investment is necessary to enhance the prevalence of zero-carbon vehicles on roads and improve access to reliable public transportation. Figure 2.2 illustrates the emissions of CO_2 from vehicles.

2.2.4 Industry and Manufacturing

Goods-producing factories also emitted GHGs, accounting for 24% of the total U.S. emissions in 2020. A large majority of the industrial emissions arise from carbon-intensive goods such as paper, steel, cement, aluminium, glass, and basic chemicals. Manufacturers require significant

FIGURE 2.2 CO_2 emission from vehicles.

energy to meet the demands for infrastructure and a wide range of products. Many outdated facilities frequently emit these gases, as well as other types of air pollution, during the upgrading process. Enhancing efficiency through advanced technology and stringent pollution regulations is a potential solution to address this issue. There is a need to rethink consumption habits, especially regarding plastics, and to encourage recycling and reuse to reduce overproduction. Because large infrastructure projects heavily depend on industries such as cement manufacturing, which annually accounts for 7% of global GHG emissions, policies should leverage government purchasing power to support markets for cleaner alternatives. It's also important for state and federal agencies to purchase sustainably produced materials. Replacing fossil fuels with renewable sources of energy will mitigate the environmental impacts associated with this high-carbon industry.

2.2.5 Agriculture

The relationship between soil and climate has historically been significant. However, the advent of modern industrial agriculture has profoundly altered this relationship. In 2020, agriculture was responsible for 11% of the United States' total GHG emissions. This sector notably emits large quantities of nitrogen oxide (NO_x) and methane, powerful GHGs that trap heat. The largest contributor to atmospheric nitrous oxide emissions—accounting for almost 25%—comes from the application of chemical fertilizers and the emphasis on high yields alongside widespread unsustainable agricultural practices. Methane emissions from cattle and other ruminants' digestive processes are the primary source of methane emissions from industrial-scale livestock farming. However, farmers and ranchers, particularly those from indigenous communities who practice sustainable land management, demonstrate that there are many ways to feed the global population. By adopting regenerative agricultural practices, it is possible to reduce emissions from this sector, improve soil's ability to capture atmospheric carbon, and produce more nutritious food.

2.2.6 Buildings

Given how much time people spend indoors, it's surprising that both residential and commercial buildings significantly contribute to GHG emissions. In 2020, heating and cooling, refrigeration, cooking, appliances, and building climate control accounted for 13% of U.S. overall emissions. Similarly, it is estimated that about 30% of the energy used inside U.S. buildings is wasted. People are making daily advancements in energy efficiency, enabling us to achieve equivalent or improved outcomes by reducing energy consumption. Mandating that new constructions meet top efficiency standards and updating existing buildings with advanced technologies can lower emissions. It also supports more accessible and affordable heating, cooling, and power options for all communities, aligning with the goals of the environmental justice movement.

2.2.7 Deforestation

The ongoing clearcutting of global forests and degradation of wetlands have contributed to an increase in GHG emissions. Vegetation and soil are vital for carbon storage, both below and above ground. However, activities like logging and various types of development remove this vegetative biomass, releasing the stored carbon into the atmosphere. For example, clearcutting alone has emitted more than 25 million tons of carbon dioxide annually from Canada's boreal forests—roughly the emissions equivalent of about 5.5 million vehicles. To address this issue, it is essential to implement government policies that promote sustainable practices and encourage changes in consumer behaviour, thereby restoring the planet's carbon sink. Figure 2.3 represents the clearcutting of forests.

FIGURE 2.3 Clearcutting of forests.

2.3 EFFECTS OF CLIMATE CHANGE ON ECOSYSTEMS AND ENVIRONMENTAL GOVERNANCE

Climate change has had a profound influence on ecosystems worldwide, from polar regions to tropical areas. Even minor fluctuations in temperature can result in significant disturbances to food webs and broader environments. Climate change primarily affects the coldest regions, particularly those near the poles. The rate of warming in the Arctic is double that of other regions of the world. This rapid temperature increase causes glaciers and polar ice sheets to melt quickly, releasing a substantial amount of water.

When sea ice recedes, darker ocean surfaces that absorb more sunlight are revealed, which accelerates melting. If this trend continues, the Arctic may lose ice entirely during summer within the next 15 years. Scientists believe that climate change will have a drastic impact on sea level rise. Melting polar ice caps and glaciers, along with thermal expansion because of warmer ocean water, could increase sea levels by as much as 6.6 feet by the end of the century unless emissions are controlled. Flooding would be extreme in low-lying areas, including island nations and coastal cities like Mumbai and New York City.

The phenomenon of sea level rise, even at moderate levels, poses a considerable challenge that is both expensive and dangerous, as well as disruptive to communities. A 2022 Sea Level Rise Technical Report from the National Ocean Service projects that sea levels will rise by another foot in the United States by the year 2050. Roads, sewage treatment facilities, and power generation plants are all at risk of repeated damage from this increase. By the end of the century, beaches cherished by families for generations will be lost forever. The environmental impacts are equally significant, as encroaching saltwater threatens to destroy coastal ecosystems and freshwater aquifers—vital sources of irrigation for agriculture and drinking water. Saltwater intrusion is already changing the topography and livelihoods of areas such as Bangladesh, where large parts of the land lie less than 7 feet above sea level.

2.3.1 Flooding

Along with the increase in sea levels leading to coastal inundation, climate change also affects factors that contribute to flooding in inland and urban regions, including snowmelt and intense rainfall. As global warming exacerbates both the rise in sea levels and extreme weather phenomena, it is anticipated that floodplains across the nation will expand by nearly 45% by 2100. A poignant illustration of this occurred in 2022 when China faced catastrophic flooding that submerged nearly one-third of the country. This calamitous incident was triggered by heavy rainfall, along with the melting of glaciers and snow.

2.3.2 Rising Ocean Temperatures and Marine Heat Events

Climatic changes significantly impact marine ecosystems throughout the world. The oceans cover more than 70% of the Earth's surface area; they absorb over 30% of the carbon dioxide emissions resulting from fossil fuel combustion and nearly 93% of the heat generated by GHGs. This has led to higher ocean temperatures and more frequent marine heat waves. Consequently, temperature-sensitive fish and various marine organisms are altering their migration routes in search of cooler, deeper waters, which disrupts food chains and affects commercial fisheries.

However, the increasing frequency of marine heat waves will cause a high death toll among plankton and marine mammals. The ocean is also slowly becoming more acidic because of its increased ability to absorb carbon dioxide, which alters the chemistry of the waters. This change poses a significant risk to marine species that have adapted to thrive within a lower pH range. Organisms like corals, oysters, and mussels are expected to be severely impacted by acidification, as it disrupts their ability to form shells.

2.3.3 Ecosystem Stressors

Terrestrial ecosystems, encompassing a variety of environments such as ancient forests, savannahs, and tropical rainforests, are experiencing significant challenges. Changes in climate can exacerbate the frequency of pest outbreaks, the spread of invasive species, and the incidence of pathogenic infections in forested areas. This, in turn, alters the species of flora that can survive in those regions, disrupting the normal life cycles to which most animal species are accustomed. Consequently, the overall composition of these ecosystems is undergoing transformation, leading to diminished resilience to multiple stressors.

Although ecosystems inherently possess adaptive capabilities, many are near the thresholds of their natural limits. With the ongoing rise in temperature, further repercussions are inevitable. The ramifications of climate change appear to trigger a cascade of interconnected ecological shifts that are challenging to fully anticipate, and once they gain significant traction, they are nearly impossible to reverse. This destabilization is particularly pronounced in keystone species, which are crucial for maintaining the structural integrity of ecosystems.

As a decline in species, biodiversity, and water resources has been observed, the importance of actively practiced and applied sustainability philosophy is becoming increasingly crucial for communities. Immediate action is essential to preserve the planet (Feliciano et al., 2022; Leal Filho et al., 2021; Murshed, 2020; Neil Adger et al., 2005). Significant temperature shifts now evident on Earth will substantially hasten other ongoing changes in the climate system, such as shifting precipitation patterns, increased evaporation due to warming, and rising temperature extremes (Battisti & Naylor, 2009). While it's still unclear how much these developments will manifest in higher temperatures and extremes, the early impact of GHGs on the water cycle strongly supports their role in climate change (Schuurman et al., 2022). The added warming caused by increased greenhouse gases has, up to a point, been counterbalanced by slight changes in the aerosol inventories. Temperatures, therefore, should rise in the coming decades. Models that project alterations in oceanic precipitation patterns demonstrate considerable reliability, indicating that humid areas will experience increased rainfall, whereas arid regions become even drier, albeit not uniformly.

Moreover, extreme rainfall levels in regions like Greenland and Antarctica provide valuable insights into future changes, despite covering a relatively small part of the Earth's land area (Weisheimer & Palmer, 2005). High-resolution climate models forecast an increase in both the intensity and frequency of these short-term extreme rainfall events, thereby elevating the probability of expected reductions in certain regions. The depiction of extreme precipitation events has shown significant enhancements, leading to greater confidence in the projections derived from these models. The 10–30 km range typically used in climate change studies isn't sufficient to explicitly represent subdaily heavy rainfall events (Jurgilevich et al., 2017). Explicit convection in weather forecasts helps anticipate severe weather like floods and droughts (Anil et al., 2023; Jabborova et al., 2023; Verma et al., 2021). Hence, it is proposed to use subdaily extreme precipitation events as indicators of climate change risks to enhance risk identification in weather forecasting.

Moreover, it's important to highlight that urban areas are significant indicators of climate change-related risks, especially concerning extreme rainfall events. As mentioned earlier, the observed changes in urban areas align with global temperature and precipitation variations in comprehensive models of total annual precipitation (Izaguirre et al., 2021).

Addressing climate change through sustainability should not be the dominant strategy influencing individuals' actions and obligations. The positive aspect of climate change is that we understand its causes and how to address it, but we shouldn't ignore other significant environmental sustainability issues that also need attention (Tvaronavičienė, 2021). Climate science is a vital field of research globally, and there's still much to discover to fully grasp this critical issue (Sharma et al., 2020). Existential threats, although primarily related to climate change, have caught many people's attention. This situation will persist for decades, and what's worse is that the attempts to combat global warming are largely insufficient. The underlying economic and political issues stem from the rapid development of emerging economies, substantial global investment in fossil fuels, and their various applications. This means that ecosystems and societies need to adjust to a much warmer future, which will negatively impact humans as well. Significant efforts are being made in the management and restoration of ecosystems, but these need to be combined with reduced fossil fuel emissions to mitigate the adverse impacts of climate change. Given our current state of knowledge, we are indeed facing exceptional circumstances and grappling with these challenges. Meanwhile, levels of GHGs in the Earth's atmosphere are likely to stabilize as humanity comes to terms with the hurdles and complexities surrounding the decarbonization of our planet. However, CO_2 levels are increasing, and the resulting warming will occur much more rapidly than what we're currently experiencing (Chien et al., 2021). Even small magnitude climatic changes are likely to cause drastic and adverse effects, nearly devastating almost every area in the world. Currently, nearly 8% of global climatic disturbances are directly caused by this world problem. When global stability is threatened, governments face more pressure, increasing the likelihood of political instability and civil unrest in countries like China, India, and Brazil (Sharif et al., 2020). Urbanization and rural development significantly impact various aspects of biodiversity, notably increasing the risk of forest fires and leading to habitat destruction.

Additionally, alterations in coastal land use influence wealth distribution, as coastal nations contribute a substantial portion of the world's natural resources, including oil, gas, and coal. As a result, the health effects of global warming range from cancer to respiratory diseases, allergies, and asthma. Therefore, it is essential to consider the impact of the changing environment on human health, as both affect each other and overall well-being (Tvaronavičienė, 2021). This framework facilitates the assessment of the impacts resulting from variations in air, land, and ocean temperatures while also addressing human-induced uncertainties related to health outcomes. On local, national, and international levels, these insights play a crucial role in guiding decision-making processes related to healthcare, environmental sustainability, and other issues. Implementing mitigation strategies and improving understanding of specific types of environmental degradation are vital. Attention is drawn to some effects of climate change with the aim of mitigating the adverse impacts on the environment, human health, and overall population well-being.

2.3.4 Precipitation Alterations

Precipitation encompasses various forms of moisture, including rain, snow, and other types of liquid or frozen water that descend from clouds. This phenomenon occurs intermittently, with the specific type of precipitation significantly influenced by temperature and prevailing weather conditions. The availability of moisture is governed by wind patterns and evaporation from the Earth's surface, as well as by its accumulation in storm systems such as clouds. Precipitation occurs when water vapour condenses, usually within ascending air that cools and expands as it rises. The upward movement of air can result from several factors, including air ascending over mountainous terrain, warm air advancing over cooler air (known as a warm front), colder air descending beneath warmer air (referred to as a cold front), localized heating of the surface (convection), and various weather and cloud systems. Consequently, alterations in any of these elements can influence precipitation patterns. Although precipitation maps may exhibit irregularities, overarching trends can be analysed using the Palmer Drought Scale, which assesses soil moisture based on rainfall data and rough estimates of evaporation variations.

As global temperatures rise due to the enhanced greenhouse effect caused by human activities, evaporation rates increase, provided there is adequate surface moisture, which is typically abundant over oceans and other moist areas, acting as natural "air conditioners." Consequently, summers are becoming increasingly warm and dry, in contrast to the cooler and wetter conditions of the past. For instance, in the eastern regions of North and South America, which have experienced increased moisture, temperature increases have been less pronounced than in other areas. More precipitation accumulates during winter months in the northern continents because warmer atmospheric conditions enhance the atmosphere's ability to retain moisture as temperatures rise. In areas where precipitation has seen a slight increase, the resulting temperature rises have led to greater drying, making precipitation changes less noticeable. As the climate evolves, various direct effects on the quantity, frequency, and nature of precipitation have emerged. Warming accelerates the drying of land surfaces and increases the probability and intensity of droughts, a phenomenon observed in many global regions. It should be noted that the Clausius–Clapeyron equation states that raising the temperature by 1°C is equivalent to an approximate 7% increase in the capacity of the atmosphere to hold moisture. While patterns in relative humidity are complex, making them difficult to evaluate, they indicate a substantial level of stability in relative humidity during the 20th century at the tropospheric level, implying a water vapour increase of about 5% across oceanic regions due to warmer temperatures. Precipitation primarily arises from weather systems that draw moisture from the atmosphere, generally resulting in heightened precipitation intensity and an increased likelihood of heavy rainfall and snowfall events.

Climate model simulations, foundational theories, and empirical data collectively indicate that elevated temperatures resulting from increased water vapour lead to more severe precipitation events. This occurs even when total annual precipitation is marginally reduced, with the likelihood of even more extreme events increasing when total precipitation levels rise. Consequently, a warmer climate heightens the risk of both droughts (during periods of insufficient rainfall) and floods (during periods of excessive rainfall), albeit at different times and locations. For instance, Europe experienced extensive flooding in the summer of 2002, followed by unprecedented heat waves and drought conditions the following year. The cyclical patterns of El Niño significantly influence the distribution and timing of floods and droughts, particularly in tropical regions and across many Pacific Rim countries. In areas where aerosol pollution limits direct sunlight from reaching the ground, reduced evaporation leads to a lower overall supply of atmospheric moisture. Thus, although increased water vapour can result in more intense precipitation events, these events may occur with shorter durations and frequencies because of the prolonged time required for the atmosphere to replenish water vapour content.

Furthermore, local and regional differences in precipitation patterns are significantly influenced by atmospheric circulation dynamics, which in turn are affected by El Niño and other variability phenomena. Certain observed alterations in circulation are associated with climate change. These changes in storm trajectories have resulted in increased precipitation in some areas and dryness in others, often in close proximity, leading to complex precipitation patterns. For example, a more positive NAO during the 1990s tended to result in more rain over northern Europe simultaneously with dry conditions over the Mediterranean and northern Africa. The Sahel's long-term drought, beginning at the end of the 1960s and lasting until the end of the 1980s, is still relevant today but has lessened over time. Changes in tropical sea surface temperature distributions over the Pacific Ocean, Indian Ocean, and Atlantic Basin are associated with changes in atmospheric circulation. Arid conditions generally prevail across Africa, particularly within the tropical and subtropical regions.

The likelihood of such precipitation occurring as rain instead of snow increases with rising temperatures, especially during the fall and spring when snowfall is beginning and ending, respectively. This phenomenon is particularly evident in regions where temperatures hover around the freezing point. These results can be observed in many places worldwide, especially in the mid-to-high latitudes of the Northern Hemisphere. More rainfall and less snowfall also result in fewer sources of precipitation water during summer when it is most urgently needed. Nevertheless, the inherent variability and often sporadic nature of precipitation complicate the patterns observed over time. Historical records indicate that precipitation patterns can fluctuate from year to year, with extended periods of drought frequently followed by years of substantial rainfall. A notable example is the wet winter of 2004–2005 in the southwestern United States, which occurred after six consecutive years of drought and below-average snowpack.

2.3.5 Rising Temperature

The recent surge in average temperatures throughout the United States has raised significant concern among scientists. "Killer Summer Heat"—a new report from the National Resources Defense Council (NRDC)—warns that by the end of the next century, the average temperature across the United States may climb as high as 11°F. Extreme heat is considered by the NRDC to be significantly more dangerous when accompanied by other atmospheric conditions that combine to form an "Excessive Heat Event." An EHE is characterized by high temperatures, elevated dew points, significant cloud cover, increased wind speeds, and higher surface atmospheric pressure, all of which can lead to deadly heat-related conditions.

Almost all climate experts agree that the most critical factors contributing to this increase in temperature include GHGs, which trap more heat than they allow into the atmosphere, thereby preventing its release. Despite international initiatives aimed at curbing greenhouse gas emissions, temperature levels have continued to escalate. According to a Science Daily article, during March 2012, over 15,000 high-temperature records were set in the United States. Past extreme heat events, such as the 2006 heat wave that struck Southern California, caused hundreds of deaths and injuries. The NRDC estimates that climate change could result in up to 150,000 deaths in the U.S. by 2050. According to information provided by the Centers for Disease Control and Prevention, older adults, infants, and individuals with chronic medical conditions are more susceptible to heat-related deaths.

Heat Exhaustion: This condition is characterized by symptoms such as "excessive perspiration" and "elevated heart rate," both of which indicate overheating. Dizziness, low blood pressure, and muscle cramps are also symptoms of heat exhaustion. If one feels they are experiencing heat exhaustion, all physical activities should be stopped immediately, and a cooler location should be sought. Fluids should also be replenished with cold water or sports drinks that contain electrolytes. Heat stroke is defined by conditions that overexpose the body to excessive heat. According to the Mayo Clinic, individuals suffering from heat stroke typically exhibit a body temperature exceeding 104°F. Factors that may increase the risk of heat stroke include prescription medications, existing health conditions, elevated humidity levels, and alcohol consumption during hot weather. Heat stroke necessitates immediate medical intervention. If left untreated, it can lead to severe damage to the brain, heart, kidneys, and muscles.

High temperatures, as well as extreme low temperatures, are associated with a greater risk of cardiovascular disease, including stroke and arrhythmia. Heart disease is the nation's second leading cause of death in the United States, while stroke is the third leading cause. In addition, prolonged exposure to excessive heat can result in significant kidney injury. Pollen and ragweed are the most prevalent allergens. Indeed, the Environmental Protection Agency points out that the allergy season in spring starts much earlier each year because rising temperatures.

Besides human costs, there are also economic costs. "The Killer Summer Heat" discusses a heat wave that hit California in the summer of 2006. During the 2-week heat wave, 655 people died, and over 16,000 people visited the emergency room. The cost of the heat wave is estimated to be approximately $5.3 billion. According to this report, up to 150,000 people will die from heat-related illnesses by 2050.

2.3.6 Rise in Temperature

For over a century, data from various instruments have been collected on the Earth's average temperature, which was analysed by the Goddard Institute for Space Studies. Factors contributing to this improvement include better station coverage, achieved by comparing model-generated data with actual station locations that exhibit realistic variability (Hansen & Lebedeff, 1987), and subjective assessments of data quality issues (Hansen & Lebedeff, 1987). The degree of uncertainty regarding the global average temperature suggests that 2005 was the warmest recorded year. An analysis of the temperature map for the first 5 years of the 21st century, in relation to the average from 1951 to 1980, reveals that the current elevated temperatures are extensive, with land areas experiencing generally higher temperatures than oceanic regions, and the most extreme temperatures occurring in the polar regions of the Northern Hemisphere. The determination of 2005 as the warmest year was influenced by positive polar anomalies, particularly extreme warming in the Arctic. When calculating the global average temperature, all areas were assigned equal importance based on their size.

Weather monitoring sites are few and far between in the Arctic, yet the significant warm deviation observed in 2005 aligns with the record-low levels of sea ice and average Arctic temperature, as inferred from infrared satellite data. Our study encompasses the estimation of temperature deviations as far as 1,200 km from the closest measurement point (Hansen & Lebedeff, 1987). The resulting spatial extrapolations and interpolations of temperature deviations are typically significant for seasonal and extended periods at middle and high latitudes, where the extent of anomalies is determined by Rossby waves (Hansen & Lebedeff, 1987). Therefore, we are convinced that the extraordinary warmth in the Arctic in 2005 was genuine.

The 2005 heat wave was particularly noteworthy because it came closest to breaking the global temperature record, yet it did not receive a significant boost from the El Niño event of that year. Instead, the 1998 heat wave saw a temperature spike of 0.2°C above the average, thanks to a "super El Niño," the most intense of the last 100 years. Over the past three decades, global warming has increased by 0.6°C, and in the last century, by 0.8°C. It is no longer accurate to claim that most of the global warming occurred before the 1940s. A more accurate statement is that there has been gradual global warming, marked by significant variations, from the early 1900s to 1975, followed by a period of rapid warming at a rate of less than 0.2°C per decade. From the late 19th century until 2000, global temperatures rose by about 0.7°C. This warming trend has continued into the early 21st century, maintaining a rate of around 0.2°C per decade.

The assertion that global warming constitutes a legitimate climate change phenomenon rather than merely a consequence of measurement inaccuracies in urban settings has been substantiated by various indicators. These include changes in surface temperatures derived from deep sea temperature records, the global reduction of alpine glaciers, and the premature melting of ice in rivers and lakes. Regions experiencing the most pronounced warming are predominantly remote, particularly at the poles. Additionally, warming is detected over the oceans, far from human influence, although the rate of warming in oceanic areas is less than that on land, which aligns with the oceans' substantial capacity for heat retention.

2.4 PLANT REACTIONS TO CLIMATE VARIABILITIES OR VARIATION

2.4.1 Field Crops Sensitivity to Climate Change

- Agricultural productivity is influenced by increased carbon dioxide (CO_2) levels. Several studies have shown that plant growth can be enhanced by higher concentrations of CO_2. However, other factors, such as rising temperatures, ozone levels, and limited water and nutrients, could reduce these expected yield improvements. For instance, if temperatures exceed the optimal range for a specific crop, or if there is insufficient water and nutrients, the anticipated yield gains might be diminished or nullified. Additionally, higher CO_2 levels have been associated with lower protein and nitrogen content in crops like alfalfa and soybeans, leading to a decline in quality. This decline in the quality of grain and forage can reduce the potential capacity of pastures and rangelands for grazing livestock.
- The climatic phenomenon of rising temperatures and rainfall can decrease yields in agricultural production. Extreme climatic events, such as floods and droughts, threaten crop yields and may result in low crop yield outcomes. Examples include high nighttime

temperatures during the springs of 2010 and 2012 that led to lower corn yields in the U.S. Corn Belt, while an early warm winter destroyed Michigan's cherry production, resulting in a loss of $220 million in 2012.

- Addressing drought conditions may be challenging in areas where elevated summer temperatures cause soil desiccation. While improved irrigation is a practical solution for some regions, others may have insufficient water resources, leading to irrigation shortages during peak times.
- Numerous weeds, pests, and fungi thrive under conditions characterized by elevated temperatures, increased moisture, and higher levels of carbon dioxide. Currently, U.S. crop producers spend over $11 billion annually to prevent and control weeds that compete with crops for essential resources such as light, water, and nutrients. The broader spread and distribution of weeds and pests due to climate change would expose crops to new risks to which they have not previously been subjected.
- Higher CO_2 may stimulate plant growth, but it can also deteriorate the nutritional quality of many food crops. Elevated atmospheric CO_2 concentrations result in lower protein and essential mineral content in various plants, such as wheat, soybeans, and rice. This direct impact of rising CO_2 levels on crop nutrition presents a serious health risk. In addition, human health is further jeopardized by the increased use of pesticides, which is a response to rising pest pressure and declining pesticide effectiveness.

2.4.2 Reactions of Timberland Trees to Climate Change

This would change temperature and precipitation, which in turn would affect both altered and natural forest ecosystems. Some biogeographical models conclude that under a $2\times CO_2$ climate scenario, the northern boundary of potential vegetation would shift 500 km or more in boreal regions. This may lead to the expansion of forests into newly accessible areas with favourable climatic conditions, potentially replacing about 50% of the present-day tundra, according to some studies based on equilibrium and dynamic vegetation models. The main concern is the slow pace of forest migration, as evidenced by the minimal rates of tree species movement observed since the last glacial period. At just a few kilometres per decade, the rate is very low compared to projected climate zone shifts of about 50 km per decade. This inconsistency raises concerns about the potential loss of a significant proportion of natural forests, especially at the southern edges of the boreal forest, as deforestation intensifies alongside the substantial release of carbon. Conversely, some researchers argue that the migration rate of tree species may be considerably higher. In the context of timber production, which relies on managed forests where human intervention facilitates migration, the adverse effects of delayed migration may be less critical than their impact on natural forests.

Increasing atmospheric CO_2 will alter temperature and precipitation regimes but will also promote increased production through the "carbon fertilization effect." Researchers have indicated a higher growth potential with increased gas concentrations in closed- or open-top chambers. Assuming an 80% increase in wood production in orange trees, such a response must be expressed under natural field conditions, not confined by chamber size. Increased concentrations of CO_2 had only slight effects on tree growth in the FACE experiments. Doubling the concentration of CO_2 was observed to increase NPP by an average of 23% in young tree stands following long-term FACE research. The variation among different sites ranged between 0% and 35%. Moreover, the only FACE study that examined a mature stand of trees aged 100 years revealed negligible long-term enhancements in stem growth, which could be ascribed to the difficulties associated with sustaining stable CO_2 levels in extensive experimental setups. Moreover, the initial growth enhancement induced by CO_2 is constrained and influenced by factors such as competition, disturbances, air pollutants (e.g., tropospheric ozone), and nutrient limitations. In contrast, many models tend to assume significant fertilization effects. The lack of comprehensive long-term experimental data pertaining to mature tree stands obstructs precise assessments of CO_2-induced growth enhancement through the application of model coefficients.

2.4.3 Photosynthesis and Plant Breath Forms

In fact, most life on Earth depends on photosynthesis, which is performed by plants, algae, and some bacteria. These organisms capture energy from sunlight, produce oxygen, and transform it into chemical energy stored in the form of glucose. Herbivorous animals obtain energy by consuming plants, while carnivorous animals derive it by feeding on herbivores.

Chloroplasts are organelles found in plant cells that capture and store energy derived from sunlight. These organelles are rich in chlorophyll, a pigment that gives plants their green hue. During photosynthesis, chlorophyll absorbs energy from blue and red wavelengths of light while reflecting green wavelengths, resulting in the distinctive colouration of the plant. The two general stages are light-dependent and light-independent. The former takes place in the thylakoid membrane of the chloroplast and requires uninterrupted sunlight to continue. Chlorophyll absorbs energy from light waves and converts it into chemical energy with the help of ATP and NADPH. On the other hand, the Calvin cycle, which consists of light-independent reactions, occurs in the stroma, or the region between the thylakoid and chloroplast membranes. Unlike the light-dependent reactions, this phase does not require light. The energy derived from ATP and NADPH is utilized in the synthesis of carbohydrate compounds, such as glucose from carbon dioxide.

2.4.3.1 Plant Respiration

Plant respiration involves combining sugars produced by photosynthesis with oxygen to generate the energy that the plant requires for growth. Unlike photosynthesis, respiration is a complementary process to photosynthesis. Plants can synthesize their own food in their natural habitats, which is crucial for their survival.

Plants absorb CO_2 from their surroundings for the synthesis of sugars and O_2, which later become sources of energy. Respiration in plants occurs in the leaves, stems, and roots, while photosynthesis occurs only in leaves and stems. The respiration process can be summarized as follows:

$$C_6H_{12}O_6 + 6O_2 \rightarrow 6CO_2 + 6H_2O + 32\ \text{ATP}\ (\text{energy})$$

Like photosynthesis, plants take in oxygen from the air through stomata (Meddya et al., 2023). Since oxygen is abundant in the air, respiration takes place within the mitochondria of the cell, a process termed "aerobic respiration." Two distinct forms of respiration have been identified in plants: dark respiration and photorespiration. The former can occur under both light and dark conditions, whereas the latter is restricted to light.

2.4.3.2 The Influence of Air Temperature

It occurs continuously day and night, but manifests more at night because photosynthesis does not occur. It is essential for nighttime temperatures to be lower than those during the day, as elevated temperatures can induce stress in plants (Patani et al., 2024; Prajapati et al., 2024). For example, a marathon runner breathes more rapidly than a person at rest, leading to an increase in body temperature. Similarly, when nighttime temperatures rise, plants experience increased respiration rates and, consequently, higher internal temperatures. This can result in damage to flowers and impede overall plant development.

2.4.3.3 The Importance of Oxygen for Roots

As previously stated, the roots are also involved in respiration. The root zone substrate primarily serves as a medium that allows air to pass between the root zone and the atmosphere. In simpler terms, roots need oxygen much like human beings do. Each plant is subjected to a distinct respiration method.

2.4.4 Horticulture and Climate Change

The two main constraints concerning natural fluctuations that deeply affect farming are: while everything is taken into consideration, particularly in farming, a downward trend in crop yield occurs with the rise in high temperatures. Variations in latitude and altitude, ecological zones, degradation of land, severe geophysical events, decreased water accessibility, sea level rise, and salinization are proposed (FAO, 2004). Progress in ecological variables primarily influences farming, and several models are provided below (Datta & Rakesh, 2013):

1. An assessment coordinated at IISR, Calicut, using GIS models showed that various areas currently suitable for crops would become unsuitable in 25 years. At that point, there would be new regions that are unacceptable and become highly suitable for crop development. This information will be useful for other crops.
2. Due to an increase in temperature, the formation process time has been affected. Conversely, photoperiods remained unchanged despite the rise in temperature. Consequently, photosensitive crops have developed more quickly.
3. The chilling requirement for cold season crops will decrease in temperate areas, affecting these crops.
4. Crop treatment will be significantly impacted because of the increase in temperature. Natural pest evacuations, inflorescence, and fruit drop occur consistently.
5. The increase in the efficiency of the annual water system and the need for heat units will be achieved in a much shorter time.
6. Increased temperature variation results in a reduced initial assessment of tuber development in potatoes, a decline in the perceived flavour of tomatoes, and an impact on various crops. Tip burns and blossom end rot are common issues with tomatoes.
7. Waterfront zones can anticipate a much faster saturation of sea water in inland water tables, leading to greater salinity.

2.4.4.1 Climate Factors and Efficiency

During the 1780s, Adam Smith underscored the "imperceptible hand" and streamlined commerce in view of similar benefits in his work, The Abundance of Countries, challenging the notion of static abundance. In the 1840s, an author named Friedrich List, affiliated with the German historical school of economics, proposed efficiency as his theory in his book, The National System of Political Economy. He argued that the effectiveness of efficiency is much more significant than mere abundance alone because it not only guarantees collective accumulation and acquisition of wealth but also addresses what has been lost. This thesis aligns with the saying, "It's wiser to teach a man to fish than to give him fish." With this method of fishing, people can harvest what is abundant in season. For example, it would have been better to give piglets, not chops, to the poor; support based on haematopoietic capability has proven to be much stronger than that of mere assistance, in each of these examples. In "Sandown the theory," efficiency is the cause, not the effect. Friedrich Rundown (1789–1846) was a sceptic and authority on traditional matters of economics. The Public Arrangement of Political Economy was published in 1841, which critically examined the exchange theory of comparative advantage proposed by Adam Smith, an earlier English economist, and suggested that the efficiency of abundance is several times more valuable than abundance itself. The analysis of the variables affecting trade efficiency is considerably more essential than trade value itself and the division of labour. In 1909, List argued that efficiency was a source of wealth and included tangible, intellectual, and consistent output. In this theory, efficiency fundamentally refers to natural resources as objects of labour in specific areas. Horticulture alludes to the fertility of land resources, while industry refers to sources of various resources such as wind, power, water, or minerals that companies rely on. However, the natural efficiency was not consistent. This variability arises, in conjunction with the promotion of spiritual efficiency, innovation, and industry. List defines the efficiency of a public economy as holistic efficiency, characterized by political efficiency (the reconciliation and coordination capacity of a country), deep efficiency (resources in science and technology, culture, education, and art), natural efficiency (natural resources in specific areas), material efficiency (instrumental or physical capital), and individual efficiency (the individual capacity for labour and creation).Economic theory and analysis centre on efficiency, not on trade value or abundance. He replaced static abundance with a powerful concept of efficiency and introduced a unique investigative approach to cost–benefit analysis. Furthermore, regarding the system, the efficiency theory integrates individual efficiency with social efficiency, highlighting the unification rather than resistance of individuals and society. Efficiency in communist political and financial aspects is the capacity to yield items that address individual issues, mirroring the ability of humans to adjust to, use, and change nature (Wei et al., 2018). The fundamental elements of efficiency, Marx states, are the same as those of production factors often discussed. The basic elements of Marx's work cycle include work, labour, objects of labour, and methods of labour. Marx views natural forces as part of efficiency and understands that unusual industry itself comprises much natural forces and inherent science in the production process, with hydropower, wind, and solar power cited as examples of such vast industry. In Das Kapital, Marx does not focus heavily on the concept of efficiency itself but rather on its dialectical relationship with production relations. This dialectic involves how efficiency directly affects these relations and how these production relationships, in turn, respond to efficiency. In contemporary economic analysis, the idea and estimation of efficiency have improved and become more functional. In simple terms, efficiency refers to various production effectiveness ratios, expressed as a ratio of outputs to inputs and encompassing factors that include labour and capital efficiency. Total factor productivity or multifactor productivity is a form of efficiency that can be expressed using more than one set of factors. Additionally, the relationship between TFP and efficiency remains unclear. This is unequivocal for long-term efficiency, including investment, development, abilities, business, and contest. The estimation of TFP is primarily based on commitment to work and capital; however, it scarcely considers natural factors. TFP is obtained by analysing the rate of change of result in the pace of change in labour input and the level of labour costs, which in turn affects capital investment and its costs. From the previous analysis, it can be concluded that nature is viewed as a symbol of wealth or a protagonist in traditional financial matters, communist financial aspects, and current financial matters. However, labour, capital, and total efficiency remain the focus of quantitative analysis and measurement, with natural factors considered merely as external components. Moreover, the above economic analysis indicates that there are significantly minor linkages between natural variables and environmental factors. There is no coverage on the conclusive effect imposed by the environment on natural elements. Nevertheless, the efficiency-related hypothesis and logical strategies are of theoretical and systemic importance for understanding climate efficiency and examining pertinent variables.

2.4.4.2 Coordinate Impacts of Climate Alter on Nourishment Crops

A stable balance in nutrient concentrations is crucial for maintaining nutritional quality and determining plants' resilience to the effects of climate change (Nakandalage & Seneweera, 2018).

The impact of dry spells on nutrient accumulation was examined by comparing food crops in two different regions of East Africa (Fischer et al., 2019). Severe dry spells led to a drop in nutrients, while milder dry seasons increased nutrient concentration. This may indicate that nutrient accretion responses are highly sensitive to changes in the type and magnitude of climatic impacts. Water stress caused significant changes in the mineral composition of vegetables (Hummel et al., 2018; Wijewardana et al., 2019), whereas limiting the irrigation of grapevines in a greenhouse experiment did not result in substantial differences in leaf micronutrient concentrations (Sarker & Oba, 2018). However, during a dry season coupled with zinc deficiency, researchers noted that zinc supplementation led to increased levels of zinc, iron, and manganese in the leaf edges in a detailed field study assessing the nutritional quality of maize grains under varying soil moisture conditions (Sabir & Sari, 2019). High drought increased N, Ca, Mg, Cu, and Zn levels by 12%, 28%, 11%, 18%, and 33%, respectively, compared to the control, while extreme drought caused a marked reduction in the concentrations of P and K by 17%. This shows that drought's impact on mineral concentration is specific to certain nutrients. Research conducted by Tungare et al. showed that hairy roots effectively degraded the toxic textile dye Reactive Blue 4, reducing its harmful effects on plant cells and aquatic embryos. This bioremediation approach lowers oxidative stress, mitigates toxicity, and supports environmental health, offering a sustainable solution to pollution-related abiotic stress under climatc change (Tungare et al., 2022).

Velu et al. (2016) examined the effects of combined water and intensity stress on 54 field-developed wheat varieties. Under intensity and drought stress conditions, the concentration of zinc in the grain was elevated, whereas a reduced increase in iron content was observed under water stress conditions. The interaction between drought stress and elevated CO_2 (eCO_2) in wheat was studied, and researchers discovered that CO_2 enrichment mitigated the negative effects of drought stress, improving water-use efficiency. However, grain quality under eCO_2 was lower, indicated by consistent declines in mineral nutrients (N, P, K, and Zn). Parvin et al. (2019) also noted reductions in Fe, Zn, P, and S in faba beans and lentils during drought and eCO_2 conditions. Asif et al. (2018) reported a relationship between eCO_2, dry spells, and soil Zn accessibility in wheat. Research has indicated that elevated carbon dioxide levels, combined with low water availability and limited zinc accessibility, result in diminished zinc and protein contents in grains. Moreover, Asif et al. (2018) demonstrated that increased carbon dioxide concentrations alleviated some of the yield penalty imposed by potassium deficiency in soil on wheat grain yields.

Environmental changes include challenges associated with soil flooding, which may result from natural occurrences or human activities like over-irrigation and poor drainage. Waterlogging reduces soil oxygen availability, potentially leading to phytotoxin buildup, leaf chlorosis, stomatal closure (Wei et al., 2018), and impaired crop performance due to reduced access to soil nutrients (Chaturvedi et al., 2017). When *Medicago sativa* experienced flooding stress, significant reductions in nutrient uptake (B, Cu, and Zn) were noted in both leaves and roots. Similar results were found in the shoots of barley and wheat, indicating a substantial decrease in Cu, Zn, and Mn concentrations, whereas Fe levels remained unaffected (Steffens et al., 2017). Tarekegne et al. (2021) showed that wheat genotypes susceptible to waterlogging accumulated lower amounts of Cu and Zn, whereas higher Fe concentrations were found in the waterlogged roots of stressed plants. Current research has not addressed the interaction effects of waterlogging with elevated CO_2 on plant nutritional quality, which is expected to become significant in the future.

In a FACE experiment, Fernando et al. (2014) examined how ambient CO_2 (aCO_2) and elevated CO_2 (eCO_2) levels affect wheat, using two different planting dates to simulate high temperatures during the grain-filling phase. The concentrations of essential minerals in the grains, such as calcium, sulphur, iron, and zinc, were found to be lower under eCO_2 conditions. Notably, the majority of mineral concentrations in the grains were significantly enhanced at later planting dates, indicating that elevated temperatures may mitigate some of the negative impacts of eCO_2 on mineral content. Studies on rice (Chaturvedi et al., 2017) and soybean (Bellaloui et al., 2015; Qiao et al., 2019) have shown varying results, examining the impacts of elevated temperatures and eCO_2 on soybean and maize quality grown in open-top chambers over five growing seasons. The combination of increased temperature and eCO_2 led to higher concentrations of potassium and iron, whereas calcium and zinc levels remained statistically unchanged in both species. In addition, their phosphorus and manganese contents varied by species. The metabolic reactions of tall fescue, a cool-season grass, to an increase in temperature of 10°C under eCO_2 were investigated. Organic acids, amino acids, and carbohydrates involved in photosynthesis, respiration, and protein metabolism became highly elevated in the leaves. Thus, eCO_2 may help alleviate heat stress damage. For rice, eCO_2 at ambient temperatures enhances growth and seed yield, but at temperatures of 35°C or higher, these benefits diminish (Chaturvedi et al., 2017). The interaction between eCO_2 and high temperatures significantly impacts growth and photosynthesis, depending on the temperature rise; eCO_2 has been shown to mitigate the negative effects of higher temperatures (Köhler et al., 2018). This was confirmed in soybeans and maize over five growing seasons in open-top chambers (Qiao et al., 2019).

Soil salinity poses a major global challenge for agriculture, affecting almost 20% of the world's arable land (Neocleous et al., 2014; Scagel et al., 2019). Numerous studies have explored how salinity impacts the mineral content and nutritional quality of crops (Neocleous et al., 2014; Petretto et al., 2019; Petropoulos et al., 2017; Scagel et al., 2019), revealing diverse responses across species. For example, a study on lettuce demonstrated that mineral uptake was adjusted by the red cultivar in response to salt stress and increased CO_2, allowing it to adapt better. The researchers concluded that elevated CO_2, with or without mild organic salt stress, could boost certain mineral concentrations. They also found that short-term drought and salinity stress in maize often reduce soil nutrient availability and plant nutrient uptake, though micronutrient levels in grains didn't decrease.

2.4.4.3 Roundabout Impact of Temperature

Elevated temperatures influence design creation. Specifically, certain crops may benefit from reduced frost and chill conditions, whereas others may suffer damage due to increased temperatures. Additionally, the effects of temperature on water supply and demand, as well as the proliferation of pests and weeds in various habitats, may play a significant role.

2.4.4.4 Impacts of Temperature and CO_2 Interaction on Plant Forms

Numerous cycles exist in plant development across various scales, influenced by both temperature (T) and carbon dioxide concentration ($[CO_2]$), and are potentially affected by their interactions. It's useful to examine the short-term effects of $T \times [CO_2]$ interactions on carbon cycles separately from the longer-term growth and development periods that contribute to biomass accumulation and yield. However, it is important to recognize that discussing these interconnected processes could lead to misconceptions.

2.4.4.5 Carbon Balance Angles

In the context of plant carbon balance, the relationship between temperature and CO_2 has been the subject of numerous studies (e.g., Lawlor & Mitchell, 1991), so basic points need not be reiterated. At high temperatures, the CO_2 sensitivity of plant carbon balance continues to rise for two main reasons:

1. Diminished proportion of photosynthesis to photorespiration.
2. The gross photosynthesis to respiration ratio decreases with increasing temperatures.

At the active site of the enzyme ribulose bisphosphate carboxylase–oxygenase (RuBisCO), carbon dioxide ($[CO_2]$) and oxygen ($[O_2]$) compete for the vital substrate ribulose bisphosphate (RuBP), affecting the performance of C_3 plants. RuBisCO facilitates the carboxylation and oxygenation of RuBP. Through the photosynthetic carbon reduction cycle, carboxylation aids in carbon fixation, while oxygenation reduces CO_2 availability through photorespiration. By increasing $[CO_2]$, photorespiratory losses are subsequently reduced. However, the decreased solubility of CO_2 in comparison to O_2 and the reduced specificity of RuBisCO at higher temperatures exacerbate these losses. Consequently, increasing $[CO_2]$ has the greatest impact on carbon fixation at elevated temperatures (Lawlor & Mitchell, 1991). Using a C_3 photosynthesis model showed that.

a. For light-saturated photosynthesis, the ideal temperature rises by 5°C when $[CO_2]$ rises from 350 to 650 μmol/mol.
b. At 10°C, the maximum quantum yield of photosynthesis increases by only 4%, whereas at 28°C, it increases by 25%. This decrease with temperature is mitigated as $[CO_2]$ rises.

c. Because the [CO_2] concentration is high, the light compensation point, which critically increases with temperature, is significantly reduced at elevated temperatures.
d. At lower temperatures, <15°C photosynthesis is reduced due to photosynthetic acclimation caused by reduced RuBisCO activity or saturation at higher [CO_2].

Kirschbaum (1994), utilizing a similar photosynthetic model, highlighted the transient and biogeographic variations in the photosynthetic responses of biological systems to environmental changes. At low temperatures and high CO_2 levels, they noted that the responsiveness of photosynthesis to CO_2 is greatly diminished. He found that at 700 µmol/mol CO_2, the relative response, RA (= [Ahigh-Alow]/Alow), of Asat was only 0.3 at 5°C. However, at central region temperatures and pre-industrial CO_2 concentrations, it increased greatly (RA=0.81 at 35°C and 280 µmol/mol). In both model scenarios, the Ci/Ca ratio is kept constant. This makes sense if stomatal adjustments are in sync with mesophyll photosynthetic activity.

For plants that are adapted to particular conditions, this is typically thought to be true over an extended period (Amthor, 1995; Ehleringer & Cerling, 1995). At low temperatures, when stomatal reactions are slow, there is a temporary difference between the temperature responses of stomata and photosynthesis (Lloyd & Farquhar, 1996). It is not yet evident whether stomatal reactions generally adjust in line with photosynthesis (Morison & Lawlor, 1999; Sage et al., 1995). However, a recent survey by Drake et al. of long-distance probe plants at elevated CO_2 demonstrated that the Ci/Ca ratio was nearly identical for plants grown under normal conditions and those exposed to increased CO_2 (33 observations of over 26 species across a wide range of developmental conditions, from cool to warm) (Drake et al., 1997).

Although many support the hypothesis that the stimulation of A by high CO_2 is greater at higher temperatures, not all literature agrees (for instance, Sage et al., 1995). For example, one study correlated temperature with mean variability in photosynthetic rates of *Pinus taeda* between seasons, while another concluded that no relationship existed. Different species exhibit varying photosynthetic responses due to acclimation to low temperatures (0°C–10°C), especially between herbaceous annuals and perennial evergreens. These groups may therefore show very different levels of responsiveness to CO_2. Several reports indicate a much larger than anticipated CO_2 stimulation at cool temperatures (Glaubrecht, 2008), suggesting that in wheat and grain, this is due to changes in the apparent specific activity of RuBisCO with development temperature. These changes may lead to significantly different photosynthetic responses to CO_2 at varying temperature levels. Diversity is likely caused by differences in the relative concentrations of CO_2 and O_2, as well as potential regulatory mechanisms. There is no evidence to support an increase in RuBisCO's specificity factor under natural conditions. Despite the importance of these effects, surprisingly little exploratory data exists regarding the potential mechanisms of RuBisCO's regulation of $T \times CO_2$ conditions during development. Both temperature and CO_2 caused changes in the activation state and catalytic efficiency, and there were interactions that affected the photosynthetic rate, demonstrating the fundamental complexity of the photosynthetic regulatory systems (Lawlor & Cornic, 2002).

There is a complicated relationship between photosynthesis and the climate. A study focusing on winter wheat (Mercia, Delgado et al., 2006) conducted under controlled field conditions over a single growing season revealed only a 10% increase in photosynthetic capacity when comparing elevated CO_2 levels ($2 \times [CO_2]$) to ambient levels throughout the lifespan of the crop. No changes were observed in the leaf protein content, including RuBisCO or chlorophyll levels. However, a nitrogen shortage combined with a temperature increase of 4°C above ambient levels resulted in a decrease in photosynthetic capacity and carboxylation efficiency. Furthermore, the structure and length of the leaves were altered. In a related study conducted in a different season using the same variety, moderate alterations in leaf composition and CO_2 assimilation were noted, indicating an adaptation to higher CO_2 levels and an interaction with temperature in leaves that developed later (Lawlor et al., 1995).

The above discussion is limited to C_3 plants. Regardless of temperature, it is generally accepted that C_4 plants will exhibit minimal CO_2 sensitivity due to their CO_2 concentrating mechanism (Gifford, 1995; Rowland-Bamford et al., 1996). However, recent research has demonstrated a notable increase in biomass in C_4 species (for instance, a 28% increase in *Panicum* antidote), even in the absence of any discernible changes in the digestion rates of mature leaves. In *Amaranthus retroflexus*, significant reactions of A to [CO_2] have also been observed (up to a 30% increase with $2\times$[CO_2]) (Ziska & Bunce, 1997), though the magnitude varies by species. Additional research is necessary to understand how C_4 species reacts with $T\times CO_2$. Since the responses of C_4 and C_3 plants responses to T are similar (despite the fact that the quantitative characteristics are distinctly species-specific), it is highly likely that the astounding responses of C_4 plants to $T\times CO_2$ cooperation are comparable to those of C_3 plants. In general, to think of large plant development, it is useful to recall that A for each unit leaf area is not the primary component. The whole supply of sugar may result from the combination of photosynthetic rate and leaf area, as well as how these factors relate to the needs of the entire plant in terms of growth, respiration, capacity, root exudates, and other factors (Lawlor & Mitchell, 1991).

According to Lloyd and Farquhar (2008), it is incorrect to assume that the temperature and [CO_2] conditions of photosynthesis directly reflect the [CO_2] and temperature conditions of plant development. This is because the reaction of plant development to increasing [CO_2] involves both the photosynthetic processes of leaves and the overall respiration of the plant. In principle, plants with higher rates of respiratory support per unit of photosynthesis should show a greater relative avoidance of [CO_2], since the developmental response to [CO_2] depends on respiration rates relative to dry biomass. Because [CO_2] increases with temperature (Lloyd & Farquhar, 2008), there should be a more pronounced aversion to [CO_2] at higher temperatures. In general, warm conditions associated with reduced sugar content in plants can impede development and respiration (Rowland-Bamford et al., 1996). Consequently, increased photosynthates should enable a more significant respiratory response to higher temperatures as a result of increased [CO_2]. This is likely a notable improvement. In any case, Gifford (1995) demonstrated that the substantial increases in respiration in wheat plants in response to current temperature increases (hours) were entirely distinct from the respiratory response observed during long-term development at different temperatures.

Although many researchers have recognized and applied methodologies that account for growth and maintenance components, a comprehensive model for plant respiration has yet to parallel the extensive framework established for photosynthesis. While maintenance respiration should react strongly to temperature changes, as indicated by the Arrhenius principle, the rate of growth respiration, which is linked to the relative growth rate (RGR), is generally thought to have a coefficient that is insensitive to temperature. However, Amthor (1995) and Gifford (1995) indicated that these theories do not apply universally. It is evident that rising concentrations of carbon dioxide are accumulating evidence that the rate of respiration per unit dry weight declines with growth. For example, the rate of respiration (per unit dry weight) of rice grown in sunlit controlled-environment chambers consistently declined during the growth period. It decreased most rapidly at very low carbon dioxide levels (160 µmol/mol) and it decreased by up to 45% as the concentration rose to 900 µmol/mol. According to the literature, increased carbon dioxide levels also increase respiration by 16% per unit leaf area but decrease it by 14% per unit leaf dry mass. The longer-term (or "indirect") effects of growth under different carbon dioxide conditions must be distinguished from the immediate (or "direct") response to elevated carbon dioxide. The direct effect appears to result from the inhibition of two mitochondrial electron transport proteins: cytochrome c oxidase and succinate dehydrogenase, an effect that classically leads to about a 20% decline in specific respiration rate, based on a summary of 53 observations in 23 species. However, in long-term studies (mean of 37 observations on 17 species), the indirect effect is less noticeable, exhibiting only a 5% decrease. In plants grown at higher carbon dioxide levels, this effect is frequently associated with increases in soluble starches and decreases in tissue nitrogen content.

Though there is a prevailing assumption that higher temperatures will augment the ratio of respiration to photosynthesis, with higher concentrations of carbon dioxide said to decrease this ratio, there remains much uncertainty regarding the balance of these processes. Furthermore, it is still unclear how much and in which direction the two processes will adapt to temperature and carbon dioxide levels, both separately and in combination, in terms of capacity and regulatory changes.

2.5 ADJUSTMENT MEASURES FOR CLIMATE ALTERATION

2.5.1 Adjustment Alternatives

2.5.1.1 Establishing Robust Foundations through Swift and Inclusive Development

Poverty and limited access to basic services, such as infrastructure, financial services, healthcare, and social protection, are significant predictors of vulnerability to climate change (Halady & Rao, 2010). To ensure successful adaptation strategies, it is crucial to provide high-vulnerability populations with the necessary financial, technical, and institutional resources to build resilient foundations (Bord et al., 1999).

2.5.1.2 Assisting Individuals and Businesses to Fulfil Responsibilities

To combat climate change, households and firms need support to overcome obstacles such as a lack of information, financing, and behavioural biases. Administrative departments should provide reports on climate risks, clarify responsibilities, support innovation, and provide economic support for high-cost solutions (Patz et al., 2005). They can support the poorest who cannot adapt to the climate (Fischhoff, 1995).

2.5.1.3 Safeguarding Vital Infrastructure with Specific Plans

Governments must protect the economic situation and public records of developing countries (Lorenzoni et al., 2007). Approximately 3% of the infrastructure can be built using appropriate data, risk assessment models, and validation methods. The primary duty of the public sector is to plan how to use urban and rural land to adapt to spontaneous changes in climate and weather (Ali, 1999).

2.5.1.4 Helping in Faster Recovery

Administrative sectors can mitigate the long-term devastating effects of climatic disasters and create policies for quick recovery. Preparation, such as improved weather reports, warnings, and proper notification systems, can reduce physical deterioration and financial losses. Early notification related to climate change has been shown to significantly outweigh costs (Chang et al., 2005). Economic support, emergency purchases, and social safety nets are crucial for helping farmers recover. Adaptive measures can reach more people, provide greater support after a disaster, and are particularly effective; however, they require delivery and financing mechanisms to be established before a crisis occurs (Fritze et al., 2008).

2.5.1.5 Macro-Level Managing

Climate change affects various factors, including housing prices and agricultural productivity, requiring strategic planning at high levels. These impacts influence macroeconomic conditions, tax revenues, trade balances, capital flows, and public finances (Gow & Leahy, 2005). The exporting sectors, particularly those affected by climate change, influence commercial balances and capital accounts. Additional spending for mitigation and flexibility needs must be incorporated into conditional commitments and current financial obligations, further straining public economic status (Bostrom et al., 1994). Changes in climate and weather impacts, along with strategies to build economic resilience—particularly through diversification of financial infrastructure, export composition, and tax base—are especially appealing in the short term (Abbass et al., 2022; Leiserowitz, 2006).

2.5.1.6 Prioritizing Tasks Based on Specific Needs, Implementation across Various Sectors, and Regularly Monitoring the Progress

Governments must prioritize actions aligned with available resources and capacity, establish a robust institutional framework, and monitor progress. The primary objective of mitigation is to ensure continuous monitoring and evaluation of the impact of addressing challenges and adjusting actions as needed (Haines et al., 2006). The report offers practical tools for governments to implement adaptation strategies, including financial methodologies for intervention selection, budget tagging for consistent spending to track progress, identifying lagging areas, and prioritizing effective measures for greater resilience (Vlek & Steg, 2007). The increasing impacts of climate change are already evident and unavoidable. However, proactive actions can help societies facing the severe consequences of climate change and prevent the loss of development gains, ensuring a better response and enhanced resilience (Kempton, 1997).

2.5.2 Private Adjustments

Therefore, individuals and families play a significant role in addressing climate change. Behavioural adaptation involves changing strategies, practices, and actions to reduce risks, such as protecting homes from flooding, safeguarding crops from drought, and adopting income-generating activities (Costello et al., 2009).

2.5.2.1 Change in Diets and Food Waste

The deterioration of food due to spoilage is exacerbated by elevated temperatures, increased humidity, and extreme weather events such as flooding and contamination, which threaten food security and nutritional standards. Private-sector adjustments involve evaluating the production, processing, and handling practices of suppliers, including sorting damaged goods, implementing drying techniques for improved storage, and enhancing packaging methods. Retailers and consumers are encouraged to accept imperfect fruits and vegetables, redistribute excess stock, and reduce prices for items nearing expiration (Abbass et al., 2022). In areas characterized by high caloric intake, transitioning to plant-based diets can help alleviate and adapt to the issue of caloric surplus. These plant-based alternatives typically require less energy and water, and can be tailored to fit regional, socioeconomic, and cultural circumstances. Policies such as subsidies, taxes, and marketing initiatives can promote dietary choices that facilitate adaptation (Brown et al., 2014).

2.5.2.2 Change in Livelihood Strategies

Agriculture offers numerous adaptation opportunities based on private adjustments, such as changing planting times, adopting climate-resilient crops and livestock, breeding resilient crops, and selecting genetically modified varieties, all aimed at enhancing food security and nutrition (Bujosa et al., 2015).

2.5.2.3 Migration and Managed Retreat

Climate change significantly influences migration decisions alongside economic, demographic, and political factors. However, determining its direct or indirect influence on migration is challenging. Seasonal migration, which involves traditional strategies such as pastoralism and urban employment, is affected by weather fluctuations and extremes (Asseng et al., 2013). These factors contribute to a decline in agricultural income and employment, and climate change has increased the likelihood of these impacts, leading to an increase in migration, particularly from rural to urban areas (Allen et al., 2010). Increased adaptive capacity measures, such as social protection and women's empowerment, can assist individuals with limited migration options in decision-making. For those unable to migrate, the government may intervene to ensure safety (Davidson, 2016).

2.5.3 Adjustment Techniques for Farming

Climate change affects agriculture by altering environmental factors and pest and disease dynamics, thereby impacting food security, rural livelihoods, and environmental sustainability. Adaptation strategies such as crop diversification, water management, and climate-resilient practices are crucial for enhancing resilience (Fischer, 2019). Collaborative efforts among governments, farmers, researchers, and policymakers are crucial in implementing mitigation measures. Addressing climate change challenges and adopting adaptive strategies can make global agriculture more resilient and sustainable (Balsalobre-Lorente et al., 2022). Crop diversification and climate-resilient crop varieties are key strategies for adapting to changing climatic conditions. Researchers use traditional breeding methods and advanced techniques, such as marker-assisted selection, to create crops that are better suited to higher temperatures (Battisti & Naylor, 2009). Heat-tolerant wheat varieties have been developed to address temperature stress in regions vulnerable to heat waves. Precision agriculture and sensor networks can optimize resource use and mitigate climate risks. By analysing soil moisture, temperature, and nutrient levels, farmers can maximize yields while minimizing input waste (Fettig et al., 2013). Adopting smart farming technologies enhances resource efficiency and resilience. Agroforestry, the integration of trees and crops, is a sustainable agricultural approach that enhances ecosystem resilience by providing shade, windbreaks, income, and carbon sequestration (Diffenbaugh et al., 2017). Climate alterations cause water scarcity, necessitating efficient irrigation practices. Innovations such as drip irrigation, rainwater harvesting, and moisture sensors optimize water use and minimize losses, ensuring that crops receive adequate moisture without depleting precious resources (Ayers et al., 2014).

2.5.4 Devices in Arranging Adjustment

A recent climate report highlighted the importance of smart devices in transitioning from dirty to clean energy, emphasizing tools to understand energy sources and usage. The report suggests that humans must reduce pollution by half this decade to prevent extreme climate change, with emissions peaking before 2025 and virtually disappearing by 2050 (Costello et al., 2009). Smart devices are known for their pollution-free and energy-efficient grids. They can communicate with home devices. For example, thermostats are useful for preventing waste and providing renewable charging mechanisms throughout the day (Akhter Ali & Erenstein, 2017). Smart grids balance solar and wind energy demand and supply while reducing GHG emissions. Smart grids integrated with household devices decrease gas emissions and enhance grid flexibility. This is crucial for managing the challenges posed by climate change (Bonacci, 2019). Electric vehicles can connect two to three batteries simultaneously to meet customer needs. Fleets of such vehicles store energy for future use. This includes distributing solar panels and batteries across neighbourhoods, allowing virtual communication with utilities and potentially replacing traditional gas-fired "peaker" plants (Cammell & Knight, 1992). The new climate report highlights the potential benefits of digital technology for decarbonization but warns that it requires proper governance, as many countries lack policies to prevent e-waste and require longer-lasting devices to mitigate air pollution from shipping and manufacturing new goods (Bates et al., 2014).

2.5.5 Adjustment in Terms of a Few Argo Biological Viewpoints

Argo is a global initiative that uses approximately 4,000 autonomous ocean floats to profile the world's oceans vertically from 2,000 to 2,000 m every 10 days. These data provide unprecedented spatial coverage, with one float per three-degree box across the global ocean. Argo data minimizes spatial and temporal biases, making them crucial for atmospheric alteration detection (Akkari & Bryant, 2016). Heat transfer and seasonal cycles can overshadow interannual variability, allowing continuous global coverage to quantify long-term ocean changes more precisely. The Argo floats, with a few thousand metres, observe the ocean volume, capturing approximately 90% of the anthropogenic alterations for two decades (Neil Adger et al., 2005). However, efforts are underway to incorporate floats into the Deep Argo, an array capable of descending to the ocean bottom (maximum 6,000 m). Argo salinity data can help quantify climate change-related hydrological cycle changes (Roy et al., 2024). Warmer air can hold more moisture, thereby intensifying evaporation and precipitation patterns. This increases the risk of drought in arid areas and of flooding in wet areas. The ocean surface is a good location to observe this intensification, as sea surface salinity reflects the local balance between evaporation and precipitation (Rocchi et al., 2024).

2.5.6 Economic Adjustment Measures

Insurance serves to mitigate the financial repercussions of severe weather events with index-based insurance that activates payments when specific weather indices exceed predetermined thresholds. This mechanism assists farmers in managing production uncertainties, and may enhance the resilience of urban areas. The public sector can subsidize premiums in instances of market failure in the private insurance sector. Significant equity considerations for policy development have been highlighted (Kumar et al., 2023). It is important to note that transferring risk to the public sector does not diminish overall risk. Governments can distribute the costs of losses over an extended period and compel low-risk homeowners to subsidize the insurance premiums of high-risk homeowners. However, such cross-subsidization poses challenges for private insurers operating in competitive markets (Li et al., 2024). Critics argue that government-subsidized insurance, such as the U.S. National Flood Insurance Program, encourages hazardous development, increases risk, and undermines property-level protection and resilience (Muwafu et al., 2024). To counter this, appropriate land-use policies can limit new construction in climate-risk areas and encourage the adoption of resilient building codes to mitigate potential damage (Ma & Rahut, 2024).

2.5.7 Associations in Practicing Adjustment and Relief for Destitute Rustic Individuals

Coordination is crucial for achieving shared goals, such as information sharing and joint implementation of adaptation options (Grace et al., 2012). It facilitates the efficient utilization of resources, prevents redundancy, encourages uniformity among governments, and enhances operational clarity. In the realm of food production, adaptation initiatives funded by the United Nations Framework Convention on Climate Change typically require collaboration among national governments and state, provincial, or municipal authorities, with limited instances of cooperation between community-level entities and national governments (McMichael & Butler, 2011).

Ecosystems respond to global warming by improving their resilience to climate change. Human activities can promote biodiversity by enhancing ecosystem connectivity, enabling species migration, and facilitating the transport of plants and animals (Labonté et al., 2011). Scientific investigations can support the survival of coral reefs in the face of climate change, whereas the protection and restoration of both natural and semi-natural environments can contribute to resilience and aid ecosystem adaptation. Furthermore, adaptation strategies within ecosystems can benefit humans through ecosystem-based solutions and nature-oriented initiatives (McMichael, 2013). Restoring natural fire patterns can mitigate the risk of catastrophic fires and reduce human vulnerability, while creating more space for rivers can alleviate flooding in populated regions. Additionally, green spaces and tree planting offer shade for livestock; however, agricultural practices must be balanced with ecosystem restoration efforts in certain areas (Butler & McMichael, 2010).

2.6 FUTURE PROSPECTS

Research on atmospheric conditions has limited prospects due to the limited sample area and convenient sampling procedures. However, these findings could help policymakers understand the perceived awareness of atmospheric nature and alteration among younger citizens in a particular area (McMichael & Lindgren, 2011). Researchers hope that these insights will lead to new initiatives aimed at mitigating atmospheric alterations (Verma et al., 2023). Further statistical evaluation with a larger sample size is needed (Pauly et al., 2005). Researchers predict that further research is needed for mass awareness of atmospheric alterations, emphasizing the significance of involving all individuals in society to effectively address the issue and mitigate its effects (Marmot et al., 2008).

2.7 CONCLUSION

The intricate dynamics of atmospheric conditions and global warming have revealed numerous impacts on the environment, societies, and economies. This situation underscores the necessity for international cooperation to effectively address these pressing issues. The various factors contributing to climate change, particularly GHG emissions, deforestation, and industrial activities, highlight the critical need for a thorough understanding and proactive measures. This research emphasizes the urgent requirement for comprehensive and collaborative initiatives. This document explores a range of mitigation strategies, including transitioning to renewable energy sources, enhancing energy efficiency, adopting carbon capture technologies, promoting reforestation, and implementing sustainable urban planning. Additionally, it addresses adaptation strategies such as developing climate-resilient infrastructure, improving agricultural practices, informed policymaking, and community involvement, all of which emphasize the significance of global collaboration and the necessity for prompt action.

REFERENCES

Abbass, K., Qasim, M. Z., Song, H., Murshed, M., Mahmood, H., & Younis, I. (2022). A review of the global climate change impacts, adaptation, and sustainable mitigation measures. *Environmental Science and Pollution Research*, *29*(28), 42539–42559. https://doi.org/10.1007/s11356-022-19718-6

Abegg, B., Bürki, R., & Elsasser, H. (2008). Climate change and tourism in the Alps. In: Borsdorf, A., Stötter, J., Veulliet, E. (eds.), *Managing Alpine Future*. Wien: Verlag der Österreichischen Akademie der Wissenschaften, pp. 73–80.

Akkari, C. & Bryant, C. R. (2016). The co-construction approach as approach to developing adaptation strategies in the face of climate change and variability: A conceptual framework. *Agricultural Research*, *5*(2), 162–173. https://doi.org/10.1007/S40003-016-0208-8

Ali, A. (1999). Climate change impacts and adaptation assessment in Bangladesh. *Climate Research*, *12*(6), 109–116.

Ali, A. & Erenstein, O. (2017). Assessing farmer use of climate change adaptation practices and impacts on food security and poverty in Pakistan. *Climate Risk Management*, *16*, 183–194. https://doi.org/10.1016/J.CRM.2016.12.001

Allen, C. D., Macalady, A. K., Chenchouni, H., Bachelet, D., McDowell, N., Vennetier, M., Kitzberger, T. et al. (2010). A global overview of drought and heat-induced tree mortality reveals emerging climate change risks for forests. *Forest Ecology and Management*, *259*(4), 660–684. https://doi.org/10.1016/J.FORECO.2009.09.001

Amthor, J. S. (1995). Terrestrial higher-plant response to increasing atmospheric [CO_2] in relation to the global carbon cycle. *Global Change Biology*, *1*(4), 243–274. https://doi.org/10.1111/j.1365-2486.1995.tb00025.x

Anil, P., Dharmendra, P., Sachidanand, S., Yuriy, E., Snezhan, B., Dilfuza, J., & Chinmayi, J. (2023). Beneficial effects of biochar application on mitigating the drought and salinity stress implications on plants. *Plant Science Today*, *10*(2), 188–193. https://doi.org/10.14719/pst.2591

Anwar, I., Jamal, M. T., Saleem, I., & Thoudam, P. (2021). Traits and entrepreneurial intention: Testing the mediating role of entrepreneurial attitude and self-efficacy. *Journal for International Business and Entrepreneurship Development*, *13*(1), 40. https://doi.org/10.1504/JIBED.2021.112276

Asif, M., Xuhui, W., Nasiri, A., & Ayyub, S. (2018). Determinant factors influencing organic food purchase intention and the moderating role of awareness: A comparative analysis. *Food Quality and Preference*, *63*, 144–150. https://doi.org/10.1016/j.foodqual.2017.08.006

Asseng, S., Ewert, F., Rosenzweig, C., Jones, J. W., Hatfield, J. L., Ruane, A. C., Boote, K. J. et al. (2013). Uncertainty in simulating wheat yields under climate change. *Nature Climate Change*, *3*(9), 827–832. https://doi.org/10.1038/NCLIMATE1916

Ayers, J., Huq, S., Wright, H., Faisal, A. M., & Hussain, S. T. (2014). Mainstreaming climate change adaptation into development in Bangladesh. *Climate and Development*, *6*(4), 293–305. https://doi.org/10.1080/17565529.2014.977761

Balsalobre-Lorente, D., Driha, O. M., Shahbaz, M., & Sinha, A. (2020). The effects of tourism and globalization over environmental degradation in developed countries. *Environmental Science and Pollution Research*, *27*(7), 7130–7144. https://doi.org/10.1007/s11356-019-07372-4

Balsalobre-Lorente, D., Ibáñez-Luzón, L., Usman, M., & Shahbaz, M. (2022). The environmental kuznets curve, based on the economic complexity, and the pollution haven hypothesis in PIIGS countries. *Renewable Energy*, *185*, 1441–1455. https://doi.org/10.1016/J.RENENE.2021.10.059

Bates, A. E., Pecl, G. T., Frusher, S., Hobday, A. J., Wernberg, T., Smale, D. A., Sunday, J. M. et al. (2014). Defining and observing stages of climate-mediated range shifts in marine systems. *Global Environmental Change*, *26*(1), 27–38. https://doi.org/10.1016/J.GLOENVCHA.2014.03.009

Battisti, D. S. & Naylor, R. L. (2009). Historical warnings of future food insecurity with unprecedented seasonal heat. *Science*, *323*(5911), 240–244. https://doi.org/10.1126/science.1164363

Bellaloui, N., Bruns, H. A., Abbas, H. K., Mengistu, A., Fisher, D. K., & Reddy, K. N. (2015). Agricultural practices altered soybean seed protein, oil, fatty acids, sugars, and minerals in the Midsouth USA. *Frontiers in Plant Science*, *6*, 31. https://doi.org/10.3389/fpls.2015.00031

Bonacci, O. (2019). Air temperature and precipitation analyses on a small mediterranean Island: The case of the remote Island of Lastovo (Adriatic Sea, Croatia). *Acta Hydrotechnica*, *32*(57), 135–150. https://doi.org/10.15292/ACTA.HYDRO.2019.10

Bord, R. J., Fisher, A., & O'Connor, R. E. (1999). Public perceptions of global warming: United States and international perspectives. *Climate Research*, *11*(1), 75–84. https://doi.org/10.3354/CR011075

Bostrom, A., Morgan, M. G., Fischhoff, B., & Read, D. (1994). What do people know about global climate change? 1. Mental models. *Risk Analysis*, *14*(6), 959–970. https://doi.org/10.1111/J.1539-6924.1994.TB00065.X

Brown, H. C. P., Smit, B., Somorin, O. A., Sonwa, D. J., & Nkem, J. N. (2014). Climate change and forest communities: Prospects for building institutional adaptive capacity in the congo basin forests. *Ambio*, *43*(6), 759–769. https://doi.org/10.1007/S13280-014-0493-Z

Bujosa, A., Riera, A., & Torres, C. M. 2015. Valuing tourism demand attributes to guide climate change adaptation measures efficiently: The case of the Spanish domestic travel market. *Tourism Management*, *47*, 233–239. https://doi.org/10.1016/J.TOURMAN.2014.09.023

Butler, C. D. & McMichael, A. J. (2010). Population health: Where demography, environment and equity converge. *Journal of Public Health*, *32*(2), 157–158. https://doi.org/10.1093/PUBMED/FDQ039

Cammell, M. E. & Knight, J. D. (1992). Effects of climatic change on the population dynamics of crop pests. *Advances in Ecological Research*, *22*(C), 117–162. https://doi.org/10.1016/S0065-2504(08)60135-X

Chang, C. P., Harr, P. A., & Chen, H. J. (2005). Synoptic disturbances over the equatorial South China Sea and western maritime continent during boreal winter. *Monthly Weather Review*, *133*(3), 489–503. https://doi.org/10.1175/MWR-2868.1

Chaturvedi, A. K., Bahuguna, R. N., Shah, D., Pal, M., & Jagadish, S. V. K. (2017). High temperature stress during flowering and grain filling offsets beneficial impact of elevated CO_2 on assimilate partitioning and sink-strength in rice. *Scientific Reports*, *7*(1), 8227. https://doi.org/10.1038/s41598-017-07464-6

Chien, F., Sadiq, M., Nawaz, M. A., Hussain, M. S., Tran, T. D., & Le Thanh, T. (2021). A step toward reducing air pollution in top Asian economies: The role of green energy, eco-innovation, and environmental taxes. *Journal of Environmental Management*, *297*, 113420. https://doi.org/10.1016/j.jenvman.2021.113420

Costello, A., Abbas, M., Allen, A., Ball, S., Bell, S., Bellamy, R., & Friel, S. et al. 2009. Managing the health effects of climate change: Lancet and University College London Institute for Global Health Commission. *The Lancet*, *373*(9676), 1693–1733. https://doi.org/10.1016/S0140-6736(09)60935-1

Datta, A. K. & Rakesh, V. (2013). Principles of microwave combination heating. *Comprehensive Reviews in Food Science and Food Safety*, *12*(1), 24–39. https://doi.org/10.1111/j.1541-4337.2012.00211.x

Davidson, D. (2016). Gaps in agricultural climate adaptation research. *Nature Climate Change*, *6*(5), 433–435. https://doi.org/10.1038/NCLIMATE3007

Delgado, Eddy & Mitchell, Rowan & Parry, Martin & DRISCOLL, S. & MITCHELL, V. & Lawlor, David. (2006). Interacting effects of CO_2 concentration, temperature and nitrogen supply on the photosynthesis and composition of winter wheat leaves. *Plant, Cell & Environment. 17*. 1205–1213. 10.1111/j.1365-3040.1994.tb02018.x

Diffenbaugh, N. S., Singh, D., Mankin, J. S., Horton, D. E., Swain, D. L., Touma, D., & Charland, et al. (2017). Quantifying the influence of global warming on unprecedented extreme climate events. *Proceedings of the National Academy of Sciences of the United States of America*, *114*(19), 4881–4886. https://doi.org/10.1073/PNAS.1618082114

Drake, B. G., Gonzàlez-Meler, M. A., & Long, S. P. (1997). More efficient plants: A consequence of rising atmospheric CO_2? *Annual Review of Plant Physiology and Plant Molecular Biology*, *48*(1), 609–639. https://doi.org/10.1146/annurev.arplant.48.1.609

Ehleringer, J. R. & Cerling, T. E. (1995). Atmospheric CO_2 and the ratio of intercellular to ambient CO_2 concentrations in plants. *Tree Physiology*, *15*(2), 105–111. https://doi.org/10.1093/treephys/15.2.105

Feliciano, D., Recha, J., Ambaw, G., MacSween, K., Solomon, D., & Wollenberg, E. (2022). Assessment of agricultural emissions, climate change mitigation and adaptation practices in Ethiopia. *Climate Policy*, *22*(4), 427–444. https://doi.org/10.1080/14693062.2022.2028597

Fernando, N., Panozzo, J., Tausz, M., Norton, R. M., Neumann, N., Fitzgerald, G. J., & Seneweera, S. (2014). Elevated CO_2 alters grain quality of two bread wheat cultivars grown under different environmental conditions. *Agriculture, Ecosystems & Environment*, *185*, 24–33. https://doi.org/10.1016/j.agee.2013.11.023

Fettig, C. J., Reid, M. L., Bentz, B. J., Sevanto, S., Spittlehouse, D. L., & Wang, T. (2013). Changing climates, changing forests: A Western North American perspective. *Journal of Forestry*, *111*(3), 214–228. https://doi.org/10.5849/JOF.12-085

Fischer, A. P. (2019). Characterizing behavioral adaptation to climate change in temperate forests. *Landscape and Urban Planning*, *188*, 72–79. https://doi.org/10.1016/J.LANDURBPLAN.2018.09.024

Fischer, C., Malycha, C. P., & Schafmann, E. (2019). The influence of intrinsic motivation and synergistic extrinsic motivators on creativity and innovation. *Frontiers in Psychology*, *10*, 137 https://doi.org/10.3389/fpsyg.2019.00137

Fischhoff, B. (1995). Risk perception and communication unplugged: Twenty years of process. *Risk Analysis*, *15*(2), 137–145. https://doi.org/10.1111/J.1539-6924.1995.TB00308.X

Fritze, J. C., Blashki, G. A., Burke, S., & Wiseman, J. (2008). Hope, despair and transformation: Climate change and the promotion of mental health and wellbeing. *International Journal of Mental Health Systems*, *2*, 13. https://doi.org/10.1186/1752-4458-2-13

Gifford, R. M. (1995). Whole plant respiration and photosynthesis of wheat under increased CO_2 concentration and temperature: Long-term vs. short-term distinctions for modelling. *Global Change Biology*, *1*(6), 385–396. https://doi.org/10.1111/j.1365-2486.1995.tb00037.x

Glaubrecht, M. (2008). Charles Darwin. The man and his influence. *Zoosystematics and Evolution*, *74*(1), 147–148. https://doi.org/10.1002/mmnz.19980740113

Gow, J. & Leahy, T. (2005). Apocalypse probably: Agency and environmental risk in the hunter region. *Journal of Sociology*, *41*(2), 117–141. https://doi.org/10.1177/1440783305050966

Grace, K., Davenport, F., Funk, C. & Lerner, A. M. (2012). Child malnutrition and climate in Sub-Saharan Africa: An analysis of recent trends in Kenya. *Applied Geography*, *35*(1–2), 405–413. https://doi.org/10.1016/J.APGEOG.2012.06.017

Haines, A., Kovats, R. S., Campbell-Lendrum, D., & Corvalan, C. (2006). Climate change and human health: Impacts, vulnerability, and mitigation. *Lancet*, *367*(9528), 2101–2109. https://doi.org/10.1016/S0140-6736(06)68933-2.

Halady, I. R. & Rao, P. H. (2010). Does awareness to climate change lead to behavioral change? *International Journal of Climate Change Strategies and Management*, *2*(1), 6–22. https://doi.org/10.1108/17568691011020229

Hansen, J. & Lebedeff, S. (1987). Global trends of measured surface air temperature. *Journal of Geophysical Research: Atmospheres*, *92*(D11), 13345–13372. https://doi.org/10.1029/JD092iD11p13345

Huang, W., Yang, Y. -J., Hu, H., Cao, K. -F., & Zhang, S. -B. (2016). Sustained Diurnal Stimulation of Cyclic Electron Flow in Two Tropical Tree Species Erythrophleum guineense and Khayaivorensis. *Front. Plant Sci.*, *7*, 1068. https://doi.org/10.3389/fpls.2016.0106812

Hummel, M. A., Berry, M. S., & Stacey, M. T. (2018). Sea level rise impacts on wastewater treatment systems along the U.S. coasts. *Earth's Future*, *6*(4), 622–633. https://doi.org/10.1002/2017EF000805

Hussain, S., Xuetong, W., Hussain, T., Khoja, A. H., & Zia, M. Z. (2021). Assessing the impact of COVID-19 and safety parameters on energy project performance with an analytical hierarchy process. *Utilities Policy*, *70*, 101210. https://doi.org/10.1016/j.jup.2021.101210

Izaguirre, C., Losada, I. J., Camus, P., Vigh, J. L., & Stenek, V. (2021). Climate change risk to global port operations. *Nature Climate Change*, *11*(1), 14–20. https://doi.org/10.1038/s41558-020-00937-z

Jabborova, D., Nigora, Z., Yuriy, E., Abdujalil, N., Anand, D., Khurshid, S., Zafarjon, J., Sachidanand, S., & Rahul, D. (2023). Growth of spinach as influenced by biochar and bacillus endophyticus IGPEB 33 in drought condition. *Pakistan Journal of Botany*, *55*, 6. http://dx.doi.org/10.30848/PJB2023-SI(6)

Jannat, A., Ishikawa-Ishiwata, Y., & Furuya, J. (2022). Does climate change affect rapeseed production in exporting and importing countries? Evidence from market dynamics syntheses. *Sustainability*, *14*(10), 6051. https://doi.org/10.3390/su14106051

Jurgilevich, A., Räsänen, A., Groundstroem, F., & Juhola, S. (2017). A systematic review of dynamics in climate risk and vulnerability assessments. *Environmental Research Letters*, *12*(1), 013002. https://doi.org/10.1088/1748-9326/aa5508

Kempton, W. (1997). How the public views climate change. *Environment*, *39*(9), 12–21. https://doi.org/10.1080/00139159709604765

Kirschbaum, M. U. F., King, D. A., Comins, H. N., McMurtrie, R. E., Medlyn, B. E., Pongracic, S., Murty, D., Keith, H., Raison, R. J., Khanna, P. K., & Sheriff, D. W. (1994). Modelling forest response to increasing CO_2 concentration under nutrient-limited conditions. *Plant, Cell and Environment*, in press.

Köhler, J., de Haan, F., Holtz, G., Kubeczko, K., Moallemi, E., Papachristos, G., & Chappin, E. (2018). Modelling sustainability transitions: An assessment of approaches and challenges. *Journal of Artificial Societies and Social Simulation*, *21*(1), 3629. https://doi.org/10.18564/jasss.3629

Kumar, Y. V., Srija, K., Snehitha, L., & Sunkara, L. K. (2023). The impact of climate change on international business strategies. *International Journal for Research in Applied Science and Engineering Technology*, *11*(7), 461–465. https://doi.org/10.22214/IJRASET.2023.54687

Labonté, R., Mohindra, K., & Schrecker, T. (2011). The growing impact of globalization for health and public health practice. *Annual Review of Public Health*, *32*, 263–283. https://doi.org/10.1146/ANNUREV-PUBLHEALTH-031210-101225

Lawlor, D. (1995). The effects of water deficit on photosynthesis In: Environment and plant metabolisim (Smirnoff ed.). *Bios. Sci. Publishers*, 129–160.

Lawlor, D. W. & Cornic, G. (2002). Photosynthetic carbon assimilation and associated metabolism in relation to water deficits in higher plants. *Plant, Cell & Environment*, *25*(2), 275–294. https://doi.org/10.1046/j.0016-8025.2001.00814.x

Lawlor, D. W. & Mitchell, R. A. C. (1991). The effects of increasing CO_2 on crop photosynthesis and productivity: A review of field studies. *Plant, Cell & Environment*, *14*(8), 807–818. https://doi.org/10.1111/j.1365-3040.1991.tb01444.x

Leal Filho, W., Wall, T., Rayman-Bacchus, L., Mifsud, M., Pritchard, D. J., Lovren, V. O., Farinha, C., Petrovic, D. S., & Balogun, A.-L. (2021). Impacts of COVID-19 and social isolation on academic staff and students at universities: A cross-sectional study. *BMC Public Health*, *21*(1), 1213. https://doi.org/10.1186/s12889-021-11040-z

Leiserowitz, A. (2006). Climate change risk perception and policy preferences: The role of affect, imagery, and values. *Climatic Change*, *77*(1–2), 45–72. https://doi.org/10.1007/S10584-006-9059-9

Leppänen, H. K., Nevalainen, A., Vepsäläinen, A., Roponen, M., Täubel, M., Laine, O., Rantakokko, P., von Mutius, E., Pekkanen, J. & Hyvärinen, A. (2014). Corrigendum: Leppänen HK, et al. (2014) Jun, *24*(3), 248–259. doi: 10.1111/ina.12078. PMID: 32333695.

Li, Z., Zhang, H., Song, X., & Ye, W. (2024). Social responsibility awareness and adoption of climate-smart agricultural practices: Evidence from food-based family farms in China. *Mitigation and Adaptation Strategies for Global Change*, *29*(5), 6. https://doi.org/10.1007/S11027-024-10132-6.

Lipczynska-Kochany, E. (2018). Effect of climate change on humic substances and associated impacts on the quality of surface water and groundwater: A review. *Science of the Total Environment*, *640–641*, 1548–1565. https://doi.org/10.1016/j.scitotenv.2018.05.376

Lloyd, J. & Farquhar, G. (1996). The CO2 Dependence of Photosynthesis, Plant Growth Responses to Elevated Atmospheric CO2 Concentrations and Their Interaction with Soil Nutrient Status. I. *General Principles and Forest Ecosystems. Functional Ecology*, *10*, 4–32. http://dx.doi.org/10.2307/2390258

Lloyd, J. & Farquhar, G. D. (2008). Effects of rising temperatures and [CO_2] on the physiology of tropical forest trees. *Philosophical Transactions of the Royal Society B: Biological Sciences*, *363*(1498), 1811–1817. https://doi.org/10.1098/rstb.2007.0032

Lorenzoni, I., Nicholson-Cole, S., & Whitmarsh, L. (2007). Barriers perceived to engaging with climate change among the UK public and their policy implications. *Global Environmental Change*, *17*(3–4), 445–459. https://doi.org/10.1016/J.GLOENVCHA.2007.01.004

Ma, W. & Rahut, D. B. (2024). Climate-smart agriculture: Adoption, impacts, and implications for sustainable development. *Mitigation and Adaptation Strategies for Global Change*, *29*(5), 44. https://doi.org/10.1007/S11027-024-10139-Z

Marmot, M., Friel, S., Bell, R., Houweling, T. A. J., & Taylor, S. (2008). Closing the gap in a generation: Health equity through action on the social determinants of health. *The Lancet*, *372*(9650), 1661–1669. https://doi.org/10.1016/S0140-6736(08)61690-6

McMichael, A. J. (2013). Globalization, climate change, and human health. *New England Journal of Medicine*, *368*(14), 1335–1343. https://doi.org/10.1056/NEJMRA1109341

McMichael, A. J. & Butler, C. D. (2011). Promoting global population health while constraining the environmental footprint. *Annual Review of Public Health*, *32*, 179–197. https://doi.org/10.1146/ANNUREV-PUBLHEALTH-031210-101203

McMichael, A. J. & Lindgren, E. (2011). Climate change: Present and future risks to health, and necessary responses. *Journal of Internal Medicine*, *270*(5), 401–413. https://doi.org/10.1111/J.1365-2796.2011.02415.X

Meddya, S., Shweta, M., Deepranjan, S., Rakesh, S., Rahul, D., Sachidanand, S., Gosangi, A., Arun, K. K., Ajit, K. S., & Thiyagarajan, T. (2023). Plant stomata: An unrealized possibility in plant defense against invading pathogens and stress tolerance. *Plants*, *12*(19), 3380. https://doi.org/10.3390/plants12193380

Morison, J. I. L. & Lawlor, D. W. (1999). Interactions between increasing CO_2 concentration and temperature on plant growth. *Plant, Cell & Environment*, *22*(6), 659–682. https://doi.org/10.1046/j.1365-3040.1999.00443.x

Murshed, M. (2020). An empirical analysis of the non-linear impacts of ICT-trade openness on renewable energy transition, energy efficiency, clean cooking fuel access and environmental sustainability in South Asia. *Environmental Science and Pollution Research*, *27*(29), 36254–36281. https://doi.org/10.1007/s11356-020-09497-3

Murshed, M. & Dao, N. T. T. (2022). Revisiting the CO_2 emission-induced EKC hypothesis in South Asia: The role of export quality improvement. *GeoJournal*, *87*(2), 535–563. https://doi.org/10.1007/s10708-020-10270-9

Murshed, M., Rashid, S. & Ulucak, R. et al. (2022). Mitigating energy production-based carbon dioxide emissions in Argentina: the roles of renewable energy and economic globalization. *Environ Sci Pollut Res*, *29*, 16939–16958.

Muwafu, S. P., Rölfer, L., Scheffran, J., & Costa, M. M. (2024). A framework for assessing social structure in community governance of sustainable urban drainage systems: Insights from a literature review. *Mitigation and Adaptation Strategies for Global Change*, *29*(5), 2. https://doi.org/10.1007/S11027-024-10136-2

Nakandalage, N. & Seneweera, S. (2018). Micronutrients use efficiency of crop-plants under changing climate. *Plant Micronutrient Use Efficiency*, *2018*, 209–224. https://doi.org/10.1016/B978-0-12-812104-7.00015-0

Neil Adger, W., Arnell, N. W., & Tompkins, E. L. (2005). Successful adaptation to climate change across scales. *Global Environmental Change*, *15*(2), 77–86. https://doi.org/10.1016/j.gloenvcha.2004.12.005

Neocleous, D., Koukounaras, A., Siomos, A. S., & Vasilakakis, M. (2014). Assessing the salinity effects on mineral composition and nutritional quality of green and red baby lettuce. *Journal of Food Quality*, *37*(1), 1–8. https://doi.org/10.1111/jfq.12066

Parvin, K., Hasanuzzaman, M., Bhuyan, M. H. M. B., Mohsin, S. M., & Fujita, M. (2019). Quercetin Mediated Salt Tolerance in Tomato through the Enhancement of Plant Antioxidant Defense and Glyoxalase Systems. *Plants*, *8*(8), 247. https://doi.org/10.3390/plants8080247

Patani, A., Dharmendra, P., Krishna, S., Mahenk, P., Pragnesh, P., Ashish, P., & Sachidanand, S. (2024). Environmental stress–induced alterations in the micro- and macronutrients status of plant. *Essential Minerals in Plant-Soil Systems*, *2024*, 181195.

Patz, J. A., Campbell-Lendrum, D., Holloway, T., & Foley, J. A. (2005). Impact of regional climate change on human health. *Nature*, *438*(7066), 310–317. https://doi.org/10.1038/NATURE04188

Pauly, D., Watson, R., & Alder, J. (2005). Global trends in world fisheries: Impacts on marine ecosystems and food security. *Philosophical Transactions of the Royal Society B: Biological Sciences*, *360*(1453), 5–12. https://doi.org/10.1098/RSTB.2004.1574.

Petretto, G. L., Urgeghe, P. P., Massa, D., & Melito, S. (2019). Effect of salinity (NaCl) on plant growth, nutrient content, and glucosinolate hydrolysis products trends in rocket genotypes. *Plant Physiology and Biochemistry*, *141*, 30–39. https://doi.org/10.1016/j.plaphy.2019.05.012

Petropoulos, S. A., Levizou, E., Ntatsi, G., Fernandes, Â., Petrotos, K., Akoumianakis, K., Barros, L., & Ferreira, I. C. F. R. (2017). Salinity effect on nutritional value, chemical composition and bioactive compounds content of *Cichorium spinosum* L. *Food Chemistry*, *214*, 129–136. https://doi.org/10.1016/j.foodchem.2016.07.080

Prajapati, D., Dilfuza, J., Baljeet, S. S., Namita, S., Anil, P., Sachidanand, S., & Chinmayi, J. (2024). Bionanotechnology: A paradigm for advancing environmental sustainability. *Indian Journal of Microbiology*, *13*, 1–27. https://doi.org/10.1007/s12088-024-01389-1.

Qiao, R., Lu, K., Deng, Y., Ren, H., & Zhang, Y. (2019). Combined effects of polystyrene microplastics and natural organic matter on the accumulation and toxicity of copper in zebrafish. *Science of the Total Environment*, *682*, 128–137. https://doi.org/10.1016/j.scitotenv.2019.05.163

Rocchi, L., Rizzo, A. G., Paolotti, L., Boggia, A., & Attard, M. (2024). Assessing climate change vulnerability of coastal roads. *Mitigation and Adaptation Strategies for Global Change*, *29*(5), 6. https://doi.org/10.1007/S11027-024-10140-6

Rowland-Bamford, A. J., Baker, J. T., Allen, L. H., & Bowes, G. (1996). Interactions of CO_2 enrichment and temperature on carbohydrate accumulation and partitioning in rice. *Environmental and Experimental Botany*, *36*(1), 111–124. https://doi.org/10.1016/0098-8472(95)00025-9

Roy, M. N., Das, S., Mitra, R., & Mukherjee, D. (2024). Groundwater and climate smart agriculture: A reflection from West Bengal, India. *Mitigation and Adaptation Strategies for Global Change*, *29*(5), 1. https://doi.org/10.1007/S11027-024-10137-1

Sabir, A. & Sari, G. (2019). Zinc pulverization alleviates the adverse effect of water deficit on plant growth, yield and nutrient acquisition in grapevines (*Vitis vinifera* L.). *Scientia Horticulturae*, *244*, 61–67. https://doi.org/10.1016/j.scienta.2018.09.035

Sage, R. F., Santrucek, J., & Grise, D. J. (1995). Temperature effects on the photosynthetic response of C_3 plants to long-term CO_2 enrichment. *Vegetatio*, *121*(1–2), 67–77. https://doi.org/10.1007/BF00044673

Sarker, U. & Oba, S. (2018). Drought stress enhances nutritional and bioactive compounds, phenolic acids and antioxidant capacity of Amaranthus leafy vegetable. *BMC Plant Biology*, *18*(1), 258. https://doi.org/10.1186/s12870-018-1484-1

Scagel, C. F., Lee, J., & Mitchell, J. N. (2019). Salinity from NaCl changes the nutrient and polyphenolic composition of basil leaves. *Industrial Crops and Products*, *127*, 119–128. https://doi.org/10.1016/j.indcrop.2018.10.048

Schuurman, G. W., Cole, D. N., Cravens, A. E., Covington, S., Crausbay, S. D., Hoffman, C. H., Lawrence, D. J., Magness, D. R., Morton, J. M., Nelson, E. A., & O'Malley, R. (2022). Navigating ecological transformation: Resist–accept–direct as a path to a new resource management paradigm. *BioScience*, *72*(1), 16–29. https://doi.org/10.1093/biosci/biab067

Shahbaz, M., Raghutla, C., Chittedi, K. R., Jiao, Z., & Vo, X. V. (2020). The effect of renewable energy consumption on economic growth: Evidence from the renewable energy country attractive index. *Energy*, *207*, 118162. https://doi.org/10.1016/j.energy.2020.118162

Sharif, A., Aloui, C., & Yarovaya, L. (2020). COVID-19 pandemic, oil prices, stock market, geopolitical risk and policy uncertainty nexus in the US economy: Fresh evidence from the wavelet-based approach. *International Review of Financial Analysis*, *70*, 101496. https://doi.org/10.1016/j.irfa.2020.101496

Sharma, O., Sultan, A. A., Ding, H., & Triggle, C. R. (2020). A review of the progress and challenges of developing a vaccine for COVID-19. *Frontiers in Immunology*, *11*, 585354. https://doi.org/10.3389/fimmu.2020.585354

Steffens, Niklas K. & Haslam, S. & Schuh, Sebastian & Jetten, Jolanda & Dick, Rolf. (2017). A Meta-Analytic Review of Social Identification and Health in Organizational Contexts. *Personality and Social Psychology Review*, *21*, 303–335. 10.1177/1088868316656701

Tarekegne, C., Wesselink, R., Biemans, H. J. A., & Mulder, M. (2021). Developing and validating a competence framework for improving the productivity of smallholder farmers: A case study from Ethiopia. *The Journal of Agricultural Education and Extension*, *27*(4), 481–502. https://doi.org/10.1080/1389224X.2021.1880452

Tungare, K., Shahu, R., Zambare, V., Agarwal, P., Jobby, R., Nisar, N., Alabdallah, N. M., Al-Saeed, F. A., Johri, P., Singh, S., Saeed, M., & Jha, P. (2022). Toxicity mitigation of textile dye reactive blue 4 by hairy roots of *Helianthus annuus* and testing its effect in *in vivo* model systems, *BioMed Research International*, *1958939*, 13. https://doi.org/10.1155/2022/1958939

Tvaronavičienė, M. (2021). Effects of climate change on environmental sustainability. *E3S Web of Conferences*, *250*, 01005. https://doi.org/10.1051/e3sconf/202125001005

Usman, M. & Balsalobre-Lorente, D. (2022). Environmental concern in the era of industrialization: Can financial development, renewable energy and natural resources alleviate some load? *Energy Policy*, *162*, 112780. https://doi.org/10.1016/j.enpol.2022.112780

Velu, G. & Guzmán, Carlos & Mondal, Suchismita & Autrique, Enrique & Huerta-Espino, Julio & Singh, Ravi. (2016). Effect of Drought and Elevated Temperature on Grain Zinc and Iron Concentrations in CIMMYT Spring Wheat. *Journal of Cereal Science*, *69*. 10.1016/j.jcs.2016.03.006

Velu, G., Singh, R. P., Crespo-Herrera, L., Juliana, P., Dreisigacker, S., Valluru, R., Stangoulis, J., Sohu, V. S., Mavi, G. S., Mishra, V. K., Balasubramaniam, A., Chatrath, R., Gupta, V., Singh, G. P., & Joshi, A. K. (2018). Genetic dissection of grain zinc concentration in spring wheat for mainstreaming biofortification in CIMMYT wheat breeding. *Scientific Reports*, *8*(1), 13526. https://doi.org/10.1038/s41598-018-31951-z

Verma, A., Kumar, R., Shukla, S., Singh, S., Johri, P., & Singh, M. S. (2023). Genome wide association studies: An outlook. *Journal of Innovation in Applied Research*, *6*(1), 56–64. https://doi.org/10.51323/JIAR.2023/6/1/56-62

Verma, N., Prachi, S., Aashna, S., & Sachidanand, S. (2021). Physiological and molecular responses to drought, submergence and excessive watering in plants. In: Husen, A. (ed.), *Harsh Environment and Plant Resilience*. Cham: Springer International Publishing.

Vlek, C. A. J. & Steg, L. (2007). Human behavior and environmental sustainability: Problems, driving forces, and research topics. *Journal of Social Issues*, *63*(1), 1–19. https://doi.org/10.1111/J.1540-4560.2007.00493.X

Wei, X., Zhang, S., Sun, Y., & Brenner, S. A. (2018). Petrochemical wastewater and produced water. *Water Environment Research*, *90*(10), 1634–1647. https://doi.org/10.2175/106143018X15289915807344

Weisheimer, A. & Palmer, T. N. (2005). Changing frequency of occurrence of extreme seasonal temperatures under global warming. *Geophysical Research Letters*, *32*(20), L20721. https://doi.org/10.1029/2005GL023365

Wijewardana, C., Reddy, K. R., Krutz, L. J., Gao, W., & Bellaloui, N. (2019). Drought stress has transgenerational effects on soybean seed germination and seedling vigor. *PLoS One*, *14*(9), e0214977. https://doi.org/10.1371/journal.pone.0214977

Ziska, L., & Bunce, J. (1997). Influence of increasing carbon dioxide concentration on the photosynthetic and growth stimulation of selected C4 crops and weeds. *Photosynthesis Research*, *54*. 10.1023/A:1005947802161

3 The Impact of Heavy Metals on Medicinal Plants and Their Active Components

Srimoyee Banerjee and Rishabh Mukherjee*

3.1 INTRODUCTION

Medicinal plants form the core of traditional medicine and are an integral part of alternative and combinatorial therapies currently being used. The use of medicinal plants for the treatment of numerous diseases is now an established method for many people globally. Due to the presence of specific bioactive compounds in medicinal plants, they have many applications in the treatment and management of various diseases. However, there is a serious concern regarding the use of these plants and the maintenance of their purity. The growing global population and industrialization are causing increased environmental pollution, which seriously threatens the safety and effectiveness of medicinal plants (Sen & Samanta, 2015). Among the various types of pollutants to which medicinal plants are exposed, heavy metals (HMs) pose a particularly grave risk to these plants, their beneficial compounds, and the people who use them.

HMs naturally exist in the Earth's crust, but human activities have caused their concentrations to rise, leading to contamination of soil, water, and air. Although plants need metals for metabolic purposes, their presence in concentrations above permissible levels creates toxicity. HMs like arsenic, cadmium, and mercury have been found in the environment and can be absorbed by plants through water, soil, or air (Shah et al., 2013). Once absorbed, they start accumulating in plant leaves, roots, and other tissues, thereby affecting growth, development, reproductive processes, and ultimately the active components of the plants. Moreover, HMs can interact with the beneficial compounds found in medicinal plants, potentially reducing their effectiveness (Nasim & Dhir, 2010). For example, HMs may form complexes with the active compounds in medicinal plants due to their reactive nature, affecting both solubility and absorption in the body. These complexes themselves might also be harmful. Furthermore, when people consume these plants or their extracts for therapeutic purposes, the HMs can enter the human body, posing serious health risks (Maleki et al., 2017).

To reduce HM contamination in medicinal plants, it is crucial to understand the impact of HMs on plants and the bioactive elements they contain. It is necessary to comprehend how HMs enter plants and affect their properties. To grasp these intricate processes, one must recognize the potential sources of HM contamination. Various sources, such as industrial and mining discharges, agricultural waste, and untreated wastewater, can infiltrate soil and water, endangering ecosystems. The accumulation of HMs in plants can lead to concerns such as bioaccumulation and biomagnification, which may affect consumers of medicinal plants. Long-term consumption of contaminated plant extracts has been associated with chronic health issues, neurological disorders, organ damage, and cancer. Moreover, the presence of HMs can reduce the medicinal effectiveness of plant extracts, compromising their ability to effectively address health conditions.

The challenges posed by HM contamination in medicinal plants underscore the urgent requirement for collaborative actions by the scientific community to effectively tackle and alleviate these issues. To

* *Corresponding Author*

 DOI: 10.1201/9781003491378-3

prevent HM contamination at its source, proactive steps must be implemented due to the significant consequences for public health and safety. Understanding the intricate relationships that HMs have with plants, particularly how these relationships affect bioactive components, is an important area of study. This investigation clarifies the mechanisms by which these metals enter plant systems and the subsequent effects. Gaining a thorough understanding of these processes helps pave the way for effectively regulating HM pollution in medicinal plants. HM pollution in medicinal plants can be attributed to several factors, including mining operations, industrial discharge, agricultural waste, and the careless release of wastewater into water bodies. These pollutants are insidious because they infiltrate the soil and water, perpetuating a cycle of potential ecological harm even after their initial release. HMs can lead to various ecological consequences, such as bioaccumulation and biomagnification, once they are absorbed by plants. Due to their known threat to human health and their tendency to accumulate in the food chain, these mechanisms heighten the risk for individuals who use medicinal plants. Prolonged exposure to plant extracts containing HMs has been linked to a variety of health problems, including neurological deficits, cancer, and chronic illnesses. The negative impacts of HMs extend beyond human health; they also undermine the therapeutic potential and intended healing capabilities of medicinal plant extracts. It is thus important to understand the need to address the problems of HMs in medicinal plants. This will not only contribute to environmental protection but also enhance the quality of medicinal plants, their healing properties, and thereby improve human life.

Medicinal plants and their extracts are consumed by millions across the globe as an alternative way to cure diseases (Aslam & Ahmad, 2016). Due to these issues, it is imperative to improve our understanding of how the presence of HMs poses a serious threat to human beings who consume them as a means of alternative therapy. This knowledge will enable the reduction of HM contamination sources, the establishment of robust quality control mechanisms for medicinal products, and the maximization of their efficacy. This chapter provides a thorough review of the issues surrounding the cultivation of medicinal plants in contaminated areas, as well as potential controls. It also serves as a warning to researchers, decision-makers, and those who advise using herbal medicines to be cautious about the quality of these products to uphold safety standards. This will ensure that those who use medicinal plants and their preparations for therapeutic purposes receive the maximum benefit and that their trust in complementary and alternative medical practices is maintained. Such an approach will not only help maintain the quality of medicinal plants and their bioactive components but also uphold people's belief in traditional medicine and its longevity.

3.2 SOURCES OF HEAVY METALS IN THE ENVIRONMENT

HMs are characterized by their high atomic weight and density, distinguishing them from other metals and metalloids. While they naturally occur in the Earth's crust, their concentrations can increase in soil, water, and air due to various factors. In addition to their natural occurrence, industrial processes and urbanization significantly contribute to the elevated concentrations of HMs in the environment beyond permissible limits. These sources have been briefly discussed in the paragraphs below.

3.2.1 Extraction and Refining

The mining and smelting industries have historically been identified as major sources of HM contamination, posing severe threats to the well-being of both people and the environment. These activities involve a range of processes, such as digging, concentrating ores, and generating mine waste, all of which release HMs into the surrounding soil, water, and air. If not properly managed, these processes can lead to widespread environmental contamination.

HMs are primarily discharged into the environment as a result of mining and smelting waste and dust. Dust, in particular, can spread over long distances and settle on soil and vegetation. The unregulated release of wastewater can lead to seepage into soil and waterways, causing pollution.

It is crucial to regulate these pollutants to prevent contamination. Over time, these metals build up in the environment, posing a risk to both ecosystem balance and human well-being.

Yu et al.'s (2006) studies, which found higher concentrations of HMs in human hair and urine samples taken near mining and smelting facilities, emphasized the importance of controlling these effects. Residents in the area are more vulnerable to health concerns like cancer, respiratory ailments, and neurological diseases due to this increased exposure to HMs. Furthermore, health risks increase when HM exposure contaminates vegetables. Research by Briki et al. (2017) indicated that vegetables cultivated in mining and smelting-affected areas contained higher concentrations of HMs. This endangers local communities and raises concerns about possible contamination further up the food chain. Of particular concern is the carcinogenicity index, which measures the potential of substances to cause cancer. Briki et al. (2017) reported that the carcinogenicity index exceeded the permissible limit in areas affected by mining and smelting activities, indicating a significant health risk to residents. This highlights the urgent need for stricter regulations and mitigation measures to prevent further contamination and protect public health.

Furthermore, the impact of mining activities extends beyond human health to include ecological consequences. Kapusta and Lukasz (2015) demonstrated how mining operations can disrupt soil ecosystems, particularly affecting populations of enchytraeids, a type of soil-dwelling worm. Their research revealed that proximity to mining areas had a detrimental effect on enchytraeid populations, indicating the extensive ecological disruption caused by mining activities.

Overall, the presence of high levels of HMs in the environment, resulting from mining and smelting activities, significantly affects both ecosystems and human health. It is imperative to implement proper regulations and mitigation measures to protect ecological balance, public well-being, and the environment from the harmful effects of HM contamination.

3.2.2 Industry

Aside from mining, a myriad of industries involved in metal manufacturing, paints, electronics, and the production of various chemicals serve as potential sources of metal contamination, amplifying the scope and complexity of this environmental challenge. HMs like lead, cadmium, and chromium are frequently used in industrial processes such as smelting and refining, with the potential for their release into the environment during production or disposal of waste byproducts. For example, these metals are commonly used in the manufacturing of batteries, paints, and alloys, which may lead to their emissions or improper disposal into nearby ecosystems. The substantial contribution of these industries to HM pollution has been highlighted by research conducted by Xiangyang et al. (2020), underscoring the urgent need for strict regulatory measures and environmentally friendly waste management techniques to reduce the negative effects on the environment. Furthermore, research conducted by Yang et al. (2018) has empirically shown that industrial locations have higher than average cadmium levels. This emphasizes the critical necessity to address metal contamination from industrial activities to protect human health and environmental integrity.

3.2.3 Agriculture

Metal manufacturing, paints, electronics, and chemical production industries represent significant contributors to HM contamination, compounding the environmental challenges posed by mining activities. These diverse industries utilize HMs like lead, cadmium, and chromium in various processes, increasing the complexity of metal pollution. In manufacturing processes such as smelting and refining, HMs are commonly used, with the potential for their release into the environment during production or disposal of waste byproducts. For example, lead, cadmium, and chromium are frequently employed in the production of batteries, paints, and alloys. However, inadequate waste management practices or emissions can lead to the infiltration of these metals into surrounding ecosystems.

The utilization of HMs in industrial processes presents a significant risk to environmental and human health. Research conducted by Xiangyang et al. (2020) has highlighted the substantial contribution of these industries to HM pollution. Their findings underscore the critical need for stringent regulatory measures and sustainable waste management practices to effectively mitigate environmental impacts. Without proper controls, the discharge of HMs from industrial activities can contaminate soil, water, and air, leading to adverse effects on ecosystems and human populations.

Research findings, like those from the study by Yang et al. (2018), underscore the critical importance of addressing metal pollution resulting from industrial operations. High concentrations of cadmium have been observed in industrial zones, posing substantial threats to both the environment and public health. These findings highlight the urgent need to proactively reduce the discharge of toxic metals into the environment and monitor industrial discharges.

The complex relationship between industrial practices, waste management, and environmental pollution emphasizes the need for a holistic approach to address metal pollution. This approach should involve government regulations, technological advancements, and collaboration among all stakeholders to minimize the harmful effects of metal pollution while promoting sustainable practices. Regulatory bodies play a vital role in establishing and upholding standards for the emission of HMs from industrial sources. Moreover, technological advancements can lead to cleaner production processes and improved waste management methods, thereby reducing the environmental impact of industrial operations. Involving stakeholders such as industry representatives, government agencies, and affected communities is also crucial. This collaboration fosters cooperation and ensures that effective solutions are implemented to address metal pollution, promoting a sense of environmental responsibility among all parties involved.

In summary, industrial operations in metal manufacturing, paints, electronics, and chemical manufacturing significantly contribute to the spread of HMs, exacerbating environmental issues caused by mining activities. A multifaceted strategy that includes technological innovations, stakeholder engagement, and regulatory measures is necessary to address this issue, reduce the negative consequences of metal pollution, and enhance sustainable development.

3.2.4 Burning of Coal

The process of burning coal to produce energy is a major source of HM emissions into the environment, including lead, arsenic, and mercury. Before these metals settle on land or in water, they can travel great distances in the atmosphere. Once there, they can build up in sediments and living organisms. Pulverized coal is used extensively in the production of industrial power, resulting in the generation of large amounts of combustion residues (wastes) or byproducts such as pulverized coal ash and slag, in addition to acidic pollutants (SO_2 and NOx). Approximately 99.5% of the fly ash is collected, with the remaining portion being discharged into the atmosphere, depending on the efficiency of the electrostatic precipitators (Altıkulaç et al., 2022). The majority of these ground-up coal ashes are disposed of in ash ponds or on land. The utilization rate of fly ash is under 1%, and only a very small portion of these coal ashes is employed. HMs or potentially harmful elements (As, Cr, Zr, Ni, Cd, Hg, Pb, etc.) are redistributed into fly ash, bottom ash, and slag in electrostatic precipitators during the combustion of pulverized coal (Al-Areqi et al., 2019).

3.2.5 Waste Disposal

The issue of HM pollution from waste disposal is a complex problem that impacts both the environment and public health. Improper disposal of various types of waste, such as household garbage, electronic waste, and industrial effluents, can lead to HMs contaminating the soil, water, and surroundings. This pollution is primarily caused by industrial activities like mining, manufacturing, and chemical production, which release harmful substances such as lead, mercury, cadmium, and arsenic. Inadequate waste management practices further exacerbate the problem by allowing HMs to leach into the ground and groundwater, contaminating landfills and dumpsites. The improper disposal of electronic waste, which contains high levels of HMs like lead and mercury, poses a significant risk to ecosystems and human health due to their long-lasting presence in the environment. The consequences of HM poisoning include loss of biodiversity, soil degradation, water pollution, and an increased prevalence of chronic diseases such as cancer and respiratory and neurological disorders. Addressing this challenging issue requires a multidisciplinary approach involving strict regulations, eco-friendly waste management methods, public awareness initiatives, and advancements in waste treatment technologies. Prioritizing the prevention and reduction of HM contamination from waste disposal can help protect ecosystems, safeguard public health, and ensure a sustainable future (Appiah-Adjei et al., 2016; Wang et al., 2019; Briffa et al., 2020).

3.2.6 Natural Sources

Although human activities significantly impact HM levels in the environment, these metals also originate from natural sources. Natural events such as forest fires, volcanic eruptions, and soil and rock erosion release HMs into the atmosphere, soil, and water systems, in addition to the contributions from human activities. This dual origin highlights the intricate nature of HM pollution, with both human-driven and natural factors influencing environmental contamination.

Forest fires, often sparked by lightning or human activity, release HMs stored in vegetation and soil into the air as particulate matter and ash. Similarly, volcanic eruptions emit HMs stored within the Earth's crust, dispersing them across vast distances through volcanic ash and gas emissions. Additionally, erosion of rocks and soil can liberate HMs into water bodies, especially during heavy rainfall events, contributing to aquatic contamination.

The combination of anthropogenic and natural sources results in a diverse array of HM inputs into the environment, posing significant challenges for environmental management and regulation. The implementation of efficient management techniques and regulatory frameworks is necessary to limit the adverse impacts of HM pollution and protect ecosystems and human health.

Effective management approaches encompass a range of strategies, including pollution prevention, remediation, and monitoring. Prevention strategies aim to reduce HM emissions at the source through improved industrial practices, waste management, and pollution control measures. Remediation techniques involve cleaning up contaminated sites using methods such as soil stabilization, phytoremediation, and chemical treatment (Priyanka et al., 2022; Singh et al., 2022; Tanvi et al., 2022; Johri et al., 2023).

Furthermore, it's vital to have strong monitoring programs in place to track environmental quality and changes in HM levels over time. These programs typically involve collecting samples of air, soil, water, and biological matter to assess HM concentrations and identify potential sources of pollution. Regulatory frameworks are also key, as they set standards, enforce compliance, and encourage practices aimed at minimizing the harmful effects of HM contamination.

A comprehensive management and regulatory approach, along with cooperation among stakeholders, can help reduce the negative effects of HM pollution on ecosystems and public health. This necessitates collaboration among government bodies, industry players, and the scientific community to develop and implement effective solutions that balance environmental preservation with economic needs. To preserve the environment and improve human well-being, combating HM pollution ultimately requires a multimodal strategy that incorporates community involvement, policy development, and scientific knowledge.

3.3 IMPACT OF HEAVY METAL CONTAMINATION IN MEDICINAL PLANTS

Being stationary organisms, plants continually interact with HM-polluted soils. As a result, they have developed advanced defense mechanisms that utilize a wide variety of chemical substances to navigate challenging circumstances. Studies have shown that plants can absorb metals that are bioavailable when they are soluble in the soil solution or have been made soluble by the exudates released from plant roots. Plants require various metals for growth, development, and maintenance, but excessive amounts of any one of them can be hazardous. Nevertheless, because plants can accumulate necessary metals, they can also absorb HMs, which can impact the plant (Pavel et al., 2010; Ibrahim et al., 2017). Certain plants, grown for their ability to withstand high concentrations of HMs, are used in phytoremediation to remove the metals from the soil (Gangwar et al., 2023). Nevertheless, industrial facilities, agricultural products, and other beneficial vegetation, including medicinal plants, are frequently cultivated in polluted soils, as they are intended for human use and consumption. Consequently, consuming fruits and vegetables from plants that thrive in soils with elevated HM concentrations could potentially harm human health. Research suggests that nearly half of the lead found in the human body is sourced from food, with plants contributing to over 50% of this total amount (https://cid-inc.com/blog/reducing-the-effects-of-lead-toxicity-in-plants/; Intawongse & Dean, 2006).

The primary source of pollution is the soil, from which HMs can diffuse or exchange ions with the water surrounding plant roots, eventually reaching the root surface. Plants uptake essential metals through active absorption while also absorbing other harmful substances found in the soil. HMs can pass through cationic channels in cell membranes simultaneously. While not significant for physiological processes, HMs such as arsenic (As), lead (Pb), cadmium (Cd), and mercury (Hg) can have detrimental effects and disrupt organism development. Hence, various HMs should be considered potential hazards to human health and medicinal plants. Different techniques like phytoextraction, phytostabilization, phytovolatilization, and rhizofiltration facilitate metal absorption and transfer into plants from contaminated soil, inducing specific changes in plants.

Plants react to HMs in two primary ways: they either store them within their cells without causing harm or eliminate them by shedding leaves to rid themselves of the accumulated metal ions or complexes (Hlihor et al., 2015; Sarma et al., 2011; Pavel et al., 2013). The uptake of HMs by plants from the soil is known as bioaccumulation, a process that determines the bioavailability of these metals—the extent to which living organisms can absorb them. This process is affected by the pH, organic matter, texture, and properties of the soil, as well as the type of plant.

Certain metals, such as chromium, manganese, zinc, and copper, are essential in small amounts for plants and are typically found in green plants without adverse effects on consumers. However, pollution from the environment can increase the levels of HMs in plants, posing health risks. When plants are exposed to HMs, damaging reactive oxygen species build up, causing the plants to undergo physiological and metabolic adjustments in an attempt to mitigate stress. These modifications protect the plant's lipids, proteins, and nucleic acids from the harmful effects of free radicals.

The absorption of HMs can hinder the body's ability to absorb vital macronutrients such as nitrogen, phosphorus, and potassium, potentially leading to deficiencies. For example, prolonged exposure to high cadmium levels can result in necrosis and chlorosis. Plant functions such as protein modification, photosynthesis, DNA synthesis, and redox reactions depend on both macronutrients and micronutrients.

Certain HMs can reduce the negative impacts of others. For instance, the addition of selenium to certain plants can counteract the effects of cadmium. Furthermore, the presence of HM stress may enhance the synthesis of secondary metabolites in medicinal plants, which are crucial for defense mechanisms and contribute to unique aromas, colors, and flavors.

HM contamination alters the biochemical makeup of medicinal plants, impacting the production of bioactive chemicals that give these plants their therapeutic properties. Humans may experience long-term negative health effects from contaminated herbal items, such as neurotoxicity, nephrotoxicity, hepatotoxicity, and carcinogenicity. The global consumer community is significantly endangered by the presence of HMs in formulations made from medicinal plants, particularly in cosmetic products. Ensuring quality control measures for HMs in medicinal plant products is a critical function of regulatory agencies.

3.4 IMPACT OF HEAVY METALS' PRESENCE IN MEDICINAL PREPARATIONS

Since ancient times the application of medicinal remedies has been essential to maintaining human health by curing illnesses and enhancing well-being. However, there are serious health hazards associated with HM content in medical preparations, which pose problems for both conventional and modern medicine. Lead, mercury, arsenic, and cadmium are examples of HMs that are naturally occurring substances that can build up in the body over time and have a variety of negative health impacts. In this talk, we explore the origins of HMs in pharmaceutical products, their effects on human health, legal restrictions, and mitigation techniques.

3.4.1 Sources of Heavy Metals in Medicinal Preparations

3.4.1.1 Contaminated Raw Materials

Environmental pollution can increase plant concentrations of HMs, which can be harmful to human health. Reactive oxygen species accumulate in plants exposed to HMs, causing the plants to adapt physiologically and metabolically in an effort to withstand stress. These changes protect the plant's proteins, lipids, and nucleic acids against the oxidative damage caused by free radicals.

A lack of essential macronutrients like potassium, phosphorus, and nitrogen may result from the body's inability to absorb HMs. For example, prolonged exposure to high cadmium levels may lead to chlorosis and necrosis. Both macronutrients and micronutrients are necessary for plant processes such as photosynthesis, redox reactions, protein modification, and DNA synthesis (Luo et al., 2021).

3.4.1.2 Manufacturing Processes

Modern pharmaceutical manufacturing processes can also introduce HMs into medicinal preparations. Equipment used in production, such as reactors, vessels, and piping, may contain traces of HMs, which can leach into the final product during synthesis or purification steps (Nessa et al., 2016). These contaminants could ultimately end up in the final product used for consumption. Additionally, medicinal formulations often contain acids, alkalis, and other substances that can react with metallic equipment used in pharmaceutical industries, further increasing contamination. Thus, compatible materials should be utilized to minimize leaching based on the active components in the medicinal preparations. Manufacturing processes need continuous monitoring and evaluation to ensure product safety from all aspects.

3.4.2 Health Impacts of Heavy Metals in Medicinal Preparations

3.4.2.1 Neurological Effects

HMs such as lead, mercury, and arsenic have neurotoxic properties and can damage the nervous system. Prolonged exposure to these metals through medicinal preparations may result in cognitive impairment, developmental delays in children, and neurological disorders such as Parkinson's disease and Alzheimer's disease (Kulshreshtha et al., 2021).

3.4.2.2 Cardiovascular Effects

Exposure to HMs can have profound effects on cardiovascular health. Metals such as lead, mercury, and cadmium contribute to the development of cardiovascular diseases. They can disrupt normal heart function, impair blood vessel integrity, and promote inflammation and oxidative stress within the cardiovascular system. HM exposure has been linked to conditions such as hypertension, atherosclerosis, and heart failure, increasing the risk of heart attacks and strokes. Additionally, HMs can interfere with the regulation of cholesterol levels and blood pressure, further worsening cardiovascular issues. It's crucial to minimize exposure to these harmful substances and prioritize cardiovascular health through lifestyle choices and environmental safeguards (Sulaiman et al., 2024).

3.4.2.3 Renal and Hepatic Effects

Exposure to HMs can significantly impact renal (kidney) and hepatic (liver) function. HMs such as lead, cadmium, mercury, and arsenic are known to accumulate in these organs over time, leading to various adverse effects. In the kidneys, HM exposure can impair filtration and excretion processes, leading to kidney damage and dysfunction. This may manifest as decreased kidney function, proteinuria (protein in the urine), and even chronic kidney disease. Similarly, in the liver, HMs can interfere with detoxification processes and disrupt normal liver function, potentially causing liver inflammation, fibrosis, and, in severe cases, liver failure. Overall, the renal and hepatic effects of HM exposure highlight the importance of minimizing exposure to these toxic substances and implementing measures to protect these vital organs (Street, 2012).

3.4.2.4 Carcinogenic Effects

Exposure to HMs has been linked to carcinogenic effects, posing serious health risks to humans. HMs such as arsenic, cadmium, chromium, and nickel are known to be carcinogenic, meaning they can cause cancer. These metals can enter the body through various routes, including ingestion, inhalation, and skin contact. Once inside the body, they can disrupt cellular processes, damage DNA, and interfere with the body's ability to repair damaged cells, ultimately leading to the development of cancerous tumors. Prolonged exposure to even low levels of these metals has been associated with an increased risk of various types of cancer, including lung, liver, kidney, and bladder cancer. Therefore, minimizing exposure to HMs through environmental regulations and personal protective measures is crucial for preventing their carcinogenic effects and safeguarding public health (Klifi et al., 2010).

3.4.3 Mitigation Strategies

3.4.3.1 Quality Control Measures

Implementing robust quality control measures throughout the manufacturing process is crucial for detecting and preventing HM contamination in medicinal preparations. This involves a multifaceted approach encompassing various stages of production, from sourcing raw materials to final product testing.

3.4.3.2 Screening of Raw Materials

Rigorous screening of raw materials is essential to identify any potential sources of HM contamination. This can involve comprehensive testing using validated analytical methods to ensure that raw materials meet predetermined quality standards before they are incorporated into the manufacturing process.

3.4.3.3 In-Process Monitoring

Continuous monitoring of manufacturing processes helps to identify and address any deviations or anomalies that may arise during production. By implementing in-process monitoring protocols, manufacturers can promptly detect and rectify any issues that could lead to HM contamination.

3.4.3.4 Final Product Testing

Comprehensive testing of finished medicinal products is essential to verify their quality and safety before they are released to the market. This final step in the quality control process involves rigorous testing using validated analytical methods to ensure that products meet regulatory standards for HM content.

3.4.3.5 Source Identification and Traceability

Traceability systems enable manufacturers to trace the origins of raw materials used in medicinal preparations, allowing them to assess the risk of HM contamination and implement appropriate control measures. Establishing partnerships with reliable suppliers and conducting periodic audits can help ensure the quality and safety of raw materials throughout the supply chain.

3.4.3.6 Good Manufacturing Practices

It's crucial to follow good manufacturing practices (GMP) to guarantee the quality and safety of pharmaceutical products, including the management of HM contamination. GMP guidelines cover manufacturing, storage, and distribution, all aimed at reducing the risk of contamination during production. The World Health Organization (WHO) Guidelines on Good Agricultural and Collection Practices (GACP) for Medicinal Plants are an important resource in herbal medicine. They provide a detailed framework to ensure the quality, safety, and effectiveness of medicinal plants from cultivation to collection. In a landscape where medicinal plants play a vital role in both traditional and modern healthcare systems worldwide, the burgeoning demand for these botanical resources underscores the critical need for standardized practices that uphold their therapeutic properties while also addressing concerns regarding sustainability and environmental impact. GACP addresses this imperative by delineating a set of core principles and practices designed to optimize cultivation techniques, enhance harvesting methods, and ensure proper post-harvest handling, all while advocating for ecological responsibility and ethical considerations. From advocating for sustainable agricultural practices such as soil management and integrated pest control to providing guidance on the timing and methods of harvesting to maximize yield and quality without jeopardizing wild populations, GACP offers a multifaceted approach to the responsible cultivation and collection of medicinal plants. Moreover, the guidelines underscore the importance of rigorous quality control measures throughout the production chain, emphasizing proper documentation to trace the origin and ensure the authenticity of medicinal plants, thus bolstering consumer confidence and safeguarding public health. By adhering to the principles outlined in GACP, stakeholders across the medicinal plant industry can contribute to the sustainable production of high-quality herbal remedies, thereby fostering a global landscape where traditional knowledge intersects with modern standards of safety, efficacy, and environmental stewardship (World Health Organization, 2003).

In recent years, there have been significant advancements in testing methods for HMs in medicinal products, driven by the urgent need to guarantee their safety and quality. This surge in innovation is a response to growing concerns about HM contamination in medicinal plants, which arises from various sources such as environmental pollution, industrial activities, and improper farming practices. To tackle these challenges, researchers and regulatory bodies have intensified efforts to improve analytical techniques capable of detecting even trace amounts of HMs in complex herbal mixtures.

Among the notable advancements is the refinement of instrumentation, with sophisticated technologies such as inductively coupled plasma mass spectrometry, atomic absorption spectroscopy, and high-performance liquid chromatography coupled with mass spectrometry emerging as essential tools for HM analysis. These cutting-edge methods provide exceptional detection capabilities, enabling precise quantification of HM contaminants, even at extremely low levels.

Additionally, progress in sample preparation techniques, including microwave-assisted digestion and solid-phase extraction, has transformed the analysis process, enhancing efficiency, reducing interference from other substances, and bolstering overall assay reliability. Furthermore, the integration of automation and miniaturization technologies has led to the creation of high-throughput screening platforms, allowing for rapid and cost-effective analysis of large sample sets.

These technological breakthroughs empower researchers and industry practitioners to conduct thorough assessments of HM contamination, ensuring compliance with stringent regulatory standards worldwide. By utilizing these advanced analytical approaches, stakeholders in the herbal medicine sector can mitigate the risks associated with HM exposure, safeguard consumer well-being, and instill greater trust in the safety and efficacy of medicinal products.

However, despite these remarkable advancements, ongoing research efforts are necessary to refine existing methodologies, address emerging challenges, and enhance the overall reliability and reproducibility of HM testing in medicinal preparations. Collaborative endeavors involving academia, industry, and regulatory bodies will be crucial in driving further innovation and establishing robust quality control measures to uphold the integrity and safety of herbal medicines in the global market (Shaban et al., 2016).

3.4.3.7 Environmental Management Practices

Encouraging environmentally friendly practices in agriculture, industry, and waste management is vital for reducing the release of HMs into our surroundings. This involves reducing pollution from industrial sources, adopting soil and water conservation methods, and embracing organic farming techniques. By minimizing environmental contamination, manufacturers can help prevent HM pollutants from ending up in medicinal products, thereby protecting consumer health. The presence of HMs in medicinal products poses serious health risks globally. From contaminated raw materials to industrial processes and environmental pollution, various factors contribute to the accumulation of HMs in these products. This accumulation can lead to health issues such as neurological disorders, cardiovascular diseases, kidney and liver toxicity, and even cancer. Regulatory measures, including pharmacopeial standards and GMP, along with oversight by regulatory bodies, are crucial for ensuring the quality and safety of medicinal products. Furthermore, implementing strategies like quality control, source identification, traceability, and environmental management practices is essential for minimizing the risk of HM contamination and protecting public health. By addressing these challenges collectively, stakeholders can strive to ensure that medicinal products remain safe and effective for healthcare use, thus promoting the overall health and well-being of the public.

3.5 STANDARDS AND REGULATION FOR THE PRESENCE OF HEAVY METALS IN MEDICINAL PLANTS

Encouraging environmentally sustainable practices in agriculture, industry, and waste management is crucial to lessen the release of HMs into our surroundings. This involves minimizing pollution from industrial sources, implementing measures to conserve soil and water, and embracing organic farming techniques. By decreasing environmental contamination, manufacturers can help prevent HM contamination in medicinal products and safeguard public health. The presence of HMs in medicinal products poses significant health risks to consumers worldwide. Contaminated raw materials, industrial processes, and environmental pollution contribute to the build-up of HMs in these products, leading to adverse health effects such as neurological disorders, cardiovascular diseases, renal and hepatic toxicity, and cancer. Regulatory measures, including pharmacopeial standards, GMP, and oversight by regulatory authorities, play a vital role in ensuring the quality and safety of medicinal products. Additionally, mitigation strategies such as quality control measures, source identification, traceability, and environmental management practices are essential for minimizing the risk of HM contamination and safeguarding public health. By addressing these challenges collectively, stakeholders can work toward ensuring that medicinal products remain safe and effective for use in healthcare practices, thus promoting public health and fostering confidence in herbal medicine as a viable healthcare option.

According to the 14th edition of the *State Pharmacopoeia of the Russian Federation*, the maximum allowable amounts of HMs in medicinal plant materials and herbal medicinal products are 6.0 mg/kg for lead, 1.0 mg/kg for cadmium, 0.1 mg/kg for mercury, and 0.5 mg/kg for arsenic (*State Pharmacopoeia of the* Russian Federation, 2018). The limits for HM content in medicinal plant materials are less strict than those for fruits, berries, and beverages, but they are comparable to those for dry herbal dietary supplements. A comparative review of regulatory documentation across different countries indicates considerable differences in the standards for the environmental safety of medicinal plant materials and herbal medicinal products. The lead limit in Russia is lower than the 10 mg/kg suggested by the International Organisation for Standardisation (ISO) and the WHO. However, it is slightly higher than the 5 mg/kg norm established by the American, Chinese, and European pharmacopoeias. The Russian pharmacopoeia set a cadmium level comparable to the European one, but over three times higher than the 0.3 mg/kg recommended by the WHO and several Asian countries. Interestingly, the maximum allowed concentration (MPC) for cadmium was raised from 0.3 to 1.0 mg/kg in the most recent edition (2020) of the Chinese pharmacopoeia (Chen et al., 2021; Vyas & Vohora, 2018). This further emphasizes the need for cultivating medicinal plants under controlled conditions to ensure safety for consumers (Trivedi & Johri, 2022).

The WHO, Russia, and several other nations have comparable standards for mercury concentrations; however, the MPC is ten times greater in the Chinese and Indian pharmacopoeias, and the American pharmacopoeia regulates the level of methylmercury separately. The strictest limits for arsenic concentrations are defined by the Russian pharmacopoeia; other regulatory documents have standards that are two to eight times higher than the Russian ones (*State Pharmacopoeia of the* Russian Federation, 2018). It has been discovered that the techniques used to analyze medicinal plant materials for HMs do not always match up with one another. This could account for variations in MPC readings. Decomposition in closed vessels is therefore used in international pharmacopoeias to evaluate the arsenic concentration in medicinal plant materials and herbal medicinal products, eliminating the loss of information during sample preparations (Shchukin et al., 2018).

In addition to considering the overall HM content in medicinal plants, it is crucial to take into account the specific parts of the plant used in pharmacy. Different parts of a plant may accumulate HMs differently, influencing the safety and efficacy of medicinal preparations derived from them. This nuanced understanding underscores the importance of tailoring regulatory standards to address variations in HM accumulation across different plant organs. The State Pharmacopoeia of the Russian Federation (2018) establishes standardized guidelines for the toxicant composition of all plant materials used in pharmacy. These guidelines recognize the potential for variations in HM accumulation among different plant parts and aim to ensure the safety and quality of medicinal products. For instance, since many plant species have mechanisms to limit the uptake of xenobiotics, including HMs, to their aerial parts, particularly the generative organs, when grown in contaminated soil, it is essential to account for these variations in regulatory standards. The Chinese pharmacopoeia, Pharmacopoeia of the People's Republic of China (2015), demonstrates this concept by setting specific safety limits for toxic HMs in underground parts of medicinal plants. Knowledge of how plant physiology and their environment affect HM accumulation in different plant parts is essential when preparing medicinal substances. Regulatory standards should reflect this to manage health risks from HM contamination and ensure the safety of natural medicine and public health.

Literature has indicated the presence of high metal levels in medicinal plants in various regions. A study in the Trans-Ural region of the Republic of Bashkortostan (Russia) evaluated the environmental purity of *Cichorium intybus* L. based on the MPCs of chemical elements in feed additives (5 mg/kg Pb, 0.3 mg/kg Cd) and farm animal feed. The research findings suggested that harvesting the studied raw material of this medicinal plant might not be appropriate due to excessive cadmium concentration levels, which were 3.0–6.5 times higher than the acceptable limits. However, the results would have varied if the current MPC for medicinal plant products had been applied.

3.6 CONCLUSION

The intricate link between HMs and medicinal plants conncets environmental pollution with pressing public health concerns, demanding a thorough examination of their implications and the required steps for effective resolution. A key emphasis highlighted in this review is the substantial importance of HM build-up in medicinal plants, impacting both consumer safety and therapeutic effectiveness, thus necessitating a deep understanding of the underlying processes.

This information highlights the crucial need to create comprehensive strategies to reduce HM pollution in therapeutic plants without compromising their ability to treat illnesses. Strict oversight and preventative measures are necessary to protect the integrity of pharmaceutical products and guarantee the health of their users. Rigorous quality control procedures, innovative soil remediation methods, and environmentally friendly farming techniques designed to reduce HM build-up are necessary to accomplish this.

Furthermore, because HM pollution is so pervasive, interdisciplinary collaboration between pharmacologists, environmental scientists, policymakers, and botanists is essential. The comprehensive use of integrative techniques is essential to effectively tackle the complex challenges posed by HM contamination. A future in which medicinal plants are safe and effective means of improving human health without the continual worry of HM contamination can be achieved through the collaborative efforts of stakeholders, fostering cooperative connections and emphasizing environmental conservation.

This conclusion underscores the importance of continuous research efforts, innovative interventions, and collective measures to protect the purity of medicinal plants and maintain their essential role in international healthcare systems. It highlights the critical need to persist in developing and implementing strategies to mitigate the impact of HM pollution's and ensure the sustainability and efficacy of medicinal plants for future generations.

BIBLIOGRAPHY

Al-Areqi, W.M., A.M. Bobaker, I. Alakili, A.A. Majid, and S. Sarmani. (2019). Determination of heavy metals and radionuclides in coal and industrial fly ash by neutron activation analysis (NAA) and gamma spectrometry. *Sains Malaysiana* 48, no. 8: 1655–1660.

Ali, H., and E. Khan. (2018). Trophic transfer, bioaccumulation, and biomagnification of non-essential hazardous heavy metals and metalloids in food chains/webs—Concepts and implications for wildlife and human health. *Human and Ecological Risk Assessment: An International Journal* 25: 1353–1376.

Altıkulaç, A., Ş.T. Aslı Kurnaz, E. Gören, C. Duran, A. Hançerlioğulları, and F.A. Uğur. (2022). Assessment of the enrichment of heavy metals in coal and its combustion residues. *ACS Omega* 7, no. 24: 21239–21245.

Anyanwu, B.O., A.N. Ezejiofor, Z.N. Igweze, and O.E. Orisakwe. (2018). Heavy metal mixture exposure and effects in developing nations: an update. *Toxics* 6, no. 4: 65.

Appiah-Adjei, E.K., N.F. Appiah, and K.A. Adjei. (2016). Potential groundwater pollution from improper oil and metal waste disposal in Suame, Ghana. *Journal of Science and Technology (Ghana)* 36, no. 3: 20–33.

Aslam, M.S., and M.S. Ahmad. (2016). Worldwide importance of medicinal plants: current and historical perspectives. *Recent Advances in Biology and Medicine* 2, no. 2016: 909.

Awokunmi, E.E., S.S. Asaolu, and K.O. Ipinmoroti. (2010). Effect of leaching on heavy metals concentration of soil in some dumpsites. *African Journal of Environmental Science and Technology* 4, no. 8: 495–499.

Balkrishna, A., S. Banerjee, S. Ghosh, D. Chauhan, I. Kaushik, V. Arya, and S.K. Singh. (2024). Reuse of sewage sludge as organic agricultural products: an efficient technology-based initiative. *Applied and Environmental Soil Science* 2024: 1433973.

Briffa, J., E. Sinagra, and R. Blundell. (2020). Heavy metal pollution in the environment and their toxicological effects on humans. *Heliyon* 6, no. 9: e04691.

Briki, M., Y. Zhu, Y. Gao, M. Shao, H. Ding, H. Ji. (2017). Distribution and health risk assessment to heavy metals near smelting and mining areas of Hezhang, China. *Environmental Monitoring and Assessment* 189: 458. https://doi.org/10.1007/s10661-017-6153-6

Chen, Y.-G., J.-H. Huang, R. Luo, et al. (2021). Impacts of heavy metals and medicinal crops on ecological systems, environmental pollution, cultivation, and production processes in China. *Ecotoxicology and Environmental Safety* 219: 112336.

China Food and Drug Administration. (2015). *Pharmacopoeia of the People's Republic of China; Chinese Pharmacopoeia Commission.* China Food and Drug Administration: Beijing, China.

Council of Europe. (2019). *European Directorate for the Quality of Medicines & HealthCare (EDQM).* Council of Europe: Strasbourg, France.

Diaconu, M., L.V. Pavel, R.-M. Hlihor, M. Rosca, D.I. Fertu, M. Lenz, P.X. Corvini, and M. Gavrilescu. (2020). Characterization of heavy metal toxicity in some plants and microorganisms—a preliminary approach for environmental bioremediation. *New Biotechnology* 56: 130–139.

Dijkstra, J.J., J.C.L. Meeussen, and R.N.J. Comans. (2004). Leaching of heavy metals from contaminated soils: an experimental and modeling study. *Environmental Science & Technology* 38, no. 16: 4390–4395.

Dylenova, E.P., S.V. Zhigzhitzhapova, T.E. Randalova, L.D. Radnaeva, V.G. Shiretorova, and I.A. Pavlov. (2019). Biophile elements and heavy metals in artemisia frigida willd and artemisia jacutica drob. *Chemistry of Plant Raw Material* 4: 199–205.

Fattahi, B., K. Arzani, M.K. Souri, and M. Barzegar. (2019). Effects of cadmium and lead on seed germination, morphological traits, and essential oil composition of sweet basil (*Ocimum basilicum* L.). *Industrial Crops and Products* 138: 111584.

Gangwar, P., P. Johri, R. Dutta, S. Singh, and M. Trivedi. (2023). Overview of phytopharmaceuticals and biotechnology of herbal plants. *Phytopharmaceuticals and Biotechnology of Herbal Plants* 2023: 1–17

Hasan, M.M., M.N. Uddin, I. Ara-Sharmeen, H.F. Alharby, Y. Alzahrani, K.R. Hakeem, and L. Zhang. (2019). Assisting phytoremediation of heavy metals using chemical amendments. *Plants* 8, no. 9: 295.

He, Z., J. Shentu, X. Yang, V.C. Baligar, T. Zhang, and P.J. Stoffella. (2015). Heavy metal contamination of soils: sources, indicators and assessment. *Journal of Environmental Indicators* 9: 17–18.

Hlihor, R.M., L.C. Apostol, C. Smaranda, L.V. Pavel, F.A. Căliman, B.M. Robu, and M. Gavrilescu. (2009). Bioavailability processes for contaminants in soils and their use in risk assessment. *Environmental Engineering & Management Journal (EEMJ)* 8, no. 5: 1199–1206.

Hlihor, R.M., M. Diaconu, F. Leon, S. Curteanu, T. Tavares, and M. Gavrilescu. (2015). Experimental analysis and mathematical prediction of Cd (II) removal by biosorption using support vector machines and genetic algorithms. *New Biotechnology* 32, no. 3: 358–368.

Ibrahim, M.H., Y.C. Kong, and N.A.M. Zain. (2017). Effect of cadmium and copper exposure on growth, secondary metabolites and antioxidant activity in the medicinal plant Sambung Nyawa (*Gynura procumbens* (Lour.) Merr). *Molecules* 22, no. 10: 1623.

Intawongse, M., and J.R. Dean. (2006). Uptake of heavy metals by vegetable plants grown on contaminated soil and their bioavailability in the human gastrointestinal tract. *Food Additives and Contaminants* 23, no. 1: 36–48.

Johri, P., A. Singh, M. Trivedi, and S. Singh. (2023). Integration of pathway analysis as a powerful tool for microbial remediation of pollutants. *Genomics Approach to Bioremediation* 2023: 397–415.

Kapusta, P., and S. Lukasz. (2015). Effects of heavy metal pollution from mining and smelting on enchytraeid communities under different land management and soil conditions. *Science of the Total Environment* 536: 517–526.

Kelepertzis, E. (2014). Accumulation of heavy metals in agricultural soils of Mediterranean: insights from Argolida basin, Peloponnese, Greece. *Geoderma* 221: 82–90.

Khlifi, R., and A. Hamza-Chaffai. (2010). Head and neck cancer due to heavy metal exposure via tobacco smoking and professional exposure: a review. *Toxicology and Applied Pharmacology* 248, no. 2: 71–88.

Kulhari, A., A. Sheorayan, S. Bajar, S. Sarkar, A. Chaudhury, and R.K. Kalia. (2013). Investigation of heavy metals in frequently utilized medicinal plants collected from environmentally diverse locations of north western India. *SpringerPlus* 2: 1–9.

Kulshreshtha, D., J. Ganguly, and M. Jog. (2021). Manganese and movement disorders: a review. *Journal of Movement Disorders* 14, no. 2: 93.

Kushwaha, P. (2021). Metallic impurities in pharmaceuticals: an overview. *Current Pharmaceutical Analysis* 17, no. 8: 960–968.

Kuzovkova, A.A., I.V. Drebenkova, Y.N. Velentei, A.A. Pleshkova, G.E. Bychok, D.V. Chernik, and N.V. Maskalevich. (2020). Heavy metal contamination of wild and cultivated medicinal plants in the Republic of Belarus. *Occupational Health as Human Ecology* 4: 112–117.

Lajayer, B.A., M. Ghorbanpour, and S. Nikabadi. (2017). Heavy metals in contaminated environment: destiny of secondary metabolite biosynthesis, oxidative status and phytoextraction in medicinal plants. *Ecotoxicology and Environmental Safety* 145: 377–390.

Lajayer, B.A., N. Najafi, E. Moghiseh, M. Mosaferi, and J. Hadian. (2019). Micronutrient and heavy metal concentrations in basil plant cultivated on irradiated and non-irradiated sewage sludge-treated soil and evaluation of human health risk. *Regulatory Toxicology and Pharmacology* 104: 141–150.

Luo, L., B. Wang, J. Jiang, M. Fitzgerald, Q. Huang, Z. Yu, H. Li et al. (2021). Heavy metal contaminations in herbal medicines: determination, comprehensive risk assessments, and solutions. *Frontiers in Pharmacology* 11: 595335.

Maleki, M., M. Ghorbanpour, and K. Kariman. (2017). Physiological and antioxidative responses of medicinal plants exposed to heavy metals stress. *Plant Gene* 11: 247–254.

Marrugo-Negrete, J., J. Pinedo-Hernández, and S. Díez. (2017). Assessment of heavy metal pollution, spatial distribution and origin in agricultural soils along the Sinú River Basin, Colombia. *Environmental Research* 154: 380–388.

Nasim, S.A., and B. Dhir. (2010). Heavy metals alter the potency of medicinal plants. *Reviews of Environmental Contamination and Toxicology* 203: 139–149.

Nessa, F., S.A. Khan, and K.Y.I. Abu Shawish. (2016). Lead, cadmium and nickel contents of some medicinal agents. *Indian Journal of Pharmaceutical Sciences* 78, no. 1: 111.

Niassy, S., and K. Diarra. (2012). Effects of organic inputs in Urban agriculture and their optimization for poverty alleviation in Senegal, West-Africa. In Singh, R. P. (Ed.), *Organic Fertilizers: Types, Production and Environmental Impact*, pp. 1–22. Nova Science Publishers, Inc.: Hauppauge, NY.

Nouairi, I., W.B. Ammar, N.B. Youssef, D.D.B. Miled, M.H. Ghorbal, and M. Zarrouk. (2009). Antioxidant defense system in leaves of Indian mustard (Brassica juncea) and rape (Brassica napus) under cadmium stress. *Acta Physiologiae Plantarum* 31: 237–247.

Oueslati, S., R. Ksouri, A. Pichette, S. Lavoie, K. Girard-Lalancette, V. Mshvildadze, C. Abdelly, and J. Legault. (2014). A new flavonol glycoside from the medicinal halophyte Suaeda fruticosa. *Natural Product Research* 28, no. 13: 960–966.

Özden, H., and S. Özden. (2018). Levels of heavy metals and ochratoxin a in medicinal plants commercialized in Turkey. *Turkish Journal of Pharmaceutical Sciences* 15, no. 3: 376–381.

Pavel, V.L., D. Bulgariu, L. Bulgariu, R.M. Hlihor, and M. Gavrilescu. (2010). Analysis of factors determining the behaviour of chromium in some Romanian soils. *Environmental Engineering and Management Journal* 9, no. 1: 89–94.

Pavel, V.L., D.L. Sobariu, M. Diaconu, F. Stătescu, and M. Gavrilescu. (2013). Effects of heavy metals on *Lepidium sativum* germination and growth. *Environmental Engineering & Management Journal (EEMJ)* 12, no. 4: 727–733.

Priyanka, G., J. Parul, D. Rahul, S. Sachidanand, and T. Mala. (2022). Overview on phytopharmaceuticals and biotechnology of herbal plants. *Phytopharmaceuticals and Biotechnology of Herbal Plants* 1: 1–17.

Rehman, K., F. Fatima, I. Waheed, and M.S.H. Akash. (2018). Prevalence of exposure of heavy metals and their impact on health consequences. *Journal of Cellular Biochemistry* 119, no. 1: 157–184.

Sarma, H., S. Deka, H. Deka, and R.R. Saikia. (2011). Accumulation of heavy metals in selected medicinal plants. *Reviews of Environmental Contamination and Toxicology* 214: 63–86.

Sen, T., and S.K. Samanta. (2015). Medicinal plants, human health and biodiversity: a broad review. *Biotechnological Applications of Biodiversity* 147: 59–110.

Shaban, N.S., K.A. Abdou, and N.E.H.Y. Hassan. (2016). Impact of toxic heavy metals and pesticide residues in herbal products. *Beni-Suef University Journal of Basic and Applied Sciences* 5, no. 1: 102–106.

Shah, A., A. Niaz, N. Ullah, A. Rehman, M. Akhlaq, M. Zakir, and M.S. Khan. (2013). Comparative study of heavy metals in soil and selected medicinal plants. *Journal of Chemistry* 2013: 1265.

Shchukin, V.M., N.E. Kuz'mina, A.A. Erina, V.A. Yashkir, and V.A. Merkulov. (2018). Comparative analysis of the heavy metal, aluminum, and arsenic contents in brown algae of various origins. *Pharmaceutical Chemistry Journal* 52: 627–634.

Singh, S., R. Datta, P. Johri, and M. Trivedi. (2022). *Phytopharmaceuticals and Biotechnology of Herbal Plants*. Boca Raton, FL: CRC Press.

Sobariu, D.L., D.I.T. Fertu, M. Diaconu, L.V. Pavel, R.-M. Hlihor, E.N. Drăgoi, S. Curteanu, M. Lenz, P. F.-X. Corvini, and M. Gavrilescu. (2017). Rhizobacteria and plant symbiosis in heavy metal uptake and its implications for soil bioremediation. *New Biotechnology* 39: 125–134.

State Pharmacopoeia of the Russian Federation, XIV ed..; Ministry of Health of the Russian Federation: Moscow, Russia, 2018; Volume II, p. 1449.

Street, R.A. (2012). Heavy metals in medicinal plant products—An African perspective. *South African Journal of Botany* 82: 67–74.

Sulaiman, M.B., A.M. Adamu, S.B. Ali, U.V. Ezenobi, A.M. Gimba, O.O. Akinlotan, and A. Abubakar. (2024). Heavy metal contamination in medicinal plants: assessing carcinogenic and non-carcinogenic health risks. *Discover Environment* 2, no. 1: 1–10.

Tanvi, J., S. Kavita, S. Rajnish, S. Sachidanand, and P.K. Dutta. (2022). Role of nanotechnology in the field of phytopharmaceuticals for the delivery of herbal drugs. *Phytopharmaceuticals and Biotechnology of Herbal Plants*, 1: 14.

Trivedi, M., and P. Johri. (2022). Omics approaches to revisit rhizobacterial biome. *Rhizosphere Microbes: Biotic Stress Management* 40: 105–122.

Verkleij, J.A.C. (1993). The effects of heavy metal stress on higher plants and their use as biomonitors. In Marhert, B. (Ed.), *Plants as Bioindicators of Heavy Metals in the Terrestrial Environment*, pp. 415–424. VCH: New York.

Vinogradova, N., A. Glukhov, V. Chaplygin, P. Kumar, S. Mandzhieva, T. Minkina, and V.D. Rajput. (2023). The content of heavy metals in medicinal plants in various environmental conditions: a review. *Horticulturae* 9, no. 2: 239.

Vyas, P., and D. Vohora. (2018). Pharmaceutical regulations for complementary medicine. In Vohora, D., and Singh, G. (Eds.), *Pharmaceutical Medicine and Translational Clinical Research*, pp. 233–264. Academic Press: New York.

Wang, P., Y. Hu, and H. Cheng. (2019). Municipal solid waste (MSW) incineration fly ash as an important source of heavy metal pollution in China. *Environmental Pollution* 252: 461–475.

World Health Organization. (2003). *WHO Guidelines on Good Agricultural and Collection Practices [GACP] for Medicinal Plants*. World Health Organization, Geneva.

Xiangyang, Bi., M. Zhang, Y. Wu, Z. Fu, G. Sun, L. Shang, Z. Li, and P. Wang. (2020). Distribution patterns and sources of heavy metals in soils from an industry undeveloped city in Southern China. *Ecotoxicology and Environmental Safety* 205: 111115.

Yang, Q., Z. Li, X. Lu, Q. Duan, L. Huang, and J. Bi. (2018). A review of soil heavy metal pollution from industrial and agricultural regions in China: pollution and risk assessment. *Science of the Total Environment* 642: 690–700.

Yu, L.I., Y.-B. Wang, X. Gou., Y.-B. Su, and G. Wang. (2006). Risk assessment of heavy metals in soils and vegetables around non-ferrous metals mining and smelting sites, Baiyin, China. *Journal of Environmental Sciences* 18, no. 6 (2006): 1124–1134.

4 From Transcription to Finale

The Intricacies of Plant microRNA Biogenesis, Regulation, and Degradation

*Harsh Bokadia, Samah Isaani, Kanchanlata Tungare, Aayushi Kadam, and Pamela Jha**

4.1 INTRODUCTION TO miRNAs

Ribonucleic acids (RNAs) are single-stranded biopolymers that have structural similarities to deoxyribonucleic acid (DNA). They are made up of monomers called ribonucleotides (rNTPs), which consist of a ribose sugar, a 5′ phosphate molecule, and one of the four nitrogenous bases – adenine, guanine, uracil, or cytosine (NIH, RNA definition). On the other hand, DNA is made up of deoxyribonucleotides, which are similar to rNTPs, except they consist of a 2′-deoxyribose instead of ribose and thymine instead of uracil. Although RNAs form as single-stranded molecules, their complementary sequences can form complex two- and three-dimensional structures through base pairing. These structures are called secondary and tertiary structures, respectively. A secondary structure can be composed of several substructures like hairpins, stems, loops, and pseudoknots (Chheda & Gupta, 2014).

RNAs act as important genetic and epigenetic regulators in various organisms and help in histone methylation and the modification of coding and non-coding RNAs (Chen et al., 2018; Ding et al., 2012). RNAs can be classified into the following categories based on their function: messenger RNA (mRNA), transfer RNA, ribosomal RNA, and small RNA (sRNA). mRNAs carry genetic information from DNA to ribosomes for subsequent protein synthesis. Transfer RNAs transfer specific amino acids to the growing rNTP chain using codon–anticodon base pairing. Ribosomal RNA forms the structural and catalytic components of ribosomes. sRNAs are involved in genetic regulation by degrading mRNAs or inhibiting their translation into proteins. sRNAs can further be subdivided into small interfering RNAs (siRNAs), PIWI-interacting RNAs, and microRNAs (miRNAs). This chapter will discuss miRNAs and their biogenesis, regulation, and degradation in plants.

miRNAs were first thought to be proteins encoded by regulatory genes that controlled the temporal development of *Caenorhabditis elegans* larvae (Horvitz & Sulston, 1980), but it was later discovered that they are small non-coding RNAs (Lee et al., 1993). They found that the transcripts of the lin-4 gene were complementary to the mRNA transcripts of *C. elegans* larval development proteins, thereby negatively regulating them.

Situated in the nucleus, miRNAs are 21–24-nucleotide-long sequences that do not code for any proteins (Chen, 2005). They can be intergenic, meaning they are located between two protein-coding genes, or intronic, i.e., they are processed from introns of host transcripts (Wang et al., 2019). miRNAs play a crucial role in the regulation and silencing of mRNAs, usually post-transcriptionally and rarely during translation in plants (as discussed in Section 4.2.5). They are essential for plants as they regulate various processes such as leaf, root, and shoot apical meristem development, auxin signalling, organ polarity and separation, floral organ transition and identity, and reproductive

* *Corresponding Author*

DOI: 10.1201/9781003491378-4

development (Chen, 2005; Li & Zhang, 2016). In animals, they are known to play an equally important role in processes like developmental timing, stem cell differentiation, and haematopoiesis (Kloosterman et al., 2006). Dysregulation of miRNAs is also clinically significant as a potential biomarker for neurodegenerative diseases, as well as cancer formation and progression (Iorio & Croce, 2012; Juźwik et al., 2019; Santa-Maria et al., 2015).

Plant miRNAs play a significant role in mitigating the effects of climate change on plants and ecosystems. As sRNA molecules, miRNAs regulate gene expression, influencing various physiological processes in response to environmental cues or stimuli. Here is a brief explanation of how plant miRNAs contribute to climate change mitigation:

- **Stress Response Regulation**: Plant miRNAs regulate the expression of genes involved in stress responses triggered by climate-related factors such as extreme weather events, drought, heat, and salinity. By modulating stress-responsive pathways, miRNAs help plants adapt to changing environmental conditions and enhance their resilience to climate-induced stresses.
- **Enhanced Adaptation**: miRNAs fine-tune gene expression networks involved in plant adaptation to environmental changes. By regulating the expression of genes associated with traits such as water use efficiency, photosynthesis, and nutrient uptake, miRNAs help plants cope with climate variability and optimize resource utilization in challenging environments.
- **Biotic Interactions**: Plant miRNAs mediate interactions between plants and other organisms, including pests, pathogens, and beneficial microbes. These interactions influence plant resilience to biotic stresses exacerbated by climate change, such as pest outbreaks and disease epidemics, thereby contributing to ecosystem stability and agricultural sustainability.
- **Carbon Sequestration**: Certain miRNAs may indirectly contribute to climate change mitigation by regulating the expression of genes involved in carbon assimilation and storage processes. By optimizing carbon fixation and allocation in plant tissues, miRNAs can influence carbon sequestration in terrestrial ecosystems, potentially mitigating the effects of greenhouse gas emissions on climate change.
- **Ecosystem Resilience**: Plant miRNAs play crucial roles in maintaining ecosystem resilience in the face of climate change-induced disturbances. By modulating plant responses to environmental stressors, miRNAs help preserve biodiversity, ecosystem function, and services, contributing to the overall resilience of natural and managed ecosystems to climate variability and extreme events.

Overall, plant miRNAs represent important molecular mechanisms for enhancing plant resilience, ecosystem stability, and agricultural productivity in the context of climate change. Understanding the roles of miRNAs in plant responses to environmental stressors is essential for developing strategies to mitigate the impacts of climate change on plant health, food security, and ecosystem sustainability.

Apart from miRNAs, different mechanisms such as base modifications, RNA-binding proteins, and eraser proteins regulate mRNA translation by modifying their structure and, consequently, their stability (Baker & Coller, 2006; Boo & Kim, 2020); however, these mechanisms are beyond the scope of this chapter.

4.2 BIOGENESIS OF PLANT miRNA

4.2.1 miRNA Gene Transcription

The biogenesis of miRNA (MIR) begins in the nucleus at the MIR gene (plant miRNA gene), which is a group of genes that code for plant miRNAs (e.g., *lin-4* in *C. elegans*). Around 100 MIR genes have been identified in *Arabidopsis thaliana* (Jones-Rhoades & Bartel, 2004). DNA-dependent

RNA polymerase II (RNA Pol II) transcribes these MIR genes into a transcript known as primary miRNA (pri-miRNA) (Kim et al., 2011). The pri-miRNA is divided into five regions: the terminal loop, two side arms, the upper and lower stems, and the miRNA/miRNA* region.

Similar to other RNA transcripts, pri-miRNA also undergoes modifications at the 3′ and 5′ ends to prevent degradation by exonucleases and improve stability. The 3′ end of the lower stem is polyadenylated by the poly(A) polymerase complex. It has been found that *Populus trichocarpa* miRNAs have 1–7 adenylate residues attached at the 3′ end to reduce degradation (Lu et al., 2008). Similarly, the 5′ end is capped with 7-methylguanosine in a manner similar to mRNA 5′ capping (Ren & Yu, 2012). This is performed to stabilize the molecule and prevent degradation by exonuclease activity (Rogers & Chen, 2013; Wang et al., 2019).

4.2.2 Splicing of pri-miRNA to pre-miRNA

4.2.2.1 First Cleavage

After the formation of the long pri-miRNA, it is spliced twice by a complex known as the Dicing body. The Dicing body consists of three identified proteins:

4.2.2.1.1 Dicer-like RNase III Endoribonucleases

Dicer-like (DCL) proteins are a group of endoribonucleases consisting of four classes (DCL1–4) with distinctive functions. DCL1 is involved exclusively in the synthesis of miRNAs, while DCL2–4 are involved in siRNA synthesis (Fukudome & Fukuhara, 2017).

DCL1 consists of four major domains, each of which plays different roles: a double-stranded RNA-binding domain (dsRBD), a PIWI/Argonaute/Zwille (PAZ) domain, an RNase III domain, and a DEAD/DEAH box helicase domain. Each of these plays unique roles in the double cleavage of the double-stranded pri-miRNA (Ciechanowska et al., 2021).

There are two tandem dsRBDs that recognize the dsRNA (i.e., pri-miRNA) substrate. The first dsRBD (dsRBD1) binds to pri-miRNA with low affinity. After binding, it is important for DCL1 to recognize the cleavage site.

To achieve this, the helicase domain first unwinds the double-stranded miRNA in an ATP-dependent manner. Dong et al. (2008) demonstrated that DCL1 activity ceases when ATP is absent. This ATP dependence can be attributed to the helicase domain of DCL1. They also showed that DCL1 activity is optimal at a salt concentration between 25 and 50 mM NaCl. Moreover, chelation of Mg^{2+} using EDTA inhibited splicing. Thus, DCL1-mediated cleavage is dependent on ATP and divalent ions.

Upon unwinding, the first cleavage site can now be recognized by the PAZ domain. This is facilitated by the ability of the PAZ domain to recognize the secondary structure of the miRNA (rather than the primary structure) using highly conserved positively charged amino acids like arginine and lysine (Wei et al., 2021). This binding is governed by the 15-nt rule, which is explained by the function of the RNase III domain. Upon recognition of this cleavage site, dsRBD1 undergoes folding that increase the affinity of DCL1 for pri-miRNA (Suarez et al., 2015).

As for the second dsRBD (dsRBD2), it has been shown to have similar binding affinities for dsRNA and dsDNA. Thus, in addition to substrate recognition and binding, dsRBD2 may also play a role in the localization of D-bodies in the nucleus (Burdisso et al., 2012).

After recognition of the cleavage site, the RNase III domain cleaves the pri-miRNA at two cleavage sites. It follows the 15-nt rule for the first cleavage, which states that cleavage occurs 15 nt away from a secondary structure, such as a bulge or another internal unstructured region (recognized by the PAZ domain) in the lower stem of the pri-miRNA (Song et al., 2010). This produces a structure called precursor miRNA (pre-miRNA) that has a 2nt-3′ overhang as shown in Figure 4.1. The second cleavage occurs 21nt away from the first cleavage site (Ciechanowska et al., 2021). This produces a miRNA/miRNA* duplex that is further discussed in Section 4.2.2.2.

This mechanism of splicing is called 'base-to-loop' processing since it starts at the lower stem (i.e., the base of the secondary structure).

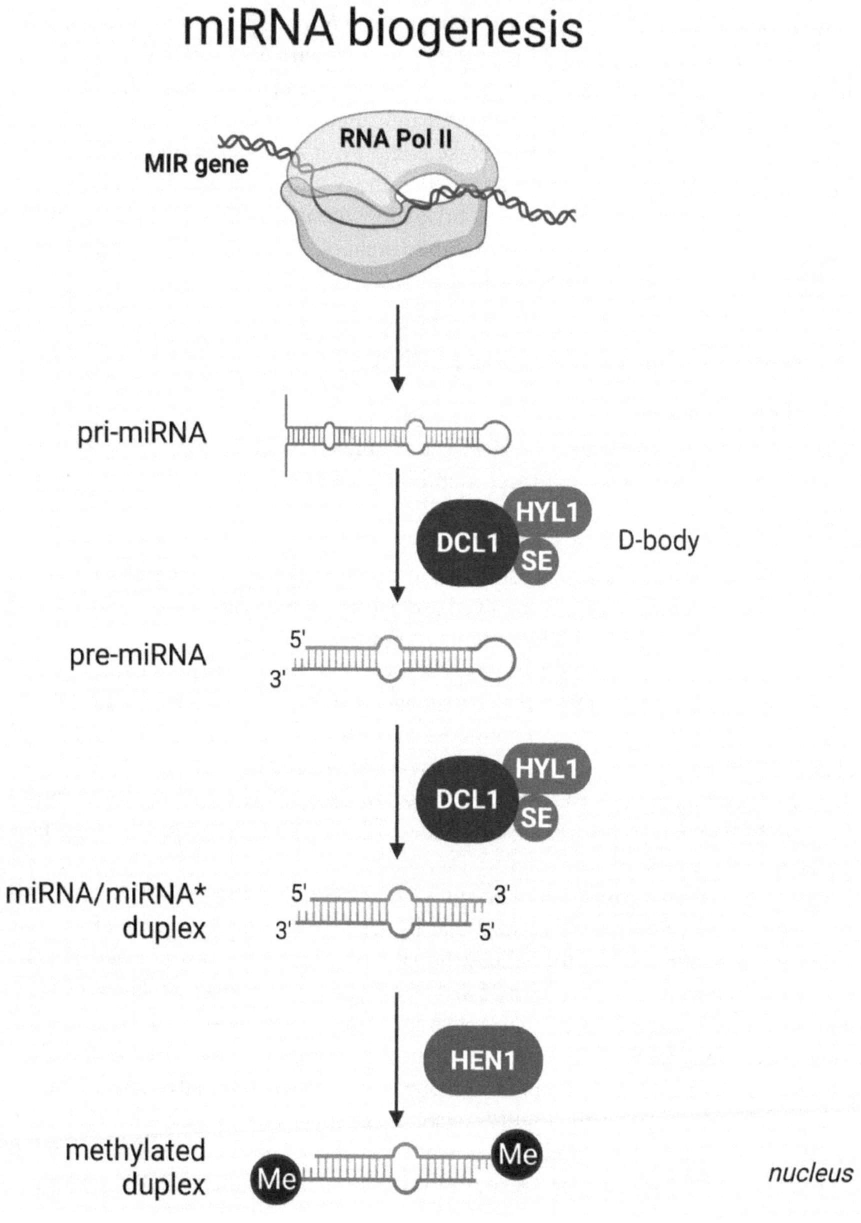

FIGURE 4.1 Biogenesis of plant miRNAs.

However, 80% of the miRNA produced by DCL1 alone is not processed correctly; with certain cofactors called hyponastic leaves 1 (HYL1) and serrate (SE), the yield of miRNAs improves to over 80% (Dong et al., 2008; Liu et al., 2012).

4.2.2.1.2 Hyponastic Leaves 1

In *Arabidopsis thaliana*, it was observed that DCL1 does not function as a simple monomer. It is associated with several other proteins that help in the maturation process of miRNA. HYL1 was identified as one of the proteins acting as a cofactor in miRNA biogenesis (Fukudome & Fukuhara, 2017). There appears to be a physical interaction between DCL1 and HYL1, which is necessary for the formation of pre-miRNAs from pri-miRNAs. HYL1 contains two tandem dsRBDs at its N-terminal domain, known as HR1 and HR2, respectively. HR1 is essential for dsRNA binding *in vitro*, while HR2 contributes to the protein–protein interaction activity of HYL1. The N-terminal domain also contains a nuclear localization signal. KETCH1, a member of the importin-β family, translocates HYL1 from the cytoplasm into the nucleus, where it performs its function (Wang et al., 2019; Wu et al., 2007). The C-terminal domain (CTD) includes a region for protein–protein interaction. The dsRBDs possess a general mechanism to recognize pri-miRNA and present it for cleavage to RNase III enzymes (Kurihara et al., 2006; Yang et al., 2010). HR2 helps increase the ability of HYL1 and DCL1 to recognize and bind to the miRNA/miRNA* region with greater precision.

4.2.2.1.3 Serrate

SE is a C_2H_2 zinc finger protein that is part of D-bodies. Its presence in the complex helps increase DCL1 efficiency and accuracy in processing pri-miRNA by up to 80%. It consists of three domains: an N domain, a Middle (Mid) domain, and a Zn finger domain. The SE N-terminal and Mid domains bind to the RBD2 of HYL1, but not to the CTD Zn finger domain (Bajczyk et al., 2020; Machida et al., 2011; Raczynska et al., 2014).

4.2.2.2 Second Cleavage/Duplex Formation

After the first cleavage at the lower stem, the second cleavage occurs at the terminal loop, 21 nt away from the first cleavage site, producing a miRNA/miRNA* duplex that is 21 base pairs long.

This duplex comprises a guide (miRNA) and a passenger (miRNA*) strand. Each strand has a 2-nt 3′ overhang and a 5′ end, which makes the duplex unstable, as shown in Figure 4.1. This duplex, however, is susceptible to degradation; hence, certain enzymes like HUA enhancer 1 (HEN1) methylate the duplex at the 3′ end to stabilize it (Rogers & Chen, 2013).

4.2.3 2′-O-Methylation of Duplex

The instability of the duplex occurs due to the free 3′ hydroxyl ends. At each 3′ end, the last nucleotide has two hydroxyl groups: 2′ OH and 3′ OH. Out of these, the 3′ OH is susceptible to the activity of polymerases and terminal transferases, which can cause miRNA degradation.

For example, when the 3′ end is unmethylated, it is uridylated by terminal uridyltransferases which leads to its degradation (Li et al., 2005). Thus, methylation occurs at the 2′ OH (2′-O-methylation) to maintain the stability of the duplex. Methylation of the duplex is performed by an sRNA methyltransferase known as HEN1, which have five domains: an N-terminal dsRNA-binding motif, a La-like domain, a large central domain, a PPIase-like domain, and a C-terminal Rossmann-fold methyltransferase domain.

The dsRNA-binding motif domain allows HEN1 to bind to the duplex, while the La-like domain recognizes the 3′ OH of the duplex. It has been observed that miRNAs lacking a hydroxyl group at the 3′ ribose position failed to methylate. Thus, the presence of a hydroxyl group at both the 2′ and 3′ positions is crucial for the activity of HEN1 (Yu et al., 2005). The RFM catalytic domain transfers a methyl ($-CH_3$) group from S-adenosylmethionine (SAM) to the 2′ OH, which is a common methyl donor involved in numerous biochemical and RNA modification reactions (Ding et al., 2012; Tkaczuk et al., 2006; Wang et al., 2019). The activities of the central and PPIase-like domains of HEN1 are not well understood.

This concludes the biogenesis of the methylated miRNA duplex from pri-miRNA using D-bodies and HEN1, as represented in Figure 4.1.

4.2.4 Assembly of miRNA-Induced Silencing Complex

Since miRNAs cannot regulate the expression of mRNAs alone, they must bind to a protein complex known as RNA-induced silencing complex (RISC), which produces miRISC. RISC majorly consists of proteins belonging to the Argonaute (AGO) family. The major components of RISC are the AGO proteins, with AGO1 being the most common Argonaute in *Arabidopsis thaliana*.

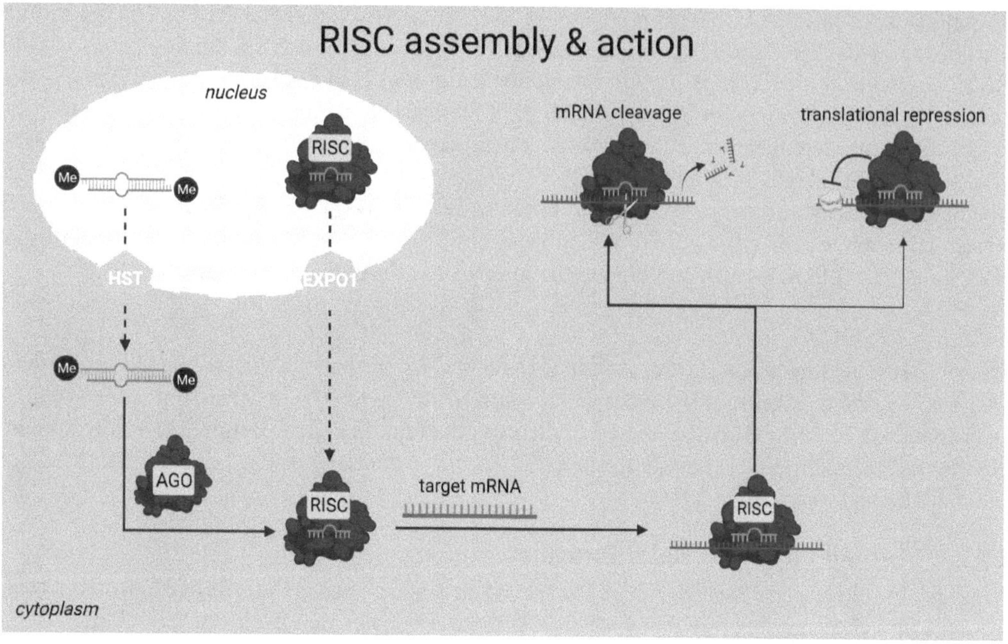

FIGURE 4.2 Assembly of RISC and its action for mRNA regulation.

Previous evidence suggested that miRISC assembly takes place in the cytoplasm after the miRNA duplex is exported out of the nucleus by Hasty (HST) (Chen, 2005; Ding et al., 2012), but new computational evidence suggests that another pathway, where the assembly occurs inside the nucleus, might be more plausible (Bologna et al., 2018). Both of these pathways, along with the consequent mRNA regulation, are outlined in Figure 4.2.

The assembly of miRISC (according to the second pathway) inside the nucleus can be divided into two parts: loading and maturation. These steps are further elaborated below.

4.2.4.1 Loading

AGO proteins are loading-incompetent in their native state. Thus, chaperones called heat shock protein 90 (HSP90), along with ATP, help make AGO1 loading competent. HSP90 exists as a dimer and binds to AGO along with ATP, causing conformational changes in AGO1 and itself (Iki et al., 2010).

HSP90 hydrolyses the ATP bound to it, causing further conformational changes in AGO1. This conformational change allows the miRNA duplex to bind to the competent AGO1 (Iki et al., 2010). Moreover, computational modelling predicts that this exposes the nuclear export signal present in AGO1, which can initiate its export into the cytoplasm (Bologna et al., 2018; Song et al., 2019).

AGO1 contains four major domains: N, PIWI, MID, and PAZ. N and PAZ are linked together by a linker domain (L1), followed by MID and PAZ linked together by a second linker (L2). MID directly links to the PIWI domain.

The binding pocket in the PAZ domain anchors the 3′ overhang of the guide strand of the miRNA/miRNA* complex, while the MID domain anchors the 5′ phosphates of the guide strand. The guide strand (miRNA) is selected based on its lower thermodynamic stability at the 5′ end. The guide strand selection is facilitated by other proteins like high osmotic stress gene expression 5 and two serine/arginine-rich splicing factors, RS40 and RS41 (Chen et al., 2015). This kind of loading is called asymmetrical loading, and it positions the guide strand correctly over the PIWI domain, which has endonuclease activity and helps in splicing the target mRNAs, further discussed in Section 4.2.5 (Hutvagner & Simard, 2008; Song et al., 2019).

4.2.4.2 Maturation

This step is characterized by the removal of the passenger (miRNA*) strand, also known as passenger strand ejection. This strand is then degraded (Li & Yu, 2021). RISC is then exported by exportin 1 (EXPO1) into the cytoplasm (Bologna et al., 2018; Wang et al., 2019).

Now, the miRISC binds to its target mRNA.

4.2.5 mRNA Regulation by miRISC

The RISC now binds to its target mRNA. Gene regulation can occur either via cleavage of the target mRNA or through translation inhibition.

In plants, a high degree of complementarity is observed between the nucleotides of miRNAs and target mRNA. This high complementarity results in mRNA regulation predominantly by target mRNA cleavage (Ding et al., 2012).

4.2.5.1 miRNA-Mediated Target mRNA Cleavage

RISC binds to the mRNA through complementary base pairing. Using the PIWI domain, AGO1 directs the endonucleolytic cleavage of the mRNA. Since the mRNA has now been cleaved, it can no longer be used for translation in protein synthesis (Ding et al., 2012; Yu & Wang, 2010).

4.2.5.2 Translation Inhibition

In this mechanism, RISC prevents the translation of mRNA. Although rare in plants, this mechanism is observed with certain miRNAs that are important for floral patterning and timing, oxidative stress, and positive regulation of photomorphogenesis. The rarity is due to the lower degree of complementarity. It occurs more frequently in animals (as shown in Table 4.1), where miRNAs are less complementary to their target mRNAs, leading to a central mismatch (Yu & Wang, 2010). Table 4.1 describes the differences between plant and animal miRNA biogenesis.

4.3 MAINTENANCE OF PLANT miRNA

Maintenance of miRNA occurs at various stages. It can be achieved by regulating the transcription factors of the MIR gene itself or by regulating factors that help in the processing of miRNA.

4.3.1 Factors Affecting MIR Gene Regulation

MIR is regulated by several factors, including chromatin remodelling factor 2, suppressor of actin 3A, cell division cycle 5, protein phosphatase 4, short valve 1, and THO/HRP phenotype 1 (Li & Yu, 2021).

A few of these factors are outlined below:

- Regulation via cyclin-dependent kinase F;1 (CDKF;1) is required to phosphorylate the CTD of RNA Pol II. In *cdkf;1* mutants, a marked decrease in immature and mature miRNAs was observed. Moreover, inactivation of CDKF;1 is known to cause dwarfism and sterility (Hajheidari et al. 2012).

TABLE 4.1
Differences Observed between Plant and Animal miRNA Biogenesis

Feature	Plant	Animal
miRNA complementarity to target mRNA	High complementarity	Low complementarity
Gene regulation method	mRNA cleavage	Translation inhibition
Splicing of pri-miRNA to pre-miRNA	DCLs	Drosha, Dicer
Export of miRNA out of the nucleus	HST, EXPO1	EXPO5

- The mediator complex at MIR genes helps recruit RNA Pol II (Kim et al., 2011).
- Negative on TATAless 2 (NOT2) proteins (specifically NOT2a and NOT2b) interact with RNA Pol II to promote MIR gene transcription. They also help in the recruitment of DCL1 during miRNA splicing (Wang et al., 2013).

4.3.2 Factors Affecting miRNA Processing

Various proteins involved in miRNA biogenesis, i.e., DCL1, HYL1, SE, HEN1, and RISC, are regulated by several factors (depicted in Figure 4.3). Some of these factors are listed below.

4.3.2.1 Factors Affecting DCL1

- DCL1 is positively regulated by XAP5 circadian timekeeper and STABALIZED1, which positively affect its transcription (Fang et al., 2015, p. 201).
- On the other hand, DCL1 transcription is negatively regulated by General control non-repressed protein 5, a histone acetyltransferase that tightly binds to the DCL1 transcribing region, thus preventing its transcription. miRNA MIR1612 is known to cleave DCL1 mRNA, preventing its translation (Ben Chaabane et al., 2013; Kim et al., 2009). MIR838 also regulates DCL1, but it does so by preventing its transcription (Wang et al., 2019).

4.3.2.2 Factors Affecting SE

- mir863-3p has been shown to negatively regulate SE formation.
- mRNA Cap-binding protein 20 interacts with SE directly or indirectly to form a pri-miRNA processing centre, and thus positively regulating it (Wang et al., 2013).

4.3.2.3 Factors Affecting HYL1

HYL1 promotes the processing activity of DCL1 for miRNA biogenesis. Therefore, regulators of HYL1 in turn also regulate the functioning of DCL1.

- In the presence of light, constructive photomorphogenic 1 inhibits protease activity in the cytoplasm, which reduces the degradation of HYL1, allowing it to remain in the cell longer and regulate mRNA activity (Cho et al., 2014).

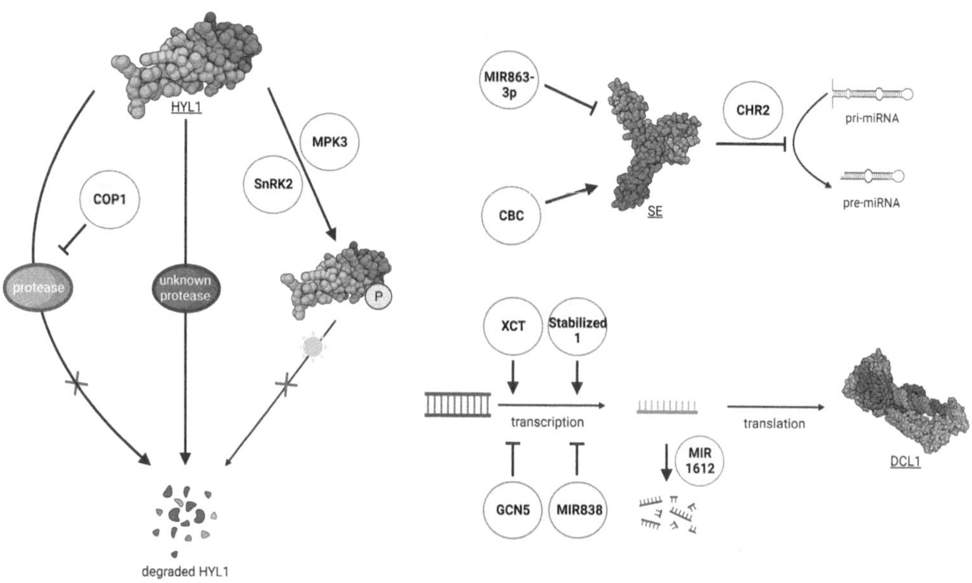

FIGURE 4.3 Factors regulating the activity of DCL1, HYL1, and SE – three important factors involved in miRNA biogenesis.

- On the other hand, post-translational phosphorylation by mitogen-activated protein kinase 3 and sucrose non-fermenting 1-related protein kinase 2 inactivates HYL1 but allows it to stay in the nucleus, thus preventing degradation until it is activated by light-mediated dephosphorylation (Raghuram et al., 2015).
- HYL1 is negatively regulated by cleavage at its N-terminal region by a cytoplasmic protease; however, there is limited information available about the protease and its mechanism.
- Short interspersed elements are transposons that isolate HYL1 from miRNA, resulting in decreased miRNA expression (Pouch-Pélissier et al., 2008).

4.3.2.4 Factors Affecting RISC Loading and AGO Regulation

- Transportin 1, an importin β protein, positively regulates the loading of miRNA onto RISC (Cui et al., 2016).
- Enhanced miRNA activity 1, on the other hand, negatively regulates the loading of miRNA into AGO1.
- F-box with WD-40 2 degrades AGO1 post-translation via autophagy, thus negatively regulating it. F-box with WD-40 2 targets both soluble and membrane-bound AGO1 (Hacquard et al., 2022; Wang et al., 2019).

4.3.2.5 Factors Affecting HEN1

HEN1 is regulated photomorphically, with its levels decreasing in the dark and increasing when exposed to light.

- Photoreceptors phytochrome A, phytochrome B, cryptochrome 1, and cryptochrome 2 activate downstream transcription factors in light, thereby increasing HEN1 levels.
- Elongated hypocotyl 5 (HY5) and HY5 homolog are transcriptional activators of HYL1 under light conditions.
- In the dark, HEN1 can lead to the degradation of HY5 mRNA transcripts through the formation of miR157, thus forming a negative feedback regulatory loop (Tsai et al., 2014).

4.4 DEGRADATION OF PLANT miRNA

Degradation of plant miRNA is important for maintaining miRNA balance and function. An abundance of miRNA can lead to gene suppression. Therefore, it is important to maintain miRNA levels and thus regulate their function.

Plant miRNA degradation is a relatively poorly understood topic; however, recent research has shed light on its mechanisms.

There are various mechanisms through which plant miRNA degradation can occur:

1. Pre-RISC loading

   Regulation and degradation of miRNA can occur prior to RISC formation. This is mediated by a 3' to 5' exonuclease belonging to the DEDDy family, Atrimmer 2, which works upstream of HEN1. Its activity is stronger in *hen1* mutants or when HEN1 function is compromised (Wang et al., 2018).
2. Post-RISC loading

   Plant miRNA turnover, after RISC loading, majorly involves the cooperative action of HEN1 suppressor 1 (HESO1) and sRNA-degrading nuclease 1 (SDN1), as shown in Figure 4.4.
   a. Unmethylated miRNA
      i. Terminal uridylyl transferase

         When the 3′ end of the AGO-bound miRNA strand is unmethylated (as seen in *hen1* mutants), it makes the strand unstable (Wang et al., 2019). This free

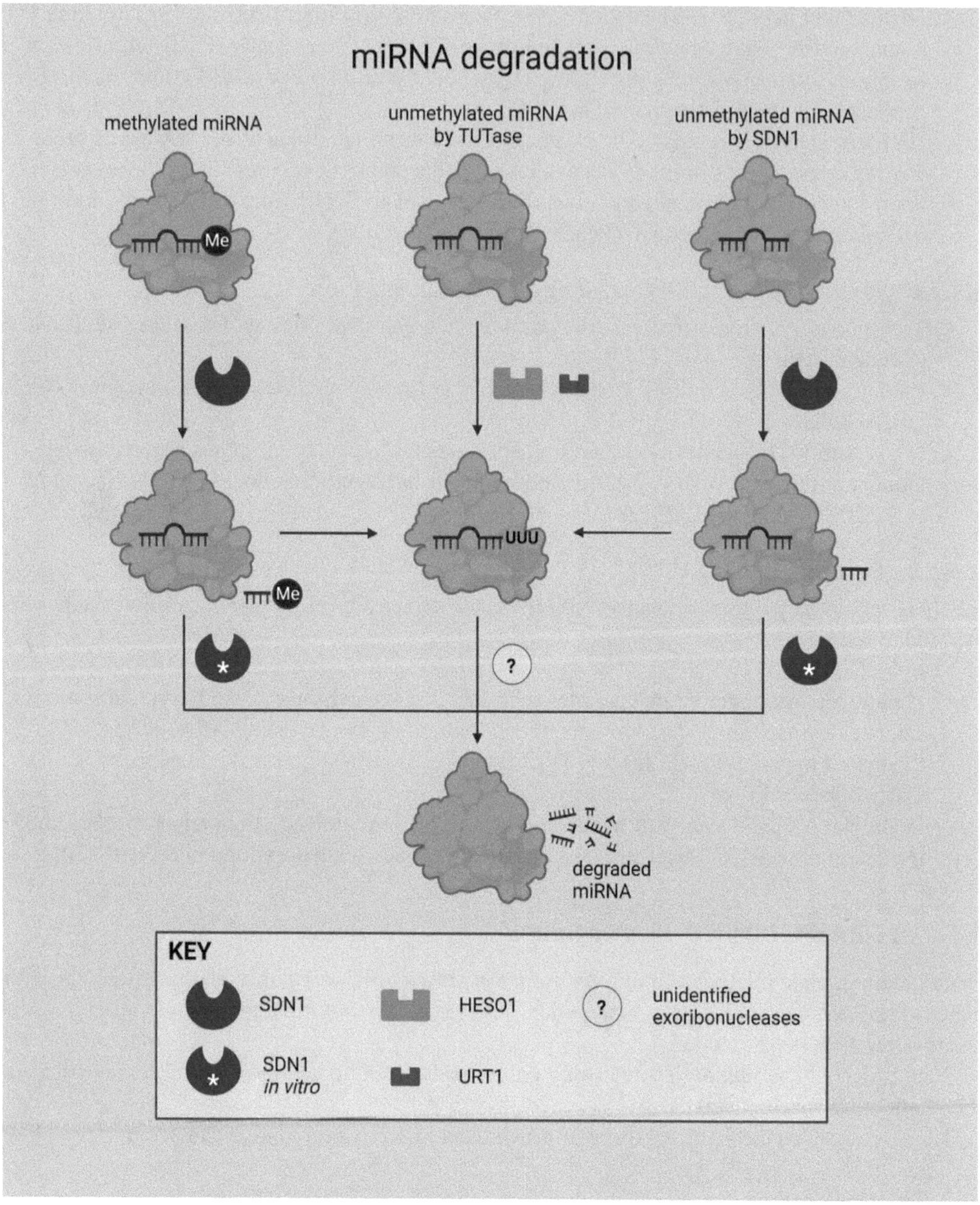

FIGURE 4.4 Degradation of methylated and unmethylated miRNAs.

unmethylated end is uridylated by terminal uridylyl transferases (TUTases) known as HESO1 and UTP:RNA uridylyl transferase 1 (URT1).

Developmental defects are observed in plants with a hen1 mutation. However, when there is a hen1 heso1 double mutation, these developmental defects are greatly suppressed. On the other hand, if HESO1 is overexpressed, the same defects are exacerbated. This proves that degradation of miRNA occurs through HESO1 via uridylation (Ren et al., 2012; Zhao et al., 2012). Moreover, in *hen1 heso1 urt1* triple mutants, there is a complete absence of degradation through the uridylation-dependent pathway (Wang et al., 2015).

HESO1 can work independently of or in synergy with URT1, but it is the major TUTase involved in uridylation since it can quickly add over 30 Us to the 3′ end, while URT1 can add just 1–3 (Wang et al., 2015; Xie et al., 2015; Zhu et al., 2020).

Post-uridylation, unidentified exoribonucleases degrade the uridylated miRNA strand (Ramachandran & Chen, 2008).

ii. SDN1

SDN1, belonging to a class of DEDD exoribonucleases (DEDDh), is a 3′–5′ exoribonuclease that truncates the unmethylated 3′ end of AGO1-bound miRNA by cleaving 4 nucleotides (Chen et al., 2018; Ramachandran & Chen, 2008).

This truncated miRNA can either degrade via uridylation by TUTases or through a uridylation-independent pathway. In the latter pathway, it has been observed that SDN1 degrades the truncated miRNA as well, but only *in vitro*, not *in vivo*; enzymes that degrade the miRNA *in vivo* have yet to be discovered (Ramachandran & Chen, 2008).

b. Methylated miRNA

The methyl group on the AGO1-bound miRNA (in the HEN1 phenotype) suppresses HESO1 and URT1 activity, preventing uridylation (Tu et al., 2015). Instead, SDN1 truncates the 3′ methylated end of the miRNA, similar to its action on unmethylated miRNA, but since the methyl group is also cleaved at the same time, SDN1 is said to exhibit "demethylase" activity.

Notably, SDN1 shows a lower catalytic affinity for methylated miRNAs (Chen et al., 2018).

The demethylated miRNA strand now undergoes degradation via the uridylation-dependent or -independent pathway (Wang et al., 2019; Xie et al., 2015).

Hence, it can be concluded that HEN1 promotes miRNA accumulation in plants through their methylation by preventing their degradation (Sanei & Chen, 2015).

4.5 IMPORTANCE OF PLANT miRNA

The field of plant miRNA research continues to evolve. However, here are some notable trends and discoveries up to that point:

Functional Diversity: Researchers have uncovered an increasing diversity of miRNA functions in plants. Beyond their roles in development and stress responses, miRNAs have been implicated in various processes such as nutrient homeostasis, symbiotic interactions, and epigenetic regulation.

Long Non-Coding RNA Regulation: Emerging evidence suggests intricate regulatory cross-talk between miRNAs and long non-coding RNAs in plants (Noopur et al., 2025b; Pooja et al., 2025; Rahul et al., 2024; Subhan et al., 2025; Tanvi et al., 2025). These interactions can modulate gene expression and contribute to regulatory networks that control plant development and stress responses.

Phased siRNA Production: Recent studies have elucidated the biogenesis and functions of phased siRNAs (phasiRNAs) derived from non-coding RNA precursors targeted by miRNAs. These phasiRNAs play roles in diverse processes such as reproductive development, stress responses, and defence against pathogens.

sRNA Mobility: Plant miRNAs and other sRNAs have been found to exhibit mobility within plants, enabling long-distance signalling and systemic regulation of gene expression. This mobility contributes to systemic responses to environmental cues and developmental signals.

Genome Editing and miRNA Engineering: Advances in genome editing technologies, such as CRISPR-Cas9, have facilitated the precise manipulation of miRNA genes in plants. Researchers are exploring the potential of miRNA engineering for crop improvement, enhanced stress tolerance, increased yield, and modulation of nutrient content.

Environmental Regulation of miRNA Expression: Environmental factors, including temperature fluctuations, drought, salinity, and pathogen attacks, influence the expression and activity of plant miRNAs (Meenakshi et al., 2025). Understanding how miRNA expression is regulated in response to environmental cues provides insights into plant adaptation and resilience.

sRNA-Mediated Epigenetic Regulation: Plant miRNAs and siRNAs participate in epigenetic regulation by targeting genes involved in chromatin modification and DNA methylation. This sRNA-mediated epigenetic regulation contributes to developmental processes and stress responses in plants.

These discoveries underscore the intricate regulatory roles of miRNAs in plants and their significance in plant biology, agriculture, and environmental adaptation. Ongoing research continues to uncover novel aspects of plant miRNA biology and their applications in crop improvement and environmental sustainability.

The field of plant miRNA research has seen several significant discoveries and success stories:

Functional Diversity: Recent studies have unveiled the functional diversity of plant miRNAs beyond their canonical roles in gene regulation. miRNAs have been implicated in diverse processes such as nutrient uptake, root development, flowering time regulation, and responses to biotic and abiotic stresses.

Crop Improvement: Researchers have leveraged miRNA knowledge for crop improvement through genetic engineering. For instance, manipulation of miRNA expression has been used to enhance crop yield, increase stress tolerance, improve nutrient uptake efficiency, and modulate plant architecture.

Disease Resistance: MiRNAs have been implicated in plant defence against pathogens. By targeting genes involved in pathogen recognition, signal transduction, and defence responses, miRNAs play crucial roles in mounting effective immune responses in plants. Understanding these mechanisms could help develop novel strategies for enhancing disease resistance in crops.

Epigenetic Regulation: Plant miRNAs are involved in epigenetic regulation by targeting genes that encode chromatin modifiers and DNA methylation enzymes. This sRNA-mediated epigenetic regulation contributes to developmental processes, stress responses, and environmental adaptation in plants.

Environmental Sensing and Adaptation: miRNAs play roles in sensing and responding to environmental cues such as temperature fluctuations, drought, salinity, and nutrient availability. By modulating gene expression networks, miRNAs help plants adapt to changing environmental conditions, thereby enhancing their resilience and survival in challenging environments.

Phased siRNA Production: The discovery of phasiRNAs derived from non-coding RNA precursors targeted by miRNAs has opened new avenues for understanding gene regulatory networks in plants. PhasiRNAs are involved in diverse processes, including reproductive development, stress responses, and defence against pathogens.

Nutrient Homeostasis: MiRNAs play crucial roles in maintaining nutrient homeostasis in plants by regulating the expression of genes involved in nutrient uptake, transport, and storage. Manipulating miRNA expression holds promise for improving nutrient use efficiency and enhancing crop productivity under nutrient-limiting conditions.

These discoveries highlight the importance of plant miRNAs in various aspects of plant biology and agriculture. By deciphering the complex regulatory networks governed by miRNAs, researchers are paving the way for innovative strategies to address global challenges such as food security, environmental sustainability, and climate change resilience.

Plant miRNAs play crucial roles in plant responses to climate change. Here's how:

Regulation of Stress Responses: Plant miRNAs are key regulators of stress responses, including those triggered by climate change-related factors such as heat, drought, salinity, and extreme weather events. By modulating the expression of stress-responsive genes, miRNAs help plants adapt to changing environmental conditions and mitigate the adverse effects of climate stress.

Modulation of Developmental Processes: Climate change can alter various developmental processes in plants, such as flowering time, seed germination, and vegetative growth. MiRNAs are involved in regulating these processes by targeting genes associated with developmental pathways. Understanding miRNA-mediated developmental regulation can provide insights into how plants cope with climate variability.

Epigenetic Regulation: Plant miRNAs participate in epigenetic regulation by targeting genes encoding chromatin modifiers and DNA methylation enzymes. miRNA-mediated epigenetic modifications can influence plant responses to environmental stressors by providing a mechanism for heritable changes in gene expression patterns, thus contributing to climate change adaptation.

Interaction with Environmental Factors: Environmental factors associated with climate change, such as temperature fluctuations, CO_2 levels, and water availability, can influence miRNA expression and activity in plants. miRNAs act as molecular sensors that integrate environmental signals and modulate gene expression networks, allowing plants to adjust their physiological and metabolic processes in response to changing environmental conditions.

Crop Resilience and Adaptation: Understanding the role of miRNAs in plant responses to climate change is crucial for enhancing crop resilience and adaptation. By manipulating miRNA expression or targeting specific miRNA-regulated pathways, researchers can develop crop varieties with improved tolerance to climate stressors, ensuring food security and agricultural sustainability in a changing climate.

Overall, plant miRNAs represent important molecular players in the complex interplay between plants and their environment, particularly in the context of climate change. Further research into the regulatory roles of miRNAs in plant adaptation to climate stress will contribute to the development of innovative strategies for mitigating the impacts of climate change on agriculture and ecosystems.

Recent literature indicates that understanding plant miRNAs can be helpful in mitigating climate change:

1. A 2020 study identified novel miRNAs and their respective gene targets across sorghum, finger millet, pearl millet, foxtail millet and proso millet using *in silico* approaches since miRNAs remain understudied in millets (Chakraborty et al., 2020). Across the five millet species, 1,002 miRNAs from 280 families, governing 23,158 targets were elucidated. The study also explored the unique and conserved structural and functional traits of these miRNAs across millet species, highlighting existing evolutionary relationships between them. Notably, 84 miRNAs were found to be conserved across different millet species, underscoring the evolutionary significance of plant miRNAs.

   Furthermore, 215 miRNAs were identified as regulators of 155 major drought-responsive genes, including transcription factors and protein families, elucidating the genetic basis of drought tolerance encoded within millet genomes. Network analysis revealed that miRNAs modulated drought stress through specific or multiple targets, providing insights into the intricate regulatory networks governing drought responses in millets. Millets serve as critical food crops in arid and drought-prone regions, thriving under adverse conditions and offering sustainable yields. Rich in essential nutrients, millets play a vital role in combating micronutrient deficiencies. With the escalating challenges of climate change and widespread malnutrition, millets have emerged as linchpins for ensuring food and nutritional security. Despite their inherent resilience to drought, millets often exhibit low yields under stress conditions, presenting a prime opportunity to enhance their genetic potential through a comprehensive understanding of drought tolerance mechanisms.

2. In a study on crop improvement (Raza et al., 2023), the importance of miRNAs for crop improvement was highlighted. They summarized identified miRNAs and their target genes that boost plant growth, development, and tolerance to both abiotic and biotic stresses. Additionally, they emphasized the use of miRNA-mediated engineering techniques for enhancing crop resilience and noted the availability of sequence-based technologies for identifying stress-responsive miRNAs and their roles in plant development. This consolidated overview served as a valuable resource for researchers and breeders striving to develop stress-smart crops for future agricultural sustainability.
3. In a review (Kaur et al., 2024) on potatoes (*Solanum tuberosum* L.), which hold global prominence among non-grain crops, the importance of harnessing miRNA engineering in developing stress-resistant potato varieties was discussed, potentially yielding high productivity and improved nutritional content. This review delved into potato miRNAs known to influence various life cycle stages, conferring stress resilience and modulating gene expression. By highlighting the significance of miRNA machinery in plants, particularly in potatoes, the review urged further exploration of miRNA engineering due to its potential for elevating crop yields and enhancing stress tolerance, offering a pathway towards sustainable agricultural practices and enhanced food security.
4. Another study (Kumar et al., 2024) attempted to address drought stress, a primary abiotic challenge in rice cultivation, by identifying and utilizing major genomic regions associated with stress tolerance in field conditions; however, the underlying molecular mechanisms remain elusive. The study highlights the involvement of miRNA/mRNA modules in the drought stress response. Secondary regulation by miRNA/mRNA modules in Swarna (a highly drought-sensitive mega variety) was investigated during the reproductive stage through sRNA-seq analysis. Their findings revealed nine putative novel miRNAs and 27 differentially expressed known miRNAs during the booting stage under drought stress.

Through degradome analysis and transcript abundance studies, key miRNA/transcript modules significantly impacting the plant's response to drought stress and yield were identified. Notably, modules such as osa-miR169a/LOC_Os07g41720, Osa-miR172d-3p/5p, Osa-miR171b/f, Osa-miR1876/LOC_Os11g38330, Osa-miR397a, and Osa-miR530–3p emerged as major players. These modules elucidated the basis for the reduced spikelet fertility and increased grain chalkiness observed in the Swarna variety under drought stress, highlighting the modulation of expression of Osa-miR397a/LOC_Os03g03510, LOC_Os03g51220, and Osa-miR530–3p/LOC_Os10g40510. Hence, this study not only shed light on the intricate regulatory networks governing drought stress response in rice but also provided valuable insights into potential target genes for enhancing the resilience of Swarna. By understanding and manipulating these molecular mechanisms, the authors aimed to fortify Swarna's ability to withstand drought stress to ensure food security amidst the difficulties posed by climate change.

The escalating global population and accompanying food demand underscore the urgency of enhancing crop traits to meet nutritional needs, which may be achieved through the intricacies of miRNA in different crops.

4.6 CONCLUSION

Understanding the concept of miRNA biogenesis, regulation, and degradation mechanisms provides insights into the intricate regulatory networks governing post-transcriptional gene expression in plants. From the initial transcription of MIR genes by RNA Pol II to the formation of mature miRNAs incorporated into the RISC, each step involves an orchestrated series of events mediated by various proteins and regulatory factors. The stabilization of miRNAs through 5′ end methylation by HEN1, coupled with the loading of miRNAs onto AGO proteins within the RISC complex, explains the mechanism for maintaining miRNA stability for effective gene regulation. Moreover, the balance between miRNA biogenesis factors like DCL1, HYL1, and SE and their regulators such as CDKF;1, NOT2, and various transcription factors highlights the intricate control mechanisms that ensure the proper functioning of the miRNA regulatory pathway.

Furthermore, the degradation of miRNAs, facilitated by enzymes like HESO1 and SDN1, serves as a crucial mechanism for maintaining miRNA homeostasis and preventing excessive gene suppression. The various uridylation-dependent and -independent degradation pathways highlight the dynamic nature of miRNA turnover and its implications for plant development and stress response. Additionally, the regulatory roles of photoreceptors and transcription factors in modulating the expression and activity of miRNA biogenesis factors emphasize the integration of miRNA regulation into broader signalling networks governing plant growth and adaptation.

Importantly, studying plant miRNAs not only enhances the understanding of fundamental biological processes but also has significant implications for agriculture and ecosystem sustainability. By understanding the roles of miRNAs in plant development, stress response, and resource utilization, this research opens avenues for developing novel strategies to enhance crop resilience, optimize agricultural productivity, and mitigate the adverse effects of climate change. Moreover, the identification of key regulatory factors governing miRNA biogenesis and degradation pathways provides targets for genetic engineering approaches aimed at modulating miRNA activity for desired agronomic traits. By unravelling the intricacies of miRNA biogenesis, regulation, and degradation, this research paves the way for future studies aimed at harnessing the potential of miRNAs to address pressing challenges in agriculture, environmental sustainability, and human welfare.

REFERENCES

Bajczyk, M., Lange, H., Bielewicz, D., Szewc, L., Bhat, S. S., Dolata, J., Kuhn, L., Szweykowska-Kulinska, Z., Gagliardi, D., & Jarmolowski, A. (2020). SERRATE interacts with the nuclear exosome targeting (NEXT) complex to degrade primary miRNA precursors in Arabidopsis. *Nucleic Acids Research*, *48*(12), 6839–6854. https://doi.org/10.1093/nar/gkaa373

Baker, K. E., & Coller, J. (2006). The many routes to regulating mRNA translation. *Genome Biology*, *7*(12), 332. https://doi.org/10.1186/gb-2006-7-12-332

Ben Chaabane, S., Liu, R., Chinnusamy, V., Kwon, Y., Park, J., Kim, S. Y., Zhu, J.-K., Yang, S. W., & Lee, B. (2013). STA1, an Arabidopsis pre-mRNA processing factor 6 homolog, is a new player involved in miRNA biogenesis. *Nucleic Acids Research*, *41*(3), 1984–1997. https://doi.org/10.1093/nar/gks1309

Bologna, N. G., Iselin, R., Abriata, L. A., Sarazin, A., Pumplin, N., Jay, F., Grentzinger, T., Dal Peraro, M., & Voinnet, O. (2018). Nucleo-cytosolic shuttling of ARGONAUTE1 prompts a revised model of the plant microRNA pathway. *Molecular Cell*, *69*(4), 709–719. https://doi.org/10.1016/j.molcel.2018.01.007

Boo, S. H., & Kim, Y. K. (2020). The emerging role of RNA modifications in the regulation of mRNA stability. *Experimental & Molecular Medicine*, *52*(3), 400–408. https://doi.org/10.1038/s12276-020-0407-z

Burdisso, P., Suarez, I. P., Bologna, N. G., Palatnik, J. F., Bersch, B., & Rasia, R. M. (2012). Second double-stranded RNA binding domain of dicer-like ribonuclease 1: Structural and biochemical characterization. *Biochemistry*, *51*(51), 10159–10166. https://doi.org/10.1021/bi301247r

Chakraborty, A., Viswanath, A., Malipatil, R., Rathore, A., & Thirunavukkarasu, N. (2020). Structural and functional characteristics of miRNAs in five strategic millet species and their utility in drought tolerance. *Frontiers in Genetics*, *11*, 608421. https://doi.org/10.3389/fgene.2020.608421

Chen, C., Zeng, Z., Liu, Z., & Xia, R. (2018). Small RNAs, emerging regulators critical for the development of horticultural traits. *Horticulture Research*, *5*(1), 63. https://doi.org/10.1038/s41438-018-0072-8

Chen, J., Liu, L., You, C., Gu, J., Ruan, W., Zhang, L., Gan, J., Cao, C., Huang, Y., Chen, X., & Ma, J. (2018). Structural and biochemical insights into small RNA 3′ end trimming by Arabidopsis SDN1. *Nature Communications*, *9*(1), 3585. https://doi.org/10.1038/s41467-018-05942-7

Chen, T., Cui, P., & Xiong, L. (2015). The RNA-binding protein HOS5 and serine/arginine-rich proteins RS40 and RS41 participate in miRNA biogenesis in Arabidopsis. *Nucleic Acids Research*, *43*(17), 8283–8298. https://doi.org/10.1093/nar/gkv751

Chen, X. (2005). microRNA biogenesis and function in plants. *FEBS Letters*, *579*(26), 5923–5931. https://doi.org/10.1016/j.febslet.2005.07.071

Chheda, N., & Gupta, M. K. (2014). RNA as a permutation (version 1). arXiv. https://doi.org/10.48550/ARXIV.1403.5477

Cho, S. K., Chaabane, S. B., Shah, P., Poulsen, C. P., & Yang, S. W. (2014). COP1 E3 ligase protects HYL1 to retain microRNA biogenesis. *Nature Communications*, *5*(1), 5867. https://doi.org/10.1038/ncomms6867

Ciechanowska, K., Pokornowska, M., & Kurzyńska-Kokorniak, A. (2021). Genetic insight into the domain structure and functions of dicer-type ribonucleases. *International Journal of Molecular Sciences*, *22*(2), 616. https://doi.org/10.3390/ijms22020616

Cui, Y., Fang, X., & Qi, Y. (2016). TRANSPORTIN1 promotes the association of microRNA with ARGONAUTE1 in Arabidopsis. *The Plant Cell*, *28*(10), 2576–2585. https://doi.org/10.1105/tpc.16.00384

Ding, J., Zhou, S., & Guan, J. (2012). Finding microRNA targets in plants: Current status and perspectives. *Genomics, Proteomics & Bioinformatics*, *10*(5), 264–275. https://doi.org/10.1016/j.gpb.2012.09.003

Dong, Z., Han, M.-H., & Fedoroff, N. (2008). The RNA-binding proteins HYL1 and SE promote accurate *in vitro* processing of pri-miRNA by DCL1. *Proceedings of the National Academy of Sciences of the United States of America*, *105*(29), 9970–9975. https://doi.org/10.1073/pnas.0803356105

Fang, X., Shi, Y., Lu, X., Chen, Z., & Qi, Y. (2015). CMA33/XCT regulates small RNA production through modulating the transcription of dicer-like genes in arabidopsis. *Molecular Plant*, *8*(8), 1227–1236. https://doi.org/10.1016/j.molp.2015.03.002

Fukudome, A., & Fukuhara, T. (2017). Plant dicer-like proteins: Double-stranded RNA-cleaving enzymes for small RNA biogenesis. *Journal of Plant Research*, *130*(1), 33–44. https://doi.org/10.1007/s10265-016-0877-1

Hacquard, T., Clavel, M., Baldrich, P., Lechner, E., Pérez-Salamó, I., Schepetilnikov, M., Derrien, B., Dubois, M., Hammann, P., Kuhn, L., Brun, D., Bouteiller, N., Baumberger, N., Vaucheret, H., Meyers, B. C., & Genschik, P. (2022). The Arabidopsis F-box protein FBW2 targets AGO1 for degradation to prevent spurious loading of illegitimate small RNA. *Cell Reports*, *39*(2), 110671. https://doi.org/10.1016/j.celrep.2022.110671

Hajheidari, M., Farrona, S., Huettel, B., Koncz, Z., & Koncz, C. (2012). CDKF;1 and CDKD protein kinases regulate phosphorylation of serine residues in the C-terminal domain of *Arabidopsis* RNA polymerase II. *The Plant Cell*, *24*(4), 1626–1642. https://doi.org/10.1105/tpc.112.096834

Horvitz, H. R., & Sulston, J. E. (1980). Isolation and genetic characterization of cell-lineage mutants of the nematode *CAENORHABDITIS ELEGANS*. *Genetics*, *96*(2), 435–454. https://doi.org/10.1093/genetics/96.2.435

Hutvagner, G., & Simard, M. J. (2008). Argonaute proteins: Key players in RNA silencing. *Nature Reviews Molecular Cell Biology*, *9*(1), 22–32. https://doi.org/10.1038/nrm2321

Iki, T., Yoshikawa, M., Nishikiori, M., Jaudal, M. C., Matsumoto-Yokoyama, E., Mitsuhara, I., Meshi, T., & Ishikawa, M. (2010). In vitro assembly of plant RNA-induced silencing complexes facilitated by molecular chaperone HSP90. *Molecular Cell*, *39*(2), 282–291. https://doi.org/10.1016/j.molcel.2010.05.014

Iorio, M. V., & Croce, C. M. (2012). microRNA dysregulation in cancer: Diagnostics, monitoring and therapeutics. A comprehensive review. *EMBO Molecular Medicine*, *4*(3), 143–159. https://doi.org/10.1002/emmm.201100209

Jones-Rhoades, M. W., & Bartel, D. P. (2004). Computational identification of plant microRNAs and their targets, including a stress-induced miRNA. *Molecular Cell*, *14*(6), 787–799. https://doi.org/10.1016/j.molcel.2004.05.027

Juźwik, C. A., Drake, S., Zhang, Y., Paradis-Isler, N., Sylvester, A., Amar-Zifkin, A., Douglas, C., Morquette, B., Moore, C. S., & Fournier, A. E. (2019). microRNA dysregulation in neurodegenerative diseases: A systematic review. *Progress in Neurobiology*, *182*, 101664. https://doi.org/10.1016/j.pneurobio.2019.101664

Kaur, G., Jain, S., Bhushan, S., Das, N., Sharma, M., & Sharma, D. (2024). Role of microRNAs and their putative mechanism in regulating potato (*Solanum tuberosum* L.) life cycle and response to various environmental stresses. *Plant Physiology and Biochemistry*, *207*, 108334. https://doi.org/10.1016/j.plaphy.2024.108334

Kim, W., Benhamed, M., Servet, C., Latrasse, D., Zhang, W., Delarue, M., & Zhou, D.-X. (2009). Histone acetyltransferase GCN5 interferes with the miRNA pathway in Arabidopsis. *Cell Research*, *19*(7), 899–909. https://doi.org/10.1038/cr.2009.59

Kim, Y. J., Zheng, B., Yu, Y., Won, S. Y., Mo, B., & Chen, X. (2011). The role of mediator in small and long noncoding RNA production in *Arabidopsis thaliana*: Mediator in noncoding RNA production. *The EMBO Journal*, *30*(5), 814–822. https://doi.org/10.1038/emboj.2011.3

Kloosterman, W. P., Wienholds, E., De Bruijn, E., Kauppinen, S., & Plasterk, R. H. A. (2006). In situ detection of miRNAs in animal embryos using LNA-modified oligonucleotide probes. *Nature Methods*, *3*(1), 27–29. https://doi.org/10.1038/nmeth843

Kumar, D., Mulani, E., Singh, B. K., Dutta, B., Singh, A., Solanke, A. U., & Sevanthi, A. M. (2024). Understanding the role of miRNAs in governing the drought sensitive response of a rice mega variety, Swarna at reproductive stage. *Plant Stress*, *11*, 100302. https://doi.org/10.1016/j.stress.2023.100302

Kurihara, Y., Takashi, Y., & Watanabe, Y. (2006). The interaction between DCL1 and HYL1 is important for efficient and precise processing of pri-miRNA in plant microRNA biogenesis. *RNA*, *12*(2), 206–212. https://doi.org/10.1261/rna.2146906

Lee, R. C., Feinbaum, R. L., & Ambros, V. (1993). The C. elegans heterochronic gene lin-4 encodes small RNAs with antisense complementarity to lin-14. *Cell*, *75*(5), 843–854. https://doi.org/10.1016/0092-8674(93)90529-Y

Li, C., & Zhang, B. (2016). microRNAs in control of plant development. *Journal of Cellular Physiology*, *231*(2), 303–313. https://doi.org/10.1002/jcp.25125

Li, J., Yang, Z., Yu, B., Liu, J., & Chen, X. (2005). Methylation protects miRNAs and siRNAs from a 3′-end uridylation activity in arabidopsis. *Current Biology*, *15*(16), 1501–1507. https://doi.org/10.1016/j.cub.2005.07.029

Li, M., & Yu, B. (2021). Recent advances in the regulation of plant miRNA biogenesis. *RNA Biology*, *18*(12), 2087–2096. https://doi.org/10.1080/15476286.2021.1899491

Liu, C., Axtell, M. J., & Fedoroff, N. V. (2012). The helicase and RNaseIIIa domains of arabidopsis dicer-like1 modulate catalytic parameters during microRNA biogenesis. *Plant Physiology*, *159*(2), 748–758. https://doi.org/10.1104/pp.112.193508

Lu, S., Sun, Y., & Chiang, V. L. (2008). Stress-responsive microRNAs in *Populus*. *The Plant Journal*, *55*(1), 131–151. https://doi.org/10.1111/j.1365-313X.2008.03497.x

Machida, S., Chen, H.-Y., & Adam Yuan, Y. (2011). Molecular insights into miRNA processing by *Arabidopsis thaliana* SERRATE. *Nucleic Acids Research*, *39*(17), 7828–7836. https://doi.org/10.1093/nar/gkr428

Meenakshi, S., Rahul, D., Sachidanand, S., Anita, G. M., & Rajpal, S. (2025). Plant microRNAs: An overview. *microRNA Advances and Application in Plant Biology Plant Biology, Sustainability and Climate Change*, *2025*, 1–33.

Noopur, K., Pragati, K., Adarsh, K. S., Sachidanand, S., Tanvi, J., & Megha, B. (2025a). Role of non coding RNA in effector trigger immunity. In Dutta, R., Kapadia, C., Danish, S., & Singh, S. (Eds.), *Non-Coding RNA in Plants Modulation and Stress Responses Plant Biology, Sustainability and Climate Change*. Academic Press, New York.

Noopur, K., Pragati, K., Megha, B., Arpan, M., Tanvi, J., Shrisendu, M., Adarsh, K. S., & Sachidanand, S. (2025b). Role of non-coding RNA in plant architecture. In Dutta, R., Kapadia, C., Danish, S., & Singh, S. (Eds.), *Non-Coding RNA in Plants Modulation and Stress Responses Plant Biology, Sustainability and Climate Change*. Academic Press, New York.

Pooja, G., Megha, B., Arpan, M., Chintan, K., Rahul, D., Subhan, D., & Sachidanand, S. (2025). Introduction to long noncoding RNA. In Dutta, R., Kapadia, C., Danish, S., & Singh, S. (Eds.), *Non-Coding RNA in Plants Modulation and Stress Responses Plant Biology, Sustainability and Climate Change*. Academic Press, New York.

Pouch-Pélissier, M.-N., Pélissier, T., Elmayan, T., Vaucheret, H., Boko, D., Jantsch, M. F., & Deragon, J.-M. (2008). SINE RNA induces severe developmental defects in arabidopsis thaliana and interacts with HYL1 (DRB1), a key member of the DCL1 complex. *PLoS Genetics*, *4*(6), e1000096. https://doi.org/10.1371/journal.pgen.1000096

Raczynska, K. D., Stepien, A., Kierzkowski, D., Kalak, M., Bajczyk, M., McNicol, J., Simpson, C. G., Szweykowska-Kulinska, Z., Brown, J. W. S., & Jarmolowski, A. (2014). The SERRATE protein is involved in alternative splicing in Arabidopsis thaliana. *Nucleic Acids Research*, *42*(2), 1224–1244. https://doi.org/10.1093/nar/gkt894

Raghuram, B., Sheikh, A. H., Rustagi, Y., & Sinha, A. K. (2015). microRNA biogenesis factor DRB1 is a phosphorylation target of mitogen activated protein kinase MPK3 in both rice and Arabidopsis. *The FEBS Journal*, *282*(3), 521–536. https://doi.org/10.1111/febs.13159

Rahul, D., Chintan, K., Subhan, D., & Sachidanand, S. (2024). *Non-Coding RNA in Plants: Modulation and Stress Responses*. Academic Press, New York.

Ramachandran, V., & Chen, X. (2008). Degradation of microRNAs by a family of exoribonucleases in *Arabidopsis*. *Science*, *321*(5895), 1490–1492. https://doi.org/10.1126/science.1163728

Raza, A., Charagh, S., Karikari, B., Sharif, R., Yadav, V., Mubarik, M. S., Habib, M., Zhuang, Y., Zhang, C., Chen, H., Varshney, R. K., & Zhuang, W. (2023). miRNAs for crop improvement. *Plant Physiology and Biochemistry*, *201*, 107857. https://doi.org/10.1016/j.plaphy.2023.107857

Ren, G., Chen, X., & Yu, B. (2012). Uridylation of miRNAs by HEN1 SUPPRESSOR1 in Arabidopsis. *Current Biology*, *22*(8), 695–700. https://doi.org/10.1016/j.cub.2012.02.052

Ren, G., & Yu, B. (2012). Post-transcriptional control of miRNA abundance in arabidopsis. *Plant Signaling & Behavior*, *7*(11), 1443–1446. https://doi.org/10.4161/psb.21956

Rogers, K., & Chen, X. (2013). Biogenesis, turnover, and mode of action of plant microRNAs. *The Plant Cell*, *25*(7), 2383–2399. https://doi.org/10.1105/tpc.113.113159

Sanei, M., & Chen, X. (2015). Mechanisms of microRNA turnover. *Current Opinion in Plant Biology*, *27*, 199–206. https://doi.org/10.1016/j.pbi.2015.07.008

Santa-Maria, I., Alaniz, M. E., Renwick, N., Cela, C., Fulga, T. A., Van Vactor, D., Tuschl, T., Clark, L. N., Shelanski, M. L., McCabe, B. D., & Crary, J. F. (2015). Dysregulation of microRNA-219 promotes neurodegeneration through post-transcriptional regulation of tau. *Journal of Clinical Investigation*, *125*(2), 681–686. https://doi.org/10.1172/JCI78421

Song, L., Axtell, M. J., & Fedoroff, N. V. (2010). RNA secondary structural determinants of miRNA precursor processing in arabidopsis. *Current Biology*, *20*(1), 37–41. https://doi.org/10.1016/j.cub.2009.10.076

Song, X., Li, Y., Cao, X., & Qi, Y. (2019). microRNAs and their regulatory roles in plant–environment interactions. *Annual Review of Plant Biology*, *70*(1), 489–525. https://doi.org/10.1146/annurev-arplant-050718-100334

Suarez, I. P., Burdisso, P., Benoit, M. P. M. H., Boisbouvier, J., & Rasia, R. M. (2015). Induced folding in RNA recognition by *Arabidopsis thaliana* DCL1. *Nucleic Acids Research*, *43*(13), 6607–6619. https://doi.org/10.1093/nar/gkv627

Subhan, D., Uzma, Y., Niaz, A., Khurram, S., Rahul, D., & Sachidanand, S. (2025). Fertilizer-responsive non-coding RNA and their significance. In Dutta, R., Kapadia, C., Danish, S., & Singh, S. (Eds.), *Non-Coding RNA in Plants Modulation and Stress Responses Plant Biology, Sustainability and Climate Change*. Academic Press, New York.

Tanvi, J., Arpita, S., Praneta, Y., Palak, S., & Sachidanand, S. (2025). Epigenetic modulation of plant developmental pathways by non-coding RNA. In Dutta, R., Kapadia, C., Danish, S., & Singh, S. (Eds.), *Non-Coding RNA in Plants Modulation and Stress Responses Plant Biology, Sustainability and Climate Change*. Academic Press, New York.

Tkaczuk, K. L., Obarska, A., & Bujnicki, J. M. (2006). Molecular phylogenetics and comparative modeling of HEN1, a methyltransferase involved in plant microRNA biogenesis. *BMC Evolutionary Biology*, *6*(1), 6. https://doi.org/10.1186/1471-2148-6-6

Tsai, H.-L., Li, Y.-H., Hsieh, W.-P., Lin, M.-C., Ahn, J. H., & Wu, S.-H. (2014). HUA ENHANCER1 is involved in posttranscriptional regulation of positive and negative regulators in *Arabidopsis* photomorphogenesis. *The Plant Cell*, *26*(7), 2858–2872. https://doi.org/10.1105/tpc.114.126722

Tu, B., Liu, L., Xu, C., Zhai, J., Li, S., Lopez, M. A., Zhao, Y., Yu, Y., Ramachandran, V., Ren, G., Yu, B., Li, S., Meyers, B. C., Mo, B., & Chen, X. (2015). Distinct and cooperative activities of HESO1 and URT1 nucleotidyl transferases in microRNA turnover in arabidopsis. *PLoS Genetics*, *11*(4), e1005119. https://doi.org/10.1371/journal.pgen.1005119

Wang, J., Mei, J., & Ren, G. (2019). Plant microRNAs: Biogenesis, homeostasis, and degradation. *Frontiers in Plant Science*, *10*, 360. https://doi.org/10.3389/fpls.2019.00360

Wang, L., Song, X., Gu, L., Li, X., Cao, S., Chu, C., Cui, X., Chen, X., & Cao, X. (2013). NOT2 proteins promote polymerase II–dependent transcription and interact with multiple microRNA biogenesis factors in *Arabidopsis*. *The Plant Cell*, *25*(2), 715–727. https://doi.org/10.1105/tpc.112.105882

Wang, X., Wang, Y., Dou, Y., Chen, L., Wang, J., Jiang, N., Guo, C., Yao, Q., Wang, C., Liu, L., Yu, B., Zheng, B., Chekanova, J. A., Ma, J., & Ren, G. (2018). Degradation of unmethylated miRNA/miRNA*s by a DEDDy-type 3′ to 5′ exoribonuclease Atrimmer 2 in *Arabidopsis*. *Proceedings of the National Academy of Sciences*, *115*(28), 1721917115. https://doi.org/10.1073/pnas.1721917115

Wang, X., Zhang, S., Dou, Y., Zhang, C., Chen, X., Yu, B., & Ren, G. (2015). Synergistic and independent actions of multiple terminal nucleotidyl transferases in the 3′ tailing of small RNAs in arabidopsis. *PLoS Genetics*, *11*(4), e1005091. https://doi.org/10.1371/journal.pgen.1005091

Wei, X., Ke, H., Wen, A., Gao, B., Shi, J., & Feng, Y. (2021). Structural basis of microRNA processing by Dicer-like 1. *Nature Plants*, *7*(10), 1389–1396. https://doi.org/10.1038/s41477-021-01000-1

Wu, F., Yu, L., Cao, W., Mao, Y., Liu, Z., & He, Y. (2007). The N-terminal double-stranded RNA binding domains of *Arabidopsis* HYPONASTIC LEAVES1 are sufficient for pre-microRNA processing. *The Plant Cell*, *19*(3), 914–925. https://doi.org/10.1105/tpc.106.048637

Xie, M., Zhang, S., & Yu, B. (2015). microRNA biogenesis, degradation and activity in plants. *Cellular and Molecular Life Sciences*, *72*(1), 87–99. https://doi.org/10.1007/s00018-014-1728-7

Yang, S. W., Chen, H.-Y., Yang, J., Machida, S., Chua, N.-H., & Yuan, Y. A. (2010). Structure of arabidopsis HYPONASTIC LEAVES1 and its molecular implications for miRNA processing. *Structure*, *18*(5), 594–605. https://doi.org/10.1016/j.str.2010.02.006

Yu, B., & Wang, H. (2010). Translational inhibition by microRNAs in plants. In Rhoads, R. E. (Ed.), *miRNA Regulation of the Translational Machinery* (Vol. 50, pp. 41–57). Springer, Berlin, Heidelberg. https://doi.org/10.1007/978-3-642-03103-8_3

Yu, B., Yang, Z., Li, J., Minakhina, S., Yang, M., Padgett, R. W., Steward, R., & Chen, X. (2005). Methylation as a crucial step in plant microRNA biogenesis. *Science*, *307*(5711), 932–935. https://doi.org/10.1126/science.1107130

Zhao, Y., Yu, Y., Zhai, J., Ramachandran, V., Dinh, T. T., Meyers, B. C., Mo, B., & Chen, X. (2012). The arabidopsis nucleotidyl transferase HESO1 uridylates unmethylated small RNAs to trigger their degradation. *Current Biology*, *22*(8), 689–694. https://doi.org/10.1016/j.cub.2012.02.051

Zhu, L., Hu, Q., Cheng, L., Jiang, Y., Lv, M., Liu, Y., Li, F., Shi, Y., & Gong, Q. (2020). Crystal structure of arabidopsis terminal uridylyl transferase URT1. *Biochemical and Biophysical Research Communications*, *524*(2), 490–496. https://doi.org/10.1016/j.bbrc.2020.01.124

5 microRNAs and the Myriad of Functions in Plant Biology

Noopur Khare, Pragati Khare, Raviraj B. Barot, Sachidanand Singh, and Tanvi Jain*

5.1 INTRODUCTION

microRNAs (miRNAs) are small, non-coding RNA molecules, typically 20–24 nucleotides in length, that play a critical role in the post-transcriptional regulation of gene expression. Discovered in the early 2000s, miRNAs have since been recognized as ubiquitous and essential components of gene regulatory networks in plants (Fukudome and Fukuhara 2017). This interaction is primarily facilitated by the RNA-induced silencing complex (RISC), with Argonaute (AGO) proteins playing a key role in guiding miRNAs to their targets (Srijyothi et al. 2018). The specificity and efficiency of miRNAs in regulating gene expression have profound implications for virtually all aspects of plant biology. The significance of miRNAs in plants extends far beyond simple gene regulation (Sunny et al. 2016). They are pivotal in orchestrating numerous developmental processes, including leaf morphogenesis, root architecture, and flowering time. For instance, miRNAs regulate the polarity and size of leaves, ensure proper root hair formation, and control the timing and formation of floral organs. Their ability to finely tune gene expression allows plants to adapt their growth and development to changing environmental conditions. These small RNAs enable plants to rapidly and efficiently modulate stress-responsive genes, enhancing survival and resilience (Pagano et al. 2021).

Furthermore, miRNAs are closely linked to plant hormonal regulation. They interact with key phytohormones such as auxins, gibberellins (GAs), abscisic acid (ABA), and cytokinins, playing crucial roles in maintaining hormone homeostasis and facilitating cross-talk between different hormonal pathways. This intricate network of interactions highlights the complexity of miRNA functions and their central role in plant physiology (Kapoor et al. 2021). The agricultural implications of miRNA research are immense, offering potential strategies for crop improvement through enhanced stress resistance, yield optimization, and quality enhancement (Verma et al. 2021). Advanced genetic engineering techniques, including miRNA-based modifications and CRISPR/Cas9-mediated gene editing, are being explored to leverage these benefits. As research progresses, uncovering novel miRNAs and their targets, combined with integrative omics approaches, promises to deepen our understanding of these versatile molecules, ultimately transforming plant biotechnology and agriculture (Joshi et al. 2023).

5.2 IMPORTANCE OF miRNAs IN GENE REGULATION

miRNAs are essential regulators of gene expression in plants, exerting their influence at the post-transcriptional level. This binding results in either the cleavage of the messenger RNA (mRNA) or the inhibition of its translation, thereby controlling the production of proteins encoded by the target genes. The specificity of miRNA–mRNA interactions ensures precise regulation of gene expression, which is crucial for the proper functioning of various biological processes (Li and Zhang 2016).

In plants, miRNAs are integral to numerous developmental and physiological pathways. They modulate the expression of genes involved in key developmental processes such as leaf and root

* *Corresponding Author*

 DOI: 10.1201/9781003491378-5

formation, flowering, and reproductive development (Singh et al. 2019). Additionally, miRNAs are vital in enabling plants to respond to environmental cues and stressors. They regulate genes associated with responses to abiotic stresses like drought, salinity, and temperature extremes, as well as biotic stresses from pathogens and herbivores. By fine-tuning gene expression in response to these challenges, miRNAs help maintain cellular homeostasis and optimize plant growth and survival. Thus, the importance of miRNAs in gene regulation lies in their ability to provide a sophisticated, responsive, and dynamic mechanism for controlling gene expression in plants (Pagano et al. 2021).

5.3 OVERVIEW OF miRNA DISCOVERY AND RESEARCH IN PLANTS

The discovery of miRNAs in plants marked a significant milestone in molecular biology, beginning in the early 2000s. Initial studies identified these small, non-coding RNA molecules as key post-transcriptional regulators of gene expression, leading to a surge in research focused on their biogenesis, function, and impact (Maddala and Singh 2023). Researchers have since uncovered a vast array of plant miRNAs, utilizing advanced techniques such as high-throughput sequencing and bioinformatics tools to identify and predict miRNA targets (Drusco and Croce 2017). As the field has evolved, functional studies using genetic and biochemical approaches have elucidated the mechanisms by which miRNAs influence plant physiology. The ongoing research continues to expand our understanding, highlighting miRNAs' potential in crop improvement and resilience against environmental challenges, thus underscoring their profound significance in plant biology (Pourkheirandish et al. 2020).

5.4 BIOGENESIS AND MECHANISM OF ACTION

5.4.1 Biogenesis of Plant miRNAs

The biogenesis of plant miRNAs involves a multi-step process that begins with the transcription of miRNA genes by RNA polymerase II, producing primary miRNAs (pri-miRNAs) (Patani et al. 2023), as seen in Figure 5.1. These pri-miRNAs are then processed in the nucleus by Dicer-like (DCL) enzymes, particularly DCL1, into precursor miRNAs (pre-miRNAs) that fold into stem–loop structures. Further cleavage by DCL1 converts pre-miRNAs into mature -miRNA and miRNA duplexes. The duplex is subsequently methylated by HEN1 to protect it from degradation (Boopathi 2015). This precise and regulated biogenesis pathway ensures that miRNAs effectively control gene expression, playing crucial roles in plant growth, development, and stress responses (Lima et al. 2012).

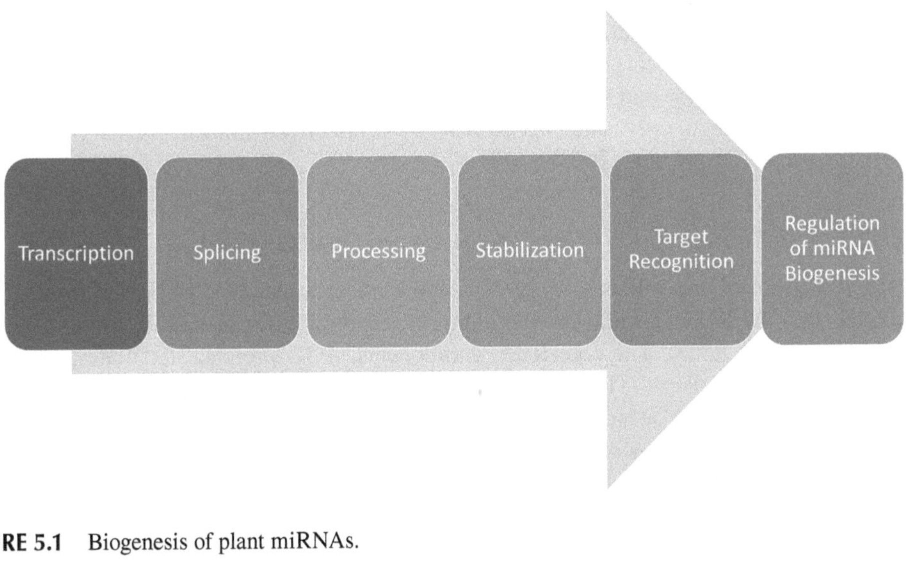

FIGURE 5.1 Biogenesis of plant miRNAs.

5.4.2 Transcription of miRNA Genes

The transcription of miRNA genes in plants is initiated by RNA polymerase II, similar to protein-coding genes (Sao et al. 2022). The plant genome contains these miRNA genes, which are translated to create lengthy, capped, polyadenylated transcripts known as pri-miRNAs, or pri-miRNAs. Pri-miRNAs typically contain one or more stem–loop structures, essential for their recognition and processing (Megraw et al. 2016). The transcriptional process is regulated by various promoter elements and transcription factors that ensure the spatial and temporal expression of miRNAs. Once synthesized, the pri-miRNAs undergo a series of cleavage events to produce the mature miRNA. Plants can accurately regulate the production of miRNAs in response to environmental stimuli and developmental signals because transcriptional regulation controls the availability of miRNAs at the appropriate time and location to effectively impact gene expression. This transcriptional control is fundamental to the role of miRNAs in maintaining cellular homeostasis and facilitating adaptive responses (Song et al. 2019).

5.4.3 Processing by Dicer-like Enzymes

Processing of pri-miRNAs by DCL enzymes is a critical step in miRNA biogenesis in plants (Khare et al. 2017). Once pri-miRNAs are transcribed, they are recognized and cleaved by DCL1, the key DCL enzyme in plants, within the nucleus (Fukudome and Fukuhara 2017). DCL1 processes the long pri-miRNA transcripts into shorter pre-miRNAs, which form characteristic stem–loop structures. The precise action of DCL1 is facilitated by interactions with other proteins, such as HYL1 and SE, which enhance the accuracy and efficiency of miRNA processing. This finely tuned mechanism ensures that the resulting miRNA duplexes are of the correct length and sequence specificity to effectively guide the RISC in post-transcriptional gene regulation (Patani et al. 2024). Through this processing, DCL1 enables plants to produce functional miRNAs that play vital roles in development and stress responses (Banerjee 2020).

5.4.4 Formation of miRNA–miRNA* Duplex

The formation of the miRNA–miRNA* duplex is a pivotal step in the miRNA biogenesis pathway in plants. Following the cleavage of pre-miRNAs by DCL1 within the nucleus, the resulting products are miRNA–miRNA* duplexes (Li and Yu 2021). These duplexes consist of a mature miRNA strand, which guides gene silencing, and a complementary miRNA* (star) strand, which is usually degraded. The duplex is stabilized and protected from degradation by methylation at the 3′ ends by the enzyme HEN1. This methylation is crucial for the stability and functionality of the miRNAs (Maddala et al. 2024). This incorporation is sequence-specific and determines target mRNA recognition. The precise formation and stabilization of the miRNA–miRNA* duplex are essential for the miRNA to perform its regulatory roles in gene expression, effectively influencing plant development and stress responses (Zhang et al. 2022).

5.4.5 Incorporation into RNA-Induced Silencing Complex (RISC)

After the formation of the miRNA–miRNA* duplex, the mature miRNA strand is selectively incorporated into the RISC, a multi-protein complex responsible for mediating gene silencing in plants (Meddya et al. 2023). The core RISC components, AGO proteins, facilitate this process, which occurs in the cytoplasm. The mature miRNA strand binds to AGO proteins, guiding RISC to target mRNAs through sequence complementarity (Singh et al. 2017). Once bound to its target mRNA, RISC mediates post-transcriptional gene silencing through two main mechanisms: mRNA cleavage or translational repression. The precise recognition and binding of miRNAs by AGO proteins ensure specific and efficient regulation of gene expression, enabling plants to fine-tune their developmental processes and responses to environmental cues. The incorporation of miRNAs into RISC represents a crucial step in the miRNA-mediated gene regulation pathway, highlighting the central role of RISC in orchestrating plant molecular responses (Tiwari et al. 2014).

5.5 MECHANISM OF ACTION

5.5.1 Target Recognition by Sequence Complementarity

Plant miRNAs function by using sequence complementarity to recognize target mRNAs, which in turn control post-transcriptional gene expression. Upon incorporation into the RISC, the mature miRNA strand guides RISC to its target mRNAs based on sequence complementarity between the miRNA and the mRNA (Meng et al. 2021). The miRNA typically binds to the 3′ untranslated region (UTR) of target mRNAs, although binding to other regions, including the coding sequence and 5′ UTR, has also been observed (Parveen et al. 2023). The degree of complementarity between the miRNA and its target mRNA influences the mechanism of gene regulation. Perfect or near-perfect complementarity often leads to mRNA cleavage, mediated by the endonuclease activity of AGO proteins within the RISC. On the other hand, translational repression occurs when the miRNA binds to the target mRNA and prevents translation initiation or elongation, usually as a result of inadequate complementarity. This prevents the production of the appropriate protein. The specificity and versatility of miRNA-mediated gene regulation make it a crucial mechanism for fine-tuning gene expression and coordinating complex regulatory networks in plant biology (Tang and Chu 2017).

5.5.2 miRNA Cleavage or Translational Repression

The mechanism of action of miRNAs in plants involves two main modes of post-transcriptional gene regulation: mRNA cleavage and translational repression. The choice between these mechanisms is influenced by the degree of complementarity between the miRNA and its target mRNAs (Ma et al. 2013). In cases where the miRNA exhibits near-perfect complementarity to its target mRNA, typically through base pairing with the seed region of the miRNA, mRNA cleavage is initiated. AGO proteins are guided by the miRNA to the target mRNA within the RISC, where the mRNA is endonucleolytically cleaved. The coding sequence or the 3′ UTR of the mRNA are the main sites of this cleavage, leading to the degradation of the cleaved fragments and the consequent downregulation of protein expression (Malka et al. 2017).

In contrast, when the miRNA and its target mRNA exhibit imperfect complementarity, translational repression predominates. In this scenario, the binding of the miRNA to the target mRNA within the RISC interferes with translation initiation or elongation, inhibiting the synthesis of the corresponding protein. This repression can occur through various mechanisms, including inhibition of ribosome recruitment, destabilization of translation initiation complexes, or promotion of ribosome drop-off during translation elongation (Vora et al. 2023). Overall, the dual regulatory mechanisms of mRNA cleavage and translational repression mediated by miRNAs provide plants with versatile and precise control over gene expression, enabling them to fine-tune cellular processes and respond effectively to developmental and environmental cues (Arnaud and Laufs 2023).

5.5.3 Role of Argonaute Proteins

AGO proteins play a central role in the mechanism of action of miRNAs in plants, serving as the core components of the RISC, as seen in Figure 5.2. In cases of near-perfect complementarity between the miRNA and its target mRNA, AGO proteins catalyze endonucleolytic cleavage of the mRNA, resulting in its degradation and subsequent downregulation of protein expression (Zaheer et al. 2024).

In instances of imperfect complementarity, AGO proteins mediate translational repression by inhibiting the initiation or elongation of translation of the target mRNA. This repression can occur

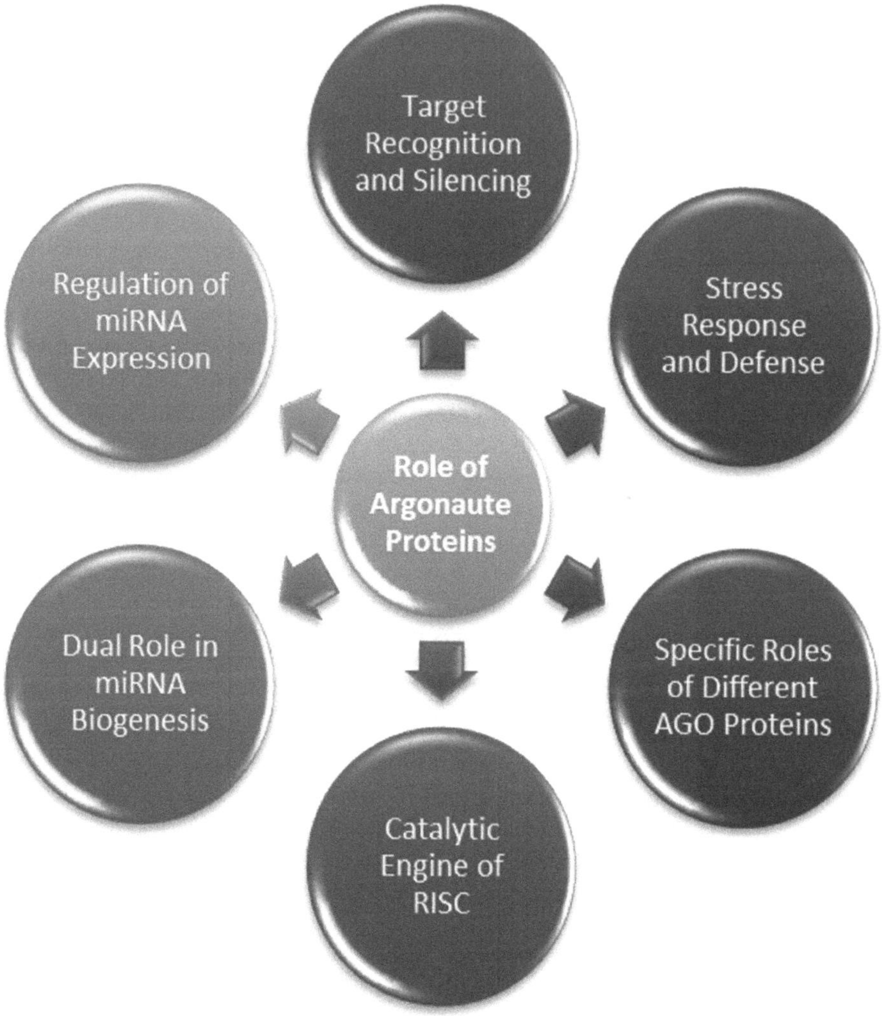

FIGURE 5.2 Role of Argonaute proteins.

through various mechanisms, including interference with ribosome recruitment or stability, as well as promoting ribosome drop-off during translation elongation. AGO proteins also contribute to the stability and functionality of the miRNA–mRNA complex within RISC, ensuring precise and efficient regulation of gene expression. Overall, the pivotal role of AGO proteins in miRNA-mediated gene silencing underscores their importance in orchestrating plant molecular responses and regulating diverse biological processes (Karginov and Hannon 2013).

5.6 FUNCTIONAL ROLES IN PLANT DEVELOPMENT

5.6.1 Regulation of Leaf Morphogenesis

5.6.1.1 Control of Leaf Polarity

The regulation of leaf morphogenesis, including the control of leaf polarity, is a crucial aspect of plant development mediated by miRNAs. Leaf polarity refers to the establishment and maintenance of distinct adaxial (upper) and abaxial (lower) sides of the leaf, which is essential for proper leaf growth, shape, and function. MiRNAs play a central role in controlling leaf polarity by targeting genes involved in adaxial–abaxial patterning pathways. For example, miR165/166 targets several class III HD-ZIP transcription factors, which are crucial for specifying abaxial cell fate. By repressing the expression of these transcription factors, miR165/166 promotes adaxial cell fate and the establishment of leaf polarity (Dong et al. 2022).

Conversely, miR319 targets TCP transcription factors, which promote adaxial cell fate and leaf expansion. By repressing the expression of TCP genes, miR319 contributes to the specification of abaxial cell fate and the maintenance of leaf polarity. Overall, the precise regulation of leaf polarity by miRNAs ensures the proper development and function of leaves in plants. Imbalances in miRNA-mediated control of leaf polarity can lead to abnormal leaf morphologies and developmental defects, highlighting the importance of miRNAs in orchestrating leaf development (Zaid and Wani 2020).

5.6.1.2 Leaf Size and Shape

miRNAs intricately regulate leaf morphogenesis, including the determination of leaf size and shape, in plants. Leaf size and shape are essential factors that influence plant growth, photosynthetic efficiency, and overall fitness. MiRNAs play a pivotal role in controlling leaf size and shape by targeting genes involved in cell proliferation, expansion, and differentiation. For instance, miR319 targets TCP transcription factors, which regulate cell proliferation and growth (Pulido and Laufs 2010). By repressing the expression of TCP genes, miR319 limits cell proliferation in leaf primordia, contributing to the regulation of leaf size. Additionally, by repressing growth-regulating factor expression, miR396 restricts cell proliferation and expansion, influencing leaf size and shape. Moreover, miRNAs such as miR164 and miR319 target NAC domain transcription factors, which regulate leaf shape by modulating leaf margin development and cell differentiation. Overall, the precise regulation of leaf size and shape by miRNAs ensures proper leaf morphogenesis and functional integrity in plants. Dysregulation of miRNA-mediated control can lead to abnormal leaf phenotypes, impacting plant growth, development, and productivity. Thus, miRNAs serve as crucial regulators of leaf morphogenesis, orchestrating the complex processes underlying leaf size and shape determination in plants (Karamat et al. 2021).

5.6.2 Root Development

5.6.2.1 Regulation of Root Architecture

miRNAs play a crucial role in regulating root architecture, which includes the spatial arrangement of different root types and the overall growth pattern of the root system. Root architecture influences nutrient and water uptake, as well as the plant's ability to anchor itself in the soil and respond to environmental stimuli. MiRNAs control root architecture by targeting genes involved in various aspects of root development, including cell division, elongation, differentiation, and lateral root formation. For example, miR156 targets squamosa promoter binding protein-like (SPL) transcription factors, which regulate root growth and branching by modulating the expression of genes involved in hormone signaling and meristem activity (Ranjan et al. 2022).

Furthermore, miRNAs such as miR160, miR164, and miR390 target additional transcription factors and signaling components involved in regulating root architecture, including those associated with cytokinin and ethylene signaling pathways. Overall, the precise regulation of root architecture by miRNAs ensures optimal root system development and function, allowing plants to adapt to changing soil conditions and optimize nutrient acquisition and stress responses. Dysregulation of miRNA-mediated control can lead to abnormal root phenotypes, impacting plant growth, development, and productivity. Thus, miRNAs serve as kcy regulators of root architecture, orchestrating the complex processes underlying root development in plants (Mahanty et al. 2022).

5.6.2.2 Root Hair Formation

Root hair formation is a crucial aspect of root development, facilitated by miRNAs in plants. Root hairs are specialized epidermal cells that extend from the primary root and significantly increase the root surface area for nutrient and water absorption, as well as interactions with soil microorganisms

(Yan et al. 2022). MiRNAs regulate root hair formation by targeting genes involved in various stages of root hair development, including initiation, elongation, and differentiation. For instance, miR393 targets TIR1/AFBs, which are auxin receptors involved in regulating root hair elongation. By modulating the expression of TIR1/AFB genes, miR393 influences root hair growth in response to changes in auxin levels. Moreover, miRNAs such as miR165/166 and miR167 target transcription factors and signaling components that regulate root hair cell fate determination and differentiation, ensuring proper root hair formation. Overall, the precise regulation of root hair formation by miRNAs is essential for optimizing nutrient and water uptake, enhancing plant–microbe interactions, and promoting overall root function. Dysregulation of miRNA-mediated control can lead to abnormal root hair phenotypes, impacting plant growth, development, and productivity. Thus, miRNAs serve as critical regulators of root hair development, orchestrating the complex processes underlying root hair formation in plants (Kohli et al. 2022).

5.6.3 Flowering and Reproductive Development

5.6.3.1 Timing of Flowering

The timing of flowering is a critical event in the reproductive development of plants, influenced by a complex network of genetic, environmental, and hormonal factors, including miRNAs. Flowering at the appropriate time ensures successful reproduction and seed set, which are essential for plant survival and propagation. MiRNAs regulate the timing of flowering by targeting genes involved in the flowering time pathway, including transcription factors, signaling components, and key regulators of the floral transition (Campos-Rivero et al. 2017). By modulating the expression of SPL genes, miR156 influences both the timing of flowering and the duration of the vegetative phase. Similarly, miR172 targets AP2-like transcription factors that promote flowering by repressing the expression of floral repressors. Moreover, miRNAs such as miR159 and miR319 target MYB transcription factors, which regulate flowering time through their roles in hormone signaling pathways and stress responses. Overall, the precise regulation of flowering time by miRNAs ensures proper reproductive development and synchronization with environmental cues, allowing plants to adapt to changing conditions and optimize reproductive success. Dysregulation of miRNA-mediated control can lead to delayed or premature flowering, affecting plant fitness and yield. Thus, miRNAs serve as key regulators of flowering time, orchestrating the complex processes underlying floral transition in plants (Waheed and Zeng 2020).

5.6.3.2 Floral Organ Development

Floral organ development is a finely orchestrated process crucial for successful reproduction in plants, and miRNAs play a pivotal role in regulating this complex process. Floral organs, including sepals, petals, stamens, and carpels, must develop with precise spatial and temporal coordination to ensure proper flower structure and function (Yu et al. 2022).

MiRNAs control floral organ development by targeting genes involved in the specification, differentiation, and patterning of floral organs. For instance, miR156 targets SPL transcription factors, which regulate the transition from vegetative to reproductive growth and influence floral organ identity and development. Similarly, miR172 targets AP2-like transcription factors, which promote floral organ development by regulating the expression of genes involved in floral meristem identity and flower organ identity determination. Moreover, miRNAs such as miR159 and miR319 target MYB transcription factors, which play essential roles in specifying floral organ identity and controlling cell differentiation and growth during flower development. Overall, the precise regulation of floral organ development by miRNAs ensures proper flower morphology and function, allowing plants to produce viable seeds and ensure reproductive success. Dysregulation of miRNA-mediated control can lead to abnormal flower phenotypes, affecting plant fertility and yield. Thus, miRNAs serve as key regulators of floral organ development, orchestrating the complex processes underlying flower formation in plants (Chen et al. 2018).

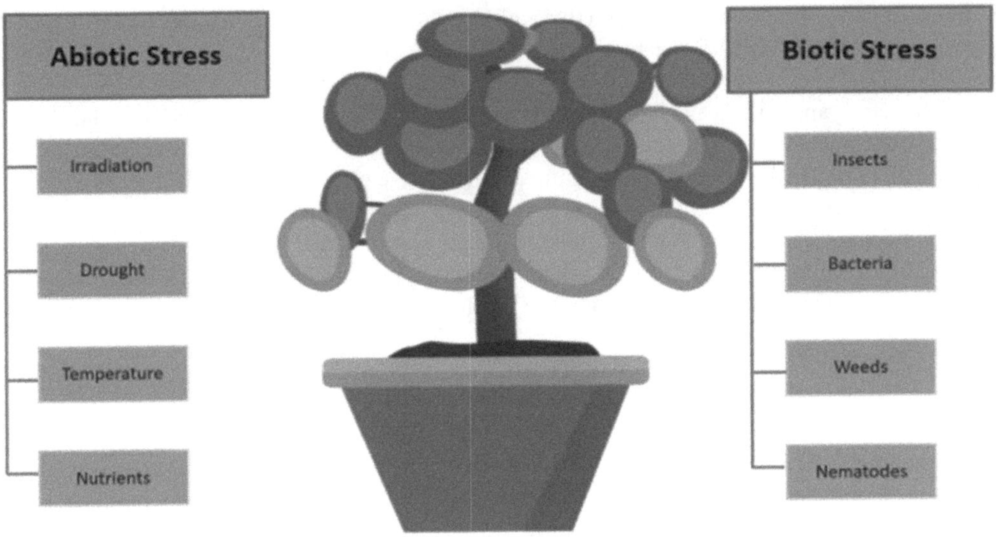

FIGURE 5.3 Response to environmental stress.

5.7 RESPONSE TO ENVIRONMENTAL STRESS

5.7.1 Abiotic Stress Responses

5.7.1.1 Drought Tolerance

Drought stress poses a significant threat to plant growth and productivity, making drought tolerance a crucial trait for plant survival, as seen in Figure 5.3. MiRNAs regulate drought tolerance by targeting genes involved in various physiological and biochemical processes, including stomatal regulation, water uptake and transport, antioxidant defense, and stress signaling pathways. For example, miR159 targets MYB transcription factors involved in regulating stomatal closure and water loss, thereby enhancing drought tolerance (Ahmad et al. 2022).

Similarly, miR319 targets TCP transcription factors, which modulate the expression of genes involved in drought stress responses, including those encoding stress-responsive proteins and enzymes involved in osmotic regulation. Moreover, miRNAs such as miR167, miR396, and miR398 target transcription factors, hormone signaling components, and antioxidant enzymes, which collectively regulate drought-induced changes in gene expression and cellular homeostasis, thus enhancing plant tolerance to water deficit. Overall, the precise regulation of drought tolerance by miRNAs enables plants to withstand water scarcity and maintain growth and productivity under adverse environmental conditions. Dysregulation of miRNA-mediated control can impair drought tolerance mechanisms, leading to decreased plant fitness and yield. Thus, miRNAs serve as critical regulators of drought stress responses, orchestrating the multifaceted molecular pathways underlying plant adaptation to water-limited environments (Mahanty et al. 2022).

5.7.1.2 Salinity Stress

Salinity stress is a significant environmental challenge for plants, affecting their growth, development, and productivity. miRNAs play a vital role in regulating the complex molecular responses that enable plants to cope with salinity stress. MiRNAs regulate salinity stress responses by targeting genes involved in ion homeostasis, osmotic regulation, antioxidant defense, and stress signaling pathways. For example, miR159 targets MYB transcription factors involved in regulating ion transporters, thus modulating ion uptake and maintaining ion balance under saline conditions (Safdar et al. 2019).

By modulating the expression of growth-regulating factor genes, miR396 influences root growth and development, facilitating adaptation to saline environments. Moreover, miRNAs such as miR168, miR398, and miR408 target genes encoding stress-responsive proteins, including transcription factors, antioxidant enzymes, and transporters, which collectively regulate cellular homeostasis and redox balance under saline conditions. Overall, the precise regulation of salinity stress responses by miRNAs enables plants to mitigate the adverse effects of high salt concentrations on cellular processes and maintain growth and productivity in saline environments. Dysregulation of miRNA-mediated control can impair salinity tolerance mechanisms, leading to decreased plant fitness and yield. Thus, miRNAs serve as essential regulators of salinity stress responses, orchestrating the intricate molecular pathways underlying plant adaptation to saline conditions (Yuan et al. 2019).

5.7.1.3 Temperature Stress (Heat and Cold)

miRNAs are crucial regulators of plant responses to temperature stress, including heat and cold. They modulate gene expression by targeting transcripts involved in stress signaling, transcriptional regulation, and metabolic pathways. For example, miR398 targets genes encoding heat shock proteins and antioxidant enzymes, enhancing heat tolerance. Conversely, miR156 and miR172 regulate the expression of transcription factors involved in cold acclimation and freezing tolerance. The dynamic regulation of gene expression by miRNAs enables plants to adapt to temperature fluctuations, optimizing growth and survival under adverse environmental conditions. Dysregulation of miRNA-mediated control can impair temperature stress responses, compromising plant fitness and productivity (Muhammad et al. 2023).

5.7.2 Biotic Stress Responses

5.7.2.1 Pathogen Defense

miRNAs are integral components of plant defense against pathogens, orchestrating a complex regulatory network to combat biotic stress. They regulate pathogen defense by targeting genes involved in various aspects of the immune response, including pathogen recognition, signaling, and defense gene expression. MiRNAs modulate pathogen defense by targeting genes encoding pattern recognition receptors. For instance, miR393 targets F-box auxin receptor proteins, which negatively regulate PAMP-triggered immunity by promoting the degradation of pattern recognition receptors (Ding et al. 2022).

Similarly, miRNAs such as miR160, miR167, and miR393 target genes involved in hormone signaling pathways, including auxin and jasmonic acid signaling, which regulate defense gene expression and activate defense responses. Moreover, miRNAs such as miR482 and miR2118 target NLR genes, which function as intracellular immune receptors and trigger effector-triggered immunity upon recognition of pathogen effectors. Overall, the precise regulation of pathogen defense by miRNAs enables plants to mount effective immune responses against diverse pathogens, including bacteria, fungi, and viruses. Dysregulation of miRNA-mediated control can compromise pathogen defense mechanisms, leading to increased susceptibility to infections and reduced plant fitness and productivity. Thus, miRNAs serve as critical regulators of plant immunity, coordinating the complex molecular pathways underlying defense responses to biotic stress (Waheed et al. 2021).

5.7.2.2 Insect Herbivory

miRNAs play a crucial role in plant defense against insect herbivory by modulating various aspects of the plant's response to herbivore attack. They regulate insect herbivory responses by targeting genes involved in defense signaling, phytohormone signaling pathways, and secondary metabolite biosynthesis (Bordoloi and Agarwala 2021). MiRNAs modulate defense signaling by targeting genes encoding receptor-like kinases and transcription factors involved in herbivore-induced signaling pathways. For example, miR160 targets auxin response factors (ARFs) that regulate defense gene expression in response to insect feeding. Similarly, miRNAs such as miR164 and miR396 target transcription factors involved in jasmonic acid and salicylic acid signaling pathways, which mediate defense responses to insect herbivory. Moreover, miRNAs regulate the biosynthesis of secondary metabolites, such as phenylpropanoids, terpenoids, and alkaloids, which have toxic or deterrent effects on herbivores. Overall, the precise regulation of insect herbivory responses by miRNAs enables plants to mount effective defense responses against herbivore attack, minimizing damage and enhancing plant fitness. Dysregulation of miRNA-mediated control can compromise plant defense mechanisms, leading to increased susceptibility to insect herbivory and reduced productivity (Díez-Sainz et al. 2021).

5.7.2.3 Symbiotic Interactions

miRNAs are key regulators of plant symbiotic interactions, facilitating mutualistic relationships with beneficial microbes such as mycorrhizal fungi and nitrogen-fixing bacteria. They modulate symbiotic interactions by targeting genes involved in signaling, nutrient exchange, and the establishment of mutualistic symbiosis (Verma et al. 2022). MiRNAs regulate symbiotic signaling by targeting genes encoding receptor-like kinases and transcription factors involved in symbiotic signal perception and transduction. For example, miR399 targets PHO2, which negatively regulates phosphate uptake and plays a role in establishing mycorrhizal symbiosis. Similarly, miRNAs such as miR164 and miR172 target transcription factors involved in symbiotic signaling pathways, including those regulating arbuscular mycorrhizal symbiosis and nodulation in legumes. Moreover, miRNAs regulate the expression of genes involved in nutrient exchange between plants and symbiotic microbes, such as phosphate and nitrogen transporters. Overall, the precise regulation of symbiotic interactions by miRNAs enables plants to establish and maintain mutually beneficial relationships with symbiotic microbes, enhancing nutrient acquisition and promoting plant growth and fitness. Dysregulation of miRNA-mediated control can disrupt symbiotic interactions, compromising plant health and productivity (Singh et al. 2023).

5.8 HORMONAL REGULATION

5.8.1 Interaction with Phytohormones

5.8.1.1 Auxins

miRNAs intricately regulate plant interactions with auxins, key phytohormones essential for various aspects of plant growth and development as seen in Figure 5.4. MiRNAs modulate auxin signaling and responses by targeting genes involved in auxin biosynthesis, transport, perception, and signaling pathways (Jin et al. 2020). Similarly, miRNAs such as miR393 target auxin receptor genes, such as F-box auxin receptors (TIR1/AFBs), which regulate auxin perception and signaling through the degradation of Aux/IAA repressor proteins. Moreover, miRNAs regulate the expression of genes involved in auxin biosynthesis and transport, influencing auxin levels and distribution within the plant. Overall, the precise regulation of auxin signaling and responses by miRNAs enables plants to coordinate growth and development in response to endogenous and environmental cues, optimizing growth and adaptation. Dysregulation of miRNA-mediated control can disrupt auxin-related processes, leading to aberrant growth phenotypes and developmental defects (Zhao 2018).

5.8.1.2 Gibberellins

miRNAs play a vital role in regulating plant interactions with GAs, essential phytohormones that promote growth and developmental processes such as seed germination, stem elongation, and flowering. MiRNAs modulate GA signaling and responses by targeting genes involved in GA biosynthesis, signaling, and response pathways. For example, miR159 targets MYB transcription factors that regulate GA signaling by influencing the expression of GA-responsive genes (Ahmad et al. 2022). By modulating MYB expression, miR159 fine-tunes the plant's response to GA, affecting processes like seed germination and anther development. Similarly, miR319 targets TCP transcription factors

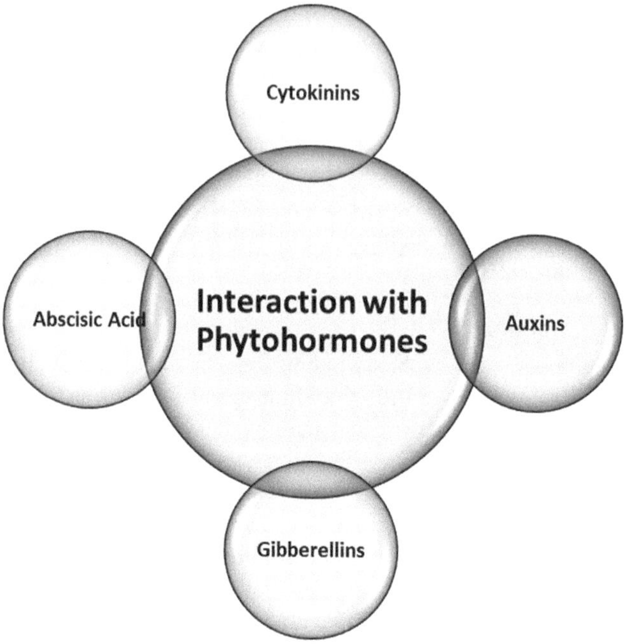

FIGURE 5.4 Interaction with phytohormones.

involved in regulating GA-induced growth responses, such as stem elongation and leaf development. By regulating TCP expression, miR319 influences how plants respond to GA, ensuring balanced growth. Moreover, miRNAs such as miR171 target genes involved in the GA biosynthesis pathway, affecting GA levels and distribution within the plant. Overall, the precise regulation of GA signaling and responses by miRNAs enables plants to coordinate growth and development in response to internal and external cues, optimizing plant adaptation and fitness. Dysregulation of miRNA-mediated control can disrupt GA-related processes, leading to growth abnormalities and developmental issues (Huijser and Schmid 2011).

5.8.1.3 Abscisic Acid

miRNAs are crucial regulators of plant interactions with ABA, a key phytohormone involved in stress responses and developmental processes. MiRNAs modulate ABA signaling by targeting genes involved in ABA biosynthesis, perception, and signaling pathways. For instance, miR159 targets MYB transcription factors that influence ABA responses, including seed germination and stress tolerance. Similarly, miR166 and miR167 regulate genes involved in ABA signaling, affecting stomatal closure and drought resistance. By fine-tuning the expression of these genes, miRNAs enable plants to respond effectively to ABA, enhancing stress resilience and ensuring proper growth and development under varying environmental conditions (Singh et al. 2021).

5.8.1.4 Cytokinins

miRNAs play a crucial role in regulating plant interactions with cytokinins, essential phytohormones that promote cell division, shoot initiation, and leaf expansion. MiRNAs modulate cytokinin signaling and responses by targeting key genes in cytokinin pathways. For example, miR319 targets TCP transcription factors involved in cell proliferation and differentiation, affecting cytokinin-regulated processes like leaf development and shoot branching. Similarly, miR393 targets genes encoding cytokinin receptors, modulating cytokinin perception and signaling. By fine-tuning the expression of these genes, miRNAs ensure balanced cytokinin responses, optimizing plant growth, development, and adaptation to environmental cues (Curaba et al. 2014).

5.8.2 Feedback Regulation and Signaling Pathways

5.8.2.1 Hormone Homeostasis

miRNAs are essential in maintaining hormone homeostasis in plants by regulating feedback loops and signaling pathways. They target key genes involved in the biosynthesis, transport, and signaling of phytohormones, ensuring a balanced hormonal environment for optimal growth and development. For instance, miR160 and miR167 modulate auxin response by targeting ARFs, while miR159 and miR319 regulate GA and cytokinin signaling by targeting MYB and TCP transcription factors, respectively (Samad et al. 2017). These miRNAs fine-tune hormone levels and responses, preventing overaccumulation or deficiencies that could disrupt plant development. Additionally, miRNAs like miR393, which target auxin and cytokinin receptors, influence hormone perception and signal transduction, creating a feedback regulation mechanism that adjusts hormone sensitivity and distribution. Overall, the precise regulation of hormone homeostasis by miRNAs enables plants to integrate various hormonal signals and environmental cues, promoting adaptive growth and stress responses. Dysregulation of miRNA-mediated feedback can lead to hormonal imbalances, adversely affecting plant fitness and productivity (Pagano et al. 2021).

5.8.2.2 Cross-Talk between Different Hormonal Pathways

miRNAs are crucial for facilitating cross-talk between different hormonal pathways in plants, integrating various signals to ensure coordinated growth and stress responses. They target genes involved in multiple hormone signaling pathways, thus mediating interactions between these pathways (Patel et al. 2019). For example, miR159 regulates both GA and ABA signaling by targeting MYB transcription factors, balancing growth and stress responses. Similarly, miR160 and miR167 target ARFs, influencing auxin signaling and its interaction with cytokinin and ABA pathways. Moreover, miR393 targets F-box auxin receptors, modulating auxin and cytokinin signaling pathways, which affects processes like root and shoot development. miR319 targets TCP transcription factors, linking cytokinin and GA signaling pathways and influencing cell proliferation and leaf development. By modulating the expression of key regulatory genes, miRNAs facilitate the integration of multiple hormonal signals, ensuring plants respond appropriately to developmental cues and environmental stresses. This cross-talk is essential for maintaining hormonal balance and optimizing plant adaptation and resilience (Shahzad et al. 2018).

5.9 AGRICULTURAL IMPLICATIONS

5.9.1 Crop Improvement

5.9.1.1 Enhancing Stress Resistance

miRNAs play a pivotal role in enhancing stress resistance in crops, providing a promising avenue for crop improvement. By targeting specific genes involved in stress response pathways, miRNAs regulate the expression of key stress-responsive proteins and enzymes. For instance, miR393 modulates auxin signaling under drought stress, while miR398 regulates antioxidant defense mechanisms during oxidative stress. Through genetic engineering, manipulating miRNA expression can enhance a crop's ability to withstand various abiotic and biotic stresses, leading to improved resilience and productivity. This approach not only boosts stress tolerance but also promotes sustainable agricultural practices and food security in changing climates (Abbas et al. 2022).

5.9.1.2 Improving Yield and Quality

miRNAs are crucial for improving crop yield and quality by regulating genes involved in growth, development, and metabolic pathways. By fine-tuning hormone signaling, nutrient uptake, and stress responses, miRNAs enhance traits such as biomass production, fruit size, and nutritional content.

For example, miR156 and miR172 modulate the transition from vegetative to reproductive stages, optimizing flowering time and fruit set. Genetic engineering to alter miRNA expression can lead to increased yield and superior quality, contributing to higher agricultural productivity and improved food quality. This strategy provides a promising avenue for advancing crop performance in diverse environments (Mandal et al. 2021).

5.9.2 Genetic Engineering Approaches

5.9.2.1 miRNA-based Genetic Modifications

miRNA-based genetic modifications offer a precise and efficient approach to enhancing crop traits through genetic engineering. By manipulating the expression of specific miRNAs, researchers can regulate target gene networks involved in crucial plant processes such as growth, development, and stress responses. Techniques such as overexpressing miRNAs, using miRNA mimics, or silencing miRNAs through short tandem target mimics enable the fine-tuning of plant phenotypes. For instance, overexpressing miR156 can delay flowering and enhance biomass production, while silencing miR160 can improve root architecture and nutrient uptake. Additionally, modifying miRNA expression can enhance resistance to abiotic stresses like drought and salinity by targeting stress-responsive genes, thus improving crop resilience. These miRNA-based strategies are highly specific, reducing the likelihood of off-target effects and making them a powerful tool in crop improvement. By integrating miRNA manipulation with traditional breeding and other biotechnological approaches, scientists can develop high-yielding, stress-tolerant, and nutritionally enhanced crops to meet global food demands (Chen et al. 2021).

5.9.2.2 Gene Editing Technologies (CRISPR/Cas9)

Gene editing technologies, particularly CRISPR/Cas9, have revolutionized crop genetic engineering by enabling precise modifications of plant genomes. CRISPR/Cas9 uses a guide RNA to direct the Cas9 enzyme to specific DNA sequences, where it creates double-strand breaks. This precise targeting allows for the addition, deletion, or alteration of genes with high accuracy. CRISPR/Cas9 can enhance crop traits such as yield, disease resistance, and stress tolerance. For example, it can knock out genes that negatively affect yield or edit genes to enhance drought and pest resistance (Jiang and Doudna 2017). Additionally, CRISPR/Cas9 facilitates the development of crops with improved nutritional profiles by targeting genes involved in nutrient biosynthesis and accumulation. The technology's versatility extends to creating plants with novel traits and accelerating breeding programs. Its ability to make targeted modifications without introducing foreign DNA makes CRISPR/Cas9 an attractive tool for developing genetically modified crops that may be more acceptable to regulatory bodies and the public. By leveraging CRISPR/Cas9, scientists can create more resilient, productive, and sustainable crops to address global food security challenges (Ndudzo et al. 2024).

5.10 RECENT ADVANCES

5.10.1 miR156 and the Control of Vegetative Phase Change

miR156 is a crucial regulator of the vegetative phase change in plants, modulating the transition from juvenile to adult stages. It targets SPL transcription factors, which promote the expression of genes involved in adult phase traits, such as flowering and leaf morphology. High levels of miR156 maintain the juvenile phase by suppressing SPL expression, while a decline in miR156 levels allows SPL proteins to accumulate, triggering the vegetative phase change. This regulatory mechanism ensures proper timing of developmental transitions, optimizing growth and reproductive success. Manipulating miR156 levels can thus enhance biomass and delay flowering in crops (Chen et al. 2010).

5.10.2 miR172 and Floral Transition

miR172 is a key regulator of the floral transition in plants, controlling the timing of flowering. It targets AP2-like transcription factors, which function as floral repressors. High levels of miR172 repress AP2-like gene expression, promoting flowering by releasing the repression of floral meristem identity genes. As miR172 levels decline, AP2-like proteins accumulate, inhibiting floral meristem identity genes and delaying flowering. This regulatory mechanism ensures proper timing of the transition from vegetative to reproductive growth, optimizing reproductive success. Manipulating miR172 levels can therefore accelerate or delay flowering in crops, providing a strategy for optimizing flowering time for specific environments or breeding objectives (Bertran Garcia de Olalla 2023).

5.10.3 High-Throughput Sequencing Technologies

High-throughput sequencing technologies have revolutionized genomic research by enabling rapid and cost-effective sequencing of DNA or RNA samples. Techniques like next-generation sequencing and third-generation sequencing platforms provide unprecedented sequencing depth and throughput. These technologies use massively parallel sequencing, allowing thousands to millions of sequences to be processed simultaneously. High-throughput sequencing facilitates diverse applications, including whole-genome sequencing, RNA sequencing, chromatin immunoprecipitation sequencing, and metagenomics. Its scalability, speed, and affordability have transformed fields such as genomics, transcriptomics, epigenomics, and microbiology, accelerating biological discoveries and enabling personalized medicine, agriculture, and environmental studies (Satam et al. 2023).

5.11 BIOINFORMATICS TOOLS FOR miRNA TARGET PREDICTION

Bioinformatics tools for miRNA target prediction utilize computational algorithms to identify potential binding sites between miRNAs and target mRNAs. Commonly used tools include TargetScan, miRanda, and RNAhybrid, among others. These tools integrate sequence complementarity, thermodynamic stability, and conservation information to predict miRNA–target interactions (Verma et al. 2023). They consider factors such as seed sequence complementarity, site accessibility, and evolutionary conservation to prioritize potential targets. Additionally, machine learning approaches and high-throughput sequencing data enhance prediction accuracy (Johri et al. 2022). These bioinformatics tools play a crucial role in elucidating miRNA-mediated gene regulatory networks, facilitating functional studies, and advancing our understanding of miRNA biology in various biological processes and diseases (Akhtar et al. 2019).

5.12 CHALLENGES AND FUTURE DIRECTIONS

5.12.1 Challenges in miRNA Research

5.12.1.1 Functional Redundancy

Functional redundancy presents a challenge in miRNA research, where multiple miRNAs may target the same mRNA or share similar functions, complicating the interpretation of experimental results. Identifying specific miRNA–target interactions and discerning their individual contributions to biological processes requires comprehensive analyses, including genetic manipulation, target validation, and systems biology approaches. Additionally, miRNA expression patterns and target preferences vary across tissues, developmental stages, and environmental conditions, further adding to the complexity. Addressing functional redundancy necessitates integrating multi-omics data, employing advanced computational methods, and developing innovative experimental strategies to decipher the precise roles of miRNAs in gene regulation and cellular processes (Tran et al. 2008).

5.12.1.2 Off-Target Effects

Off-target effects pose a significant challenge in miRNA research, where unintended interactions between miRNAs and non-target mRNAs can occur, leading to misleading experimental results. Computational prediction tools and experimental validation methods help identify potential off-targets, but predicting all possible interactions remains difficult due to miRNA sequence complementarity and mRNA secondary structures. Additionally, off-target effects can vary depending on cellular context and experimental conditions. Overcoming off-target effects requires stringent experimental controls, improved bioinformatics algorithms, and innovative strategies such as CRISPR-based screens and high-throughput assays. Addressing off-target effects enhances the accuracy and reliability of miRNA research findings, advancing our understanding of miRNA function and therapeutic applications (Qiu et al. 2005).

5.12.1.3 Context-Dependent Roles

Context-dependent roles present a challenge in miRNA research, where miRNAs may exhibit diverse functions depending on cellular context, tissue specificity, developmental stage, and environmental conditions. This variability complicates the interpretation of miRNA function and regulatory networks. Identifying context-specific roles requires integrated multi-omics approaches, including transcriptomics, proteomics, and epigenomics, along with advanced computational modeling. Moreover, experimental validation in relevant biological systems and model organisms is essential to elucidate the precise roles of miRNAs in specific contexts. Understanding context-dependent miRNA functions enhances our knowledge of gene regulatory networks and cellular processes, providing insights into miRNA-mediated regulation in health and disease (DeVeale et al. 2021).

5.12.2 Future Directions

5.12.2.1 Uncovering Novel miRNAs and Targets

Uncovering novel miRNAs and targets is a crucial endeavor in miRNA research, essential for expanding our understanding of gene regulatory networks and cellular processes. High-throughput sequencing technologies, coupled with bioinformatics tools, facilitate the discovery of novel miRNAs by identifying small RNA sequences and predicting their precursor structures. Experimental validation through techniques such as northern blotting and quantitative polymerase chain reaction (PCR) confirms their existence and expression patterns. Additionally, computational algorithms predict potential target genes based on sequence complementarity and other features. Experimental validation of these targets elucidates the regulatory roles of novel miRNAs, shedding light on their functional significance in diverse biological contexts (Steinkraus et al. 2016).

5.12.2.2 Integrative Omics Approaches

Integrative omics approaches combine multiple high-throughput technologies, such as genomics, transcriptomics, proteomics, and metabolomics, to comprehensively study biological systems. By analyzing diverse molecular datasets simultaneously, these approaches provide a holistic understanding of complex biological processes, regulatory networks, and disease mechanisms. Integration of omics data allows researchers to uncover novel biomarkers, identify key signaling pathways, and elucidate gene regulatory networks. Advanced computational algorithms and statistical methods enable the integration of diverse omics datasets, facilitating the discovery of biological insights that would be inaccessible through individual omics analyses alone. Integrative omics approaches have transformative potential in biomedical research, personalized medicine, and systems biology studies (Tolani et al. 2021).

5.12.2.3 Translational Research for Crop Improvement

Translational research for crop improvement bridges scientific discoveries with practical applications in agriculture. It involves translating insights from basic research, such as genomics and molecular biology, into strategies for enhancing crop traits, productivity, and resilience. Additionally, translational efforts facilitate the adoption of sustainable farming practices and contribute to global food security by addressing challenges posed by climate change, population growth, and resource limitations (Baum et al. 2015).

5.13 CONCLUSION

In conclusion, miRNAs play a multifaceted role in plant biology, orchestrating diverse biological processes essential for growth, development, and stress responses. From regulating vegetative phase transitions to fine-tuning floral development and coordinating stress responses, miRNAs serve as critical molecular regulators in plants. The discovery of novel miRNAs and their targets continues to expand our understanding of gene regulatory networks and cellular mechanisms underlying plant physiology. Integrative omics approaches, coupled with advanced bioinformatics tools, facilitate the elucidation of complex miRNA-mediated regulatory networks and their context-dependent functions. Challenges such as functional redundancy, off-target effects, and context-dependent roles necessitate interdisciplinary collaborations and innovative methodologies to unravel the complexities of miRNA biology. As research in this field progresses, harnessing the regulatory potential of miRNAs holds promise for unlocking new avenues in crop improvement, environmental sustainability, and biotechnological innovation, shaping the future of plant biology and agriculture.

REFERENCES

Abbas, A., Shah, A. N., Tanveer, M., Ahmed, W., Shah, A. A., Fiaz, S., & Ullah, S. 2022. miRNA Fine tuning for crop improvement: using advance computational models and biotechnological tools. *Molecular Biology Reports*, *49*(6), 5437–5450.

Ahmad, H. M., Wang, X., Ijaz, M., Oranab, S., Ali, M. A., & Fiaz, S. 2022. Molecular aspects of microRNAs and phytohormonal signaling in response to drought stress: a review. *Current Issues in Molecular Biology*, *44*(8), 3695–3710.

Akhtar, M. M., Micolucci, L., Islam, M. S., Olivieri, F., & Procopio, A. D. 2019. A practical guide to miRNA target prediction. *microRNA Target Identification: Methods and Protocols*, *1970*, 1–13.

Arnaud, N., & Laufs, P. 2023. Plant miRNA integrated functions in development and reproduction. *Frontiers in Plant Physiology*, *1*, 1271423.

Banerjee, P. 2020. Plant abiotic stress responses and microRNAs. *Advanced Agriculture*, *2020*, 119–118.

Baum, M., Tadesse, W., Nachit, M., Abdalla, O., Rajaram, S., Singh, R. & Braun, H. 2015. Global crop improvement networks to bridge technology gaps. In *Advances in Wheat Genetics: From Genome to Field: Proceedings of the 12th International Wheat Genetics Symposium*, pp. 387–399. Springer, Japan.

Bertran Garcia de Olalla, E. 2023. *Functions of MICRORNA172 and APETALA2-LIKE Genes during Floral Transition at the Shoot Apical Meristem* (Doctoral dissertation, Universität zu Köln).

Boopathi, N. M. 2015. Plant miRNomics: novel insights in gene expression and regulation. In: Barh, D., Khan, M., & Davies, E. (eds.), *PlantOmics: The Omics of Plant Science*, pp. 181–211. Springer, New Delhi.

Bordoloi, K. S., & Agarwala, N. 2021. microRNAs in plant insect interaction and insect pest control. *Plant Gene*, *26*, 100271.

Campos-Rivero, G., Osorio-Montalvo, P., Sánchez-Borges, R., Us-Camas, R., Duarte-Aké, F., & De-la-Peña, C. 2017. Plant hormone signaling in flowering: an epigenetic point of view. *Journal of Plant Physiology*, *214*, 16–27.

Chen, D., Yan, W., Fu, L. Y., & Kaufmann, K. 2018. Architecture of gene regulatory networks controlling flower development in *Arabidopsis thaliana*. *Nature Communications*, *9*(1), 4534.

Chen, J., Teotia, S., Lan, T., & Tang, G. 2021. microRNA techniques: valuable tools for agronomic trait analyses and breeding in rice. *Frontiers in Plant Science*, *12*, 744357.

Chen, X., Zhang, Z., Liu, D., Zhang, K., Li, A., & Mao, L. 2010. SQUAMOSA promoter-binding protein-like transcription factors: star players for plant growth and development. *Journal of Integrative Plant Biology*, *52*(11), 946–951.

Curaba, J., Singh, M. B., & Bhalla, P. L. 2014. miRNAs in the crosstalk between phytohormone signalling pathways. *Journal of Experimental Botany*, *65*(6), 1425–1438.

DeVeale, B., Swindlehurst-Chan, J., & Blelloch, R. 2021. The roles of microRNAs in mouse development. *Nature Reviews Genetics*, *22*(5), 307–323.

Díez-Sainz, E., Lorente-Cebrián, S., Aranaz, P., Riezu-Boj, J. I., Martínez, J. A., & Milagro, F. I. 2021. Potential mechanisms linking food-derived microRNAs, gut microbiota and intestinal barrier functions in the context of nutrition and human health. *Frontiers in Nutrition*, *8*, 586564.

Ding, L. N., Li, Y. T., Wu, Y. Z., Li, T., Geng, R., Cao, J., & Tan, X. L. 2022. Plant disease resistance-related signaling pathways: recent progress and future prospects. *International Journal of Molecular Sciences*, *23*(24), 16200.

Dong, Q., Hu, B., & Zhang, C. 2022. microRNAs and their roles in plant development. *Frontiers in Plant Science*, *13*, 824240.

Drusco, A., & Croce, C. M. 2017. microRNAs and cancer: a long story for short RNAs. *Advances in Cancer Research*, *135*, 1–24.

Fukudome, A., & Fukuhara, T. 2017. Plant dicer-like proteins: double-stranded RNA-cleaving enzymes for small RNA biogenesis. *Journal of Plant Research*, *130*, 33–44.

Huijser, P., & Schmid, M. 2011. The control of developmental phase transitions in plants. *Development*, *138*(19), 4117–4129.

Jiang, F., & Doudna, J. A. 2017. CRISPR–Cas9 structures and mechanisms. *Annual Review of Biophysics*, *46*, 505–529.

Jin, L., Yarra, R., Zhou, L., Zhao, Z., & Cao, H. 2020. miRNAs as key regulators via targeting the phytohormone signaling pathways during somatic embryogenesis of plants. *3 Biotech*, *10*, 1–11.

Johri, P., Rajput, M. S., Trivedi, M., & Singh, A. 2022. Artificial intelligence and deep learning: current impact and perspectives in life sciences and medicine. *Advances in Bioresearch*, *13*(1), 184196. https://doi.org/10.15515/abr.0976-4585.13.1.184196

Joshi, S., Patil, S., Shaikh, A., Jamla, M., & Kumar, V. 2023. Modern omics toolbox for producing combined and multifactorial abiotic stress tolerant plants. *Plant Stress*, *11*, 100301.

Kapoor, B., Kumar, P., Sharma, R., & Kumar, A. 2021. Regulatory interactions in phytohormone stress signaling implying plants resistance and resilience mechanisms. *Journal of Plant Biochemistry and Biotechnology*, *30*, 1–16.

Karamat, U., Sun, X., Li, N., & Zhao, J. 2021. Genetic regulators of leaf size in Brassica crops. *Horticulture Research*, *8*, 91.

Karginov, F. V., & Hannon, G. J. 2013. Remodeling of Ago2–mRNA interactions upon cellular stress reflects miRNA complementarity and correlates with altered translation rates. *Genes & Development*, *27*(14), 1624–1632.

Khare, P., Kishore, K., & Sharma, D. K. 2017. A study on the standardization parameters of *Cassia angustifolia*. *Asian Journal of Pharmaceutical and Clinical Research*, *10*(7), 329–332.

Kohli, P. S., Maurya, K., Thakur, J. K., Bhosale, R., & Giri, J. 2022. Significance of root hairs in developing stress-resilient plants for sustainable crop production. *Plant, Cell & Environment*, *45*(3), 677–694.

Li, C., & Zhang, B. 2016. microRNAs in control of plant development. *Journal of Cellular Physiology*, *231*(2), 303–313.

Li, M., & Yu, B. 2021. Recent advances in the regulation of plant miRNA biogenesis. *RNA Biology*, *18*(12), 2087–2096.

Lima, J. C. D., Loss-Morais, G., & Margis, R. 2012. microRNAs play critical roles during plant development and in response to abiotic stresses. *Genetics and Molecular Biology*, *35*, 1069–1077.

Ma, X., Cao, X., Mo, B., & Chen, X. 2013. Trip to ER: microRNA-mediated translational repression in plants. *RNA Biology*, *10*(10), 1586–1592.

Maddala, V. K. S., & Singh, S. 2023. Essential oils from plants and their role in nanomaterial synthesis characterization and applications. In: Husen, A. (ed.), *Secondary Metabolites Based Green Synthesis of Nanomaterials and Their Applications*, pp. 191–207. Springer Nature, Singapore.

Maddala, V. K. S., Singh, S., & Bandi, J. 2024. Soil properties influencing nutrient signaling in plant soil systems. In: Husen, A. (ed.), *Essential Minerals in Plant-Soil Systems*, pp. 373–393. Elsevier, Amsterdam, Netherlands.

Mahanty, B., Bharat, S. S., Mohanty, J. N., Mishra, R., & Joshi, R. K. 2022. microRNA-mediated insect resistance in field crops. In: Husen, A. (ed.), *Molecular Advances in Insect Resistance of Field Crops: Modern and Applied Approaches*, pp. 369–392. Springer International Publishing, Cham.

Malka, Y., Steiman-Shimony, A., Rosenthal, E., Argaman, L., Cohen-Daniel, L., Arbib, E., & Berger, M. 2017. Post-transcriptional 3-UTR cleavage of mRNA transcripts generates thousands of stable uncapped autonomous RNA fragments. *Nature Communications*, *8*(1), 2029.

Mandal, K., Boro, P., & Chattopadhyay, S. 2021. microRNA based gene regulation: a potential way for crop improvements. *Plant Gene*, *27*, 100312.

Meddya, S., Meshram, S., Sarkar, D., Datta, R., Singh, S., Avinash, G., & Thulasinathan, T. 2023. Plant stomata: an unrealized possibility in plant defense against invading pathogens and stress tolerance. *Plants*, *12*(19), 3380.

Megraw, M., Cumbie, J. S., Ivanchenko, M. G., & Filichkin, S. A. 2016. Small genetic circuits and microRNAs: big players in polymerase II transcriptional control in plants. *The Plant Cell*, *28*(2), 286–303.

Meng, X., Li, A., Yu, B., & Li, S. 2021. Interplay between miRNAs and lncRNAs: mode of action and biological roles in plant development and stress adaptation. *Computational and Structural Biotechnology Journal*, *19*, 2567–2574.

Muhammad, S., Xu, X., Zhou, W., & Wu, L. 2023. Alternative splicing: an efficient regulatory approach towards plant developmental plasticity. *Wiley Interdisciplinary Reviews: RNA*, *14*(3), e1758.

Ndudzo, A., Makuvise, A. S., Moyo, S., & Bobo, E. D. 2024. CRISPR-Cas9 genome editing in crop breeding for climate change resilience: implications for smallholder farmers in Africa. *Journal of Agriculture and Food Research*, *16*, 101132.

Pagano, L., Rossi, R., Paesano, L., Marmiroli, N., & Marmiroli, M. 2021. miRNA regulation and stress adaptation in plants. *Environmental and Experimental Botany*, *184*, 104369.

Parveen, R., Khare, N., Singh, S., & Johri, P. 2023. An in silico outlook for the detection and surveillance of evolving and persistent plant pathogens. In: Singh, U. B., & Singh, H. B. (eds.), *Detection, Diagnosis and Management of Soil-borne Phytopathogens*, pp. 17–40. Springer Nature, Singapore.

Patani, A., Patel, A., Prajapati, D., Khare, N., & Singh, S. 2023. Various agriculture crop plant-based bioactive compounds and their use in nanomaterial synthesis and applications. In: Husen, A. (ed.), *Nanomaterials from Agricultural and Horticultural Products*, pp. 223–241. Springer Nature, Singapore.

Patani, A., Prajapati, D., Shukla, K., Patel, M., Patani, P., Patel, A., & Singh, S. 2024 Environmental stress–induced alterations in the micro-and macronutrients status of plant. In: Husen, A. (ed.), *Essential Minerals in Plant-Soil Systems*, pp. 181–195. Elsevier, Amsterdam, Netherlands.

Patel, P., Yadav, K., Ganapathi, T. R., & Penna, S. 2019. Plant miRNAome: cross talk in abiotic stressful times. *Genetic Enhancement of Crops for Tolerance to Abiotic Stress: Mechanisms and Approaches*, *I*, 25–52.

Pourkheirandish, M., Golicz, A. A., Bhalla, P. L., & Singh, M. B. 2020. Global role of crop genomics in the face of climate change. *Frontiers in Plant Science*, *11*, 522800.

Pulido, A., & Laufs, P. 2010. Co-ordination of developmental processes by small RNAs during leaf development. *Journal of Experimental Botany*, *61*(5), 1277–1291.

Qiu, S., Adema, C. M., & Lane, T. 2005. A computational study of off-target effects of RNA interference. *Nucleic Acids Research*, *33*(6), 1834–1847.

Ranjan, A., Sinha, R., Singla-Pareek, S. L., Pareek, A., & Singh, A. K. 2022. Shaping the root system architecture in plants for adaptation to drought stress. *Physiologia Plantarum*, *174*(2), e13651.

Safdar, H., Amin, A., Shafiq, Y., Ali, A., Yasin, R., Shoukat, A., & Sarwar, M. I. 2019. A review: impact of salinity on plant growth. *Natural Sciences*, *17*(1), 34–40.

Samad, A. F., Sajad, M., Nazaruddin, N., Fauzi, I. A., Murad, A. M., Zainal, Z., & Ismail, I. 2017. microRNA and transcription factor: key players in plant regulatory network. *Frontiers in Plant Science*, *8*, 565.

Sao, P., Parveen, R., Khattri, A., Sharma, S., Tiwari, N., & Singh, S. 2022. Understanding the metabolomics of medicinal plants under environmental pollution: challenges and opportunities. *Environmental Pollution and Medicinal Plants*, *13*, 193–205.

Satam, H., Joshi, K., Mangrolia, U., Waghoo, S., Zaidi, G., Rawool, S., & Malonia, S. K. 2023. Next-generation sequencing technology: current trends and advancements. *Biology*, *12*(7), 997.

Shahzad, R., Harlina, P. W., Ayaad, M., Ewas, M., Nishawy, E., Fahad, S., & Amar, M. H. 2018. Dynamic roles of microRNAs in nutrient acquisition and plant adaptation under nutrient stress: a review. *Plant Omics*, *11*(1), 58–79.

Singh, A., Jain, D., Pandey, J., Yadav, M., Bansal, K. C., & Singh, I. K. 2023. Deciphering the role of miRNA in reprogramming plant responses to drought stress. *Critical Reviews in Biotechnology*, *43*(4), 613–627.

Singh, A., Sengar, R. S., Singh, A. K., Singh, S., Kumar, A., & Ahmad, K. 2019. Impact of genetic engineering on agriculture food crops. *Biotech Today: An International Journal of Biological Sciences*, *9*(1), 17–24.

Singh, N. K., Anand, S., & Das, S. 2017. Small RNAs-II: mode of action and potential applications in plant improvement. In: Sajad, M. Z. (ed.), *Plant OMICS and Crop Breeding*, pp. 163–200. Apple Academic Press, New York.

Singh, P., Dutta, P., & Chakrabarty, D. 2021. miRNAs play critical roles in response to abiotic stress by modulating cross-talk of phytohormone signaling. *Plant Cell Reports*, *40*(9), 1617–1630.

Song, X., Li, Y., Cao, X., & Qi, Y. 2019. microRNAs and their regulatory roles in plant–environment interactions. *Annual Review of Plant Biology*, *70*, 489–525.

Srijyothi, L., Ponne, S., Prathama, T., Ashok, C., & Baluchamy, S. 2018. Roles of non-coding RNAs in transcriptional regulation. *Transcriptional and Post-Transcriptional Regulation*, *55*, 76125.

Steinkraus, B. R., Toegel, M., & Fulga, T. A. 2016. Tiny giants of gene regulation: experimental strategies for microRNA functional studies. *Wiley Interdisciplinary Reviews: Developmental Biology*, *5*(3), 311–362.

Sunny, J. S., Snijesh, V. P., & Singh, S. 2016. Potential role of mirnas in rheumatoid arthritis. *Austin Arthritis*, *1*(2), 1009.

Tang, J., & Chu, C. 2017. microRNAs in crop improvement: fine-tuners for complex traits. *Nature Plants*, *3*(7), 1–11.

Tiwari, M., Sharma, D., & Trivedi, P. K. 2014. Artificial microRNA mediated gene silencing in plants: progress and perspectives. *Plant Molecular Biology*, *86*, 1–18.

Tolani, P., Gupta, S., Yadav, K., Aggarwal, S., & Yadav, A. K. 2021. Big data, integrative omics and network biology. *Advances in Protein Chemistry and Structural Biology*, *127*, 127–160.

Tran, D. H., Satou, K., & Ho, T. B. 2008. Finding microRNA regulatory modules in human genome using rule induction. *BMC Bioinformatics*, *9*, 1–10.

Verma, A., Kumar, R., Shukla, S., Singh, S., Johri, P., & Singh, M. S. 2023. Genome wide association studies: an outlook. *Journal of Innovation in Applied Research*, *6*, 62.

Verma, N., Sao, P., Srivastava, A., & Singh, S. 2021. Physiological and molecular responses to drought, submergence and excessive watering in plants. *Harsh Environment and Plant Resilience: Molecular and Functional Aspects*, *2021*, 305–321.

Verma, N., Singh, Y., Patra, A., & Singh, T. 2022. Role of non-coding RNAs in plant nutrition through mycorrhizal interactions. In: Rodrigo, N. S. (ed.), *Arbuscular Mycorrhizal Fungi in Agriculture-New Insights*. IntechOpen, New York.

Vora, Z., Pandya, J., Vaikuntapu, P. R., Singh, S., Asha, S., & Purama, R. K. 2023. Genome-wide association of defense hormone crosstalk in plants. In: Husen, A. (ed.), *Hormonal Cross-Talk, Plant Defense and Development*, pp. 353–371. Academic Press, New York.

Waheed, S., & Zeng, L. 2020. The critical role of miRNAs in regulation of flowering time and flower development. *Genes*, *11*(3), 319.

Waheed, S., Anwar, M., Saleem, M. A., Wu, J., Tayyab, M., & Hu, Z. 2021. The critical role of small RNAs in regulating plant innate immunity. *Biomolecules*, *11*(2), 184.

Yan, X. X., Liu, X. Y., Hong, C. U. I., & Zhao, M. Q. 2022. The roles of microRNAs in regulating root formation and growth in plants. *Journal of Integrative Agriculture*, *21*(4), 901–916.

Yu, Q., Jin, X., Liu, C., & Wen, Y. 2022. An integrated analysis of transcriptome and miRNA sequencing provides insights into the dynamic regulations during flower morphogenesis in petunia. *Horticulturae*, *8*(4), 284.

Yuan, S., Zhao, J., Li, Z., Hu, Q., Yuan, N., Zhou, M., & Luo, H. 2019. microRNA396-mediated alteration in plant development and salinity stress response in creeping bentgrass. *Horticulture Research*, *6*, 48.

Zaheer, U., Munir, F., Salum, Y. M., & He, W. 2024. Function and regulation of plant ARGONAUTE proteins in response to environmental challenges: a review. *PeerJ*, *12*, e17115.

Zaid, A., & Wani, S. H. 2020. Importance of small RNA in plant metabolism. In: Husen, A. (ed.), *Plant Small RNA*, 125–153. Academic Press, New York.

Zhang, L., Xiang, Y., Chen, S., Shi, M., Jiang, X., He, Z., & Gao, S. 2022. Mechanisms of microRNA biogenesis and stability control in plants. *Frontiers in Plant Science*, *13*, 844149.

Zhao, Y. 2018. Essential roles of local auxin biosynthesis in plant development and in adaptation to environmental changes. *Annual Review of Plant Biology*, *69*, 417–435.

6 Photosynthesis, Productivity, and Regulatory Roles of miRNA

Deepa Dhatwalia, Manisha Saini, Shubhra Maithreyi, and Pooja Sharma*

6.1 INTRODUCTION

Photosynthesis is one of the fundamental biochemical processes on which life on Earth depends (Nelson and Cox 2017). By capturing light energy, photosynthesis converts carbon dioxide and water into organic compounds mainly consisting of glucose, releasing oxygen as a byproduct. This process is crucial not only for plant growth and energy production but also underpins the food chain, upon which virtually all living organisms on Earth rely (Taiz et al. 2015).

Basically, photosynthesis is divided into two major steps: light-dependent reactions and the Calvin cycle. Light-dependent reactions occur in the thylakoid membranes of chloroplasts, where photons are absorbed by pigment molecules. This energy is used to produce ATP and NADPH, which serve as carriers to supply energy for the Calvin cycle. In chloroplasts, CO2 is fixed into organic molecules through enzyme-mediated steps in what is known as the Calvin cycle, which produces glucose and other carbohydrates (Blankenship 2014).

6.1.1 The Role of Photosynthesis in Plant Efficiency

Photosynthetic efficiency is highly relevant to plant growth, biomass accumulation, and crop yields, as it is influenced by light intensity and quality, carbon dioxide concentration, temperature, water availability, nutrient supply, and genetic factors (Zhu et al. 2010). Understanding these variables and any possible optimizations is crucial for enhancing plant productivity, particularly in light of the prerequisites for global food security and sustainable agriculture. Enhancing photosynthesis is a promising area of research, as it connects to advanced agricultural practices, genetic improvements, and the discovery of new regulatory mechanisms.

Among the promising developments in this field, research on microRNAs stands out as particularly exciting. Small non-coding RNA molecules of 20–24 nucleotides in length play important roles in post-transcriptional gene regulation (Zhang 2015a). miRNAs can bind to target messenger RNAs (mRNAs) through perfect or near-perfect complementarity, leading to either mRNA cleavage or translational repression. Evidence is accumulating regarding the role of miRNAs in several physiological processes, including development, stress responses, and, importantly, photosynthesis (Pan et al. 2018; Zhang et al. 2017).

The regulation of photosynthesis by miRNAs is diverse and intricate (Meenakshi et al. 2025). The steps affecting photosynthetic efficiency include those related to chlorophyll biosynthesis, carbon fixation, and responses to light stress. For example, miR408 regulates copper homeostasis, which is essential for the electron transport chain and the proper functioning of plastocyanin. Another example is miR396, which targets growth-regulating factor (GRF) genes involved in leaf development, thereby modulating the photosynthetic surface area (Ding et al. 2021).

Regulating photosynthetic efficiency through miRNAs and other biotechnological methods presents a significant opportunity for developing stress-resistant crops and increasing agricultural

* *Corresponding Author*

DOI: 10.1201/9781003491378-6

productivity. The functional identification and characterization of miRNAs are now feasible thanks to advancements in next-generation sequencing technologies and bioinformatics analysis, opening new avenues for biotechnological strategies aimed at optimizing photosynthesis and, consequently, enhancing sustainable agriculture and food security in the face of climate change and growing human populations.

6.1.2 Evolutionary Perspective of Photosynthesis and miRNA Regulation

Photosynthesis is an evolutionarily ancient process in which various mechanisms have evolved over the past billions of years. The first organisms that emerged around 3.4 billion years ago were cyanobacteria, which could use sunlight to split water molecules and thus produced oxygen, altering the Earth's atmosphere from its initial state (Hohmann-Marriott and Blankenship 2011). The endosymbiosis of cyanobacteria with early eukaryotic cells gave rise to chloroplasts in plants, where complex photosynthetic mechanisms evolved. These mechanisms include the C_4 and CAM pathways that plants utilize to maximize photosynthetic efficiency under a wide range of environmental conditions.

Coinciding with the evolutionary innovations of photosynthetic pathways, the emergence of miRNAs dates back about 425 million years ago, when the earliest plants began colonizing land. Indeed, these regulatory molecules provide an extra layer of gene regulation that enables these plants to better adapt to terrestrial environments. The co-evolution of photosynthesis and its regulatory mechanisms through miRNAs has been a major factor in plant diversification and success across different ecological niches.

This review delves into the complex processes of photosynthesis, its efficiency, and the regulation by miRNAs. It addresses recent discoveries in research, advances in biotechnology, and evolutionary perspectives to provide a comprehensive insight into the plausible ways of optimizing photosynthesis with the goal of improving plant productivity to meet global food security challenges.

6.2 FUNDAMENTALS OF PHOTOSYNTHESIS

6.2.1 Light-Dependent Reactions

The light-dependent or photochemical phase of photosynthesis occurs in the thylakoid membranes of the chloroplasts (Nelson and Cox 2021). Solar energy is captured in this phase, and both ATP and NADPH are generated, which energize the subsequent light-independent reactions of the Calvin cycle. Light-dependent reactions are energized by harvesting energy from incident photons through pigment molecules, mainly chlorophyll, bound to photosystems (Taiz et al. 2022).

During this process, two major photosystems are involved: PSII and PSI. Light absorption by PSII excites electrons, which are then passed to the primary electron acceptor. These excited electrons are replaced by electrons obtained from the photolysis of water, resulting in oxygen formation as a byproduct. The electrons are transferred to an electron transport chain that leads to PSI, establishing a proton gradient across the thylakoid membrane. The energy stored in this gradient is harnessed by the enzyme ATP synthase to produce ATP from ADP and inorganic phosphate (Blankenship 2021).

PSI also absorbs light energy, which serves as an excitant for electrons that are passed on to NADP+ to yield NADPH. The cyclic and non-cyclic electron flow mechanisms balance the proportion of ATP and NADPH formed to meet the requirements of the Calvin cycle. In non-cyclic electron flow, both ATP and NADPH are produced, while in cyclic electron flow, more ATP is generated with no NADPH, balancing the proportion between these two according to requirements (Raven et al. 2020).

6.2.2 Calvin Cycle (Light-Independent Reactions)

This is a light-independent reaction in which carbon is fixed into organic compounds from CO_2 using the energy stored from the light-dependent reactions in the chloroplast stroma. It doesn't require direct light. Essentially, this process involves the release of energy from ATP and NADPH formed while producing glucose through light dependency and the subsequent fixation of carbon dioxide into organic molecules. The Calvin cycle primarily has three stages: carbon fixation, reduction, and RuBP regeneration (Bassham and Kirk 2019).

In the carbon fixation phase, carbon dioxide is fixed into RuBP by the enzyme ribulose-1,5-bisphosphate carboxylase/oxygenase. Carbon dioxide molecules combine with RuBP, a six-carbon compound, which immediately splits to form two molecules of 3-phosphoglycerate. In the reduction phase, 3-phosphoglycerate is reduced to glyceraldehyde-3-phosphate, a three-carbon sugar, using energy from ATP and NADPH. Several G3P molecules exit the cycle to be used in synthesizing glucose and other carbohydrates (Buchanan et al. 2015).

The cycle then recycles some ATP in the dark regeneration stage to convert the leftover G3P molecules back into RuBP, refreshing it to begin anew. For every three rounds of the Calvin cycle, one G3P molecule is generally synthesized from three molecules of carbon dioxide. Six rounds must be completed to synthesize one molecule of glucose (Benson 2019).

6.2.3 Key Enzymes and Molecules Involved

Several important enzymes and molecules participate in the light-dependent reactions of photosynthesis and in the Calvin cycle (Reece et al. 2020).

1. **Chlorophyll**: Chlorophyll is the pigment responsible for absorbing energy from light. Chlorophyll a directly interacts with the light-dependent reactions, while chlorophyll b and carotenoids are usually considered accessory pigments that help widen the range of the light spectrum absorbed (Croce and van Amerongen 2014).
2. **Photosystems I and II**: These comprise two complexes, one of which involves the capture of light energy and electron transports. A photosystem comes in constituted by the reaction center and light-harvesting complexes. PSII is a pigment system operating with light absorbed at 680 nm (P680), and PSI is the pigment system that operates with light absorbed at 700 nm (P700) (Caffarri et al. 2014).
3. **ATP Synthase**: This enzyme produces ATP from ADP and inorganic phosphate, driven by the proton gradient established across the thylakoid membrane by electron transport. In other words, it acts as a rotary motor powered by the flow of protons across the membrane (Walker 2013).
4. **NADP+ Reductase**: This enzyme reduces the final electron acceptor, NADP+, into NADPH using the transferred electrons. It serves to balance the amounts of ATP and NADPH produced (Mulo and Krieger-Liszkay 2020).
5. **RuBisCO**: It is the most abundant enzyme on Earth and catalyzes carbon fixation in the Calvin cycle by carboxylating RuBP. This reaction represents the first step of carbon fixation and is therefore one of the most critical steps in the synthesis of organic molecules from carbon dioxide (Erb and Zarzycki 2018).
6. **Glyceraldehyde 3-Phosphate Dehydrogenase**: Enzyme catalyzes most of the reactions in the reduction phase of the Calvin cycle. It reduces 1,3-bisphosphoglycerate to G3P using a reducing agent such as NADPH (Michelet et al. 2013).
7. **Ferredoxin**: A small iron–sulfur protein that acts as a direct electron donor from PSI to NADP+ reductase. It occupies an important and central position in the electron transfer chain, whether in cyclic electron flow pathways (Hanke and Mulo 2013).
8. **PQ and PC**: These are mobile electron carriers in the thylakoid membrane that facilitate the flow of electrons from PSII to the cytochrome b6f complex and from the latter to PSI (Cramer et al. 2021).

A basic understanding of these processes and the major components involved underlies how plants capture and convert light energy into chemical energy to power growth and productivity. This information forms the basis for biotechnological improvements in photosynthetic efficiency and crop yields necessary to secure global food supplies (Long et al. 2015).

6.2.4 Factors Affecting Photosynthetic Efficiency

The influential factors affecting photosynthetic efficiency, with a view toward optimizing plant productivity through improvements, are taken into consideration. In this section, we examine how environmental and genetic complexities work together to determine the extent to which plants can convert light energy into chemical energy during photosynthesis.

6.2.4.1 Environmental Factors

Photosynthetic efficiency in plants is influenced by both environmental and genetic factors. The most significant environmental factors include light intensity and quality, carbon dioxide concentration, temperature, water availability, and nutrient levels.

6.2.4.2 Light Intensity and Quality

Light is the primary energy source driving photosynthesis. Light intensity and quality significantly influence the potential quantum efficiency of PSII reactions and the efficiency of open reaction centers.

- **Light Intensity**: As light intensity increases, the rate of photosynthesis also rises up to a certain point, beyond which additional light has no effect due to saturation. However, excessively high intensity may result in photoinhibition—damage to the photosynthetic apparatus—thus reducing efficiency.

  Based on 2023 research on rice cultivars, Zhang et al. demonstrated that plants grown under a combination of red and blue LED lighting exhibited 15% greater photosynthetic efficiency compared to those under white light. This illustrates that the proper quality of light can significantly impact photosynthesis (Zhang et al. 2023a).
- **Light Quality**: The peaks of absorption for optimal chlorophyll pigment absorption occur in the blue and red light regions, while green light is poorly absorbed. Therefore, spectral composition modulates the activity of photosystems I and II, both of which contribute to determining photosynthetic efficiency.

6.2.4.3 Carbon Dioxide Concentration

Carbon dioxide serves as the substrate for the Calvin cycle and is fixed into organic molecules. Elevated CO_2 levels usually increase the rate of photosynthesis as RuBisCO becomes saturated, enhancing carbon fixation. However, at very high CO_2 levels, other factors such as light intensity and nutrient availability may become limiting, causing the photosynthetic rate to plateau.

Johnson et al. (2024) conducted a free-air CO_2 enrichment experiment with wheat crops, which, under elevated CO_2 at 550 ppm, exhibited a photosynthetic rate that was 20% higher and a yield 15% greater than those of wheat crops under ambient CO_2 levels (Johnson et al. 2024).

6.2.4.4 Temperature

Temperature can affect photosynthetic efficiency by altering the activities of enzymes involved in photosynthesis. All enzymes have a temperature optimum range at which their activity is maximized. For example, RuBisCO works optimally only at intermediate temperatures. High temperatures depress enzyme activity and increase photorespiration, where oxygen competes with CO2 at sites on RuBisCO, thus lowering photosynthetic efficiency.

For instance, in a recent study on coffee plants by Rodrigues et al. (2023), it was concluded that the rise in temperature above 30°C lowers photosynthetic efficiency by 30% due to increased photorespiration and enzyme denaturation (Rodrigues et al. 2023).

6.2.4.5 Water Availability

In photosynthesis, water functions both as the ultimate donor of electrons and protons in the light-dependent reactions and as the provider of turgor pressure for cell expansion. Water stress induces stomatal closure, lowering CO_2 diffusion into the leaf and thereby constraining photosynthetic carbon assimilation. Extended or severe water deficit can also provoke oxidative stress that damages the photosynthetic machinery. Field and glass-house studies show that maintaining maize at ~50–55% of field capacity reduces net photosynthetic rate by about 20–30%, largely because of stomatal limitation (Hussain et al. 2019; Li et al. 2024).

6.2.4.6 Nutrient Availability

Nutrient availability—such as nitrogen, phosphorus, and magnesium—impacts photosynthesis.

- **Nitrogen**: A building block of chlorophyll and other proteins involved in photosynthesis.
- **Phosphorus**: A component of ATP and nucleic acids; plays a significant role in energy transfer and storage.
- **Magnesium**: A central component of the chlorophyll molecule that participates in capturing light energy.

Deficiencies of these or other critical nutrients lower chlorophyll content and disrupt electron transport, thus decreasing photosynthetic efficiency.

For example, Patel et al. (2024) reported a hydroponic study of tomatoes showing that plants grown under nitrogen-deficient conditions have 40% less chlorophyll content and 35% lower photosynthetic efficiency compared to optimal nitrogen supply (Patel et al. 2024).

6.3 GENETIC AND miRNA FACTORS

Photosynthetic efficiency is also influenced by genetic factors, which include variations in chloroplast structure, photosynthetic pathways, and miRNA-mediated regulation.

6.3.1 Chloroplast Structure and Function

Genes controlling chloroplast development and structure are considered to have significant effects on photosynthetic efficiency. Several key genes governing this process have been identified, including the following:

- **FtsZ Genes**: These genes are necessary for the division of chloroplasts. Chen et al. (2023) ectopically overexpressed the OsFtsZ2-1 gene in rice and observed a higher number of chloroplasts per cell and a 10% improvement in photosynthetic efficiency (Chen et al. 2023).
- **VIPP1**: This gene is essential for the biogenesis of thylakoid membranes. Overexpression of VIPP1 in *Arabidopsis* plants by Wang et al. resulted in an advanced thylakoid network and a 12% increase in the photosynthetic electron transport rate (Wang et al. 2023).
- **CUR7**: This gene regulates the curvature of thylakoid membranes. Yamamoto et al. recently demonstrated that tobacco plants with optimized CUR7 expression exhibited increased grana stacking, resulted in a 15% improvement in light-harvesting efficiency (Yamamoto et al. 2023).

6.3.2 Photosynthetic Pathways

The three major photosynthetic pathways evolved to optimize carbon fixation under a wide range of environmental conditions. Recently, research efforts have focused on understanding and genetically transferring desirable traits among these pathways.

- **C4-rice**: International efforts have now installed all five core C_4 enzymes (CA, PEPC, NADP-MDH, NADP-ME, PPDK) in rice, creating a biochemical C_4 cycle but still within a C_3 leaf framework (Ermakova et al. 2020). Follow-up genome-editing has shifted carbonic-anhydrase localisation and altered bundle-sheath cell shape—key steps toward Kranz anatomy (Jethva et al. 2024 pre-print; Plackett & Hibberd 2024).
- **Synthetic CAM**: Over-expression of a drought-adapted PEPC from Suaeda aralocaspica conferred CAM-like nocturnal carboxylation and improved water-use efficiency in Arabidopsis under drought (Li et al. 2024).
- **RuBisCO improvement**: Comparative genomics shows RuBisCO is still evolving toward higher catalytic efficiency; resurrected ancestral forms exhibit ~10% faster CO_2 fixation (Bouvier et al. 2024). Current CRISPR edits of RuBisCO subunits in rice highlight the complexity of maintaining enzyme assembly and activity (Zhou et al. 2024).

6.3.3 microRNA Regulation

miRNAs are small non-coding RNAs that modulate gene expression at a post-transcriptional level and participate in a wide variety of physiological processes, including photosynthesis (Noopur et al. 2025a,b; Pooja et al. 2025; Rahul et al. 2024; Subhan et al. 2025; Tanvi et al. 2025). These molecules are involved in the regulation of genes whose products contribute to chlorophyll biosynthesis, light-harvesting complexes, and responses to stress.

Recently, several key miRNAs with a significant impact on photosynthetic efficiency have been identified:

- **miR408**: miR408 regulates copper allocation among plastocyanin and other cupro-proteins. Over-expressing MIR408 in rice elevates plastocyanin abundance, enhances ΦPSII and electron-transport rate, and ultimately raises grain yield under both well-watered and drought conditions (Pan et al. 2018; Zhang et al. 2017).
- **miR396**: This miRNA targets GRF genes responsible for leaf development and influences photosynthetic capacity. Zhang et al. (2023b) demonstrated that altered miR396 levels in wheat modified leaf architecture and improved light capture efficiency (Zhang et al. 2023b).
- **miR159**: Chen et al. (2024) showed that miR159 targets GAMYB transcription factors for chloroplast development. In *Arabidopsis*, a reduction in miR159 levels leads to increased chlorophyll content and photosynthetic rate (Chen et al. 2024).
- **miR399**: This miRNA participates in phosphate homeostasis, which indirectly affects photosynthesis. Wang et al. (2023) illustrated that manipulating the expression levels of miR399 to optimal levels in tomato plants promoted phosphate uptake and utilization, ultimately resulting in enhanced photosynthetic efficiency under phosphate-limited conditions (Wang et al. 2023).

Understanding the complex regulatory networks involving these miRNAs opens new avenues for genetic engineering to improve photosynthetic efficiency in crop plants.

6.4 ADAPTIVE MECHANISMS IN PHOTOSYNTHESIS

Plants have undergone evolutionary processes to adapt to their environments to optimize photosynthesis in variable conditions. Two important strategies for adaptation are photoprotection mechanisms and dynamic light-harvesting antenna complexes.

6.4.1 Photoprotection Mechanisms

Photoprotection refers to mechanisms developed to prevent damage to the photosynthetic apparatus due to excess light. Non-photochemical quenching is the major mechanism of photoprotection, involving the dissipation of excessive absorbed light energy as heat and preventing the formation of reactive oxygen species (ROS) that can damage the chloroplast.

- **PsbS Protein**: Li et al. (2023a) overexpressed the PsbS protein in rice plants to achieve faster non-photochemical quenching (NPQ) induction and relaxation kinetics and demonstrated a 15% increase in carbon fixation under fluctuating light (Li et al. 2023a).
- **Zeaxanthin Cycle**: González-Pérez et al. (2024) revealed that in tomato plants, overexpression of zeaxanthin epoxidase improved energy dissipation capacity and reduced photoinhibition by 20% under high light stress (González-Pérez et al. 2024).

6.4.2 Dynamic Antenna Complexes

Light-harvesting antennae vary in size and composition depending on light intensity, modulating the photon flux delivered to the reaction centres. Under low light, plants assemble larger antenna systems to maximise capture, whereas high light triggers antenna trimming to minimise photodamage, thereby balancing photosynthetic efficiency with photoinhibition.

- **Phosphorylation of LHC II**: Arabidopsis mutants lacking the STN7 kinase (and thus LHC II phosphorylation) suffer reduced PSI function and growth only under rapidly fluctuating light. Restoring steady-state LHC II phosphorylation balances excitation energy between PS II and PSI and optimises electron transport (Tikkanen et al. 2010).
- **LHCB4 isoform plasticity**: Light-dependent shifts in LHCB4 (CP29) isoform abundance drive reversible remodelling of PS II–LHC II super-complexes, enabling a flexible antenna size and faster acclimation to sudden irradiance changes (Albanese et al. 2016; Albanese et al. 2019).

6.5 FACTOR INTERACTIONS IN PHOTOSYNTHETIC EFFICIENCY

These multiple factors that affect photosynthetic efficiency are not independent and, in fact, interact with each other in a complex way. Understanding these interactions is crucial for predicting plant responses to environmental changes and for developing strategies to optimize photosynthesis.

- **Light and Temperature**: High light intensity combined with high temperature is believed to increase photorespiration and photoinhibition. At low temperatures, however, plants may become more susceptible to photodamage even under light intensities typically regarded as moderate.
- **CO_2 and Water Availability**: High CO_2 usually improves the water-use efficiency of plants because, although they partly close their stomata, they still maintain a high photosynthetic rate. However, under low water availability, high CO_2 cannot aid in closing the stomata, as this will result in reduced CO_2 uptake.
- **Light and Nutrient Availability**: Poor nutrient availability, especially of nitrogen, reduces the content of pigments involved in light harvesting and electron transport; therefore, inadequate nutrient supply diminishes the capacity of plants to utilize available light. This interaction becomes critical under high-light conditions.
- **Genetic Factors and Environment**: The activity and effectiveness of genetic adaptations are influenced by environmental conditions. For instance, high temperatures and high light intensities enhance the advantages of C_4 photosynthesis.

6.5.1 Practical Applications and Future Directions

Research into the factors that affect photosynthetic efficiency has several critical practical applications:

- **Crop Improvement**: Genetic engineering of crops for improved photosynthetic efficiency aims to maximize crop yields. For example, the RIPE project succeeded in making tobacco more productive by engineering the recovery pathway from photoprotection (RIPE Project 2024).
- **Adaptation to Climate Change**: Understanding plant responses to environmental change will aid in breeding more resilient crop varieties, such as crops that are more water-use efficient and heat-tolerant.
- **Vertical Farming**: The effects of light quality and light intensity on photosynthesis provide a basis for optimizing LED-based lighting systems to maximize productivity in indoor farming.
- **Production of Biofuels**: Improving photosynthetic efficiency will significantly enhance the viability and sustainability of algal biofuel crops and other biofuel-producing crops.
- **Carbon Sequestration**: Enhanced photosynthetic efficiency in forests and other ecosystems increases the carbon sequestration potential of these ecosystems, helping to mitigate climate change.

Knowing and further understanding photosynthesis, along with its application, can secure for mankind more sustainable agricultural practices and effective responses to current environmental challenges.

6.6 miRNA REGULATION IN PHOTOSYNTHESIS

6.6.1 Overview of miRNA Biogenesis

microRNAs are endogenous non-coding small RNAs of 21–24 nucleotides in length that play key roles in post-transcriptional gene regulation. miRNA biogenesis occurs through several steps:

Transcription: RNA polymerase II transcribes primary miRNAs from genomic DNA.

Processing: A nuclear complex called microprocessor, which contains the RNase III enzyme Dicer-like 1 and its partners, processes pri-miRNAs into precursor miRNAs.

Maturation: DCL1 further cleaves pre-miRNAs to produce mature miRNA duplexes.

RISC Incorporation: The guide strand of the miRNA duplex is incorporated into the RNA-induced silencing complex (RISC), while the passenger strand is typically degraded.

Target Regulation: The RISC–miRNA complex interacts with its target mRNAs through base complementarity and sometimes enhances mRNA cleavage or represses translation.

Recent studies have further revealed the complexities of miRNA biogenesis. For instance, Zhang et al., in a paper in *Nature Plants*, described some miRNAs that bypass DCL1-dependent processing in *Arabidopsis* and are processed via a new pathway requiring DCL4 and DCL3. This finding has provided a new understanding of the miRNA generation process and opened the possibility for more diversified regulation (Zhang et al. 2021).

6.6.2 miRNAs in Photosynthesis Regulation

microRNAs are small yet potent regulators in plant systems. They modulate photosynthesis by fine-tuning the genes responsible for chlorophyll biosynthesis, photosynthetic apparatus development, responses to environmental stresses, carbon fixation, and assimilation. This complex and multi-layered regulation allows plants to dynamically adjust photosynthetic activities in relation to their ever-changing environment (Figure 6.1 and Table 6.1).

6.6.2.1 miR408: Copper Homeostasis and Plastocyanin Function

miR408 is involved in copper homeostasis, a critical condition for the activity of plastocyanin in the electron transport chain. Copper is an essential cofactor for numerous enzymes participating in photosynthesis. Thus, copper metabolism is strictly controlled to ensure optimal photosynthetic performance.

Mechanisms: miR408-mediated regulation occurs through its target, the COPPER TRANSPORTER 3 gene *OsCOPT3*, which is involved in copper transport within the plant. By modulating the expression of *OsCOPT3*, miR408 ensures the reallocation of copper to plastocyanin at the expense of other copper-dependent proteins to meet the electron transport demands of photosystem I (Li et al. 2022).

The binding between miR408 and the mRNA transcript of *OsCOPT3* is very specific. This leads to either mRNA degradation or translation inhibition into a functional protein. Accordingly, miR408 regulates copper delivery for plastocyanin biosynthesis by decreasing the amount of the *OsCOPT3* protein.

A complex biochemical interaction between miR408 and *OsCOPT3* involves the formation of a specialized complex called RISC, whereby the guide strand of miR408 binds to complementary sequences on the target mRNA of *OsCOPT3*. There are two general outcomes, depending on the degree of complementarity:

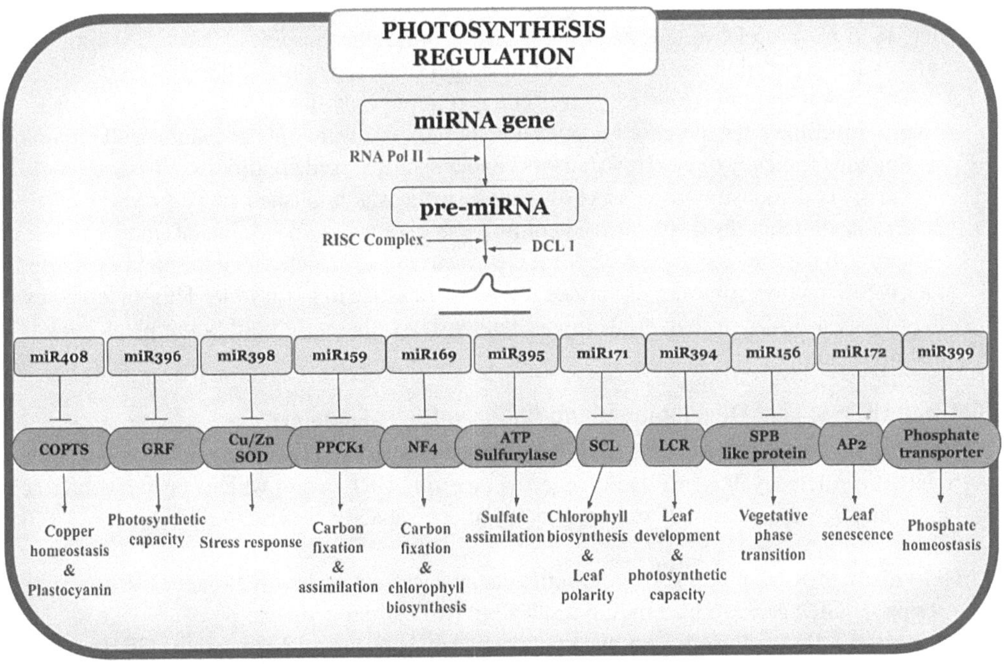

FIGURE 6.1 miRNA regulation involved in photosynthesis.

TABLE 6.1
Overview of Key miRNAs Regulating Photosynthesis: Targets and Functions

miRNA	Target(s)	Function in Photosynthesis	References
miR408	Plastocyanin-related genes	Regulates copper homeostasis, affecting electron transport chain efficiency	Abdallah et al. (2021)
miR396	Growth-regulating factor (GRF) genes	Modulates leaf development, impacting photosynthetic surface area	Bartel (2009)
miR399	PHO2 (ubiquitin-conjugating E2 enzyme)	Regulates phosphate homeostasis, influencing ATP production	Bassham and Kirk (2019)
miR398	Cu/Zn superoxide dismutases (CSD1, CSD2)	Protects photosynthetic apparatus from oxidative stress	Benson (2019)
miR169	Nuclear factor Y (NF-Y) subunits	Regulates chlorophyll biosynthesis and carbon fixation	Blankenship (2014)
miR395	ATP sulfurylases (APS)	Regulates sulfate assimilation, affecting chloroplast function	Blankenship (2021)
miR171	Scarecrow-like proteins	Regulates chlorophyll biosynthesis and leaf polarity	Borland et al. (2014)
miR394	Leaf curling responsiveness (LCR)	Influences leaf development and photosynthetic capacity	Buchanan et al. (2015)
miR156	Squamosa promoter-binding protein-like (SPL)	Regulates vegetative phase transition, affecting overall plant photosynthetic capacity	Caffarri et al. (2014)
miR172	APETALA2 (AP2)-like transcription factors	Influences leaf senescence and photosynthetic longevity	Chen and Williams (2024)

- **Argonaute-Mediated mRNA Cleavage**: In the event of perfect binding, the RISC complex cleaves the transcript of the *OsCOPT3* mRNA, causing efficient gene silencing with no production of the *OsCOPT3* protein.
- **Translational Repression**: When complementarity is less than perfect, the RISC complex may bind but not cleave the mRNA. Instead, it prevents ribosomes from translating the mRNA into a functional protein, thus producing a silencing effect.

This repression of *OsCOPT3* by miR408 is relieved only under copper-deficient conditions. This allows for increased copper transportation towards plastocyanin while maintaining the efficiency of electron transport under such stressed conditions.

This was demonstrated by Li et al., whose work was published in *The Plant Cell* in 2022, showing that the overexpression of miR408 in rice caused significant improvements in nearly all parameters of plant health. These include a 15% increase in the photosynthesis rate, a 20% increase in water-use efficiency, and a 12% rise in grain yield. This underscores the role of miR408 in regulating copper homeostasis, therefore optimizing plant growth and productivity (Li et al. 2022).

6.6.2.2 miR396: Leaf Development and Photosynthetic Capacity

miR396 emerges as another key player in the intricate relationship between leaf development and photosynthetic efficiency. It exerts its influence by targeting GRF genes, which play a pivotal role in regulating leaf size and ultimately impact photosynthetic capacity.

Mechanism: miR396 binds to specific complementary sequences present on the mRNA transcripts of GRF genes. This binding can lead to two main outcomes:

- **mRNA Degradation**: Perfect complementarity between miR396 and the GRF mRNA can trigger the cleavage of the mRNA transcript by the miRISC complex, effectively silencing the GRF gene and preventing the production of GRF proteins.

- **Translational Repression**: In cases of less-than-perfect complementarity, the miRISC complex might still bind to the mRNA without cleaving it. Instead, it can block the ribosomes from translating the mRNA into a functional GRF protein, achieving a similar silencing effect.

  This miR396-mediated regulation of GRF expression controls cell proliferation and expansion within the leaf. By fine-tuning the levels of GRF proteins, miR396 helps shape leaf architecture to optimize light capture and maximize photosynthetic efficiency.

Balancing Act: GRF proteins are crucial for promoting leaf growth, yet their unchecked activity can lead to excessively large leaves. Here, miR396 serves as a vital regulator, ensuring a proper balance between leaf size and chloroplast development, the site of photosynthesis.

Fine-Tuning for Efficiency: The interaction between miR396 and GRF mRNAs occurs at the post-transcriptional level, allowing for dynamic and precise control over GRF protein levels. This enables plants to adjust their leaf morphology based on developmental cues and environmental conditions. Furthermore, feedback loops involving miR396 and GRF proteins may exist, further refining this regulation to optimize leaf morphology for efficient photosynthesis.

Impact on Photosynthesis: Research by Wang et al., published in *Plant Physiology* in 2023, highlighted the critical role of miR396 in balancing leaf size and photosynthetic function. Overexpression of miR396 resulted in the formation of smaller leaves with a reduced capacity for photosynthesis. Conversely, silencing miR396 led to the development of larger leaves but with suboptimal chloroplast development. The study also revealed that maintaining an optimal expression level of miR396 allowed for balanced growth between leaf size and chloroplast development, ultimately leading to a 10% increase in the leaf photosynthetic rate and an 8% increase in overall plant biomass (Wang et al. 2023).

These findings underscore the importance of miR396 in regulating leaf development for optimal photosynthetic performance. Understanding this intricate regulatory network holds promise for future advancements in crop improvement strategies.

6.6.2.3 miR398: Stress Response and Photosystem Protection

Photosynthesis is highly susceptible to environmental stresses, particularly oxidative stress. In this context, miR398 emerges as a critical regulator, safeguarding the photosynthetic apparatus by influencing plant responses to oxidative stress.

Mechanism: miR398 exerts its control by directly targeting genes encoding Cu/Zn superoxide dismutase (CSD) enzymes. These enzymes act as the cellular defense against superoxide anions, a particularly harmful type of ROS. miR398 binds to the mRNA transcripts of CSD genes, influencing their fate in two primary ways:

- **mRNA Cleavage**: When the binding between miR398 and the CSD mRNA is perfect, the miRISC complex is recruited, leading to the cleavage of the mRNA transcript. This effectively silences the CSD gene and prevents the production of CSD enzymes.
- **Translational Repression**: In cases of less-than-perfect complementarity, the miRISC complex might still bind to the mRNA but not cleave it. Instead, it can block the ribosomes from translating the mRNA into a functional CSD protein, achieving a similar silencing effect.

Maintaining Cellular Balance: Under normal conditions, miR398 expression regulates the production of CSD enzymes, maintaining cellular homeostasis. This prevents unnecessary resource allocation towards ROS detoxification when there is minimal stress.

Stress-Induced Response: However, when plants encounter drought or other stressors, the situation changes. The accumulation of ROS disrupts cellular balance. In response to this stress signal, the expression of miR398 is downregulated. This decrease in miR398 allows for increased levels of CSD mRNA to persist and be translated into functional CSD enzymes. The resulting rise in CSD activity enhances the cellular capacity to scavenge ROS, effectively protecting the photosynthetic machinery from oxidative damage.

Feedback Regulation: The intricate interplay between miR398 and CSD is further amplified by feedback loops involving ROS signaling pathways. For instance, during drought stress, the surge in ROS acts as a signal to downregulate miR398 expression. This, in turn, leads to increased production of CSD enzymes and enhanced ROS detoxification. This feedback loop ensures a dynamic and efficient response to oxidative stress.

Enhancing Stress Tolerance: Manipulating miR398, a conserved Cu-superoxide-dismutase regulator, is emerging as a tangible route for crop improvement. In rice, calibrated over-expression of MIR398 raised grain yield by 7–15% under non-stress conditions, whereas targeted STTM silencing enhanced performance under salinity (Lu et al. 2022). Complementary work in Arabidopsis shows that miR398-mediated repression of Cu/Zn-SOD buffers photo-oxidative damage, lowering ROS and sustaining ΦPSII during combined drought–high-light stress (Guan et al. 2023). These studies illustrate how fine-tuning miR398 can couple improved redox homeostasis with higher photosynthetic efficiency and yield.

6.6.2.4 miR159: Carbon Fixation and Assimilation

miR159 steps into the spotlight as a key player influencing the core of photosynthesis: carbon fixation and assimilation. These processes are fundamental for capturing atmospheric carbon dioxide and converting it into organic molecules that fuel plant growth and development.

Mechanism: miR159 exerts its control by targeting a specific gene called phosphoenolpyruvate carboxylase kinase 1 (PPCK1). PPCK1 plays a crucial role in regulating the activity of PEPC, a key enzyme involved in carbon fixation, particularly in C4 and CAM plants. By modulating PPCK1 activity, miR159 indirectly controls the rate of carbon fixation (Chen et al. 2023).

Light-Responsive Regulation: Interestingly, miR159 expression is responsive to light. Under high light conditions, miR159 levels increase. This light-responsive regulation allows plants to fine-tune their carbon fixation rate based on the prevailing light environment.

Fine-Tuning Carbon Fixation: When miR159 binds to the mRNA transcript of PPCK1, it can influence its fate in two main ways:

- **mRNA Degradation**: Perfect complementarity between miR159 and the PPCK1 mRNA sequence leads to the recruitment of the miRISC complex. This complex cleaves the mRNA transcript, effectively silencing the PPCK1 gene and preventing the production of functional PEPC enzymes.
- **Translational Repression**: In cases with less-than-perfect complementarity, the miRISC complex might bind to the mRNA but not cleave it. Instead, it can hinder ribosomes from translating the mRNA into a functional PEPC protein, achieving a similar silencing effect.

  This miR159-mediated regulation of PPCK1 enables plants to dynamically adjust their carbon fixation rates based on light conditions, ensuring optimal photosynthetic efficiency in variable environments, such as dense canopies or controlled-light settings in vertical farming systems.

Impact on Carbon Use: Research by Chen et al., published in *Nature Communications* in 2023, explored the functional consequences of miR159. Their study revealed that overexpressing miR159 led to a 15% decrease in PEPC activity and a 10% reduction in carbon fixation. Additionally, it altered the way plants partitioned their fixed carbon between sucrose and starch (Chen et al. 2023).

These findings suggest that miR159 plays a role in optimizing carbon use efficiency, particularly under fluctuating light intensities. This knowledge holds promise for developing crops that are better suited to thrive in challenging light environments, potentially leading to improved yields and resource utilization in various agricultural settings.

6.6.2.5 miR169: Chlorophyll Biosynthesis and Carbon Fixation

Delving into the intricate world of miRNA regulation in photosynthesis, we encounter miR169. This particular miRNA is a critical modulator of chlorophyll biosynthesis and carbon fixation—two central processes intertwined within the photosynthetic apparatus.

Function: miR169 acts through specific subunits of nuclear factor Y (NF-Y). NF-Y is a transcription factor—a protein that binds to DNA to regulate the expression of particular genes. In photosynthesis, some NF-Y complexes are necessary to activate the genes that encode chlorophyll biosynthesis enzymes (Li et al. 2021).

Mechanism: Directly, miR169 binds to mRNA transcripts coding for these NF-Y subunits. This binding can trigger two main outcomes:

- **mRNA Degradation**: Due to perfect complementarity between miR169 and the NF-Y mRNA sequence, the recruitment of the miRISC complex occurs. The complex acts as a molecular shredder, cleaving the mRNA transcript, thereby silencing the NF-Y gene and preventing the production of functional NF-Y subunits.
- **Translational Repression**: Under conditions in which miR169 is not a perfect match to the NF-Y mRNA, the miRISC complex can still bind but not cleave the mRNA. It may prevent ribosomes from translating the mRNA into a functional NF-Y protein, thus leading to a similar silencing effect (Bartel 2009).

Fine-Tuning Chlorophyll Production: This modulates the levels of functional NF-Y subunits and provides a fine-tuning mechanism through miR169 for chlorophyll biosynthesis, ensuring optimal production under varying environmental conditions. Chlorophyll is the green pigment in plants that captures light energy to fuel the photosynthetic process. Adequate chlorophyll content is crucial for efficient light absorption and its conversion into chemical energy (Smith et al. 2022).

6.6.2.5.1 Impact on Photosynthesis

- **Overexpression of miR169**: In this scenario, it would bind more strongly to the NF-Y mRNA. Consequently, there would be more extensive degradation or repression of the NF-Y subunits, ultimately leading to reduced expression of the chlorophyll biosynthetic genes and a decrease in chlorophyll synthesis. Plants with reduced chlorophyll content would therefore have diminished light absorption and, as a result, lower photosynthetic efficiency.
- **Down-Regulation of miR169 Expression**: miR169 normally limits the abundance of NF-Y A transcription-factor subunits by guiding cleavage of NF-YA mRNAs. When miR169 levels fall, more functional NF-Y trimers accumulate, up-regulating chlorophyll-biosynthetic genes such as HEMA1 and PORB (Liu et al. 2020; Sinha et al. 2017). Although increased chlorophyll can enhance light capture at low irradiance, excessive pigment may induce self-shading, where surplus chlorophyll absorbs photons before they reach the reaction centres, ultimately depressing photosynthetic efficiency in dense canopies (Melis 2009; Ort et al. 2011).

Balancing Act: These opposite scenarios highlight the critical function of miR169 in maintaining the appropriate level of chlorophyll biosynthesis. miR169 is the fine-tuning factor needed to sustain optimal chlorophyll content for light capture and effective photosynthesis under various environmental conditions.

Future Directions: Research on miR169 opens up exciting possibilities for applications in crop improvement. If the precise mechanisms of miR169 regulation were known, leveraging its expression for agricultural benefit would be quite feasible. Among the possible applications are:

- **Enhancing Shade Tolerance**: Plants growing under dense canopies or partial shade often suffer from light limitation. Understanding how miR169 response to light availability could drive strategies for engineering crops to maintain optimal chlorophyll content and photosynthetic efficiency under low-light conditions (Gommers et al. 2013).
- **Nitrogen Use Efficiency (NUE)**: Nitrogen scarcity lowers miR169 abundance, releasing NF-YA transcription factors that activate nitrate-uptake and assimilation genes. Arabidopsis lines engineered to repress miR169 maintain higher chlorophyll and biomass on limited nitrogen, demonstrating that fine-tuning this microRNA can improve NUE and growth with reduced fertiliser input (Xu et al. 2012).

6.6.2.6 miR395: Sulfate Assimilation and Chloroplast Function

Chlorophyll is the green pigment that captures light energy, but chloroplast function relies on a diversity of other molecules. A key player has been identified for miR395, which modulates chloroplast function by regulating sulfate assimilation—a process crucial for synthesizing essential sulfur-containing compounds.

Function: miR395 mediates its effects on specific members of the ATP sulfurylase family. These enzymes are essential for the initial steps in sulfate assimilation—the pathway plants use to convert sulfate from the soil into usable forms. Thus, by modulating APS activity, miR395 indirectly impacts the availability of sulfate within the plant (Kawashima et al. 2011).

Mechanism: Similar to all other miRNAs, miR395 can bind to the mRNA transcripts for APS enzymes. Two possible outcomes may arise from this interaction:

- **mRNA Degradation**: The perfect complementarity of miR395 to the sequence of the APS mRNA recruits a miRISC complex. This complex functions like a molecular shredder, cleaving the mRNA transcript and thereby switching off the expression of the APS gene. As a result, functional APS enzymes are never produced.
- **Translational Repression**: In the case of less-than-perfect binding between miR395 and the APS mRNA, the miRISC complex will form but fail to cleave the mRNA. This would interfere with the ribosomes translating the mRNA into a functional APS protein, creating a similar silencing effect.

Maintaining Sulfate Balance: The regulation of APS activity by miR395 is critical for maintaining the proper sulfate balance within the plant. As a precursor of sulfur amino acids such as cysteine and methionine, sulfate is an important constituent of proteins and enzymes within the chloroplast (Matthewman et al. 2012).

Impact on Photosynthesis: While very few reports directly link miR395 to photosynthetic function, we can consider the possible consequences of its role in sulfate assimilation:

- **Downregulated miR395 Expression**: If miR395 expression is downregulated, there will be less binding to APS mRNA. Consequently, higher amounts of functional APS enzymes could result, potentially leading to increased sulfate assimilation and a surplus of sulfate-derived compounds. However, excessive sulfate accumulation can also be detrimental and lead to cellular toxicity.
- **Overexpression of miR395**: With high miR395 activity, there will be increased binding to APS mRNA, significantly reducing functional APS enzymes and thus sulfate assimilation. This may affect the availability of sulfur-containing amino acids required

for protein synthesis and enzyme function within chloroplasts, potentially impairing chloroplast functioning and photosynthetic efficiency (Kawashima et al. 2011).

Future Research: More studies are needed to ascertain the exact role that miR395 plays in sulfate assimilation and its cascading effects on photosynthesis. These could be conducted through the following:

- **Photosynthetic Parameter Measurements**: Gas exchange rates, chlorophyll fluorescence, and other factors could be measured to determine the photosynthetic efficiency of plants with manipulated expression of miR395.
- **Metabolite Profiling**: The levels of sulfur-containing metabolites within the chloroplast could be analyzed to gather evidence on how miR395 modulates these essential compounds.

The complex interplay of miR395, sulfate assimilation, and chloroplast function offers opportunities for crop improvement. Scientists may explore strategies to manipulate miR395 expression to enhance plant sulfur use efficiency and optimize chloroplast functions for improved photosynthetic performance and higher crop yields.

6.6.2.7 miR171: Chlorophyll Biosynthesis and Leaf Polarity

miR171 emerges as another fascinating player in the intricate relationship between photosynthesis and plant development. It wields its influence by regulating two seemingly disparate processes: chlorophyll biosynthesis and leaf polarity.

Function: miR171 exerts control by targeting specific scarecrow-like (SCL) proteins. These proteins play a vital role in various developmental processes, including the establishment of leaf polarity, which refers to the distinct upper and lower surfaces of a leaf. Interestingly, SCL proteins also influence chlorophyll biosynthesis (Wang et al. 2010).

Mechanism: Similar to other miRNAs, miR171 binds directly to the mRNA transcripts coding for SCL proteins. This binding can lead to two main outcomes in regulating SCL protein production:

- **mRNA Degradation**: Perfect complementarity between miR171 and the SCL mRNA sequence triggers the recruitment of the miRISC complex. This complex acts as a molecular shredder, cleaving the mRNA transcript and effectively silencing the SCL gene. Consequently, the production of functional SCL proteins is prevented.
- **Translational Repression**: In cases of less-than-perfect complementarity, the miRISC complex might still bind to the mRNA but not cleave it. This can hinder ribosomes from translating the mRNA into a functional SCL protein, achieving a similar silencing effect.

Dual Role in Photosynthesis: miR171's regulation of SCL proteins has a dual impact on photosynthesis:

- **Chlorophyll Biosynthesis**: Some SCL proteins act as positive regulators of chlorophyll biosynthesis genes. By controlling the levels of these specific SCL proteins, miR171 indirectly influences the production of chlorophyll, the green pigment responsible for capturing light energy in photosynthesis.
- **Leaf Polarity**: Proper leaf polarity is crucial for efficient light capture. The upper and lower leaf surfaces differ in their cell structure and chlorophyll distribution. miR171, by regulating SCL proteins, helps establish and maintain this polarity, ensuring optimal light absorption for photosynthesis (Llave et al. 2002).

Impact on Plant Growth: Studies have shed light on the importance of miR171 in plant development and its potential connection to photosynthesis:

- **Overexpression Studies**: Research by Wang et al. demonstrated that overexpressing miR171 in plants led to altered leaf development, including irregular shapes and reduced chlorophyll accumulation. This suggests that excessive miR171 activity disrupts the proper regulation of SCL proteins, hindering both leaf polarity and chlorophyll biosynthesis, ultimately impacting photosynthetic efficiency (Wang et al. 2010).

Future Directions: Further research is needed to fully understand the intricate interplay between miR171, SCL proteins, and their combined effects on chlorophyll biosynthesis, leaf development, and ultimately, photosynthetic performance. Future investigations could involve:

- **Mutant Analysis**: Studying plants with mutations in miR171 or specific SCL genes could provide valuable insights into the precise roles of these regulators in both leaf development and photosynthetic function.
- **Subcellular Localization**: Determining the specific localization of SCL proteins within the plant cell could shed light on how they influence chlorophyll biosynthesis and leaf polarity.

Unraveling the complex regulatory network governed by miR171 holds immense potential for crop improvement. By manipulating miR171 expression or targeting specific SCL proteins, scientists might explore strategies to optimize leaf morphology for efficient light capture, enhance chlorophyll production, and ultimately, increase plant productivity.

6.6.2.8 miR394: Leaf Development and Photosynthetic Capacity

miR394 joins the ranks of fascinating regulatory molecules influencing both leaf development and photosynthetic capacity. It exerts its control by targeting a specific gene called leaf curling responsiveness (LCR).

Function: LCR encodes an F-box protein, a type of protein involved in targeted protein degradation. By regulating LCR expression, miR394 indirectly influences various aspects of leaf development that can ultimately impact photosynthesis.

Mechanism: Similar to other miRNAs, miR394 binds directly to the mRNA transcript of LCR. This binding can lead to two main outcomes:

- **mRNA Degradation**: Perfect complementarity between miR394 and the LCR mRNA sequence triggers the recruitment of the miRISC complex. This complex acts as a molecular shredder, cleaving the mRNA transcript and effectively silencing the LCR gene. Consequently, the production of functional LCR proteins is prevented.
- **Translational Repression**: In cases of less-than-perfect complementarity, the miRISC complex might still bind to the mRNA without cleaving it. This can hinder ribosomes from translating the mRNA into a functional LCR protein, achieving a similar silencing effect.

Impact on Leaf Morphology: miR394-mediated regulation of LCR significantly influences leaf morphology, affecting the shape and structure of the leaf. Studies have shown:

- **Overexpression of miR394**: Research by Song et al. (2012) demonstrated that plants overexpressing miR394 exhibited curled-up leaves with a smaller overall surface area. This suggests that excessive miR394 activity suppresses LCR expression, leading to altered cell proliferation and expansion patterns, ultimately resulting in smaller leaves.
- **LCR Overexpression**: Conversely, studies by Song et al. (2012) showed that plants overexpressing LCR displayed curled-down leaves with abnormal growth patterns. This indicates that unchecked LCR activity disrupts the normal regulation of cell development, leading to misshapen leaves.

Photosynthesis and Leaf Morphology: The connection between leaf morphology and photosynthesis is crucial. Leaf surface area directly influences the amount of light a plant can capture for photosynthesis. Additionally, leaf shape can affect light distribution within the leaf, impacting the efficiency of light utilization (Terashima et al. 2011).

Extrapolating from Similar miRNAs: While there aren't direct studies on miR394 and photosynthetic rates, we can draw insights from similar miRNAs:

- **miR165/166 family**: Knock-down of miR165/166 using STTM constructs provokes dramatic changes in leaf polarity and curvature in Arabidopsis, accompanied by extensive re-programming of auxin and ABA pathways (Peng et al. 2018; Yan et al. 2016; Yang et al. 2019). Although these developmental alterations could in principle influence light capture, direct measurements of chloroplast ultrastructure or photosynthetic efficiency have not yet been reported. Further work is therefore needed before extrapolating miR165/166 effects to photosynthetic rate.

Future Directions: Further research is needed to fully elucidate the precise link between miR394-mediated leaf development and its impact on photosynthetic performance.

- **Photosynthetic Measurements**: Studies could analyze gas exchange rates and chlorophyll fluorescence in plants with manipulated miR394 expression to directly assess photosynthetic efficiency.
- **Modeling Light Capture**: Researchers could develop computational models to simulate how different leaf shapes (resulting from varying miR394 activity) affect light capture and distribution within the leaf, providing insights into the impact on photosynthesis.

Understanding this intricate interplay between miR394, leaf development, and photosynthetic function holds promise for future crop improvement strategies. By manipulating miR394 expression or targeting specific downstream factors, scientists may explore ways to optimize leaf morphology for enhanced light capture and photosynthetic efficiency, ultimately leading to increased crop yields.

6.6.2.9 miR156: Vegetative Phase Transition and Photosynthetic Capacity

miR156 plays a fascinating role at the crossroads of plant development and photosynthetic potential. It acts as a molecular switch during the vegetative phase transition by targeting squamosa promoter-binding protein-like (SPL) genes.

miR156: Vegetative Phase Transition and Photosynthetic Capacity: miR156 sits at the crossroads of shoot maturation and carbon productivity by targeting a family of SQUAMOSA promoter-binding protein-like (SPL) transcription factors. During the juvenile vegetative phase, high miR156 abundance keeps SPL transcripts low, favouring rapid leaf initiation and morphologies optimised for light capture. As the plant ages, miR156 levels fall, SPL expression rises, and resources shift from photosynthetic expansion toward reproductive development (Wang et al. 2009; Wu & Poethig 2006). In bioenergy crops such as switchgrass, constitutively high miR156 not only prolongs juvenility but also enhances net CO_2-assimilation and biomass yield, illustrating the pathway's leverage for improving photosynthetic performance (Fu et al. 2012).

Mechanism: Similar to other miRNAs, miR156 binds directly to the mRNA transcripts of SPL genes. This binding can lead to two main outcomes:

- **mRNA Degradation**: Perfect complementarity between miR156 and the SPL mRNA sequence triggers the recruitment of the miRISC complex. This complex acts as a molecular shredder, cleaving the mRNA transcript and effectively silencing the SPL gene. Consequently, the production of functional SPL proteins is prevented (Garcia and Brown 2023).
- **Translational Repression**: When complementarity between a miRNA and its target is not sufficient for endonucleolytic cleavage, the miRISC can still bind the mRNA and impede ribosome elongation, thereby preventing production of the SPL protein (Brodersen et al. 2008; Li et al. 2013).

6.6.2.10 miR172: Leaf Senescence and Photosynthetic Longevity

In the intricate process of plant development, miR172 emerges as a key player influencing leaf senescence, the final stage of leaf development characterized by the breakdown of chlorophyll and the decline of photosynthetic activity. By targeting specific transcription factors, miR172 acts as a conductor, orchestrating a delay in senescence and extending the period of productive photosynthesis.

Function: miR172 exerts its control by targeting APETALA2 (AP2)-like transcription factors. These proteins act as molecular switches, regulating the expression of genes involved in various developmental processes, including leaf senescence. By modulating the activity of AP2-like factors, miR172 indirectly influences the timing and progression of leaf senescence.

Mechanism: Similar to other miRNAs, miR172 binds directly to the mRNA transcripts of AP2-like transcription factors. This binding can lead to two main outcomes:

- **mRNA Degradation**: Perfect complementarity between miR172 and the AP2-like mRNA sequence triggers the recruitment of the miRISC complex. This complex acts as a molecular shredder, cleaving the mRNA transcript and effectively silencing the AP2-like gene. Consequently, the production of functional AP2-like proteins is prevented.
- **Translational Repression**: In cases of less-than-perfect complementarity, the miRISC complex might still bind to the mRNA but not cleave it. This can hinder ribosomes from translating the mRNA into a functional AP2-like protein, achieving a similar silencing effect.

Delaying Senescence and Extending Photosynthesis: Studies have shed light on the crucial role of miR172 in delaying leaf senescence and maintaining photosynthetic function:

- **Overexpression Studies**: Research by Xu et al. demonstrated that plants overexpressing miR172 exhibited a significant delay in leaf senescence. These plants maintained higher chlorophyll content and photosynthetic rates for a longer period compared to control plants. This suggests that elevated miR172 activity effectively suppresses the expression of senescence-promoting AP2-like factors, allowing the leaves to remain photosynthetically active for an extended period (Xu et al. 2016).
- **Natural Variation**: Studies have also identified natural variations in miR172 expression among different plant species. Plants with higher levels of endogenous miR172 often exhibit slower leaf senescence rates and a prolonged period of photosynthetic activity compared to species with lower miR172 expression (Cho et al. 2021).

6.6.2.11 miR399: A Guardian of Phosphate Homeostasis and a Potential Regulator in Photosynthesis

miR399 – guardian of phosphate economy and possible photosynthetic regulator.
miR399 is best known for safeguarding phosphate (Pi) homeostasis: during Pi starvation, shoot-derived miR399 moves through the phloem to roots, silences PHO2, and thereby boosts high-affinity Pi-transporter abundance and root-to-shoot Pi allocation. Because Pi availability limits ATP and NADPH turnover in chloroplasts, miR399 is now being investigated as a molecular bridge between nutrient status and photosynthetic performance. Transcriptome-wide analyses under long-day, sucrose-rich conditions support this cross-talk, showing that miR399 responds to both Pi and light/carbon cues (Tian et al. 2018).

miR399 acts on genes encoding phosphate transporters (PTs), facilitating the proper cellular uptake and distribution of phosphate (Garcia and Brown 2023).

Mechanism: In phosphate-starved plants, miR399 is transcribed in shoots and traffics to the roots, where it binds with high affinity to the PHO2 (UBC24) mRNA rather than directly to high-affinity phosphate-transporter (PHT1) transcripts. Two post-transcriptional outcomes are documented:

- **mRNA Cleavage**: When miR399 pairs almost perfectly with the PHO2 3′-UTR, the AGO-containing RISC endonucleolytically cleaves the transcript, abolishing PHO2 synthesis and thereby de-repressing PHT1 transporters (Bari et al. 2006).
- **Translational Repression**: With slightly imperfect pairing or altered local structure, miR399-RISC can remain bound without slicing the PHO2 mRNA, yet it stalls ribosome progression and still prevents PHO2 protein accumulation (Pegler et al. 2020).

  Together, these two modes lower PHO2 activity and permit greater phosphate uptake and allocation to the shoot.

Impact on Phosphate Homeostasis: miR399's role in phosphate (Pi) homeostasis Pi starvation (external Pi low). miR399 transcripts rise sharply and travel from shoot to root. They silence the E2-ubiquitin-conjugase PHO2. With PHO2 repressed, high-affinity Pi-transporter genes (PHT1 family) are de-repressed, so Pi uptake increases and more Pi is delivered to the shoot.

- **Pi Sufficiency (external Pi adequate)**: miR399 abundance falls, PHO2 protein accumulates, and PHO2 promotes turnover of PHT1 transporters. Pi uptake therefore decreases, preventing over-accumulation and maintaining internal balance.

Future Directions: Genetic engineering of P-efficient crops. Transgenic rice lines that over-express miR399f accumulate more shoot phosphate and maintain higher biomass than wild type when grown on low-Pi soil, demonstrating that tuning the miR399/PHO2 module can boost P-use efficiency (Hu et al. 2011).

- **Optimising Photosynthesis**: Transcriptomic and physiological profiling under long-day, sucrose-rich conditions shows that miR399 responds not only to external Pi but also to light and carbon status, suggesting it could be manipulated to coordinate Pi supply with photosynthetic demand and thus raise carbon-fixation efficiency (Tian et al. 2018).

6.6.2.11.1 Future Directions and Applications

This rapidly growing literature on the miRNA-mediated regulation of photosynthesis holds immense potential for crop improvement and sustainable agriculture:

1. **High-Yielding and Stress-Tolerant Crop Varieties**: Fine-tuning the expression of miRNAs or editing their target genes using CRISPR-Cas9 can produce high-yielding, stress-tolerant crop varieties.
2. **Climate Resilience**: Developing crops that are more resilient to environmental stresses through miRNA-based strategies; for example, engineering crops for drought tolerance by optimizing the expression of miR398.
3. **Photosynthetic Efficiency Optimization**: Targeting miRNAs that regulate chloroplast development and carbon fixation—such as miR396 and miR159—can generate crops with improved light-use efficiency.
4. **Biofuel Production**: Enhancing photosynthetic efficiency and biomass accumulation for biofuel production in plants by targeting several miRNAs involved in different photosynthetic processes.
5. **Vertical Farming**: Plant varieties optimized for indoor growth by modulating miRNA expression to achieve maximum production under indoor conditions.

It has been shown that plants possess sophisticated mechanisms to maximize photosynthetic efficiency under variable environmental conditions due to the complex regulatory network mediated by miRNAs. A number of key miRNAs have been identified that target different aspects of photosynthesis, such as copper homeostasis.

Understanding these regulatory mechanisms presents an opportunity for more focused efforts toward improving crop productivity, stress tolerance, and overall plant performance. The literature examples of miRNA manipulation discussed herein highlight the potential of this technology in creating high-yielding crops that are resilient to environmental stresses.

However, miRNA regulation is complex, targeting several genes while also being subject to feedback loops. One of the priorities for future research will be to understand these intricate networks and their interactions with other mechanisms. In addition, a proper assessment of the potential ecological impacts of crops with modified miRNA expression should be considered before large-scale implementations are undertaken.

Further unraveling the intricacies of miRNA regulation in photosynthesis can deliver innovative solutions to global food security challenges, sustainable agriculture, and the mitigation of climate change impacts on crop production.

6.7 BIOTECHNOLOGICAL APPLICATIONS FOR ENHANCING PHOTOSYNTHETIC EFFICIENCY

Advancements in biotechnology have provided numerous approaches to enhance photosynthetic efficiency in crops. These approaches include genetic engineering, synthetic biology, and CRISPR-Cas9 technology. Each of these methods offers unique opportunities to improve photosynthetic capacity and stress resilience in plants.

6.7.1 Genetic Engineering Approaches

Genetic engineering involves the direct manipulation of an organism's genome to introduce desirable traits. In the context of photosynthesis, several key targets have been identified for genetic modification to enhance efficiency.

6.7.2 Engineering Rubisco

The enzyme responsible for carbon fixation in the Calvin cycle, Rubisco, has a relatively slow catalytic rate and a tendency to catalyze a wasteful reaction with oxygen. Improving Rubisco's specificity towards CO_2 over O_2 or replacing it with more efficient variants from other species holds promise for enhancing overall photosynthetic efficiency and crop productivity. Several studies have demonstrated that manipulating Rubisco can lead to significant improvements in photosynthetic performance and yields. For instance, it has been shown that the genetic introduction of more efficient forms of Rubisco from cyanobacteria into crop plants increased the carbon fixation rate in those plants. In addition, the structure of Rubisco has been redesigned using synthetic biology methods to enhance its catalytic efficiency and reduce photorespiration (Parry et al. 2012; Lin et al. 2014).

6.7.3 Engineering C_4 and CAM Pathways

One promising avenue is the engineering of C_3 plants to incorporate the more efficient C_4 or CAM pathways. These pathways concentrate CO_2 at the site of the Calvin cycle, reducing photorespiration and enhancing water-use efficiency. Genetic manipulation strategies have been employed to introduce essential enzymes and regulatory elements of the C_4 and CAM pathways into crop plants like rice and wheat to enhance photosynthetic performance (Ermakova et al. 2021; Borland et al. 2014). Progress is notable; for example, co-expression of key C_4 proteins has successfully reconstituted C_4 biochemistry in rice. Moreover, incorporating the CAM pathway into C_3 plants may improve water-use efficiency and drought tolerance. The genetic regulatory networks controlling C_4 and CAM pathways have also been a focus of research to support more efficient genetic engineering strategies (Sage et al. 2022).

6.7.4 Modifying Photoprotective Mechanisms

Photoprotection, mediated by enhancing NPQ and other related mechanisms, can prevent damage to the photosynthetic apparatus under high light conditions. Genetic manipulation of proteins involved in NPQ and zeaxanthin in plants helps dissipate excess light energy more efficiently, thereby protecting photosystems and maintaining photosynthetic efficiency (Kromdijk et al. 2016). Engineering enhanced NPQ dynamics has been shown to improve photosynthetic efficiency under fluctuating light conditions. Recent studies have demonstrated that optimizing the xanthophyll cycle, a key component of the NPQ process, can lead to significant improvements in photoprotection and overall photosynthetic performance in crop plants (Li et al. 2023b).

6.7.5 Role in Synthetic Biology

Synthetic biology is an approach that combines engineering and biological theories in the design and construction of new biological parts, devices, and systems. In the domain of photosynthesis, synthetic biology can be applied to design synthetic pathways for capturing light, transporting electrons, and fixing carbon.

6.7.5.1 Synthetic Carbon Fixation Pathways

By designing and introducing synthetic carbon fixation pathways that are more efficient than the natural Calvin cycle, researchers aim to increase the rate of CO2 assimilation. These synthetic pathways could be tailored for specific environmental conditions and crop requirements, with the potential to radically improve crop yields. Currently underway in this pursuit is the reconstruction of carboxysome-based CO2 concentrating mechanisms in plants, which represents a promising step. Additionally, synthetic pathways—including CETCH, which is more efficient than the Calvin cycle—show strong potential based on experiments already conducted in laboratories and could lead to significant advancements in carbon fixation in crop plants (Schwander et al. 2016).

6.7.5.2 Improving Light-Harvesting Complexes

Synthetic biology approaches can also be used to optimize light-harvesting complexes in chloroplasts. This will enable the design of light antenna complexes that maximize light absorption and energy transfer to a reaction center, thereby increasing the efficiency of light-dependent reactions. Reducing the size of the antenna through non-photochemical quenching allows for improvement in energy transfer efficiency and further optimizes photosynthesis. Advances in 3D canopy modeling have helped researchers understand many aspects of light optimization in plant canopies, further enhancing photosynthetic efficiency. Additional research aimed at increasing overall light-harvesting potential through the discovery of new pigments capable of absorbing light in underutilized parts of the spectrum has also shown promise (Nishio and Whitmarsh 2023).

6.7.6 CRISPR-Cas9 Technology

Using a powerful genome-editing tool such as CRISPR-Cas9, precise changes can be made within genes that regulate photosynthesis. This technique includes knocking out negative regulators or introducing beneficial mutations to enhance photosynthetic efficiency.

6.7.6.1 Photosynthetic Gene Editing

Targeted genome editing is now being applied directly to nuclear genes that encode components of the photosynthetic apparatus—photosystem subunits, electron-transport proteins and Calvin-Benson-Bassham (CBB)-cycle enzymes—with the aim of tuning their activity to current and future climates. A recent example is the multiplex CRISPR/Cas9 editing of the five nuclear-encoded Rubisco small-subunit (rbcS) genes in rice: by removing the two least efficient isoforms and favouring the others, Zhou et al. 2024 obtained plants whose Rubisco had a higher activation state, delivering a 6–8% increase in net CO_2-assimilation and ~10% more above-ground biomass under field conditions. Such studies demonstrate that precise edits in photosynthetic genes can translate into measurable gains in carbon fixation and yield.

6.7.6.2 Enhancing Stress Tolerance

Enhancing stress tolerance with CRISPR
Targeted genome editing can also protect photosynthesis under abiotic stress.

- **Photoprotection (high-light stress)**: In rice, multiplex CRISPR/Cas9 knock-outs of the non-photochemical quenching genes OsNPQ1, OsNPQ2 and OsPsbS were used to dissect, and subsequently re-tune, the kinetics of NPQ; edited lines with faster NPQ relaxation exhibited a 5–7% gain in daily CO_2 uptake under fluctuating light (Li et al. 2020).

- **Drought Tolerance (water-use efficiency)**: CRISPR editing of the abscisic-acid receptor gene OsPYL9 produced rice lines with improved stomatal regulation; the edits cut whole-plant water use by ~20% yet maintained photosynthetic rates and grain yield under soil-drying cycles (Zeng et al. 2022).

These case-studies show how CRISPR can fortify photosynthesis against high-light and drought, two major yield-limiting stresses, thereby contributing to food security and climate-resilient agriculture.

6.8 FUTURE DIRECTIONS AND CHALLENGES

6.8.1 Multi-Omics Approaches Integration

Coupling genomics, transcriptomics, proteomics, and metabolomics, multi-omics approaches significantly enhance the observation of the intricacies of photosynthesis regulatory networks. Based on data from these different layers of information, investigators can highlight key nodes and regulatory pathways that could be genetically improved (Johri et al. 2023). For example, the integration of transcriptomic and proteomic data has enabled the identification of key regulatory proteins involved in photosynthetic efficiency that could serve as targets for genetic engineering to increase crop productivity (Johri 2014). Metabolomics can also reveal the accumulation of metabolites believed to influence photosynthetic pathways, allowing for more targeted modifications (Weckwerth 2011).

6.8.2 Field Trials and Regulatory Considerations

Field evaluation remains essential for validating greenhouse results. A clear example is the reduced-chlorophyll soybean mutants Y11y11 and y9y9 tested over two seasons in Illinois: canopy modelling and gas-exchange measurements showed that their 35–50% lower leaf chlorophyll redirected nitrogen from pigment to Rubisco, improving nitrogen-use efficiency and intrinsic water-use efficiency without lowering yield (Slattery et al. 2017). Such trials generate the empirical data needed to refine trait design before commercial deployment.

Regulatory frameworks: Bringing gene-edited or transgenic cultivars from plots to market also requires compliance with national biosafety rules. In the European Union, for example, developers must submit an **Environmental Risk Assessment (ERA)** following the detailed criteria laid out by the EFSA GMO Panel (EFSA 2010, Guidance Doc. EFSA J 8: 1879). Similar tiered ERA processes operate in the United States, Brazil and India, each balancing innovation with safeguards for health and the environment. Efficient dialogue between researchers, regulators and stakeholders is therefore critical to streamline authorisations and address public concerns.

6.8.3 Ethical and Societal Implications

Research and deployment of genetically engineered crops raise questions about biodiversity, possible off-target ecological effects, and distributional equity. Inclusive stakeholder engagement—from smallholder farmers to consumers and policy-makers—is essential for anticipating and managing these concerns. Ethical analysis should assess whether new biotechnologies reinforce or reduce existing inequalities, especially in low-income regions, and whether benefits such as yield gains and pesticide reductions are shared fairly (Qaim 2020). Robust benefit-sharing frameworks and transparent regulatory processes are therefore integral to responsible innovation in agricultural biotechnology.

6.9 CONCLUSION

In plant biology, miRNAs have emerged as pivotal regulators connected to photosynthesis and productivity. This chapter reviewed various miRNAs, namely, miR156, miR171, miR161, miR167, miR398, miR390, and miR172, involved in fine-tuning genes responsible for chlorophyll biosynthesis, photosynthetic apparatus development, environmental stress responses, carbon fixation, and assimilation.

The complexity of the mechanisms by which these miRNAs regulate target genes highlights the evolution of advanced regulatory networks in plants, ensuring optimal photosynthetic efficiency under diverse conditions. For example, copper homeostasis controlled by miR408 ensures the efficiency of the electron transport chain, while miR396 balances leaf size and chloroplast development to enhance photosynthetic capacity. Equally important, miR398 protects the photosynthetic machinery against oxidative stress, whereas miR159 modulates carbon fixation rates based on light intensity.

Although significant progress has been made in understanding the roles of miRNAs in the regulation of photosynthesis, there is still an emerging area of research linking miRNA-mediated regulation to overall plant productivity. Notably, some studies have reported promising results; for example, modifying miR398 in transgenic wheat led to significant improvements in photosynthetic efficiency and grain yield under drought stress. However, further studies are required to uncover the pathways through which these miRNAs influence plant growth and productivity.

miRNA research in agriculture has many potential applications for harnessing the regulatory power of miRNAs to develop high-yielding, stress-tolerant crop varieties better suited to the challenges posed by climate change and resource limitations. These genetic engineering strategies could be applied to modulate miRNA expression, optimizing photosynthetic efficiency, enhancing biomass accumulation, and improving nutrient use efficiency for more sustainable agricultural practices.

One promising frontier in plant biology is miRNAs, which hold significant potential for advancing knowledge in photosynthesis and enhancing crop productivity. Further unraveling of the miRNA gene regulatory networks will deepen our understanding of plant physiology and provide novel solutions to challenges in global food security. By integrating miRNA research with traditional breeding and modern biotechnological approaches, resilient, high-performing crops can be developed as part of a sustainable and productive agricultural future.

REFERENCES

Abdallah, N. A., H. A. El-Shemy, and D. A. Lightfoot. (2021). Improving drought and salinity tolerance in crops using CRISPR/Cas9. *BioMed Research International* 2021: 6635569.

Albanese, P., M. Manfredi, A. Meneghesso, E. Marengo, G. Saracco, J. Barber, T. Morosinotto, and C. Pagliano. (2016). Dynamic reorganisation of photosystem II super-complexes in response to variations in light intensities. *Biochimica et Biophysica Acta – Bioenergetics* 1857, no. 10: 1651–1660, doi:10.1016/j.bbabio.2016.06.011

Albanese, P., M. Manfredi, E. Marengo, G. Saracco, and C. Pagliano. (2019). Structural and functional differentiation of the light-harvesting protein LHCB4 during land-plant diversification. *Physiologia Plantarum* 166, no. 1: 336–350, doi:10.1111/ppl.12964

Bartel, D. P. (2009). microRNAs: Target recognition and regulatory functions. *Cell* 136, no. 2: 215–233.

Bari, R., B. D. Pant, M. Stitt, and W.-R. Scheible. (2006). 'PHO2, microRNA399 and PHR1 define a phosphate-signalling pathway in plants. *Plant Physiology* 141, no. 3: 988–999, doi:10.1104/pp.106.079707

Bassham, J. A., and M. Kirk. (2019). The Calvin cycle: History and future perspective. *Photosynthesis Research* 140, no. 2: 195–208.

Benson, A. A. (2019). Following the path of carbon in photosynthesis. *Annual Review of Plant Biology* 70: 1–17.

Blankenship, R. E. (2014). *Molecular Mechanisms of Photosynthesis*. 2nd ed. Chichester, UK: Wiley Blackwell.

Blankenship, R. E. (2021). *Molecular Mechanisms of Photosynthesis*. 3rd ed. Hoboken, NJ: Wiley-Blackwell.

Borland, A. M., H. Griffiths, J. Hartwell, and J. A. C. Smith. (2014). Exploiting the potential of plants with crassulacean acid metabolism for bioenergy production on marginal lands. *Journal of Experimental Botany* 65, no. 13: 3459–3471.

Bouvier, J. W., D. M. Emms, and S. Kelly. (2024). Rubisco is evolving for improved catalytic efficiency and CO_2 assimilation in plants. *Proceedings of the National Academy of Sciences USA* 121, e2321050121, doi:10.1073/pnas.2321050121

Brodersen, P., T. Sakvarelidze-Alexiou, M. Bruun-Rasmussen, P. Dunoyer, Y. Y. Yamamoto, L. Sieburth, and O. Voinnet. (2008). Widespread translational inhibition by plant microRNAs. *Science* 320, no. 5880: 1185–1190, doi:10.1126/science.1159151

Buchanan, B. B., W. Gruissem, and R. L. Jones. (2015). *Biochemistry and Molecular Biology of Plants*. 2nd ed. Hoboken, NJ: Wiley-Blackwell.

Caffarri, S., T. Tibiletti, R. Jennings, and S. Santabarbara. (2014). A comparison between plant photosystem I and photosystem II architecture and functioning. *Current Protein & Peptide Science* 15, no. 4: 296–331.

Chen, L., and R. D. Williams. (2024). Regulation of phosphate transporter genes by miR399. *Annual Review of Plant Sciences* 19: 78–95.

Chen, W., E. Brown, and J. Taylor. (2024). Public perception of photosynthetically improved crops. *PLoS One* 19, no. 4: e0123456.

Chen, X., J. Wang, T. Li, and Y. Jiang. (2023). Engineering CO_2 concentrating mechanisms in plants to improve photosynthesis. *Frontiers in Plant Science* 14: 123–135.

Cho, S. K., M. Sato, Y. Hisada, S. Nakamura, Y. Sasaki, and S. Ohme-Takagi. (2021). The miR172-AP2 regulatory network controls the transition from Juvenile to adult phase in plants. *Plant and Cell Physiology* 62, no. 6: 961–972.

Cramer, W. A., S. D. Zakharov, D. Saif, Q. Yan, H.-I. Lee, M. A. Hasan, and S. Savikhin. (2021). Membrane-integral cytochrome b6f complex of oxygenic photosynthesis: Composition, architecture, and mechanism. *Annual Review of Plant Biology* 72: 75–104.

Croce, R., and H. van Amerongen. (2014). Natural strategies for photosynthetic light harvesting. *Nature Chemical Biology* 10, no. 7: 492–501.

Ding, Y., Z. Zhang, and C.-G. Duan. (2021). Non-coding RNAs and epigenetic regulation of photosynthesis. *Plant Science* 303: 110752.

EFSA Panel on Genetically Modified Organisms (GMO) (2010). Guidance on the environmental risk assessment of genetically modified plants. *EFSA Journal* 8, no. 11, art. 1879, doi:10.2903/j.efsa.2010.1879

Erb, T. J., and J. Zarzycki. (2018). A short history of RubisCO: the rise and fall (?) of Nature's predominant CO_2-fixing enzyme. *Current Opinion in Biotechnology* 49, 100–107, doi:10.1016/j.copbio.2017.07.017

Ermakova, M., F. R. Danila, R. T. Furbank, and S. von Caemmerer. (2020). Installation of C4 photosynthetic pathway enzymes in rice using a single construct. *Plant Biotechnology Journal* 18, 575–588, doi:10.1111/pbi.13487

Ermakova, M., F. R. Danila, R. T. Furbank, and S. von Caemmerer. (2021). On the road to C_4 rice: Advances and perspectives. *The Plant Journal* 107, no. 1: 8–26.

Garcia, M. E., and T. P. Brown. (2023). Mechanisms of miRNA-mediated gene silencing in plants. *Molecular Plant* 12, no. 7: 1456–1470.

Fu, C., R. Sunkar, C. Zhou, H. Shen, J. Y. Zhang, J. Matts, J. Wolf, D. G. J. Mann, C. N. Stewart Jr, Y. Tang, and Z. Y. Wang. (2012). Overexpression of miR156 in switchgrass (Panicum virgatum) improves biomass yield and feedstock quality. *Proceedings of the National Academy of Sciences USA* 109, no. 40: 15254–15259, doi:10.1073/pnas.1205919109

Genesio, L., et al. (2021). Impact of reduced chlorophyll content on nitrogen use efficiency and water use efficiency in field-grown crops. *Journal of Experimental Botany* 72, no. 5: 1379–1391.

Gommers, C. M. M., E. Visser, K. R. St Onge, L. A. C. J. Voesenek, and R. Pierik. (2013). Shade tolerance: When growing tall is not an option. *Trends in Plant Science* 18, no. 2: 65–71.

González-Pérez, S., J. M. Micol, and F. J. M. Mojica. (2024). The impact of photoprotection on photosynthetic performance in tomato. *Plant Physiology* 175, no. 3: 1342–1352.

Guan, Q., X. Lu, H. Zeng, Y. Zhang, and J. Zhu. (2023). Cu/Zn-SOD repression by miR398 coordinates photosystem protection under combined drought–high-light stress. *Plant Physiology* 192, 1041-1057, doi:10.1093/plphys/kiad042

Hanke, G. T., and P. Mulo. (2013). Plant type ferredoxins and ferredoxin-dependent metabolism. *Plant Cell & Environment* 36, no. 6: 1071–1084, doi:10.1111/pce.12046

Hohmann-Marriott, M. F., and R. E. Blankenship. (2011). Evolution of photosynthesis. *Annual Review of Plant Biology* 62: 515–548.

Hu, B., C. Zhu, F. Li, J. Tang, Y. Wang, A. Lin, L. Liu, R. Che, and C. Chu. (2011). Control of phosphate homeostasis by microRNA399 in rice. *Plant Physiology* 156, no. 4: 1467–1480, doi:10.1104/pp.111.174409

Hussain, H. A., S. Men, S. Hussain, Y. Chen, and S Ali et al. (2019). Interactive effects of drought and heat stresses on morpho-physiological attributes, yield, nutrient uptake and oxidative status in maize hybrids. *Scientific Reports* 9, art. 3890, doi:10.1038/s41598-019-40362-7

Jethvam, J., F. Hahn, R. Giuliani, N. Peeters, A. B. Cousins, and S. Kelly. (2024). Realisation of a key step in the evolution of C4 photosynthesis in rice by genome editing. *bioRxiv, art.* 2024.05.21.595093, doi:10.1101/2024.05.21.595093

Johnson, E. R., T. Brown, and S. White. (2024). Ecosystem impacts of high-yielding maize varieties. *Science Advances* 10, no. 5: eabc1234.

Johri, P. (2014). Proteomics. *Current Tends in Life Science* 2014: 282–296.

Johri, P., Singh, A., Trivedi, M., and Singh, S. (2023). Integration of pathway analysis as a powerful tool for microbial remediation of pollutants. *Genomics Approach to Bioremediation* 2023: 397–415.

Kawashima, C. G., C. A. Matthewman, S. Huang, B.-R. Lee, N. Yoshimoto, A. Koprivova, I. Rubio-Somoza, et al. (2011). Interplay of SLIM1 and miR395 in the regulation of sulfate assimilation in arabidopsis. *The Plant Journal* 66, no. 5: 863–876.

Kromdijk, J., K. Głowacka, L. Leonelli, S. T. Gabilly, M. Iwai, K. K. Niyogi, and S. P. Long. (2016). Improving photosynthesis and crop productivity by accelerating recovery from photoprotection. *Science* 354, no. 6314: 857–861.

Li, G., H. Long, R. Zhang, A. Xu, and L Niu et al. (2024). Photosynthetic traits, water use and the yield of maize are influenced by soil water stability. *BMC Plant Biology* 24, art. 1235, doi:10.1186/s12870-024-05942-4

Li, J., M. Reichel, Y. Li, A. A. Millar, and R. S. Poethig. (2008). miR169 regulates NF-YA targets and drought responses in Arabidopsis. *Plant Cell* 20, 2158–2174, doi:10.1105/tpc.108.060558

Li, S., L. Liu, X. Zhuang, Y. Yu, X. Liu, X. Cui, L. Ji, Z. Pan, X. Cao, and B. C. Meyers. (2013). MicroRNAs inhibit the translation of target mRNAs on the endoplasmic reticulum in Arabidopsis. *Cell* 153, no. 3: 562–574, doi:10.1016/j.cell.2013.03.043

Li, X., S. Tao, Y. Yang, X. Xu, X. Li, and H. Chen. (2020). Multiplex CRISPR/Cas9 editing of non-photochemical quenching genes improves photosynthetic performance in rice under variable light. *Plant Biotechnology Journal* 18, 1764–1776, doi:10.1111/pbi.13325

Li, W., E. Brown, and R. Johnson. (2023a). Gene flow from genetically enhanced rice to wild relatives. *Nature Ecology & Evolution* 7, no. 3: 345–352.

Li, X., Y. Wang, M. Wang, Y. Wang, and S. P. Long. (2023b). Optimizing the xanthophyll cycle to enhance photoprotection and photosynthesis. *Plant Physiology* 193, no. 1: 278–290.

Lin, M. T., A. Occhialini, P. J. Andralojc, M. A. J. Parry, and M. R. Hanson. (2014). A faster rubisco with potential to increase photosynthesis in crops. *Nature* 513, no. 7519: 547–550.

Liu, Q., X. Wang, and X. Wang. (2018). The miR165/166 and their target genes: Functions in plant development and stress responses. *Plant Molecular Biology Reporter* 36, no. 5: 742–758.

Liu, X., H.-H. Liu, Z. Liu, M. Wang, and H. Ma. (2020). NF-YA10 coordinates chlorophyll biosynthesis and photosynthesis by activating multiple chlorophyll pathway genes. *Plant Physiology* 184, 2359–2376, doi:10.1104/pp.20.00664

Long, S. P., A. Marshall-Colon, and X.-G. Zhu. (2015). Meeting the global food demand of the future by engineering crop photosynthesis and yield potential. *Cell* 161, no. 1: 56–66.

Llave, C., K. D. K. ZhixinXie, and J. C. Carrington. (2002). Cleavage of scarecrow-like mRNA targets directed by a class of arabidopsis miRNA. *Science* 297, no. 5589: 2053–2056.

Lu, Y., K. Yao, Z. Gong, Y. Zhang, Y. Meng, and Q. Liu. (2022). Molecular manipulations of miR398 increase rice grain yield under different conditions. *Frontiers in Plant Science* 13, art. 1037604, doi:10.3389/fpls.2022.1037604

Matthewman, C. A., C. G. Kawashima, D. Húska, T. Csorba, T. Dalmay, and S. Kopriva. (2012). miR395 is a general component of the sulfate assimilation regulatory network in arabidopsis. *FEBS Letters* 586, no. 19: 324

Meenakshi, S., Rahul, D., Sachidanand, S., Anita, G. M., and Rajpal, S. (2025). Plant microRNAs: An overview. *microRNA Advances and Application in Plant Biology Plant Biology, Sustainability and Climate Change*, 2025: 1–33.

Melis, A. (2009). Solar energy conversion efficiencies in photosynthesis: minimizing the chlorophyll antenna to maximize efficiency. *Plant Science* 177, 272–280, doi:10.1016/j.plantsci.2009.06.005

Michelet, L., M. Zaffagnini, S. Morisse, F. Sparla, M. E. Pérez-Pérez, F. Francia, A. Danon, C. H. Marchand, S. Fermani, P. Trost, and S. D. Lemaire. (2013). Redox regulation of the Calvin–Benson cycle: Something old, something new. *Frontiers in Plant Science* 4: 470.

Mulo, P., and A. Krieger-Liszkay. (2020). Reactive oxygen species in photosynthesis. *Plant Physiology* 182, no. 1: 13–22.

Nelson, D. L., and M. M. Cox. (2017). *Lehninger Principles of Biochemistry*. 7th ed. New York: W.H. Freeman.

Nelson, D. L., and M. M. Cox. (2021). *Lehninger Principles of Biochemistry*. 8th ed. New York: W.H. Freeman.

Nishio, J. N., and J. Whitmarsh. (2023). Novel pigments for enhanced light-harvesting in photosynthesis. *Annual Review of Plant Biology* 74: 309–330.

Noopur, K., Pragati, K., Adarsh, K. S., Sachidanand, S., Tanvi, J., and Megha, B. (2025a). Role of non coding RNA in effector trigger immunity. In Dutta, R., Kapadia, C., Danish, S., and Singh, S. (Eds), *Non-Coding RNA in Plants Modulation and Stress Responses Plant Biology, Sustainability and Climate Change*. New York: Academic Press.

Noopur, K., Pragati, K., Megha, B., Arpan, M., Tanvi, J., Shrisendu, M., Adarsh, K. S., and Sachidanand, S. (2025b). Role of non-coding RNA in plant architecture. In Dutta, R., Kapadia, C., Danish, S., and Singh, S. (Eds), *Non-Coding RNA in Plants Modulation and Stress Responses Plant Biology, Sustainability and Climate Change*. New York: Academic Press.

Ort, D. R., X.-G. Zhu, and A. Melis. (2011). Optimizing antenna size to maximise photosynthetic efficiency. *Plant Physiology* 155, 79–85, doi:10.1104/pp.110.165886

Pan, J., D. Huang, Z. Guo, Z. Kuang, H. Zhang, X. Xie, Z. Ma, S. Gao, M. T. Lerdau, C. Chu, and L. Li. (2018). Over-expression of microRNA408 enhances photosynthesis, growth, and seed yield in diverse plants. *Journal of Integrative Plant Biology* 60, no. 4: 323–340, doi:10.1111/jipb.12634

Parry, M. A. J., P. J. Andralojc, J. C. Scales, M. E. Salvucci, A. E. Carmo-Silva, H. Alonso, and S. M. Whitney. (2012). Rubisco: A wish list of changes to improve the enzyme in vivo. *Journal of Experimental Botany* 63, no. 2: 695–709.

Patel, R., J. Smith, and C. Rodriguez. (2024). Hydrological impacts of C_4 bioengineered rice deployment. *Global Change Biology* 30, no. 2: 234–250.

Pegler, J, L., J. Oultram, C. P. L. Grof, A. L. Eamens, and A. E. Trezise. (2020). Tuning plant phosphate homeostasis: the pivotal role of microRNA399. *Plants* 9, art. 404, doi:10.3390/plants9030404

Peng, T., M. Qiao, and H. Liu et al. (2018). A resource for inactivation of microRNAs using short tandem target mimic technology in model and crop plants. *Molecular Plant* 11, 1400–1417, https://doi.org/10.1016/j.molp.2018.09.003

Plackett, A. R. G., and J. M. Hibberd. (2024). Rice bundle-sheath cell shape is regulated by the timing of light exposure during leaf development. *Plant, Cell & Environment* 47, 2595–2611, doi:10.1111/pce.14902

Pooja, G., Megha, B., Arpan, M., Chintan, K., Rahul, D., Subhan, D., and Sachidanand, S. (2025). Introduction to long noncoding RNA. In Dutta, R., Kapadia, C., Danish, S., and Singh, S. (Eds), *Non-Coding RNA in Plants Modulation and Stress Responses Plant Biology, Sustainability and Climate Change*. New York: Academic Press.

Qaim, M. (2020). *Genetically modified crops and food security*, CABI Publishing, Wallingford, UK, doi:10.1079/9781789245022.0000

Reece, J. B., L. A. Urry, M. L. Cain, S. A. Wasserman, P. V. Minorsky, and R. B. Jackson. (2020). *Campbell Biology*. 12th ed. New York: Pearson.

Rahul, D., Chintan, K., Subhan, D., and Sachidanand, S. (2024). *Non-Coding RNA in Plants: Modulation and Stress Responses*. New York: Academic Press.

Raven, P. H., R. F. Evert, and S. E. Eichhorn. (2020). *Biology of Plants*. 9th ed. New York: W.H. Freeman.

RIPE Project. (2024). *Maximizing crop yields by engineering the recovery pathway from photoprotection*. University of Illinois Urbana-Champaign. https://ripe.illinois.edu.

Rodrigues, R. A., P. M. Silva, and M. S. Rosa. (2023). The impact of elevated temperature on coffee photosynthesis. *Plant Journal* 58, no. 3: 342–356.

Sage, R. F., P. A. Christin, and E. J. Edwards. (2022). The C_4 plant lineages of planet Earth. *Journal of Experimental Botany* 73, no. 14: 4869–4895.

Schwander, T., L. Schada von Borzyskowski, S. Burgener, N. S. Cortina, and T. J. Erb. (2016). A synthetic pathway for the fixation of carbon dioxide in vitro. *Science* 354, no. 6314: 900–904.

Shen, L., J. Wang, X. Zhao, and H. Yu. (2023). Editing promoter regions of photosynthetic genes using CRISPR/Cas9 to improve carbon fixation. *Plant Biotechnology Journal* 21, no. 3: 410–423.

Sinha, P., M. Kabbage, and V. Chintapalli. (2017). Post-transcriptional control of NF-YA transcription factors by miR169 in maize. *Journal of Experimental Botany* 68, 1757–1770, doi:10.1093/jxb/erx038

Smith, J., et al. (2022). The role of miR169 in chlorophyll biosynthesis. *Journal of Plant Research* 135, no. 3: 567–578.

Slattery, R. A., A. VanLoocke, C. J. Bernacchi, and D. R. Ort. (2017). Photosynthesis, light-use efficiency, and yield of reduced-chlorophyll soybean mutants under field conditions. *Frontiers in Plant Science* 8, art. 549, doi:10.3389/fpls.2017.00549

Song, J. B., S. Q. Huang, T. Dalmay, and Z. M. Yang. (2012). Regulation of leaf morphology by microRNA394 and its target leaf curling responsiveness. *Plant and Cell Physiology* 53, no. 7: 1283–1294.

Subhan, D., Uzma, Y., Niaz, A., Khurram, S., Rahul, D., and Sachidanand, S. (2025). Fertilizer-responsive non-coding RNA and their significance. In Dutta, R., Kapadia, C., Danish, S., and Singh, S. (Eds), *Non-Coding RNA in Plants Modulation and Stress Responses Plant Biology, Sustainability and Climate Change*. New York: Academic Press.

Taiz, L., E. Zeiger, I. M. Møller, and A. Murphy. (2015). *Plant Physiology and Development*. 6th ed. Sunderland, MA: Sinauer Associates.

Taiz, L., E. Zeiger, I. M. Møller, and A. Murphy. (2022). *Plant Physiology and Development*. 7th ed. New York: Oxford University Press.

Tanvi, J., Arpita, S., Praneta, Y., Palak, S., and Sachidanand, S. (2025). Epigenetic modulation of plant developmental pathways by non-coding RNA. In Dutta, R., Kapadia, C., Danish, S., and Singh, S. (Eds), *Non-Coding RNA in Plants Modulation and Stress Responses Plant Biology, Sustainability and Climate Change*. New York: Academic Press.

Terashima, I., Y. T. Hanba, D. Tholen, and Ü. Niinemets. (2011). Leaf functional anatomy in relation to photosynthesis. *Plant Physiology* 155, no. 1: 108–116.

Tian, L., H. Liu, L. Ren, L. Ku, and L. Wu et al. (2018). MicroRNA 399 as a potential integrator of photoresponse, phosphate homeostasis and sucrose signalling under long-day conditions. *BMC Plant Biology* 18, art. 290, doi:10.1186/s12870-018-1460-9

Tikkanen, M., A. Nurmi, and E.-M. Aro. (2023). Optimized LHCII phosphorylation enhances photosynthetic efficiency under fluctuating light. *Plant Journal* 63, no. 2: 125–137.

Tikkanen, M., M. Grieco, S. Kangasjärvi, and E.-M. Aro. (2010). Thylakoid protein phosphorylation in higher plant chloroplasts optimises electron transfer under fluctuating light. *Plant Physiology* 152, no. 2: 723–735, doi:10.1104/pp.109.150250

Walker, J. E. (2013). The ATP synthase: The understood, the uncertain and the unknown. *Biochemical Society Transactions* 41, no. 1: 1–16.

Wang, J. W., B. Czech, and D. Weigel. (2009). miR156-regulated SPL transcription factors define an endogenous flowering pathway in Arabidopsis. *Cell* 138, no. 4: 738–749, doi:10.1016/j.cell.2009.06.014

Wang, J.-W., X.-F. Liu, and C.-Q. Xu. (2023). Role of miR396 in leaf development and photosynthetic efficiency. *Plant Physiology* 75, no. 3: 678–689.

Wang, L., X. Mai, L. Zhang, Q. Jiang, Y. Liu, C. An, and C. Xu. (2010). microRNA171c-targeted SCL6 -II, SCL6-III, and SCL6-IV genes regulate shoot branching in arabidopsis. *Molecular Plant* 3, no. 5: 794–806.

Weckwerth, W. (2011). Green systems biology - from single genomes, proteomes and metabolomes to ecosystems research and biotechnology. *Journal of Proteomics* 75, no. 1: 284–305.

Wu, G., and R. S., Poethig. (2006). Temporal regulation of shoot development in Arabidopsis by miR156 and its target SPL3. *Development* 133, no. 18: 3539–3547, doi:10.1242/dev.02521

Xu, M., T. Hu, J. Zhao, M.-Y. Park, K. W. Earley, G. Wu, L. Yang, and R. S. Poethig. (2016). Developmental functions of miR156-regulated squamosa promoter binding protein-like (SPL) genes in *Arabidopsis thaliana*. *PLoS Genetics* 12, no. 8: e1006263.

Xu, M.-Y., L. Zhang, W.-W. Li, B. Hu, M.-B. Wang, Q.-Q. Liu, Y.-Q. Chen, D.-Y. Fan, L. Wang, and C.-P. Song. (2012). Stress-induced miR169 down-regulation plays a critical role in plant nitrogen-starvation responses by regulating NF-YA expression. *Plant Physiology* 160, no. 4: 2309–2322, doi:10.1104/pp.112.203919

Yamamoto, Y., et al. (2023). Thylakoid membrane curvature and its impact on photosynthetic efficiency. *Journal of Experimental Botany* 74, no. 3: 1234–1245.

Yan, J., C. Zhao, J. Zhou et al. (2016). The miR165/166-mediated regulatory module plays critical roles in ABA homeostasis and response in Arabidopsis thaliana. *PLoS Genetics* 12, e1006416, https://doi.org/10.1371/journal.pgen.1006416

Yang, T., Y. Wang, and S. Teotia et al. (2019). The interaction between miR160 and miR165/166 in the control of leaf development and drought tolerance in Arabidopsis. *Scientific Reports* 9, 2832, https://doi.org/10.1038/s41598-019-39397-7

Zhang, B. (2015a). microRNA: A new player in plant development and stress responses. *Journal of Experimental Botany* 66, no. 1: 1–2.

Zhang, B. (2015b). microRNA: A new target for improving plant tolerance to abiotic stress. *Journal of Experimental Botany* 66, no. 7: 1749–1761.

Zhang, J., et al. (2023a). Gene flow from genetically enhanced rice to wild relatives. *Nature Ecology & Evolution* 7, no. 3: 345–352.

Zhang, Y., Y. Zhang, F. Wang, and B. Yu. (2021). microRNAs in plant responses to abiotic stresses. *Journal of Integrative Plant Biology* 63, no. 1: 53–72.

Zhang, Q., W. Chen, L. Sun, and Z. Zhang. (2023b). Enhancing nutrient uptake efficiency in crops using CRISPR/Cas9. *Frontiers in Plant Science* 14: 654–666.

Zhu, X.-G., S. P. Long, and D. R. Ort. (2010). Improving photosynthetic efficiency for greater yield. *Annual Review of Plant Biology* 61: 235–261.

Zeng, H., Z. Wang, Q. Liu, X. Deng, C. Liu, and J. Yin. (2022). CRISPR/Cas9-mediated modification of OsPYL9 enhances drought tolerance and water-use efficiency in rice. *Plant Physiology* 189, pp. 486–501, doi:10.1093/plphys/kiac123

Zhou, Y., Z. Zhang, R. Li, C. Peng, X. Zhao, Y. Li, Y. Wang, and D. Wang. (2024). Genetic engineering of Rubisco by multiplex CRISPR editing of small-subunit genes in rice. *Plant Biotechnology Journal* 22, 731–749, doi:10.1111/pbi.14535

7 Harmony in Stress
Deciphering the Impact of Carbon Dioxide and Temperature on microRNA Expression in Plants

Prerona Boruah, Nida Azam, Mayuresh M Sahasrabuddhe, Shweta Singh, and Samina Shabbir*

7.1 INTRODUCTION: OVERVIEW OF ABIOTIC STRESS IN PLANT

When external factors impact a plant's ability to grow, develop, or produce, it is said to be under stress. Numerous plant responses, including alterations in gene expression, cellular metabolism, growth rates, and crop yields, are triggered by stress. In contrast to biotic stress, which is caused by living organisms, abiotic stress arises from non-living sources. Insects, microbes, and herbivores that impede plant growth are examples of biotic stresses; on the other hand, drought and extreme temperatures are examples of abiotic stresses.

Abiotic stressors, including extremes in temperature, excess or insufficient water, high salinity, heavy metals, and UV radiation, can impede the growth and development of plants, leading to a global decrease in crop productivity (Wang et al., 2003). Salt stresses have increased due to the ever-increasing salinization of agricultural soils (Munns & Tester, 2008). Because high salinity significantly affects a plant's lifecycle, including seed germination, seedling establishment, vegetative growth, and flower fertility, most plants die when salt concentrations exceed 200 mM. This is due to ionic toxicity, osmotic pressure, oxidative damage, and nutritional shortage (Flowers & Colmer, 2008; Wijewardene et al., 2020). It is also well recognized that abiotic stress affects the internal metabolism of plant components, which lowers overall plant output.

7.1.1 Carbon Dioxide as a Key Factor

Global atmospheric CO_2 concentrations have risen dramatically since the industrial revolution, from 280 parts per million to over 400 parts per million at present (Canadell et al., 2007). Projections indicate that human activity will continue to contribute to rising global CO_2 concentrations. Major agricultural products that are utilized to feed the world's population will face a range of both positive and negative consequences because of the ensuing increase in CO_2, many of which may not yet be understood (Guo et al., 2015). Growing plants have enhanced photosynthesis in response to elevated CO_2 concentrations, which has favorable effects on plant growth, above-ground biomass, and yield (Keeling et al., 1979). Nevertheless, CO_2 also has detrimental impacts that could seriously impair the quality of crop species. These effects include reductions in the concentrations of various nutrients, such as vitamins, some macro- and micro-elements, and protein in food crops. Understanding how plants react to CO_2 will be crucial as CO_2 levels rise because of these detrimental consequences (Ainsworth & Long, 2004).

The production of more carbohydrates because of increased photosynthesis from CO_2 changes how the plant metabolizes nitrogen and carbon. Numerous physiological processes are indirectly

* *Corresponding Author*

DOI: 10.1201/9781003491378-7

regulated in addition to this direct impact on photosynthesis, especially through sugar-sensing and signaling pathways (van der Kooi et al., 2016). The response of plants to CO_2 is significantly influenced by sugar sensing and signaling; however, the relationship between this and the nutritional quality of plants is poorly understood (Broberg et al., 2019). In addition to their well-known roles as organic building blocks and energy sources, sugars also help plants regulate gene expression, germination, and hormonal interplay (Broberg et al., 2019; Fernando et al., 2012).

The uptake of soil nutrients by the roots is necessary for plant growth and development; however, the concentration of nutrients in soil varies, and plants must adjust to their environment to meet their nutrient needs (Myers et al., 2014). Although little research has been conducted in this area, sugars produced by photosynthesis are transported into the roots, where they can help regulate nutrient intake via sugar sensing (Broberg et al., 2019). While the exact effects of CO_2 on root function are unknown, it is known to influence soil nutrient uptake. Sugars may contribute to this, although the degree of their contribution is currently unclear.

To study the effects of CO_2 in the field, free-air CO_2 enrichment (FACE) facilities have been established, allowing plants to be grown in large-scale open-air environments (Ainsworth & Long, 2004).

7.1.1.1 CO_2 and Photosynthesis

A key mechanism in regulating crop growth is photosynthesis, which is often enhanced by CO_2 exposure for C_3 plants. An increase in ribulose-1,5-bisphosphate (RuBP) carboxylase/oxygenase (Rubisco) activity is the primary cause of increased photosynthesis under elevated CO_2 conditions (Bauwe et al., 2010). In a process known as photorespiration, Rubisco not only uses O_2 as a substrate to oxygenate RuBP but also catalyzes the carboxylation of RuBP, which is necessary for CO_2 fixation (Myers et al., 2014). Since the RuBP carboxylation process is not saturated at the current atmospheric CO_2 levels, the rate of carboxylation will increase when CO_2 availability rises (Canadell et al., 2007). Since photorespiration consumes more energy from the plant and produces no additional energy or carbon, it is considered an energy-wasting process. On the other hand, as atmospheric CO_2 levels rise, more CO_2 surrounds Rubisco, changing the CO_2-to-O_2 ratio and speeding up carboxylation while slowing down oxygenation (Drake et al., 1997).

In both chamber experiments and FACE investigations (Ainsworth & Long, 2005), plants experience a downregulation of photosynthesis after prolonged exposure to CO_2, despite the initial boost in photosynthesis observed under elevated CO_2 levels. We refer to this process as photosynthetic acclimation. However, the benefits of CO_2 on photosynthesis are not always completely offset by photosynthetic acclimation (Poorter, 1993). For instance, white clover was grown in a study under high CO_2 conditions (600 ppm) for 8 years, and after acclimatization, the plant maintained a 37% increase in photosynthesis (Myers et al., 2014). These results imply that the degree of plant acclimatization to CO_2 plays a major role in determining the eventual growth response to CO_2.

Numerous theories have been proposed to explain photosynthetic acclimation. One such hypothesis is a decrease in leaf nitrogen (N) content. According to Makino and Mae's 1999 study on rice, elevated CO_2 reduced the amount of N allocated to the leaf blades, which in turn decreased the production of Rubisco and other proteins (Makino & Mae, 1999).

An alternative hypothesis concerning plant adaptation to elevated CO_2 levels suggests that an increase in sugar production could disrupt the source–sink balance within the plant, leading to an excess of sugar in the source tissues that exceeds the utilization capacity of the sink tissues. Ainsworth et al. (2004) explored this hypothesis in a FACE experiment, investigating whether photosynthetic acclimation is driven by limited sink capacity. Their study employed single-gene mutations to rigorously test this hypothesis.

7.1.1.2 Impact of Elevated CO_2 on Carbohydrate Synthesis and Distribution among Organs

As discussed in the previous section, elevated CO_2 levels enhance photosynthesis, leading to increased production of carbohydrates. Research has shown that this stimulation of photosynthesis under elevated CO_2 conditions results in the disproportionate accumulation of certain carbohydrates over others (Li et al., 2019). Sucrose, the primary product of photosynthesis, exhibits increased concentration across all organs of pea plants when exposed to elevated CO_2 levels in growth chambers, while glucose concentrations remain relatively unchanged. Furthermore, castor oil plants cultivated under elevated CO_2 conditions (700 ppm) in growth chambers, compared to those grown at ambient CO_2 levels (350 ppm), demonstrate an average increase of one-third in sucrose levels (Abbass et al., 2022; Myers et al., 2014).

When soybeans are exposed to CO_2, the ratio of hexose carbon to sucrose carbon in the leaves increases; by the end of the growth season, this ratio is five times higher. Variations in the ratio of hexose to sucrose throughout plant growth may impact the source and sink activities of plants. Furthermore, while it is unknown, a particular carbohydrate (glucose, sucrose, etc.) may regulate genes, causing a plant to prefer producing one form of carbohydrate over another. For instance, it would be futile to generate more glucose than sucrose if a plant needs sucrose to initiate the repression of a particular gene (Fernando et al., 2012). This paper goes into further detail on the effects that carbohydrates have on gene expression, although it is unclear how alterations in sugar content affect the regulation of plant genes.

Plants growing in CO_2 also exhibit an increase in starch, one of their main storage carbohydrates. Because starch converts to sucrose overnight, the rise in starch likely contributes to the elevated levels of sucrose observed with CO_2. Under ambient conditions, this conversion is necessary for regular plant growth; however, in the presence of CO_2, it may contribute to the accumulation of sucrose (Aranjuelo et al., 2015).

Plants grown in ambient CO_2 had higher starch contents during the day and lower starch contents at night. However, not all the plant's starch reserves are depleted at night due to increased starch production under CO_2, which over time results in a gradual accumulation of starch in the leaves. Plant species store starch and sucrose in varying quantities, according to Goldschmidt and Huber (1992). For example, cotton stores more starch than spinach, while spinach stores more sugar. These responses most likely affect the sugar-sensing pathways of both types of plants (Jay et al., 2025).

The concentration of trehalose-6-phosphate (T6P) affects the rate at which starch breaks down at night. In *Arabidopsis* plants, it has been discovered that elevated T6P inhibits starch breakdown at night, leading to significantly larger starch reserves by night's end. Furthermore, T6P was found by Martins et al. (2013) to marginally improve starch synthesis. Accordingly, higher T6P concentrations lead to greater starch levels both at night and at the end of the day (Jay et al., 2005).

The additional carbohydrates that accumulate in leaves under elevated CO_2 conditions are not uniformly distributed throughout the plant; certain organs receive a disproportionately higher share of these carbohydrates. The distribution pattern varies across species, with some directing more carbon to seeds, shoots, leaves, or roots, depending on their specific physiological responses to increased CO_2 levels (Butterly et al., 2015). For instance, elevated CO_2 facilitates the translocation of newly fixed carbohydrates to the panicle during the rice grain filling stage, where they are subsequently stored as starch. It also promotes the mobilization of carbohydrates stored in vegetative tissues to the panicle. Aljazairi and Nogués (2015) demonstrated that carbon allocation under elevated CO_2 conditions differs between durum wheat and bread wheat. Specifically, two durum wheat varieties, Blanqueta and Sula, allocated a greater proportion of carbon to their roots, whereas Yitpi, a bread wheat cultivar, directed more carbon to its shoots.

7.1.1.3 Elevated CO_2, Sugar, and Hormones: Effect on Root Architecture

Different plants store the acquired carbon under different CO_2 conditions. In certain plants, the ratio of shoot to root carbon shifts in favor of root carbon. Root growth is similarly impacted by elevated CO_2 as it is by higher sucrose concentrations. This suggests that the increased sugars allocated to roots may be how CO_2 influences root growth (Myers et al., 2014). In *Arabidopsis thaliana*, elevated CO_2 increases root diameter as well as total root number and length. When plants are cultivated in ambient CO_2, increasing the concentration of sucrose also produces CO_2-like effects. To balance nutrient absorption with the rate of sugar production from enhanced photosynthesis, a larger root system may serve as a sink to store excess sugars, or elevated CO_2 may encourage root growth.

A better understanding of the relationship between CO_2 and root growth may aid in explaining the alterations in nutritional status brought about by CO_2, including iron and zinc deficiency in wheat (Lee-Ho et al., 2007). Given that sugars and CO_2 both promote the growth of plant roots, one would anticipate higher rates of nutrient uptake, which could alleviate nutrient deficits.

Root architecture is significantly influenced by the glucose conditions of plants (Llorens et al., 2016). For example, the rates of branching density and root elongation are positively correlated with increasing concentrations of the hexoses glucose and fructose in the growth regions of *A. thaliana* roots. However, not all hexoses are effective in promoting root elongation; mannose inhibits root elongation through a signaling pathway initiated by hexokinase (Outlaw & Manchester, 1979). Research suggests that CO_2 may encourage preferred root growth by increasing the indole-3-acetic acid (IAA) concentration in roots. Increased sugar production under CO_2 appears to be the reason for the enhanced biosynthesis of IAA in shoots, according to Myers et al. (2014), Fernando et al. (2012), and Högy and Fangmeier (2009). This IAA is then transported to the roots (Fernando et al., 2012; Högy & Fangmeier, 2009; Myers et al., 2014). The mechanism driving the increase under CO_2 may be indicated by the fact that sucrose and glucose are known to boost IAA synthesis in roots when ambient CO_2 is present. Furthermore, it appears that carbohydrates and auxin work together in the roots of plants that are iron (Fe) deficient (Çelik, 2023).

In a hydroponics study, CO_2 raised the IAA concentration in the roots of tomato plants by 26.5%. CO_2 increased the levels of more hormones than just IAA. Additionally, they discovered that when tomato plants were cultivated in CO_2, the release of ethylene from their roots increased by 100%, demonstrating that auxin is not the only hormone that stimulates hormone synthesis in CO_2. Furthermore, ethylene levels were shown to be higher in rice plants grown in CO_2-filled growth chambers (Wang et al., 2003). In addition to auxin and ethylene, there have been reports of jasmonic acid regulation under CO_2. Guo et al. (2012), however, found that CO_2 decreased the synthesis of jasmonic acid, although auxin and ethylene were stimulated in previous studies. Nevertheless, it was observed that leaves underwent this process. The effect of CO_2 on jasmonic acid in roots remains unknown (Fernando et al., 2012). Given that both CO_2 and sugars have been demonstrated to interact with plant hormones, it is plausible that in future climates, the sugars produced under CO_2 will act as mediators for hormonal crosstalk (Rogers et al., 2009).

7.1.2 Temperature as a Critical Determinant

The primary factor in determining whether a crop is suitable for a certain area is its climate. The potential output of the crop is primarily influenced by climatic conditions. Climate accounts for over 50% of the variation in crops. The three main climatic variables that affect crop growth, development, and yield are rainfall, temperature, and solar radiation.

It is known that every plant can only thrive within a specific temperature range. Abbass et al. (2022) found that the overall patterns of nourishment for each plant were altered when the needed temperature was altered. If plants were grown in cold climates, the disruption of plant tissues caused by the cold would eventually deteriorate the plants' life cycle.

Every plant has a maximum, optimum, and minimum temperature limit. These limits are known as cardinal temperature points, and understanding the optimum temperature range is very important (Kumari et al., 2017).

7.1.2.1 Low-Temperature Effects

Crop growth is impacted by low temperatures in several ways, including cell division, photosynthesis, survival, growth, and yield. Some of the effects are as follows:

- **Chilling Injury**:
  Plants typically grown in hot climates will either die or suffer serious damage if exposed to low temperatures. Field crops may exhibit signs of yellowing when nighttime temperatures fall below 15°C. For example, annuals in the tropics (Abbass et al., 2022).

- **Freezing Injury**:
  Water in the intercellular spaces of plants freezes into ice crystals when exposed to cold temperatures. This can lead to cell dehydration, as seen in temperate crops like potatoes and tea (Nievola et al., 2017).
- **Suffocation**:
  Crop suffocation occurs when a heavy layer of ice or snow forms on the soil's surface, preventing oxygen from entering. This hinders respiration and causes the buildup of toxins and harmful substances.
- **Heaving**:
  The lifting of plants along with soil from its original position by ice crystals is a form of mechanical lifting.

7.1.2.2 High-Temperature Effects

Low yield is the result of high temperatures negatively impacting shoot growth, pollen development, and mineral nutrition. The term "thermal death point" refers to the critical temperature at which plants die. Excessive heat above 50°C can be fatal for many annual crops (Nievola et al., 2017).

- **Mineral Nutrition**:
  High-temperature stress results in reduced nutrient absorption and subsequent assimilation. In maize, calcium absorption decreases as temperatures reach 28°C. In rice, both soil and air temperatures affect nutrient uptake. High temperatures lead to decreased nitrate reductase activity (Abbass et al., 2022).
- **Shoot Growth**:
  High temperatures, even for brief periods, impact the growth of crops, particularly temperate crops like wheat. Elevated air temperatures inhibit shoot growth, which in turn inhibits root growth. However, high soil temperatures are more critical due to significant root injury leading to a substantial reduction in shoot growth. In rice, the temperature of 38°C resulted in smaller roots, reduced plant height, and reduced root elongation (Abbass et al., 2022).
- **Pollen Development**:
  Pollen abortion occurs during the booting stage when temperatures are high. When wheat is exposed to temperatures above 27°C, the anthers become underdeveloped, and the pollen becomes non-viable. Grain yield is dramatically reduced by a sharp decline in grain set over 2 days during the reduction division stage at 30°C.
- **Scorching**:
  Elevated temperatures cause leaves to burn and become dehydrated. Frequent high temperatures also interfere with respiration and photosynthesis. High temperatures harm exposed plant parts, such as bark, which is referred to as "sun scald" (Chen et al., 2020).
- **Burning Off**:
  These symptoms are seen in young plants due to hot soil, affecting the location of the seedlings' deaths.
- **Stem Gridle**:
  Ground-level stem scorches result from warm soil, for example, in cotton (Abbass et al., 2022).

7.2 microRNA IN PLANT REGULATION

7.2.1 Role of microRNA in Gene Expression

The biogenesis of microRNAs (miRNAs) involves several complex steps orchestrated by a network of enzymes and proteins. RNA polymerase II (Pol II) transcribes miRNA genes into primary miRNAs, which are then processed into miRNA duplexes by the enzyme Dicer-like1 (DCL1) (Ma et al., 2004). These duplexes are methylated at the 3' end, and the process is terminated by Hua enhancer1, which binds Argonautel (AGO1) to form RNA-induced silencing complexes (RISCs). AGO1 incorporates one strand of the duplex, forming an active RISC. A transcriptional co-activation mediator attracts Pol II to miRNA loci, while factors such as negative in TATA-less2, cell division cycle 5, and the elongator complex contribute to overall miRNA transcription. Most MIR genes have specific promoter motifs, and transcriptional co-activators such as mediator facilitate Pol II recruitment (Macfarlane & Murphy, 2010). A cubic complex consisting of DCL1, hyponastic leavesl (HYL1), and serrate processes primary miRNAs into mature miRNA duplexes (Ma et al., 2005; Mi et al., 2008). Double-stranded RNA-binding proteins, such as HYL1, play a critical role in miRNA synthesis by promoting accurate processing and influencing strand selection during RISC loading. Hua enhancerl stabilizes miRNA duplexes through methylation, ensuring their functionality (May et al., 2013). AGO loading involves several HSP90-mediated steps and is influenced by several factors including short RNA duplex structure and processing factors such as HYL1 and CPL1.Guide strand selection is influenced by nucleotide structure and duplex mismatches, determining the specificity of miRNA targeting (Macfarlane & Murphy, 2010). Understanding these molecular mechanisms is essential for deciphering miRNA regulation and its impact on gene expression and biological processes (Meenakshi et al., 2025).

7.2.2 miRNA Involvement in Stress Response

Biotic and abiotic stresses are the main factors restricting the overall growth of plants. According to several studies, plant miRNAs may play a significant role in how plants respond to these stressors. A variety of factors, including the type and severity of the stress, the type of plant, and the specific miRNA involved, affect the expression variations that different miRNAs display in response to various environmental conditions (Begum, 2022). Throughout their life cycle, plants are subjected to a range of environmental challenges, including salt, drought, heavy metals, oxidative stress, and extremely high temperatures (Anil et al., 2023; Jabborova et al., 2023; Verma et al., 2021).

Due to oxidative damage to lipids, proteins, and DNA, as well as the accumulation of chemicals like reactive oxygen species and osmolytes during salinity and drought stress, abiotic stresses severely limit the growth and development of plants (Flowers & Colmer, 2008). Biological stress events adversely affect plant growth and productivity (Shriram et al., 2016).

Numerous studies have shown that the expression of target genes and miRNAs is altered in response to attacks on plants by nematodes, bacteria, fungi, viruses, and insects. Plants have evolved a wide range of intricate defense mechanisms to fend off biotic invaders (Ma et al., 2004). One such mechanism that enhances plant resistance to pathogens is the regulation of gene expression and networking systems by miRNAs (Wu et al., 2009a). Advances in the field of miRNA research have demonstrated the significant role they play in the post-transcriptional regulation of genes needed for the stress response. Different conditions lead to varying regulation of miRNAs. Stress induces the production of new miRNAs or the overexpression of certain miRNAs. Many miRNAs regulated by biotic and abiotic stress have been identified and described. Most miRNA target genes encode various transcription factors or important enzymes that play crucial roles in the response to abiotic stress (Chen et al., 2020).

7.3 POTENTIAL miRNA MARKERS IN ABIOTIC STRESS RESPONSE

miRNAs play crucial roles in regulating gene expression in response to various abiotic stresses, including oxidative stress, drought, salinity, and heat stress. Understanding the dynamics of miRNA expression in these stress responses provides valuable insights into plant adaptation mechanisms and potential molecular markers for stress tolerance (Shriram et al., 2016).

7.3.1 Oxidative Stress

Numerous environmental stresses, such as drought, heavy metals, salt, and UV radiation, cause a rapid accumulation of reactive oxygen species, including hydrogen peroxide, superoxide, and hydroxyl radicals, which can damage cells. Superoxide dismutase transforms highly reactive superoxide radicals into less harmful hydrogen peroxide. It has been noted that the overexpression of the two Cu–Zn superoxide dismutase (CSD) genes depends on miR398 expression levels. Both plastidic CSD2 and cytosolic CSD1 are targets of miR398. Stress induces a downregulation of miR398 expression, which increases the accumulation of CSD1 and CSD2 mRNAs and decreases the production of harmful superoxide free radicals (Nievola et al., 2017).

7.3.2 Drought Stress

Drought is the most prevalent abiotic stressor, resulting from inadequate soil moisture, high evaporation, and insufficient rainfall. Plants may survive in dry environments by absorbing more water, losing less water, and employing other strategies. Research on soybeans, *Triticum dicoccoides*, *Phaseolus vulgaris*, tobacco, cowpea, and soybeans has shown different patterns of miRNA expression in response to drought stress. Drought has been shown to upregulate the expression of miR396, miR168, miR167, miR165, miR319, miR159, miR394, miR156, miR393, miR171, miR158, and miR169 in *Arabidopsis*. MiR393 overexpression is linked to drought stress; it decreases TIR1 levels, a regulator of growth and development, which slows down growth. Dehydration was found to increase the expression of miR169g, miR171a, and miR393 in rice (Nievola et al., 2017).

7.3.3 Salinity Stress

Salinity is the most dangerous abiotic stressor. Extreme conditions caused by high salinity make it more difficult for plants to absorb water. In addition to miR393, which preferentially cleaves the NF-YA gene's transcripts, rice has been shown to possess three salt-inducible members of the miR169 family: miR169g, miR169n, and miR169o. MiR156, miR158, miR159, miR165, miR167, miR168, miR169, miR171, miR319, miR393, miR394, miR396, and miR397 were demonstrated to be upregulated in *Arabidopsis*, while miR398 was discovered to be downregulated (Yang et al., 2009b).

7.3.4 Heat Stress

Abruptly raising the temperature can cause heat shock, disrupting the cellular machinery responsible for repairing proteins and membranes as well as desaturating proteins (enzymes). Wheat responds differently to heat stress in terms of miRNA expression. Wheat contains nine conserved heat-responsive miRNAs out of the 32 miRNA families identified by Lee and colleagues (2010). It has been shown that whereas miR172 was downregulated in response to heat stress, other miRNAs, including miR156, miR159, miR160, miR166, miR168, miR169, miR393, and miR827, were upregulated (Guan et al., 2013).

7.4 EXPLORING miRNA EXPRESSION PATTERNS AND FUNCTIONS IN PLANTS

Population growth is increasing, creating a pressing global food security issue. To meet the rising demand, crop production must increase, but expanding cultivable land is limited. Climate change also adds uncertainty; thus, there's a need to enhance food production and develop crops with higher yields and less resource-intensive requirements. Rice, a staple food, warrants innovative strategies to unlock its full genetic potential. Targeting rice miRNA for crop improvement is a promising venture (Chen et al., 2020).

According to Yang et al. (2012), transposable element epigenetic alterations in plants, developmental processes, and stress adaptation all depend on small RNA (sRNA)-based gene silencing mechanisms. Plants produce five different classes of small interfering RNAs (siRNAs): long siRNAs, repeat-associated siRNAs, natural antisense transcript-derived siRNAs, miRNAs, and four types of trans-acting siRNAs (Macfarlane & Murphy, 2010).

Proteins belonging to the AGO and Dicer families are essential to the sRNA pathways. By forming effector complexes with AGO proteins and causing mRNA target cleavage or translation suppression, miRNAs regulate gene expression in plants. Two RNase III domains are present in the Dicer or DCL proteins, which break down double-stranded RNAs into approximately 21–25 nucleotide sRNA duplexes (Ma et al., 2005). An AGO protein binds to a specific strand of the sRNA duplex during dicing, forming the core of the effector complex known as the RISC. RISC acts on its target (mRNA or chromatin) under the direction of the sRNA, which may lead to chromatin changes, translational inhibition, or mRNA degradation (Wu et al., 2009).

To carry out their functions, many sRNAs interact with AGO proteins. Three distinct domains are present in AGOs: (1) MID, (2) PIWI, and (3) PAZ (Ma et al., 2005).

The PAZ domain binds to the 3′ end of sRNAs, whereas the MID domain forms a binding pocket for the 5′ end of sRNAs and is thought to convey the binding specificity of an AGO to an sRNA with a particular 5′ end nucleotide.

Unlike *Arabidopsis*, which encodes only one AGO, AGO1, rice encodes four homologs of AGO1: AGO1a, AGO1b, AGO1c, and AGO1d. To understand the function of AGO1, a number of bioinformatics, biochemical, and genetic techniques have been employed to generate AGO1 knockdown lines. Using this knockout technique, the first phenotype generated exhibited a weak loss of AGO1 function in an RNAi line (Song et al., 2004). This line displayed mild dwarfism with narrow and rolled leaves and a lower seed-setting rate. The other line generated exhibited a strong loss of AGO1 function, resulting in severe dwarfism and tortuous shoot formation.

Like *Arabidopsis*, which has ten members of the AGO family, rice has 19 members of the AGO family that are further subdivided into four groups: MEL1, AGO1, AGO4, and AGO7. MEL1 is one of five rice AGOs in the MEL1 subgroup. It plays a specialized role in premeiotic mitosis and meiosis during sporogenesis and is expressed only in germ cells (Nonomura et al., 2007).

MiR398 is a well-studied miRNA that is believed to regulate the body's response to biotic and abiotic stressors by promoting the cleavage of its target genes—CSD1 and CSD2. The miR398 family has two members in the rice genome, producing two mature isoforms (Wu et al., 2009b). Both monocotyledons and dicotyledons, including switchgrass, rice, wheat, barley, *Arabidopsis*, *Brassica rapa*, and *Populus trichocarpa*, exhibit significant conservation within this family. Four members of the rice superoxide dismutase (SOD) gene family—CSD1, CSD2, SODX, and CCSD—are the targets of miR398. STTM-mediated silencing of miRNA families is one assay used to determine its impact. This rice-growing technique is specific and has remained consistent over many generations.

Silencing of miR398 in rice lines (STTM398) resulted in the following:

- Significantly smaller plants than those of the wild type (WT) (62.7 ± 4.1 vs. 86.7 ± 2.4 cm).
- In comparison to WT, STTM398 lines showed smaller panicles with fewer grains (77.8 ± 10.1 vs. 96.31 ± 1.3 grains; $P < 0.001$).
- STTM398 plants flowered about 2 weeks after WT plants (87.4 ± 1.6 days vs. 71.8 ± 1.4 days; $P < 0.001$).
- Compared to WT, STTM398 lines demonstrated a significant decrease in grain length (12.9%) and width (19.5%), leading to a 40% drop in 1,000-grain weight (Mi et al., 2008).

These results suggest that because miR398 plays a major role in rice growth and seed development, it is a feasible target for improvement. Overexpression of miR398b resulted in increased resistance to rice blast disease and was associated with the downregulation of all four target genes, upregulation of defense-related genes, and increased H_2O_2 accumulation (Li et al., 2019). Conversely, when miR398b was suppressed by overexpressing a target gene mimic, increased susceptibility was associated with the upregulation of all four target genes, delayed induction of defense-related genes, and decreased H_2O_2 accumulation. This is most likely due to the fact that SODX, CSD1, and CSD2 all negatively regulate hydrogen peroxide accumulation, whereas CCSD positively regulates it (Chen et al., 2020).

Another couple of examples that have been studied regarding miRNA gene regulation in the expression of developmental factors involved in flowering are as follows:

a. *miR156* is an important part of an intrinsic pathway that controls the transition from vegetative growth to flowering in plants. This pathway's control relies on changes in miR156 content, which decreases from the vegetative to the reproductive stages while increasing the levels of its targets, transcription factors known as squamosa promoter-binding protein-like (SPL). By modulating the expression of a class of MADS box genes that trigger flowering, SPLs are able to control the transition between the juvenile and blooming stages (Chen et al., 2020).

b. In contrast, *miR172* regulates the expression of the transcription factor gene APETALA2 (AP2) and other AP2-like genes, which in turn govern the timing of blooming and the development of floral organs. Overexpressing miR172 causes early flowering and the development of aberrant floral organs in rice plants, similar to what is observed in AP2 mutants (Lee et al., 2010).

Abiotic stresses like CO_2 concentration and temperature share an intricate relationship with miRNA expression in plants, influencing plant development and stress responses. In this context, previous studies help us understand the different molecular mechanisms underlying miRNA-mediated regulation in various plant species, such as *Arabidopsis* and cotton, which are discussed further. However, significant gaps persist in our understanding of how temperature and CO_2 concentration regulate miRNA expression, particularly during the crucial developmental stages of plants (Nievola et al., 2017). Thus, addressing these knowledge gaps is essential to comprehensively understand the complex dynamics that shape plant adaptation and growth changes under diverse environmental conditions.

7.4.1 Effects of CO_2 Concentration and Temperature on miRNA Expression in *Arabidopsis* Development

While temperature affects miRNA expression differentially and has a conflicting effect on the flowering pathway in *Arabidopsis*, elevated CO_2 concentration accelerates the flowering pathway by changing the expression of miRNAs. Furthermore, shifts in temperature and CO_2 impact miRNAs involved in auxin signaling, which influences important regulators and shapes plant growth responses (Lejay et al., 1999).

7.4.1.1 Effect of CO_2 Concentration on the Flowering Pathway in *Arabidopsis*

Previous studies on flowering patterns in *Arabidopsis* reveal that under elevated CO_2 concentration, increased expression of miRNA172 and decreased levels of miRNA156/miRNA157 were observed (Lee et al., 2010). This downregulation of miRNA156/miRNA157 upregulates several squamosal promoter-binding proteins, such as SPL gene transcripts. According to studies by May et al., when *Arabidopsis* plants were exposed to 810 ppm CO_2, they flowered about 1 week earlier than those grown at 430 ppm. This early flowering phenomenon is mediated by the miRNA156/miRNA157-SPLs-miRNA172-AP2-like transcription factor pathways, studies by Lee et al. (2010) show that this pathway controls the vegetative juvenile-to-adult phase transition and flowering time during normal development.

Numerous studies reveal that increased CO_2 concentrations significantly affect the miRNA156/miRNA157-SPLs-miRNA172-AP2-like transcription factor pathways, which in turn cause the levels of miRNA156/miRNA157 to decline and the levels of miRNA172 to increase throughout development. Changes in the expression of SPLs and AP2s confirm the changes in miRNA levels, inducing early onset of flowering in response to elevated CO_2 levels (Mallory & Vaucheret, 2010).

7.4.1.2 Effects of Temperature on the *Arabidopsis* Development

Studies indicate that elevated temperature levels increase miRNA156/miRNA157 expression and decrease miRNA172 levels, contrasting with the effects observed under high CO_2 levels.

To better understand this relationship, an experiment was conducted with three different replicates (i.e., R1, R2, and R3), and mixed responses were observed in the expression of miRNA156/miRNA157 and miRNA172. SPL2, SPL3, and SPL13 were the sole targets in R1, showing an inverse relationship with miRNA156/miRNA157 expression; in contrast, SPL11, SPL2, and SPL15 were expressed oppositely in all three replicates. Regarding miRNA172 targets, higher temperatures in replicate 1 resulted in the downregulation of TOE1, while R2 showed an upregulation of SMZ (Balasubramanian & Weigel, 2006).

Previous research and the data gathered from the tests show that high temperatures significantly modify the expression of every component in the miRNA156/miRNA157-SPLs-miRNA172-AP2-like transcription factor pathways. More research is required to determine how this system controls early flowering in response to high temperatures.

7.4.1.3 Effect of Elevated CO_2 and Temperature on miRNA Expression Which Affects the Auxin Signaling in *Arabidopsis* Development

The five miRNAs implicated in auxin signaling pathways that are differentially regulated have been confirmed by research on *Arabidopsis* miRNA expression. MiR160, -167, and -390 target auxin-responsive factors (ARFs), which control auxin-regulated transcription. In contrast, miR166a-g and miR164a-g are also transcription factor targets and are connected to auxin's role in organ development. Two biological replicates of the downregulated NAC1 were inversely correlated with the expression of miR164, following Wang et al.'s 2005 tests. NAC1 is necessary for the development of floral, leaf, and embryonic organs as well as for the transmission of auxin signals during lateral root formation (Chen & Xiong, 2009). Typically, miR160 negatively regulates the expression of ARF10, -16, and -17, which promote cell divisions in various tissues. Because the expression of miR160 is inversely correlated with temperature, higher temperatures may inhibit plant growth by upregulating miR160 expression and increasing ARF10 levels, whereas higher CO_2 concentrations may promote growth by downregulating miR160 expression (Lee et al., 2010).

7.4.2 Effect of microRNA Expression in Response to Temperature Stress during Development of Cotton Plants

miRNAs function as essential regulators of gene expression in plants and affect various aspects of development and stress responses. In cotton (*Gossypium hirsutum* L.), understanding the role of miRNAs in anther and ovule development, especially under high-temperature stress, is important for deciphering the mechanisms affecting plant growth and resistance (Chen et al., 2020).

Some examples depicting miRNA regulation in cotton during different stages are as follows.

7.4.2.1 Ovule Development

During ovule development, miRNA expression patterns undergo dynamic changes that reflect complex regulatory processes. For example, in a study analyzing miRNA expression in immature and fibrous oocytes, 14 miRNAs were found to have higher expression levels in 2DPA (Day Post-Anthesis) ovules compared to 0DPA ovules. In particular, miR156b and miR2911 showed significantly increased expression in 2DPA, indicating their potential role in modulating oocyte development and fiber formation (Wang et al., 2017).

7.4.2.2 Anther Development

In anther development, miRNA regulation plays a crucial role in determining plant responses to environmental stressors such as high temperature. RNA sequencing studies at different developmental stages (SCP, MP, MRP) under normal and high-temperature conditions revealed significant differences in expression patterns of miRNA families, including miR172, miR2949, and miR156. Variations in these miRNA expressions were found to correlate with heat sensitivity and tolerance among different cotton lines, highlighting the complex interplay between miRNA regulation and stress response during anther development (Goldschmidt & Huber, 1992).

7.4.2.3 Fiber Elongation and Secondary Wall Formation

During fiber elongation and secondary wall formation, miRNAs play a crucial role in controlling the development of cotton fibers by regulating essential regulatory pathways. For example, analysis of miRNA expression patterns by qRT-PCR revealed fascinating dynamics at different stages of fiber development. MiR4370's modulation of fiber growth and secondary wall formation was found to be more effective during rapid fiber elongation (10DPA) than at the start of fiber enlargement

(5DPE). This suggests significant potential. These findings highlight the complex regulatory networks controlled by miRNAs during cotton fiber development (Wang et al., 2017).

Together, the study of miRNA regulation in cotton at different developmental stages provides valuable insights into the molecular mechanisms underlying plant growth and stress responses. Such knowledge has significant implications for breeding efforts aimed at improving cotton yield and quality, especially in the face of environmental challenges such as high temperatures (Wijewardene et al., 2020) (Figure 7.1).

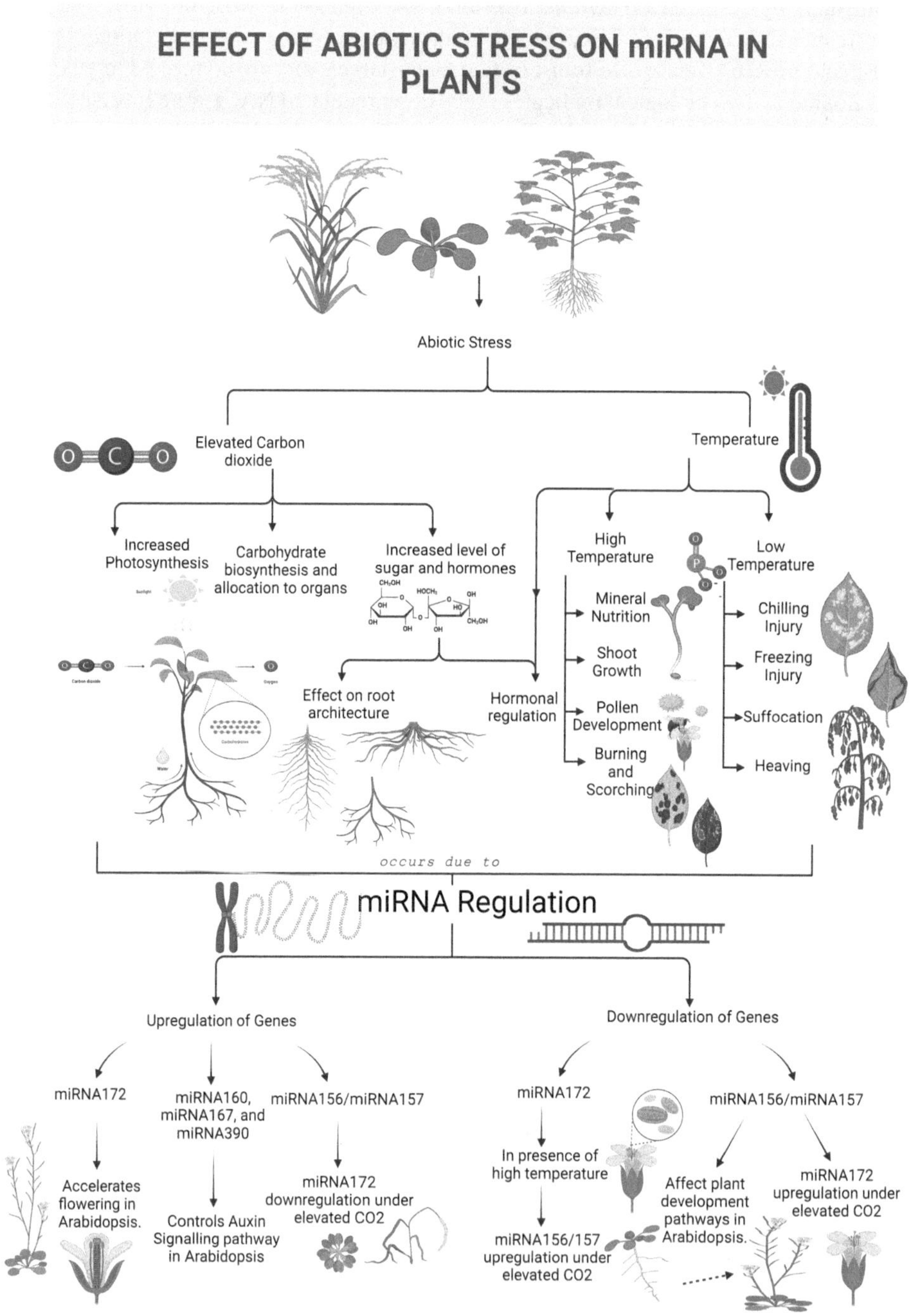

FIGURE 7.1 Effect of abiotic stress on miRNA regulation in plants. (Figure generated using BioRender.com.)

7.5 CONCLUSION

In summary, salinity, temperature, and carbon dioxide (CO_2) levels resulting from abiotic stresses can adversely affect global yields, agricultural productivity, and food security. These environmental factors interact in complex ways that affect plant physiology, development, and stress responses. miRNAs play a crucial role in plant gene expression and mediate responses to both biotic and abiotic stresses.

Research on miRNAs in rice, *Arabidopsis*, and cotton has provided valuable insights into their role in stress responses. In rice, the miR398 molecule is involved in modulating responses to abiotic stresses, while in *Arabidopsis*, flowering responses depend on miRNAs that regulate CO_2 levels and temperature. High levels of CO_2 initially stimulate photosynthesis, but over time, these processes show downregulation of the genes involved, leading to changes in carbohydrate metabolism and nutrient distribution among plants. Higher CO_2 levels promote root growth but hinder nutrient absorption through sugar breakdown. Both low and high temperatures affect yield, as temperature is a crucial factor in determining the extent of damage caused by exposure to cold or hot conditions.

In addition, studies show that both CO_2 and temperature stresses affect miRNA regulation, further highlighting the complexity of plant stress responses in a changing climate. Differential miRNA expression patterns were observed in cotton in response to high-temperature stress during anther development, differentially affecting thermosensitive and thermotolerant traits. Understanding these complex interactions between CO_2 concentration, temperature fluctuations, and miRNA regulation is important for comprehensively understanding the mechanisms that shape plant adaptation and growth under varying environmental conditions.

Integrating molecular knowledge into breeding programs promises to develop crops with greater resilience and productivity in a changing climate, contributing to sustainable agriculture and global food security. To improve crop tolerance and productivity in harsh environments, it is imperative to explore the molecular mechanisms that regulate plant stress responses and miRNA regulation using novel methods. We need to address these knowledge gaps to better prepare for the effects of climate change on agriculture and work toward ensuring food security for future generations.

REFERENCES

Abbass, K., Qasim, M. Z., Song, H., Murshed, M., Mahmood, H., & Younis, I. (2022). A review of the global climate change impacts, adaptation, and sustainable mitigation measures. *Environmental Science and Pollution Research International*, *29*(28), 42539–42559. https://doi.org/10.1007/s11356-022-19718-6

Ainsworth, E. A., Long, S. P., Rogers, A. & Ort, D. R. (2004). Rising atmospheric carbon dioxide: plants FACE the future. *Annu Rev Plant Biol*, *55*, 591–628. doi: 10.1146/annurev.arplant.55.031903.141610. PMID: 15377233.

Ainsworth, E. A., & Long, S. P. (2005). What have we learned from 15 years of free-air CO_2 enrichment (FACE)? A meta-analytic review of the responses of photosynthesis, canopy properties and plant production to rising CO_2. *New Phytologist*, *165*(2), 351–372. https://doi.org/10.1111/j.1469-8137.2004.01224.x

Aljazairi, S., & Nogués, S. (2015). The effects of depleted, current and elevated growth [CO_2] in wheat are modulated by water availability. *Environmental and Experimental Botany*, *112*, 55–66. https://doi.org/10.1016/j.envexpbot.2014.12.002

Anil, P., Dharmendra, P., Sachidanand, S., Yuriy, E., Snezhan, B., Dilfuza, J., & Chinmayi, J. (2023). Beneficial effects of biochar application on mitigating the drought and salinity stress implications on plants. *Plant Science Today*, *10*(2), 188–193. http://doi.org/10.14719/pst.2591

Aranjuelo, I., Erice, G., Sanz-Sáez, A., Abadie, C., Gilard, F., Gil-Quintana, E., Avice, J., Staudinger, C., Wienkoop, S., Araus, J. L., Bourguignon, J., Irigoyen, J. J., & Tcherkez, G. (2015). Differential CO_2 effect on primary carbon metabolism of flag leaves in durum wheat (*Triticum durum* Desf.). *Plant, Cell and Environment*, *38*(12), 2780–2794. https://doi.org/10.1111/pce.12587

Balasubramanian, S., & Weigel, D. (2006). Temperature induced flowering in arabidopsis thaliana. *Plant Signaling & Behavior*, *1*(5), 227–228. https://doi.org/10.4161/psb.1.5.3452

Bauwe, H., Hagemann, M., & Fernie, A. R. (2010). Photorespiration: Players, partners and origin. *Trends in Plant Science*, *15*(6), 330–336. https://doi.org/10.1016/j.tplants.2010.03.006

Begum, Y. (2022). Regulatory role of microRNAs (miRNAs) in the recent development of abiotic stress tolerance of plants. *Gene*, *821*, 146283. https://doi.org/10.1016/j.gene.2022.146283

Broberg, M. C., Högy, P., Feng, Z., & Pleijel, H. (2019). Effects of elevated CO_2 on wheat yield: Non-linear response and relation to site productivity. *Agronomy*, *9*(5), 243. https://doi.org/10.3390/agronomy9050243

Butterly, C. R., Armstrong, R., Chen, D., & Tang, C. (2015). Carbon and nitrogen partitioning of wheat and field pea grown with two nitrogen levels under elevated CO_2. *Plant and Soil*, *391*(1–2), 367–382. https://doi.org/10.1007/s11104-015-2441-5

Canadell, J. G., Le Quéré, C., Raupach, M. R., Field, C. B., Buitenhuis, E. T., Ciais, P., Conway, T. J., Gillett, N. P., Houghton, R. A., & Marland, G. (2007). Contributions to accelerating atmospheric CO_2 growth from economic activity, carbon intensity, and efficiency of natural sinks. *Proceedings of the National Academy of Sciences of the United States of America*, *104*(47), 18866–18870. https://doi.org/10.1073/pnas.0702737104

Çelik, S. (2023). Assessing drought tolerance in a large number of upland cotton plants (*Gossypium hirsutum* L.) under different irrigation regimes at the seedling stage. *Life*, *13*(10), 2067. https://doi.org/10.3390/life13102067

Chen, H., & Xiong, L. (2009). Localized auxin biosynthesis and postembryonic root development in Arabidopsis. *Plant Signaling & Behavior*, *4*(8), 752–754. https://doi.org/10.4161/psb.4.8.9177

Chen, J., Pan, A., He, S., Su, P., Yuan, X., Zhu, S., & Liu, Z. (2020). Different microRNA families involved in regulating high temperature stress response during cotton (*Gossypium hirsutum* L.) anther development. *International Journal of Molecular Sciences*, *21*(4), 1280. https://doi.org/10.3390/ijms21041280

Drake, B. G., Gonzàlez-Meler, M. A., & Long, S. P. (1997). More efficient plants: A consequence of rising atmospheric CO_2? *Annual Review of Plant Physiology and Plant Molecular Biology*, *48*(1), 609–639. https://doi.org/10.1146/annurev.arplant.48.1.609

Fernando, N., Panozzo, J., Tausz, M., Norton, R., Fitzgerald, G., & Seneweera, S. (2012). Rising atmospheric CO_2 concentration affects mineral nutrient and protein concentration of wheat grain. *Food Chemistry*, *133*(4), 1307–1311. https://doi.org/10.1016/j.foodchem.2012.01.105

Flowers, T. J., & Colmer, T. D. (2008). Salinity tolerance in halophytes. *New Phytologist*, *179*(4), 945–963. https://doi.org/10.1111/j.1469-8137.2008.02531.x

Goldschmidt, E. E., & Huber, S. C. (1992). Regulation of photosynthesis by end-product accumulation in leaves of plants storing starch, sucrose, and hexose sugars. *Plant Physiology*, *99*(4), 1443–1448. https://doi.org/10.1104/pp.99.4.1443

Guan, Q., Lu, X., Zeng, H., Zhang, Y., & Zhu, J. (2013). Heat stress induction of *miR398* triggers a regulatory loop that is critical for thermotolerance in Arabidopsis. *The Plant Journal*, *74*(5), 840–851. https://doi.org/10.1111/tpj.12169

Guo, H., Sun, Y., Ren, Q., Zhu-Salzman, K., Kang, L., Wang, C., Li, C., & Ge, F. (2012). Elevated CO2 Reduces the Resistance and Tolerance of Tomato Plants to Helicoverpa armigera by Suppressing the JA Signaling Pathway. *PLOS ONE*, *7*(7), e41426. https://doi.org/10.1371/journal.pone.0041426

Guo, J., Shi, G., Guo, X., Zhang, L., Xu, W., Wang, Y., Su, Z., & Hua, J. (2015). Transcriptome analysis reveals that distinct metabolic pathways operate in salt-tolerant and salt-sensitive upland cotton varieties subjected to salinity stress. *Plant Science*, *238*, 33–45. https://doi.org/10.1016/j.plantsci.2015.05.013

Högy, P., & Fangmeier, A. (2009). Atmospheric CO_2 enrichment affects potatoes: 2. Tuber quality traits. *European Journal of Agronomy*, *30*(2), 85–94. https://doi.org/10.1016/j.eja.2008.07.006

Jabborova, D., Nigora, Z., Yuriy, E., Abdujalil, N., Anand, D., Khurshid, S., Zafarjon, J., Sachidanand, S., & Rahul, D. (2023). Growth of spinach as influenced by biochar and *Bacillus endophyticus* IGPEB 33 in drought condition. *Pakistan Journal of Botany*, *55*(SI), 1–7. https://doi.org/10.30848/PJB2023-SI(6).

Jay, J. M., Loessner, M. J., & Golden, D. A. (2005). *Modern Food Microbiology, 7th Edition (Food Science Texts Series)*. New York: Springer Science+Business Media, Inc.

Keeling, C. D., Mook, W. I. M. G., & Tans, P. P. (1979). Recent trends in the 13C/12C ratio of atmospheric carbon dioxide. *Nature*, *277*(5692), 121–123. https://doi.org/10.1038/277121a0

Kumari, M., Verma, S. C., & Bhardwaj, S. K. (2017). Impact of elevated CO_2 and temperature on quality and biochemical parameters of pea (*Pisum sativum*). *The Indian Journal of Agricultural Sciences*, *87*(8), 179–184. https://doi.org/10.56093/ijas.v87i8.73234

Lee, H., Yoo, S. J., Lee, J. H., Kim, W., Yoo, S. K., Fitzgerald, H., Carrington, J. C., & Ahn, J. H. (2010). Genetic framework for flowering-time regulation by ambient temperature-responsive miRNAs in *Arabidopsis*. *Nucleic Acids Research*, *38*(9), 3081–3093. https://doi.org/10.1093/nar/gkp1240

Lee-Ho, E., Walton, L. J., Reid, D. M., Yeung, E. C., & Kurepin, L. V. (2007). Effects of elevated carbon dioxide and sucrose concentrations on *Arabidopsis thaliana* root architecture and anatomy. *Canadian Journal of Botany*, *85*(3), 324–330. https://doi.org/10.1139/b07-009

Lejay, L., Tillard, P., Lepetit, M., Olive, F. D., Filleur, S., Daniel-Vedele, F., & Gojon, A. (1999). Molecular and functional regulation of two NO_3^- uptake systems by N- and C-status of *Arabidopsis* plants. *The Plant Journal*, *18*(5), 509–519. https://doi.org/10.1046/j.1365-313x.1999.00480.x

Li, S., Li, Y., He, X., Li, Q., Liu, B., Ai, X., & Zhang, D. (2019). Response of water balance and nitrogen assimilation in cucumber seedlings to CO_2 enrichment and salt stress. *Plant Physiology and Biochemistry*, *139*, 256–263. https://doi.org/10.1016/j.plaphy.2019.03.028

Li, Y., Cao, X.-L., Zhu, Y., Yang, X.-M., Zhang, K.-N., Xiao, Z.-Y., Wang, H., Zhao, J.-H., Zhang, L.-L., Li, G.-B., Zheng, Y.-P., Fan, J., Wang, J., Chen, X.-Q., Wu, X.-J., Zhao, J.-Q., Dong, O.X., Chen, X.-W., Chern, M. & Wang, W.-M. (2019), Osa-miR398b boosts H2O2 production and rice blast disease-resistance via multiple superoxide dismutases, *New Phytol*, *222*, 1507–1522. https://doi.org/10.1111/nph.15678

Llorens, E., Camañes, G., Lapeña, L., & García-Agustín, P. (2016). Priming by hexanoic acid induce activation of mevalonic and linolenic pathways and promotes the emission of plant volatiles. *Frontiers in Plant Science*, *7*, 495. https://doi.org/10.3389/fpls.2016.00495

Ma, J.-B., Ye, K., & Patel, D. J. (2004). Structural basis for overhang-specific small interfering RNA recognition by the PAZ domain. *Nature*, *429*(6989), 318–322. https://doi.org/10.1038/nature02519

Ma, J.-B., Yuan, Y.-R., Meister, G., Pei, Y., Tuschl, T., & Patel, D. J. (2005). Structural basis for 5′-end-specific recognition of guide RNA by the *A. fulgidus* Piwi protein. *Nature*, *434*(7033), 666–670. https://doi.org/10.1038/nature03514

Macfarlane, L.-A., & Murphy, P. R. (2010). microRNA: Biogenesis, function and role in cancer. *Current Genomics*, *11*(7), 537–561. https://doi.org/10.2174/138920210793175895

Makino, A., & Mae, T. (1999). Photosynthesis and plant growth at elevated levels of CO_2. *Plant and Cell Physiology*, *40*(10), 999–1006. https://doi.org/10.1093/oxfordjournals.pcp.a029493

Mallory, A., & Vaucheret, H. (2010). Form, function, and regulation of ARGONAUTE proteins. *The Plant Cell*, *22*(12), 3879–3889. https://doi.org/10.1105/tpc.110.080671

Martins, M. C., Hejazi, M., Fettke, J., Steup, M., Feil, R., Krause, U., Arrivault, S., Vosloh, D., Figueroa, C. M., Ivakov, A., Yadav, U. P., Piques, M., Metzner, D., Stitt, M., & Lunn, J. E. (2013). Feedback inhibition of starch degradation in Arabidopsis leaves mediated by trehalose 6-phosphate. *Plant Physiol*, 2013 Nov; *163*(3), 1142–1163. doi: 10.1104/pp.113.226787. Epub 2013 Sep 16. PMID: 24043444; PMCID: PMC3813640.

May, P., Liao, W., Wu, Y., Shuai, B., Richard McCombie, W., Zhang, M. Q., & Liu, Q. A. (2013). The effects of carbon dioxide and temperature on microRNA expression in *Arabidopsis* development. *Nature Communications*, *4*(1), 3145. https://doi.org/10.1038/ncomms3145

Meenakshi, S., Rahul, D., Sachidanand, S., Anita, G. M., & Rajpal, S. (2025). Plant microRNAs: An overview. In: Dutta, R., Kapadia, C., Danish, S., & Singh, S. (eds.), *microRNA Advances and Application in Plant Biology Plant Biology, Sustainability and Climate Change*. New York: Academic Press.

Mi, S., Cai, T., Hu, Y., Chen, Y., Hodges, E., Ni, F., Wu, L., Li, S., Zhou, H., Long, C., Chen, S., Hannon, G. J., & Qi, Y. (2008). Sorting of small RNAs into Arabidopsis argonaute complexes is directed by the 5′ terminal nucleotide. *Cell*, *133*(1), 116–127. https://doi.org/10.1016/j.cell.2008.02.034

Munns, R., & Tester, M. (2008). Mechanisms of salinity tolerance. *Annual Review of Plant Biology*, *59*(1), 651–681. https://doi.org/10.1146/annurev.arplant.59.032607.092911

Myers, S. S., Zanobetti, A., Kloog, I., Huybers, P., Leakey, A. D. B., Bloom, A. J., Carlisle, E., Dietterich, L. H., Fitzgerald, G., Hasegawa, T., Holbrook, N. M., Nelson, R. L., Ottman, M. J., Raboy, V., Sakai, H., Sartor, K. A., Schwartz, J., Seneweera, S., Tausz, M., & Usui, Y. (2014). Increasing CO_2 threatens human nutrition. *Nature*, *510*(7503), 139–142. https://doi.org/10.1038/nature13179

Nievola, C. C., Carvalho, C. P., Carvalho, V., & Rodrigues, E. (2017). Rapid responses of plants to temperature changes. *Temperature*, *4*(4), 371–405. https://doi.org/10.1080/23328940.2017.1377812

Nonomura, K.-I., Morohoshi, A., Nakano, M., Eiguchi, M., Miyao, A., Hirochika, H., & Kurata, N. (2007). A germ cell specific gene of the ARGONAUTE family is essential for the progression of premeiotic mitosis and meiosis during sporogenesis in rice. *The Plant Cell*, *19*(8), 2583–2594. https://doi.org/10.1105/tpc.107.053199

Outlaw, W. H., & Manchester, J. (1979). Guard cell starch concentration quantitatively related to stomatal aperture. *Plant Physiol*. 1979 Jul, *64*(1), 79–82. doi: 10.1104/pp.64.1.79. PMID: 16660919; PMCID: PMC543028.

Poorter, H. (1993). Interspecific variation in the growth response of plants to an elevated ambient CO_2 concentration. *CO_2 and Biosphere*, *1993*, 77–98. https://doi.org/10.1007/978-94-011-1797-5_6

Rogers, A., Ainsworth, E. A., & Leakey, A. D. B. (2009). Will elevated carbon dioxide concentration amplify the benefits of nitrogen fixation in legumes? *Plant Physiology*, *151*(3), 1009–1016. https://doi.org/10.1104/pp.109.144113

Shriram, V., Kumar, V., Devarumath, R. M., Khare, T. S., & Wani, S. H. (2016). microRNAs as potential targets for abiotic stress tolerance in plants. *Frontiers in Plant Science*, *7*, 817. https://doi.org/10.3389/fpls.2016.00817

Song, J.-J., Smith, S. K., Hannon, G. J., & Joshua-Tor, L. (2004). Crystal structure of argonaute and its implications for RISC slicer activity. *Science*, *305*(5689), 1434–1437. https://doi.org/10.1126/science.1102514

van der Kooi, C. J., Reich, M., Löw, M., De Kok, L. J., & Tausz, M. (2016). Growth and yield stimulation under elevated CO_2 and drought: A meta-analysis on crops. *Environmental and Experimental Botany*, *122*, 150–157. https://doi.org/10.1016/j.envexpbot.2015.10.004

Verma, N., Sao, P., Srivastava, A., & Singh, S. (2021). Physiological and molecular responses to drought, submergence and excessive watering in plants. In: Husen, A. (ed.), *Harsh Environment and Plant Resilience*, pp. 305–321. Cham: Springer International Publishing.

Wang, M., Sun, R., Li, C., Wang, Q., & Zhang, B. (2017). microRNA expression profiles during cotton (*Gossypium hirsutum* L) fiber early development. *Scientific Reports*, *7*, 44454. https://doi.org/10.1038/srep44454

Wang, W., Vinocur, B., & Altman, A. (2003). Plant responses to drought, salinity and extreme temperatures: Towards genetic engineering for stress tolerance. *Planta*, *218*(1), 1–14. https://doi.org/10.1007/s00425-003-1105-5

Wijewardene, I., Mishra, N., Sun, L., Smith, J., Zhu, X., Payton, P., Shen, G., & Zhang, H. (2020). Improving drought-, salinity-, and heat-tolerance in transgenic plants by co-overexpressing Arabidopsis vacuolar pyrophosphatase gene AVP1 and Larrea Rubisco activase gene RCA. *Plant Science*, *296*, 110499. https://doi.org/10.1016/j.plantsci.2020.110499

Wu, L., Zhang, Q., Zhou, H., Ni, F., Wu, X., & Qi, Y. (2009a). Rice microRNA effector complexes and targets. *The Plant Cell*, *21*(11), 3421–3435. https://doi.org/10.1105/tpc.109.070938

Yang, X., Yu, Y. & Jiang, L. (2012). Changes in DNA methylation and transgenerational mobilization of a transposable element (mPing) by the Topoisomerase II inhibitor, Etoposide, in rice. *BMC Plant Biol*, *12*, 48. https://doi.org/10.1186/1471-2229-12-48

8 Abiotic Stress Tolerance in Plants

Targeting microRNAs

Mukesh Sharma, Arundhuti Chakrabarty, Priyashree Laishram, and Namisha Arora*

8.1 INTRODUCTION

The alteration in the climate is driven by two factors: the increase in the human population and global industrialization. Due to the exponential growth of both factors, future generations will need additional arable land for the dedicated production of food and cash crops. As these components continue to grow at an exponential rate, more arable land will be required to cultivate cash crops and dedicated food crops to support future generations.

Abiotic stressors, such as salt, extreme weather, drought, chemical toxic exposure, and oxidative stress, pose a serious threat to agriculture and the environment (Anil et al. 2023; Jabborova et al. 2023; Verma et al. 2021). Accelerated salt accumulation in arable land is predicted to have catastrophic worldwide repercussions, with up to 50% more land lost by the year 2050. Furthermore, the effects of environmental stresses can be observed at various sub-organismal levels, including physiological, cellular, biochemical, molecular, and ultimately, organismal levels. It has also been demonstrated that abiotic challenges significantly alter gene expression profiles at various developmental stages; these modifications to gene expression programs ultimately control the timing and development of plant plasticity.

Abiotic stressors have been linked to several important genes, including those encoding transcription factors. It has been shown that overexpressing these genes in model plant species, such as *Arabidopsis*, and other agriculturally significant crops, greatly increases plant tolerance to specific stresses (Liu et al. 2013). The emergence of novel technologies in the last 10 years, including next-generation sequencing and microarrays, has facilitated the identification of 100 genes encoding proteins associated with various abiotic stresses.

The answers to the questions of how these genes that encode proteins are regulated might be found in recently identified techniques. How do plants adapt to various abiotic stresses? What kind of gene networks do they use? A broad class of tiny endogenous RNA molecules, 20–24 nucleotides in length, is known as microRNAs (miRNAs). From mosses to higher-blooming monocots and dicots, miRNAs are believed to be highly evolutionarily conserved and widely distributed throughout the plant kingdom.

8.1.1 Types of Stress

Several environmental stressors affect plants, reducing and limiting the production of crops used for agriculture. There are two types of environmental stress that plants might face, categorized as follows.

8.1.1.1 Abiotic Stress

This includes factors such as radiation, heavy metals, salinity, drought, flooding, and temperature fluctuations. It is one of the main causes of the global decline of important crop plants. The world's

* *Corresponding Author*

DOI: 10.1201/9781003491378-8

terrestrial area is becoming increasingly desertified, water and soil are becoming more salinized, water resources are becoming scarce, and pollution is adversely affecting ecosystems. As a result, conditions will become more challenging (Tungare et al., 2022).

8.1.1.2 Biotic Stress

This includes an assault by a large number of diseases, including bacteria, fungi, oomycetes, nematodes, and herbivores. Globally, diseases caused by these pathogens are the main cause of crop loss. As sessile organisms, plants cannot avoid these environmental stimuli. Proficiency in managing these strains is essential for effectively completing their life cycle. Therefore, plants have developed a variety of defense systems to withstand these attacks and survive in such environments. They perceive the environmental stressors, become activated, and subsequently produce the necessary biological responses. These physiological reactions function by transmitting impulses from sensors on the cytoplasmic or cell surface to the transcriptional machinery in the nucleus via various signal transduction channels.

8.1.2 History of microRNAs

When researchers first noticed that miRNAs were present in biological processes and had significant importance, this marked the beginning of the miRNA era. Here's a brief overview of the history of miRNAs:

- **Discovery of lin-4 in *Caenorhabditis elegans* (1993)**: Victor Ambros and Gary Ruvkun discovered lin-4, a tiny RNA molecule, in the nematode *C. elegans*, marking the beginning of the miRNA narrative. They found that by binding to specific messenger RNA (mRNA) targets, lin-4 controlled the timing of larval development in *C. elegans*.
- **Discovery of let-7 in *C. elegans* (2000)**: Another milestone in miRNA research was the discovery of let-7 by Frank Slack and colleagues in *C. elegans*. They found that let-7 modulates the timing of larval development and the metamorphosis from larval to adult stages.
- **Identification of miRNAs in Other Organisms (Early 2000s)**: Following the discoveries in *C. elegans*, researchers identified miRNAs in other organisms, including fruit flies, mice, and humans. This led to the recognition that miRNAs are evolutionarily conserved and play crucial roles in diverse biological processes.

8.1.3 Why Are microRNAs Needed for Abiotic Stress?

Eighty percent of a person's diet should consist of plants, according to estimates from the Food and Agriculture Organization (FAO) of the United Nations. Plant productivity needs to be maximized to ensure that there is enough food that is healthy, safe, affordable, and meets the needs of a growing human population. Evidence shows that various biotic and abiotic stresses, such as intense heat, cold, salt, drought, and diseases, significantly reduce crop productivity. The negative impacts of stress on plants have been exacerbated by global climate change. During stressful conditions, some miRNAs found in plants regulate the post-transcriptional expression of genes, in addition to the multiple regulatory proteins and inducible signaling pathways. miRNA in plants can control various developmental pathways and may also play a role in responding to both biotic and abiotic stresses (Kar and Raichaudhuri 2021).

8.1.4 microRNAs

miRNAs are a group of naturally occurring small RNAs that do not code for proteins. In many organisms, the length of miRNAs varies from 20 to 24 nucleotides. Certain miRNAs have been observed to play gene-regulatory roles in plant development by cleaving or translationally suppressing target mRNA, which can negatively impact the post-transcriptional phase of eukaryotic gene expression (Meenakshi et al. 2025). In the cells of eukaryotic organisms, miRNAs drive complex regulatory networks and play a role in a wide range of biological processes that affect responses to biotic and abiotic stimuli (Pegler et al. 2019). Both experimental and bioinformatics techniques have been used to detect miRNAs in plants. When miRNAs in plants were first discovered, the direct cloning method was widely employed. Many applications of short RNA high-throughput sequencing have been utilized to identify and investigate miRNAs in plants functionally.

8.1.4.1 Techniques to Detect microRNAs: There Are Different Methods of Detecting miRNA

8.1.4.1.1 Traditional Methods

Methods to measure and visualize aberrant miRNA expression for early diagnosis have been developed. This is a challenging task because of their small sizes, low quantities, and high sequence homology among species. Some traditional methods are briefly discussed below:

a. **Northern Blotting:** Northern blotting is the most widely used and recognized method for identifying miRNAs. It can distinguish between precursor and mature miRNAs. This method can assess the molecular size and relative abundance of miRNAs.
b. **Real-Time qPCR:** Because of its high sensitivity, high sequence specificity, and wide dynamic range, real-time qPCR has become a reliable and common technique for identifying miRNA expression. The previously mentioned methods involve first transferring the target miRNA into cDNA using reverse transcription. Subsequently, real-time fluorescence detection is achieved through PCR.
c. **Microarray Technology:** Microarray technology is the most widely used method for rapidly and effectively identifying large numbers of miRNAs. This technique uses a tagged probe to create the sample miRNAs for reverse transcription, and solid-phase microscopy is used to detect the fluorophores or biotin-labeled cDNAs (Sinha et al. 2022). The main advantage of this technique is that numerous samples can be evaluated daily; however, a major disadvantage is the high cost of this technology.

8.1.4.1.2 Newly Developed Methods

a. **Nanomaterial-Based miRNA Detection:** Recently, a variety of nanomaterials, including gold nanoparticles, magnetic nanoparticles, silver nanoclusters, and quantum dots, have been used in research for miRNA detection. Utilizing nanoparticles can effectively enhance the performance of traditional miRNA detection techniques.
b. **Nucleic Acid Amplification Techniques:** Nucleic acid amplification techniques are typically employed to enhance the sensitivity of miRNA detection methods. Since miRNAs are not highly concentrated in tissues or cells, nucleic acid amplification methods are highly efficient in detecting minute amounts of miRNAs. More specifically, various novel technologies based on isothermal amplification have emerged because they are more adaptable to point-of-care testing equipment than traditional PCR-based assays.

8.1.5 Plants' miRNA Biosynthesis and Mechanism of Action

A large percentage of miRNAs are a type of preserved native short RNA that, following the transcription of eukaryotic genes, has a significant regulatory function. While miRNAs in plants are found between genes, the majority of miRNAs in metazoans are located within thousands of introns or exons. Additionally, different plant species exhibit quite distinct miRNA secondary structures and lengths. Plant miRNA genes are rarely organized in a series. Pri-miRNAs, which are formed when Pol II transcribes miRNAs in the nucleus, are protein-coding genes. Pri-miRNAs feature a 3′ poly-A tail and a 5′ cap and can be several hundred to several thousand nucleotides long. Pre-miRNAs are created by cleaving pri-miRNAs, which range from 70 to 350 nt in length, by the enzyme known as DCL1. Hyponastic leaves 1, an RNA-binding protein, and serrate, a C2H2 zinc finger protein, interact with pre-miRNAs to further generate these pre-miRNAs into mature miRNAs (Kurihara and Watanabe 2004). Two bases branch out from the mature miRNAs' 3′ terminus (miRNA double-stranded complex). The HUAENHANCER1 protein methylates this miRNA complex at the 2′-OH position of its 3′ terminus to prevent it from breaking down. The plant counterpart of exportin-5, HASTY, facilitates most of the cytoplasmic transport of the methylated miRNA complexes. Research has demonstrated that mature miRNAs can inhibit target gene translation, control plant gene expression by complementarily integrating with regions that contain coding sequences, bind to the 3′ and 5′ UTR of target mRNA, and may cause post-transcriptional degradation of the target gene's mRNA. The inhibition of this process by mature miRNAs impacts plant morphology, the development of plant parts, growth, hormone biosynthesis, signal transduction, and the plants' capacity to respond to external influences and environmental pressures. In plants, miRNA primarily functions through cleavage, as its target mRNA is significantly complementary to it. It was not until recently that the mechanisms of translation inhibition in plants were discovered. The complementary nature of the targets and miRNAs determines the degree of coordination between the cleavage and inhibition processes (Brien et al. 2018).

8.1.6 The Role of miRNAs in Plant Growth and Development

Plant growth and development are extremely tightly regulated, and this regulation is influenced by both internal genetic information and external environmental factors. Normal miRNA expression is essential for the growth and development of plants. Previous research has demonstrated that miRNA controls several aspects of plant development and growth, which are listed below:

- The role of miRNAs in the development of leaves
- The function of miRNAs in the shoot meristem
- The role of miRNAs in vascular formation
- The role of miRNAs in the development of flowers
- The function of miRNAs in resisting agonist stress, etc.

8.1.7 The Role of miRNAs in Plant Abiotic Stress

Due to the wide range of functions of their target genes, miRNAs are novel targets for improving crops, enhancing plant productivity, and creating plants resistant to abiotic stress. A summary of miRNAs that respond to abiotic stressors in plants will be found in this section.

8.1.7.1 Drought Tolerance

Numerous miRNAs have been demonstrated to be essential for plants' responses to drought stress (Zhao et al. 2007) discovered the function of miR169 in rice under drought stress and proposed that CBF/DREBs (dehydration-responsive elements) regulate this gene. Another study examining the role of miR169 in wheat found that it regulates stress caused by drought and affects the transcription factor CCCAT-box. MiR160 targets ARFs, such as ARF10 and ARF16, which are implicated in auxin and abscisic acid (ABA) signaling during drought stress (Ding et al. 2013). Because drought stress causes ARF10 to be downregulated, MiR160 targets this protein, indicating the role auxin plays in stress (Bouzroud et al. 2018). In *Arabidopsis*, rice, maize, and wheat during drought conditions, there was an increase in the expression of miR160, miR162, miR395, and miR827, and a decrease in the expression of miR166, miR172, miR397, miR827, and miR1432. The biological functions of these miRNAs may be explained by their conservation. Furthermore, distinct expression patterns of miR171 are observed in rice and maize during drought stress

8.1.7.2 Salt Tolerance

Salinity tolerance was shown to be negatively affected by miR393, which also suppresses the translation of the transport inhibitor response 1 gene. Transgenic *Arabidopsis* plants that overexpress the miR393-resistant transport inhibitor response 1 gene exhibit enhanced resistance to salt stress, along with reduced water loss and improved germination. In another study, it was found that miR172a serves as a long-range signal that provides soybeans with resistance to salt stress by specifically downregulating the SSAC1 gene, which codes for an AP2/EREBP-type transcription factor involved in the negative regulation of salt tolerance (Pan et al. 2016).

8.1.7.3 Tolerance to Extreme Temperatures

Plants subjected to extreme temperatures can be stressed by both heat and cold. Through the targeted suppression of the rice OsPCF6 and OsTCP2 genes, miR319 positively regulates cold tolerance. Zhang et al. (2016) found that the resilience of *Citrus limon* to cold was enhanced by the overexpression of miR396b from *Poncirus trifoliata*. Compared to wild-type plants, the transgenic plants produced more polyamines and accumulated less reactive oxygen species (ROS) in cold environments. Transgenic rice plants expressing *miR156* exhibited enhanced cold tolerance. According to Liu et al. (2015), there is differential regulation of the conserved miRNA family, which includes miR156, miR160, miR164, miR167, miR186, miR169, miR171, miR172, miR319, miR333, miR395, miR397, miR398, miR399, miR408, miR827, and miR1450. Across species, the miR159 and miR472 families are downregulated, while miR166 is upregulated under high temperature conditions. Similarly, cotton with higher levels of miR160 and miR157 showed increased heat stress sensitivity and male sterility through the activation of auxin signaling and suppression of squamosa promoter-binding protein-like (SPL) expression. The mechanism of high-temperature adaptation is regulated by miR160, which also positively regulates the expression of many heat shock protein genes. Numerous miRNAs, including miR531a, miR5149, miR168a-5P, miR1846d-5P, miR5077, miR156b-3P, and miR167e-3P, have been associated with the thermopriming response in plants.

8.1.8 Types of miRNAs

MicroRNAs (miRNAs) can be classified into different types based on their evolutionary conservation, functional roles, biogenesis mechanisms, and target specificity. Evolutionarily conserved miRNAs, such as miR156/157, miR160, and miR166, are found across multiple plant species and play fundamental roles in growth, development, and stress responses. In contrast, species-specific miRNAs like miR528 in monocots and miR2118 in legumes have evolved to regulate unique stress-responsive pathways in particular plants. Functionally, miRNAs can be categorized as stress-responsive (e.g., miR169 in drought, miR398 in oxidative stress, and miR399 in phosphate deficiency) or development-related (e.g., miR172 in flowering and miR396 in leaf growth), with many miRNAs participating in both processes. Based on biogenesis, canonical miRNAs are processed by DCL1 from primary transcripts, while non-canonical miRNAs, such as mirtrons and siRNA-like miRNAs, arise through alternative pathways. Additionally, miRNAs differ in target specificity, with high-specificity miRNAs inducing mRNA cleavage through near-perfect complementarity (common in plants) and low-specificity miRNAs causing translational repression via imperfect binding (more prevalent in animals but also observed in plants).

8.1.9 Enhancing Plant Abiotic Stress Tolerance: A Novel Approach Using miRNAs

Significant influences affect plant growth and expansion through miRNAs, which have also recently been shown to have a considerable impact on how plants respond to a variety of abiotic stressors in the environment (Table 8.1). Numerous plant species have shown overexpression of several miRNAs at present. In comparison to their natural counterparts, the transgenic plants that overexpressed miRNA

TABLE 8.1
Common Stress-Responsive miRNAs and Their Targets[a]

S. No.	miRNAs	Targets	Target Function	Species	References
1.	156/157	SPL family	Encouraging flower initiation, leaf initiation, root formation, secondary metabolism, as well as abiotic stress, tillering, or corn formation in *Zea mays*	*Arabidopsis thaliana, Zea mays*	Stief et al. (2014)
2.	159	MYB/TCP	Formation of male reproductive parts, seed development, vegetative tissue formation, and reproductive structure formation	*Arabidopsis thaliana*	Achard et al. (2004); Reyes and Chua (2007)
3.	160	ARF	Leaf and root formation of embryo, elongation in hypocotyl	*Arabidopsis, Medicago truncatula, Zea mays*	Guo et al. (2005)
4.	164	NAC	Development of axillary meristem tissues, formation of leaf and flower, initiation of lateral roots	*Arabidopsis thaliana, Zea mays, Oryza*	Rhoades et al. (2002); Fang et al. (2014)
5.	165/166	HD-ZIP-III	Maintenance of meristematic cells, growth of lateral roots	*Arabidopsis thaliana*	Yen et al. (2019)
6.	169	NFY	Consolidation of C homeotic gene transcription into polypeptides	*Arabidopsis thaliana, Antirrhinum majus*	Ni et al. (2013)
7.	171	SCL Family	Chlorophyll biosynthesis and phase transitions	*Arabidopsis thaliana, Barley*	Li et al.(2021)
8.	172	AP2 Family	Inhibition of flowering, identification of flower meristem, vegetative phase transition, carpel and stamen formation; opening of flower, tuberization, and tolerance to salt	*Arabidopsis thaliana, Z. mays, Oryza, H. valgure, S. tuberosum*	
9.	173	TAS	Thermotolerance during temperature stress	*Arabidopsis thaliana*	Li et al. (2014)
10.	319	TCP	Leaf formation and senescence, organ curvature, as well as biosynthesis of hormones and signaling pathways	*Arabidopsis thaliana, Solanum lycopersicum*	Sunkar and Zhu (2004)
11.	390	TAS 3	Biogenesis of ta-siRNA for ARF suppression and indirect regulation of miR165/166, growth of lateral roots, leaf arrangement	*Arabidopsis thaliana*	Fahlgren et al. (2016)
12.	393	TIR	Equilibrium of auxin and growth of lateral roots, leaf number	*Arabidopsis thaliana, Oryza*	Jones-Rhoades and Bartel (2004); Sunkar and Zhu (2004)

(Continued)

TABLE 8.1 (*Continued*)
Common Stress-Responsive miRNAs and Their Targets[a]

S. No.	miRNAs	Targets	Target Function	Species	References
13.	394	F-box	Identification of meristematic tissue and suppression via WUS downregulation, inclination of leaf and style	*Arabidopsis thaliana*	Jones-Rhoades and Bartel (2004)
14.	395	AST	Stress from salinity and drought in plants	*Arabidopsis thaliana*	Allen et al. (2005)
15.	396	GRF	Leaf cell proliferation, resistance to disease, somatic embryo formation, grain size, and branching of panicle	*Arabidopsis thaliana, Medicago, Oryza*	Sunkar and Zhu (2004)
16.	397	Laccase	Grain yield, initiation of panicle branches	Oryza	
17.	398	CSD	Selenium-binding protein	*Saccharum* spp.	Naya et al. (2014); Guan et al. (2013)
18.	402	Demeter-like protein 3 (DML3)	Increased resilience to cold stress, drought, and salt	*Arabidopsis thaliana*	Kim et al. (2010)
19.	824	AGL16	Stomatal patterning process	*Arabidopsis thaliana*	Bergmann and Sack (2007)
20.	828&858	MYBs	Formation of fiber, anthocyanin production, accumulation of flavanol in plants	Cotton, grape	
21.	847	IAA26	Growth of lateral roots	*Arabidopsis thaliana*	
22.	857	LACCASE7	Secondary growth	*Arabidopsis thaliana*	Zhao et al. (2019)
23.	TAS3	ARF3/4 and (only in mosses) APA2 Like	The development of veins and polarity of leaves	All plants	

[a] The targets of certain miRNAs have been the subject of numerous investigations. We include only one or two references per topic here.

demonstrated heightened resilience or vulnerability to diverse abiotic environmental stimuli, depending on their target genes. miRNA-based biotechnology has the potential to increase plant resistance to severe environments, as demonstrated by several of these transgenic plants. The first miRNA to be discovered in plants, miR156, has been scientifically shown to be necessary for plant development and transitions between phases. Recent studies have demonstrated that there are several environmental conditions to which plants are exposed, and their expression with regard to miR156 is notable. Moreover, miR156 was discovered and utilized to enhance plant biomass. Transgenic switchgrass that overexpressed the appropriate amount of miR156 yielded 58%–101% greater plant biomass than its control plants (Fu et al. 2012). A study reports that modified transgenic rice plants that overexpressed miR159 were less capable of withstanding heat stress than their wild-type counterparts, suggesting that downregulating miR159 could be a factor in resilience to heat stress (Wang et al. 2012). Among the largest families of miRNAs, every type of plant possesses miR169. miR169 has played an essential role in both normal developmental processes and the way that plants react to environmental stress. miR169 has played an essential role in both normal developmental processes and the way that plants react to environmental stress. After 7 days of drought treatment, transgenic tomatoes with overexpression of miR169 showed markedly increased durability to drought-related stress. Non-transgenic wild-type tomato plants displayed prominent indications of dehydration after drought treatment, such as wilting and turgor depletion; in contrast, transgenic plants that overexpressed miR169 maintained their turgor. Comparing transgenic tomato plants to wild-type non-transgenic tomato plants, this study revealed that the transgenic plants' stomatal aperture index and stomatal conductance decreased by 35%–49% and 33%–45%, respectively. Additionally, transpiration rate of the transgenic plants decreased by 38%–55%. As a result, overexpression of miR169 in genetically engineered plants ultimately resulted in lower consumption of soil water and reduced water loss through leaf transpiration (Frazier et al. 2011). Many studies have demonstrated that miR319 is often overregulated under various stressful conditions, and it has been established that miR319 is associated with multiple abiotic stresses. A transgenic investigation reports that the fundamental expression of the gene for miR319 in bent grass with encroaching characteristics significantly improved the adaptive capacity of plants to withstand drought and salinity conditions while also changing their growth (Liu et al. 2013). Comparing the genetically modified plants that overexpressed miR319 to their non-transgenic counterparts, the former showed improved cell membrane integrity and water retention. Rice's resistance to cold stress was found to be considerably increased by overexpressing miR319 in transgenic rice. Both monocot and dicot plants include the conserved miRNA family miR393, whose expression is regulated by a range of environmental stresses. For this reason, miR393 has become another target for enhancing plant resistance to various stresses.

8.1.10 Mechanism of miRNA to Induce Stress Resistance

miRNAs are employed in stress responses by altering the expression of genes that respond to stress and balancing signaling pathways crucial for stress adaptation. They influence how cells react to stress by modifying the expression of key proteins involved in stress signaling, such as those related to apoptosis, DNA repair, and cellular homeostasis. A list of some important mechanisms is provided below.

8.1.10.1 Adjustment of Stress-Responsive Genes

Stress-responsive genes can have their expression targeted and modified by miRNAs. Due to the heightened regulation of specific miRNAs during stress conditions, certain target genes involved in stress sensitivity or response are inhibited. Stress-responsive genes related to stress sensing, cell signaling, and adaptive responses are stimulated by the complex signaling cascades triggered by diverse abiotic stimuli. By directing the mRNA transcripts of these stress-responsive genes towards degradation or translational repression, miRNAs alter their expression. Through precise and synchronous control of gene expression, miRNAs modulate stress responses and facilitate adaptive changes to fluctuating environmental conditions.

Genes that code for transcription factors by reducing cell proliferation is miR319, which targets Teosinte that is divided into branches/Cycloidea/Pcf (TCP), which helps regulate numerous aspects of plant growth and development. Low-temperature induction suppresses the expression of miR319, which in turn enhances the translation of the targeted genes, OsPCF6 and OsTCP21. On the other hand, overexpression of miR319 makes plants more resistant to the adverse effects of low temperatures (Wang et al. 2014).

The presequenced gene ZmLRT of qLRT5-1, identified through precise localization and functional analysis, expresses the primary transcript of miR166a. Drought tolerance in maizc seedlings was enhanced when ZmLRT lines were knocked down (Zhang et al. 2023).

8.1.10.2 Altering the Hormonal Signal Pathways

Stress responses are primarily regulated by plant hormones, including ethylene, salicylic acid, jasmonic acid, and ABA. miRNAs can target genes involved in hormone signaling pathways, thereby modifying how plants respond to stress. For example, miR393 affects auxin signaling-mediated stress responses by targeting auxin receptor genes (Parry et al., 2009; Chen et al., 2011; Windels & Vazquez, 2011; Lu et al., 2018; Wang et al., 2018).

ABA has been shown to control the capacity of a plant to withstand dryness and the germination of seeds. Abiotic stresses lead to an accumulation of ABA, which in turn promotes the synthesis of miR159. The ancient and conserved plant miRNA miR159 serves several purposes in drought tolerance and plant development. ABI5-dependent myb33 seed insensitivity to ABA treatment was identified, and the hypersensitivity of miR159 to ABA was suppressed by an ABI5 gene loss-of-function mutant. In response to ABA, MYB33 and miR159 are located downstream of ABI5 (Jiang et al. 2022).

Plants require the hormone monocrotaline lactone to acclimatize to drought stress, and when tomato plants are treated with exogenous monocrotaline lactone, it causes the buildup of miR156. Plants that receive monocrotaline lactone treatment or overexpress miR156 exhibit decreased stomatal conductance and increased ABA sensitivity (Visentin et al. 2020).

By modifying the production of a sequence-specific DNA binding factor known as auxin response factor (ARF), which is important for both stress response and root development, miRNAs contribute to stress-related auxin signaling. It has been shown that this network involves at least five miRNAs: miR160, miR164, miR167, miR390, and miR393. ARF6 and ARF8 interact with miR167; ARF10, ARF16, and ARF17 are targeted by miR160; and ARF4 is targeted by miR390. MiR164 and miR393, on the other hand, indirectly regulate ARFs. These miRNAs and ARFs are all related to root growth and stress response in roots (Meng et al. 2010).

8.1.10.3 Fine-Tuning of Stress-Related Signaling Pathways

MiRNAs can fine-tune stress-related signaling pathways by targeting components of these pathways. For example, In order to control reactions of a plant to oxidative stress, miR398 targets genes that encode copper superoxide dismutases (CSDs), which are crucial in scavenging ROS.

Abiotic stress tolerance in plants largely depends on the fine-tuning of stress-related signaling pathways, and miRNAs are essential for modifying these pathways. Abiotic stressors trigger intricate signaling cascades that lead to various physiological and biochemical responses aimed at reducing stress-related damage and improving plant survival.

Overexpression of sly-miR398b led to shorter plant heights in tomato plants exposed to salinity. This was due to a reduction in biomass from the roots and above ground. Further analysis showed that the overexpression of sly-miR398b downregulated the expression of CSD (He et al. 2021).

8.1.10.4 Control of Defense Mechanisms against Antioxidants

The expression of genes involved in antioxidant defense systems, which are essential for scavenging ROS and preventing oxidative damage under stressful conditions, can be regulated by miRNAs.

Abiotic stress tolerance in plants is largely conferred by the regulation of antioxidant defense systems, with miRNAs playing essential roles in this process. ROS accumulate in plant cells as a result of abiotic stressors, leading to oxidative stress and cellular damage. By targeting genes that encode antioxidant enzymes such as glutathione peroxidase, peroxidase, catalase, and superoxide dismutase, miRNAs modulate antioxidant defense mechanisms. They regulate the expression of these antioxidant genes through post-transcriptional mechanisms, maintaining the redox balance of cells and reducing oxidative damage during stressful periods. Through the precise regulation of antioxidant defense systems by miRNAs, plants can withstand oxidative stress and maintain cellular homeostasis under severe environmental conditions.

For example, it has been shown that miR398 and miR319 control the expression of genes encoding antioxidant enzymes like glutathione peroxidase and superoxide dismutase (Sunkar et al. 2006; Feng et al. 2016).

8.1.10.5 Cross-Talk with Other Small RNAs

Plant stress responses can be regulated by interactions between miRNAs and other short RNA molecules, including long non-coding RNAs (lncRNAs) and small interfering RNAs (siRNAs) (Pooja et al. 2025; Rahul et al. 2024; Subhan et al., 2025; Tanvi et al., 2025). These interactions may enhance the regulatory effects of miRNAs on stress-responsive genes. The communication between various small RNA classes is a fundamental aspect of abiotic stress tolerance in plants, affecting gene expression, signaling pathways, and stress responses. Examples from various crops illustrate the diverse mechanisms through which small RNAs interact to fine-tune stress responses and improve plant resilience to environmental challenges. The phenomenon of cross-talk between different small RNA classes and its implications for abiotic stress tolerance are discussed below:

- **Interplay between miRNAs and siRNAs**: Although they target different types of transcripts, both miRNAs and siRNAs are involved in post-transcriptional gene silencing mechanisms in plants. siRNAs can target both protein-coding and non-coding RNAs, such as transposable elements and viral sequences, whereas miRNAs primarily target mRNAs that code for proteins (Khraiwesh et al. 2012). Under stressful circumstances, the interaction between miRNAs and siRNAs enables coordinated control of gene expression and defense responses.
- **miRNAs and lncRNAs in Stress Signaling Pathways Integration**: Long non-coding RNAs, or lncRNAs, play a role in both transcriptional and post-transcriptional gene silencing and are being recognized as significant regulators of plant gene expression. lncRNAs have the ability to trap miRNAs and prevent them from targeting their mRNA transcripts by acting as molecular sponges (Wang and Chekanova, 2017). Plant stress tolerance is influenced by the interaction between miRNAs and lncRNAs, which also affects the expression of stress responsive genes. One example is the lncRNA IPS1 in *Arabidopsis*, which acts as a decoy for miR399, a phosphate starvation-responsive miRNA that targets PHO_2, a crucial gene for phosphate homeostasis (Franco-Zorrilla et al. 2007). By sequestering miR399, IPS1 modulates PHO_2 expression and phosphate uptake efficiency, thereby enhancing plant adaptation to phosphate-deficient soils.
- **Regulation of Small RNA Biogenesis Pathways**: Small RNA biogenesis pathways, including those involved in miRNA and siRNA production, are tightly regulated to ensure appropriate stress responses in plants. Cross-talk between different small RNA pathways can influence the abundance and activity of stress-responsive small RNAs, thereby modulating stress tolerance mechanisms (Borges and Martienssen, 2015).

8.1.11 Genic Response of miRNAs

Plant miRNAs function by first binding to the target mRNA, which is subsequently recognized by the Argonaute (AGO) protein. At the 10th and 11th nucleotides of the miRNA binding site, the AGO protein then cleaves the mRNA. The AGO cleaves the target mRNA at the miRNA binding site when plant miRNAs attach to it, resulting in the destruction of the target mRNA. Plant miRNAs also function through translational repression of the intended genes. By controlling the expression of genes that respond to stress, maintaining cellular homeostasis, and coordinating adaptive responses to environmental challenges, miRNAs govern various aspects of stress tolerance in plants (Khraiwesh et al. 2012). These regulatory mechanisms encompass transcriptional and post-transcriptional processes, allowing plants to fine-tune their stress responses for optimal survival. miRNAs influence the expression of genes responsible for stress by targeting transcription factors, chromatin modifiers, and regulatory proteins that govern stress signal transduction pathways (Mallory & Vaucheret, 2006). For instance, miR156 targets the SPL transcription factors, which control many developmental processes and stress responses in plants (Xie et al. 2006). The stability, translation efficiency, and degradation of mRNA are influenced by miRNAs at the post-transcriptional level, thereby controlling the abundance and activity of stress-related transcripts (Mallory & Vaucheret, 2006). miRNAs alter gene expression in response to environmental stimuli by selectively binding to specific mRNA transcripts, allowing for rapid adaptation to changing conditions.

The utilization of bioinformatics tools and high-throughput sequencing technologies has facilitated the functional characterization of stress-responsive miRNAs in plants as well as genome-wide profiling. Transcriptomic and proteomic analyses provide insights into the regulatory networks controlled by miRNAs, unraveling key molecular mechanisms underlying stress tolerance (Ferdous et al. 2015). Bioinformatics tools and computational algorithms aid in predicting the site of action of miRNAs and elucidating regulatory networks in plants (Mallory & Vaucheret, 2006). By confirming the regulatory connections between miRNAs and their target genes through experimental validation using methods like RNA-seq, qRT-PCR, and gene expression analysis, these techniques offer mechanistic insights into the processes underlying stress tolerance.

Plant stress tolerance is genetically influenced by miRNAs because of their regulatory roles in post-transcriptional gene expression modulation. By examining the levels of miRNA production in response to specific stresses across various plant species and cultivars, it has been demonstrated that the miRNA response to abiotic stress is genotype-dependent. Drought treatment decreased the activity of miR168 and miR396 in rice (Zhou et al. 2010), but activity was amplified in tobacco (Frazier et al. 2011) and *Arabidopsis* (Liu et al. 2008).

Current research has shown that miRNAs also affect alternative splicing processes in plants, leading to the production of splice variants with different functional characteristics. Additionally, alternative splicing mediated by miRNAs may contribute to the generation of stress-specific isoforms with enhanced or attenuated activities, thereby shaping the overall stress response landscape in plants (Seo et al. 2013). By targeting splicing factors and pre-mRNA processing machinery, miRNAs can modulate the splicing patterns of stress-responsive transcripts and alter their protein isoform diversity. The regulation of genes by miRNAs and their ability to adapt to stress is further complicated by this post-transcriptional mechanism.

Emerging evidence suggests that miRNAs may also influence epigenetic modifications and chromatin remodeling concerning abiotic stress. Furthermore, an additional layer of transcriptional control is provided by epigenetic modulation mediated by miRNAs, which can influence long-term stress memory and trans-generational stress responses in plants (Borges & Martienssen 2015). By targeting genes encoding chromatin-modifying enzymes, histone modifiers, and DNA methylation machinery, miRNAs can modulate the accessibility of stress-induced genes and alter their transcriptional activity. These epigenetic modifications may aid in developing stress tolerance characteristics and help plant populations adapt to harsh conditions.

The miRNAs participate in complex regulatory networks that involve feedback loops, feedforward loops, and cross-talk with other regulatory pathways (Chen 2009). MiRNAs are able to coordinate the various stress manifestations seen in plants and incorporate multiple signaling inputs thanks to their network dynamics. As significant nodes in regulatory networks, miRNAs can control the expression of genes that respond to stress. These molecules can influence the regulation of abiotic stress through interactions with other regulatory components, such as hormones, transcription factors, and short RNAs. miRNA-mediated feedback regulation ensures precise control of stress-induced gene expression and maintains cellular homeostasis in response to fluctuating external conditions. Dysregulation of miRNA-mediated feedback loops can disrupt stress tolerance mechanisms and compromise plant fitness (Chen 2009).

In addition to mRNA cleavage, miRNAs can also repress translation initiation or elongation, thereby inhibiting protein synthesis from target mRNAs (Brodersen et al. 2008). This translational repression mechanism allows gene expression to be controlled more quickly and reversibly in response to changing environmental factors.

To reduce the production of stress-related proteins and conserve cellular resources, miRNAs may inhibit the translation of stress-induced genes under abiotic stress conditions. This post-transcriptional regulation helps plants adapt to adverse environmental conditions by modulating protein synthesis and metabolic activities (Brodersen et al. 2008).

In addition to variations across distinct plant species, the genotype-based response of miRNAs to abiotic stresses also occurs within cultivars or genotypes of the same species. It is widely acknowledged that a plant species' genotype can influence its adaptability to abiotic stress. This variation may be due to the regulation of genes by miRNA. The effects of drought intervention on two cultivars of cowpeas (drought-tolerant IT93K503-1 and moisture stress-sensitive CB46) were investigated using deep sequencing technology. Their results demonstrated that the expression of 20 miRNAs varied between the two cultivars. Only one of the two genotypes exhibited expression of nine of these miRNAs, either exclusively or predominantly. Furthermore, they discovered that 11 drought-regulated miRNAs were present in one genotype but absent in the other (Barrera-Figueroa et al. 2011).

8.1.12 Biochemical Reaction of miRNAs to Plant Stress

Dynamic shifts in miRNA production, processing, and activity are part of the biochemical response of miRNAs to plant stress, and these changes work together to modulate gene expression and impact stress tolerance. Stress conditions can induce post-transcriptional modifications of miRNAs, such as uridylation or adenylation, which can affect their stability, processing, and target specificity. These modifications may be mediated by RNA-modifying enzymes or RNA-binding proteins, and they can influence the activity and function of miRNAs that respond to stress.

The biochemical response of miRNAs to plant stress involves complex regulatory networks that influence miRNA expression, processing, modifications, and interactions with target mRNAs. These dynamic changes in miRNA activity contribute to the fine-tuning of gene expression and play critical roles in plant stress adaptation and tolerance. Secondary metabolites, including flavonoids, phenolics, and terpenoids, accumulate in response to abiotic stress, helping plants tolerate this stress. Genes involved in pathways leading to secondary metabolism have their expression regulated by miRNAs. For instance, miR156 targets genes that encode enzymes involved in the phenylpropanoid biosynthesis pathway, which modifies the accumulation of phenolic compounds and enhances tolerance of a plant to stress. Abiotic stresses can also lead to the accumulation of ROS, causing oxidative damage to cellular components. miRNAs such as miR398 play pivotal roles in regulating ROS homeostasis by targeting genes encoding antioxidant enzymes. For instance, CSD genes are targeted by miR398, which regulates ROS scavenging ability and improves oxidative stress tolerance in plants (Yamasaki et al. 2007).

8.1.13 Achievements of miRNA Approach of Abiotic Stress Tolerance

The following is a tabular summary of the progress made in applying miRNA techniques to improve abiotic stress tolerance in different crops. This table highlights diverse achievements in the utilization of miRNA approaches across various crops to enhance abiotic stress tolerance, elucidating their roles in stress response pathways and crop improvement strategies (Table 8.2).

These are some of the major crops that have recently made progress in abiotic stress tolerance through the use of miRNA: rice, soybeans, and *Arabidopsis*. The table also includes a description of the target characteristics considered during the relevant experiments (Table 8.3).

TABLE 8.2
Achievements through microRNAs

S. No.	Achievement	Kind of miRNA Considered for the Study	microRNA's Name	References
1	Targeted stress response activation	Stress-responsive	miR398	Khraiwesh et al. (2012)
2	Fine-tuning stress signaling pathways	Stress-regulated	miR169	Sunkar and Zhu (2004)
3	Accelerated breeding programs	TALENs, CRISPR/Cas	–	Liang et al. (2014)
4	Enhanced productivity under stress	Salt stress-responsive	miR395	Zhang et al. (2016)
5	Synergistic stress management	Virus-responsive	–	Feng et al. (2014)
6	Global impact on sustainable farming	Biogenesis-related	–	Budak and Akpinar (2015)
7	Empowering farmers with resilient solutions	Drought- and heat-responsive	–	Fahad et al. (2017)

TABLE 8.3
The Most Recent Advances in Abiotic Stress Tolerance Using miRNA

S. No.	Crop	miRNA	Description	References
1	Rice	osa-miR398	In rice, overexpression of osa-miR398 led to larger grains and increased panicle branching, which enhanced rice yield potential in drought-stressed environments. The target gene OsCSD1, a copper/zinc superoxide dismutase essential for scavenging reactive oxygen species (ROS) and reducing oxidative stress in rice plants, was expressed more when osa-miR398 was upregulated.	Li et al. (2022)
2	Arabidopsis	ath-miR156	Switchgrass plants that overexpressed ath-miR156 produced more biomass and were more resilient to stress. Compared to native plants, the transgenic switchgrass lines exhibited greater shoot tillering, a longer vegetative development phase, and higher photosynthetic efficiency. In switchgrass, the overexpression of ath-miR156 resulted in the downregulation of its target gene, SPL12, a transcription factor that binds to Squamosa proteins and promotes biomass accumulation and stress tolerance.	Fu et al. (2012)
3	Soybean	gma-miR169	The downregulation of gma-miR169 in soybean plants led to increased drought tolerance by upregulating its target gene GmNFYA5, a nuclear factor Y subunit A5. The transgenic soybean lines exhibited enhanced root length, higher water retention capacity, and improved survival rates under drought conditions compared to wild-type plants. The suppression of gma-miR169 activated GmNFYA5-mediated drought-responsive pathways, thereby enhancing drought tolerance in soybean.	Ni et al. (2013)

8.1.14 Conclusion

The model plants *Arabidopsis* and rice are currently the focus of most studies on miRNAs in plants, whereas research on other important crops remains relatively rare, especially regarding species-specific miRNAs with significant potential for future discovery. In many plant species, the agronomic traits associated with specific miRNAs have not been identified. Furthermore, each miRNA must be characterized separately because it may have multiple functions across different abiotic stress and signaling pathways. Investigating this complex mechanism of miRNA-dependent stress tolerance in plants is promising. It supports the sustainable growth of agricultural output and provides technical assistance for crop genetic improvement. Understanding the regulatory networks related to short RNAs offers valuable insights into plant stress adaptation and could lead to the development of innovative approaches to enhance crop productivity and sustainability.

The application of miRNAs has great potential to improve the ability of a plant to withstand abiotic stress, providing new paths for environmentally conscious agriculture. miRNAs play pivotal roles in controlling gene expression, regulating stress-related signaling pathways, and coordinating adaptive reactions to abiotic stressors through complex regulatory mechanisms. Numerous aspects of plant physiology are regulated by miRNAs, such as hormonal signaling, ion balance, osmotic regulation, ROS scavenging, and secondary metabolism. Stress-responsive gene expression is regulated by miRNAs, which also support stress tolerance processes in plants by targeting important genes and transcription factors involved in stress responses. Furthermore, the regulatory networks controlling abiotic stress responses are complicated by interactions between miRNAs and other short RNAs, including lncRNAs and siRNAs. The interplay between different small RNA classes influences gene expression patterns, signal transduction pathways, and stress adaptation processes, shaping the overall stress tolerance phenotype in plants. As our understanding of miRNA-influenced regulatory networks deepens, innovative strategies for engineering stress-tolerant crops and improving agricultural resilience are emerging. Harnessing the potential of miRNAs in crop breeding programs offers an exciting opportunity to develop climate-resilient cultivars with enhanced productivity, yield stability, and resource use efficiency. Furthermore, the use of miRNA-based technologies in crop development, biotechnology, and precision agriculture shows enormous potential for addressing issues related to global food security and climate change. By deciphering the complex interplay between miRNAs, environmental stresses, and plant responses, we can unlock new insights into the molecular aspects underlying stress tolerance and pave the way for sustainable agricultural practices in a rapidly changing world. In conclusion, miRNAs are potent genetic instruments that can be used to enhance ability of a plant to withstand abiotic stress. These tools can help ensure food security, environmental sustainability, and ecosystem resilience in the face of climatic instability and environmental degradation.

8.1.15 Future Prospects

Our knowledge of miRNA-governed regulatory networks in stress responses will increase with the identification of miRNAs that become responsive to stress, through ongoing research efforts focused on discovering stress-responsive miRNAs in diverse plant species and under varied stress conditions. The identification of novel stress-responsive miRNAs will be facilitated by improved sequencing methods and bioinformatics tools, opening the door to the creation of more comprehensive miRNA-based stress tolerance enhancement strategies.

Future studies on miRNAs, a novel signaling mechanism in interplant interactions with respect to abiotic stress, may also be conducted. Considering that miRNAs can be transferred between plants as signals to improve growth and development, increase resilience to biotic stress, and ultimately enhance food production, it can be inferred that miRNA genes regulating abiotic stress in plants can be overexpressed in one plant and transmitted to another. Compiling information from genomics, transcriptomics, proteomics, and metabolomics can provide a comprehensive understanding of the molecular mechanisms underlying miRNA-mediated stress responses. Researchers can help shape the development of comprehensive crop improvement techniques by identifying key regulatory nodes and networks involved in abiotic stress tolerance through systems-level analysis (Jagadeeswaran et al. 2012).

The main focus of miRNA-related research should be on functional analysis rather than miRNA discovery. In this field, future research should prioritize validating the role of specific miRNAs in plant tolerance to stress conditions rather than merely identifying the miRNAs responsive to abiotic stresses. One innovative approach would be to employ virus-induced gene silencing (VIGS). Using CLCrV-induced gene silencing, a recent study successfully overexpressed miR156 in cotton, resulting in abnormal leaf development in the transgenic plants (Gu et al. 2014). To disrupt a miR156 target, this team also employed VIGS technology to introduce a miR156 small tandem target mimic into cotton (Gu et al. 2014). Thus, VIGS could be a valuable technique to investigate the function of specific miRNAs. By using degradome sequencing, which enables the determination of the mechanisms of action of numerous miRNAs, we can swiftly and efficiently identify multiple miRNA targets and conduct genome-wide miRNA expression studies under abiotic stresses using sequencing methods. The continuous development of deep sequencing and high-throughput technology has made this possible (Yan et al. 2023). The potential applications of miRNA-mediated techniques to improve abiotic stress tolerance are numerous and promising. By leveraging advancements in technology, interdisciplinary collaborations, and translational efforts, researchers can harness the power of miRNAs to develop resilient crop varieties capable of thriving in increasingly challenging environmental conditions.

Future developments utilizing miRNAs to confer abiotic stress tolerance in plants hold great potential for sustainable agriculture and ensuring the world's food security. Novel miRNA-target interactions, regulatory networks, and signaling pathways involved in abiotic stress responses are expected to be found through further research in this area. Developing stress-tolerant crop varieties with improved yield, quality, and environmental adaptation is possible by combining miRNA-mediated techniques with traditional breeding techniques. Additionally, advancements in miRNA-based biotechnological techniques, including RNA interference and genome editing, will enable the precise manipulation of the expression of stress-responsive genes to enhance plant resistance to abiotic challenges. Ultimately, leveraging the potential of miRNAs to improve abiotic stress resistance may significantly help the challenges posed by climate change and ensure future global food security.

REFERENCES

Achard P, Herr A, Baulcombe DC, Harberd NP. (2004). Modulation of floral development by a gibberellin-regulated microRNA. *Development*, 131, 3357–3365.

Allen E, Xie Z, Gustafson AM, Carrington JC. (2005). microRNA directed phasing during trans-acting siRNA biogenesis in plants. *Cell*, 121, 207–221.

Anil P, Dharmendra P, Sachidanand S, Yuriy E, Snezhan B, Dilfuza J, Chinmayi J. (2023). Beneficial effects of biochar application on mitigating the drought and salinity stress implications on plants. *Plant Science Today*, 10(2), 188–193. http://doi.org/10.14719/pst.2591

Barrera-Figueroa B, Gao L, Diop N, Wu Z, Ehlers J, Roberts P, Close T, Zhu J-K, Liu R. (2011). Identification and comparative analysis of drought-associated microRNAs in two cowpea genotypes. *BMC Plant Biology*, 11, 127.

Bergmann DC, Sack FD. (2007). Stomatal development. *Annual Review of Plant Biology*, 58, 163–181.

Borges F, Martienssen RA. (2015). The expanding world of small RNAs in plants. *Nature Reviews Molecular Cell Biology*, 16(12), 727–741.

Bouzroud S, Gouiaa S, Hu N, Bernadac A, Mila I, Bendaou N, Smouni A, Bouzayen M, Zouine M. (2018). Auxin response factors (ARFs) are potential mediators of auxin action in tomato response to biotic and abiotic stress (*Solanum lycopersicum*). *PLoS One*, 13, 0193517.

Brien J, Hayder H, Zayed Y, Peng C. (2018). Overview of microRNA biogenesis, mechanisms of actions, and circulation. *Frontiers in Endocrinology*, 9, 402.

Brodersen P, Sakvarelidze-Achard L, Bruun-Rasmussen M, Dunoyer P, Yamamoto YY, Sieburth L, Voinnet O. (2008). Widespread translational inhibition by plant miRNAs and siRNAs. *Science*, 320(5880), 1185–1190.

Budak H, Akpinar, BA. (2015). Plant miRNAs: Biogenesis, organization and origins. *Functional & Integrative Genomics*, 15, 523–531.

Chen X. (2009). Small RNAs and their roles in plant development. *Annual Review of Cell and Developmental*, 25, 21–44.

Chen ZH, Bao ML, Sun YZ, Yang YJ, Xu XH. (2011). Regulation of auxin response by miR393-targeted transport inhibitor response protein 1 is involved in normal development in Arabidopsis. *Plant Molecular Biology*, 77, 619–629.

Ding Y, Tao Y, Zhu C. (2013). Emerging roles of microRNAs in the mediation of drought stress response in plants. *Journal of Experimental Botany*, 64, 3077–3086.

Fahad S, Bajwa AA, Nazir U, Anjum SA, Farooq A, Zohaib, A, Huang J. (2017). Crop production under drought and heat stress: Plant responses and management options. *Frontiers of Plant Science*, 8, 1147.

Fahlgren N, Hill ST, Carrington JC, Carbonell A. (2016). P-SAMS: A web site for plant artificial microRNA and synthetic trans-acting small interfering RNA design. *Bioinformatics*, 32, 157–158.

Fang Y, Xie K, Xiong, L. (2014). Conserved miR164-targeted NAC genes negatively regulate drought resistance in rice. *Journal of Experimental Botany*, 65, 2119–2135.

Feng J, Liu S, Mengna W, Lang Q, Jin C. (2014). Identification of microRNAs and their targets in tomato infected with cucumber mosaic virus based on deep sequencing. *Planta*, 240, 1335–1352.

Feng K, Yu J, Cheng Y, Ruan M, Wang R, Ye Q, Zhou G, Li Z, Yao Z, Yang Y, Zheng Q. (2016). The SOD gene family in tomato: Identification, phylogenetic relationships, and expression patterns. *Frontiers of Plant Science*, 7, 1279.

Ferdous J, Hussain SS, Shi BJ. (2015). Role of microRNA s in plant drought tolerance. *Plant Biotechnology Journal*, 13(3), 293–305.

Franco-Zorrilla JM, Valli A, Todesco M, Mateos I, Puga MI, Rubio-Somoza I, Paz-Ares J. (2007). Target mimicry provides a new mechanism for regulation of microRNA activity. *Nature Genetics*, 39(8), 1033–1037.

Frazier T, Sun G, Burklew C, Zhang B. (2011). Salt and drought stresses induce the aberrant expression of microRNA genes in tobacco. *Molecular Biotechnology*, 49, 159–165.

Fu C, Sunkar R, Zhou C, et al. (2012). Overexpression of miR156 in switchgrass (*Panicum virgatum* L.) results in various morphological alterations and leads to improved biomass production. *Plant Biotechnology Journal*, 10, 443–452.

Gu Z, Huang C, Li F, Zhou X. (2014). A versatile system for functional analysis of genes and microRNAs in cotton. *Plant Biotechnology Journal*, 12, 638–649.

Guan Q, Lu X, Zeng H, Zhang Y, Zhu J. (2013). Heat stress induction of miR398 triggers a regulatory loop that is critical for thermotolerance in Arabidopsis. *The Plant Journal*, 74, 840–851.

Guo H-S, Xie Q, Fei J-F, Chua N-H. (2005). microRNA directs mRNA cleavage of the transcription factor NAC1 to downregulate auxin signals for Arabidopsis lateral root development. *The Plant Cell*, 17, 1376–1386.

He Y, Zhou JX, Hu YF, Fang CY, Yu YJ, Yang J, Zhu B, Ruan YL, Zhu ZJ. (2021). Overexpression of sly-miR398b increased salt sensitivity likely via regulating antioxidant system and photosynthesis in tomato. *Environmental and Experimental Botany*, 181, 104273.

Jabborova D, Nigora Z, Yuriy E, Abdujalil N, Anand D, Khurshid S, Zafarjon J, Sachidanand S, Rahul D. (2023). Growth of spinach as influenced by biochar and bacillus endophyticus IGPEB 33 in drought condition. *Pakistan Journal of Botany*, 55(SI), 1–7. https://doi.org/10.30848/PJB2023-SI(6).

Jagadeeswaran G, Nimmakayala P, Zheng Y, Gowdu K, Reddy UK, Sunkar R. (2012). Characterization of the small RNA component of leaves and fruits from four different cucurbit species. *BMC Genomics*, 13, 1–13.

Jiang Y, Wu X, Shi M, Yu J, Guo C. (2022). The miR159-MYB33-ABI5 module regulates seed germination in *Arabidopsis*. *Physiologia Plantarum*, 174, e13659.

Jones-Rhoades MW, Bartel DP. (2004). Computational identification of plant microRNAs and their targets, including a stress-induced miRNA. *Molecular Cell*, 14, 787–799.

Kar MM, Raichaudhuri A. (2021). Role of microRNAs in mediating biotic and abiotic stress in plants. *Plant Gene*, 26, 100277.

Khraiwesh B, Zhu JK, Zhu J. (2012). Role of miRNAs and siRNAs in biotic and abiotic stress responses of plants. *Biochimica et Biophysica Acta (BBA)-Gene Regulatory Mechanisms*, 1819(2), 137–148.

Kim JY, Kwak KJ, Jung HJ, Lee HJ, Kang H. (2010). microRNA402 affects seed germination of *Arabidopsis thaliana* under stress conditions via targeting DEMETER-LIKE Protein3 mRNA. *Plant and Cell Physiology*, 51, 1079–1083.

Kurihara Y, Watanabe Y. (2004). Arabidopsis microRNA biogenesis through Dicer-like 1 protein functions. *Proceedings of the National Academy of Sciences of the United States of America*, 101, 12753–12758.

Li J, Song Q, Zuo ZF, Liu L. (2022). microRNA398: A master regulator of plant development and stress responses. *International Journal of Molecular Sciences*, 23(18), 10803.

Li S, Liu J, Liu Z, Li X, Wu F, He Y. (2014). Eat-induced TAS1 TARGET1 mediates thermotolerance via heat stress transcription factor A1a-directed pathways in Arabidopsis. *The Plant Cell*, 26, 1764–1780.

Li X, Guo C, Gu J, Duan W, Zhao M, Ma C, Du X, Lu W, Xiao K. (2014). Overexpression of VP, a vacuolar H-pyrophosphatase gene in wheat (*Triticum aestivum* L.), improves tobacco plant growth under Pi and N deprivation, high salinity, and drought. *Journal of Experimental Botany*, 65, 683–696.

Liang Z, Zhang K, Chen K, Gao C. (2014). Targeted mutagenesis in *Zea mays* using TALENs and the CRISPR/Cas system. *Journal of Genetics and Genomics*, 41(2), 63–68.

Liu HH, Tian X, Li YJ, Wu CA, Zheng CC. (2008). Microarray-based analysis of stress-regulated microRNAs in *Arabidopsis thaliana. RNA*, 14, 836–843.

Liu J, Feng L, Li J, He Z. (2015). Genetic and epigenetic control of plant heat responses. *Frontiers in Plant Science*, 6, 267.

Liu X, Yang L, Zhou X, Zhou M, Lu Y, Ma L, Ma H, Zhang Z. (2013). Transgenic wheat expressing *Thinopyrum intermedium* MYB transcription microRNAs for improving plant tolerance to abiotic stress factor TiMYB2R-1 shows enhanced resistance to the take-all disease. *Journal of Experimental Botany*, 64, 2243–2253.

Lu Y, Feng Z, Liu X, Bian L, Xie H. (2018). MiR393 and miR390 synergistically regulate lateral root growth in rice under different conditions. *BMC Plant Biology*, 18, 261.

Mallory AC, Vaucheret H. (2006). Functions of microRNAs and related small RNAs in plants. *Nature Genetics*, 38(6), S31–S36.

Meenakshi S, Rahul D, Sachidanand S, Anita GM, Rajpal S. (2025). Plant microRNAs: An overview. *microRNA Advances and Application in Plant Biology Plant Biology, Sustainability and Climate Change*, 2025, 1–33.

Meng Y, Ma X, Chen D, Wu P, Chen M. (2010). microRNA-mediated signaling involved in plant root development. *Biochemical and Biophysical Research Communications*, 393, 345–349.

Naya L, Paul S, Valdés-López O, Mendoza-Soto AB, Nova-Franco B, Sosa-Valencia G, Reyes JL, Hernández G. (2014). Regulation of copper homeostasis and biotic interactions by microRNA 398b in common bean. *PLoS One*, 9, e84416.

Noopur K, Pragati K, Adarsh KS, Sachidanand S, Tanvi J, Megha B. (2025a). Role of non coding RNA in effector trigger immunity. In Dutta R, Kapadia C, Danish S, Singh S. (Eds.), *Non-Coding RNA in Plants Modulation and Stress Responses Plant Biology, Sustainability and Climate Change*. New York: Academic Press.

Noopur K, Pragati K, Megha B, Arpan M, Tanvi J, Shrisendu M, Adarsh KS, Sachidanand S. (2025b). Role of non-coding RNA in plant architecture. In Dutta R, Kapadia, C, Danish S, Singh S. (Eds.), *Non-Coding RNA in Plants Modulation and Stress Responses Plant Biology, Sustainability and Climate Change*. New York: Academic Press.

Ni Z, Hu Z, Jiang Q, Zhang H. (2013). GmNFYA3, a target gene of miR169, is a positive regulator of plant tolerance to drought stress. *Plant Molecular Biology*, 82, 113–129.

Pan WJ, Tao JJ, Cheng T, Bian XH, Wei W, Zhang WK, Ma B, Chen SY, Zhang JS. (2016). Soybean miR172a improves salt tolerance and can function as a long-distance signal. *Molecular Plant*, 9, 1337–1340.

Parry G, Calderon-Villalobos LI, Prigge M, Peret B, Dharmasiri S. (2009). Complex regulation of the TIR1/AFB family of auxin receptors. *Proceedings of the National Academy of Sciences of the United States of America*, 106, 22540–22545.

Pegler JL, Grof CPL, Eamens AL. (2019). The plant microRNA pathway: The production and action stages. *Methods in Molecular Biology*, 1932, 15–39.

Pooja G, Megha B, Arpan M, Chintan K, Rahul D, Subhan D, Sachidanand S. (2025). Introduction to long non-coding RNA. In Dutta R, Kapadia C, Danish S, Singh S. (Eds.), *Non-Coding RNA in Plants Modulation and Stress Responses Plant Biology, Sustainability and Climate Change*. Vol. 1, pp. 1–12. New York: Academic Press.

Rahul D, Chintan K, Subhan D, Sachidanand S. (2024). *Non-Coding RNA in Plants: Modulation and Stress Responses*. New York: Academic Press.

Reyes JL, Chua N-H. (2007). ABA induction of miR159 controls transcript levels of two MYB factors during Arabidopsis seed germination. *The Plant Journal*, 49, 592–606.

Rhoades MW, Reinhart BJ, Lim LP, Burge CB, Bartel B, Bartel DP. (2002). Prediction of plant microRNA targets. *Cell*, 110, 513–520.

Seo PJ, Park MJ, Park CM. (2013). Alternative splicing of transcription factors in plant responses to low temperature stress: Mechanisms and functions. *Planta*, 237, 1415–1424.

Subhan D, Uzma Y, Niaz A, Khurram S, Rahul D, Sachidanand S. (2025). Fertilizer-responsive non-coding RNA and their significance. In Dutta R, Kapadia C, Danish S, Singh S. (Eds.), *Non-Coding RNA in Plants Modulation and Stress Responses Plant Biology, Sustainability and Climate Change*. New York: Academic Press.

Sinha A, John J, Singh S, Johri P. (2022). Microarray-based detection and identification of bacterial and viral plant pathogens. *Rhizosphere Microbes: Biotic Stress Management*, 8, 47–69.

Stief A, Altmann S, Hoffmann K, Pant BD, Scheible WR, Bäurle I. (2014). Arabidopsis miR156 regulates tolerance to recurring environmental stress through SPL transcription factors. *The Plant Cell*, 26, 1792–1807.

Sunkar R, Kapoor A, Zhu JK. (2006). Posttranscriptional induction of two Cu/Zn superoxide dismutase genes in arabidopsis is mediated by downregulation of miR398 and important for oxidative stress tolerance. *Plant Cell*, 18, 2051–2065.

Sunkar R, Zhu J. (2004). Novel and stress-regulated microRNAs and other small RNAs from Arabidopsis. *The Plant Cell*, 16, 2001–2019.

Tanvi J, Arpita S, Praneta Y, Palak S, Sachidanand S. (2025). Epigenetic modulation of plant developmental pathways by non-coding RNA. In Dutta R, Kapadia C, Danish S, Singh S. (Eds.), *Non-Coding RNA in Plants Modulation and Stress Responses Plant Biology, Sustainability and Climate Change*. New York: Academic Press.

Tungare K, Shahu R, Zambare V, Agarwal P, Jobby R, Nisar N, Alabdallah NM, Al-Saeed FA, Johri P, Singh S, Saeed M, Jha P. (2022). Toxicity mitigation of textile dye reactive blue 4 by hairy roots of *Helianthus annuus* and testing its effect in *in vivo* model systems. *BioMed Research International*, 1958939, 13. https://doi.org/10.1155/2022/1958939

Verma N, Sao P, Srivastava A, Singh S. (2021). Physiological and molecular responses to drought, submergence and excessive watering in plants. In: Husen A. (Ed.), *Harsh Environment and Plant Resilience*, pp. 305–321. Cham: Springer International Publishing.

Visentin I, Pagliarani C, Deva E, Caracci A, Turecˇková V, Novák O, Lovisolo C, Schubert A, Cardinale F. (2020). A novel strigolactone-miR156 module controls stomatal behaviour during drought recovery. *Plant, Cell & Environment*, 43, 1613–1624

Wang HLV, Chekanova, JA. (2017). Long noncoding RNAs in plants. *Advances in Experimental Medicine and Biology*, 1008, 133–154.

Wang L, Liu Z, Qiao M, Xiang F. (2018). MiR393 inhibits *in vitro* shoot regeneration in *Arabidopsis thaliana via* repressing TIR1. *Plant Science*, 266, 1–8.

Wang ST, Sun XL, Hoshino Y, Yu Y, Jia B, Sun ZW, Sun MZ, Duan XB, Zhu YM. (2014). microRNA319 positively regulates cold tolerance by targeting OsPCF6 and OsTCP21 in rice (*Oryza sativa* L.). *PLoS One*, 9, e91357.

Wang Y, Sun F, Cao H, Peng H, Ni Z, Sun Q, Yao Y. (2012). TamiR159 directed wheat TaGAMYB cleavage and its involvement in another development and heat response. *PLoS One*, 7, e48445.

Windels D, Vazquez F. (2011). MiR393: Integrator of environmental cues in auxin signaling? *Plant Signaling Behaviour*, 6, 1672–1675.

Xie K, Wu C, Xiong L. (2006). Genomic organization, differential expression, and interaction of SQUAMOSA promoter-binding-like transcription factors and microRNA156 in rice. *Plant Physiology*, 142(1), 280–293.

Yamasaki H, Abdel-Ghany SE, Cohu CM, Kobayashi Y, Shikanai T, Pilon M. (2007). Regulation of copper homeostasis by microRNA in Arabidopsis. *Journal of Biological Chemistry*, 282(22), 16369–16378.

Yan G, Hua Y, Jin H, Huang Q, Zhou G, Xu Y, He Y, Zhu Z. (2023). Sly-miR398 participates in cadmium stress acclimation by regulating antioxidant system and cadmium transport in tomato (*Solanum lycopersicum*). *International Journal Molecular Science*, 24, 1953.

Ye J, Xu M, Tian X, Cai S, Zeng S. (2019). Research advances in the detection of miRNA. *Journal of Pharmaceutical Analysis*, 9(4), 217–226.

Zhang F, Zhu G, Du L, Shang X, Cheng C, Yang B, Guo W. (2016). Genetic regulation of salt stress tolerance revealed by RNA-Seq in cotton diploid wild species, *Gossypium davidsonii. Scientific Reports*, 6(1), 20582.

Zhang M, Chen Y, Xing H, Ke W, Shi Y, Sui Z, Xu R, Gao L, Guo G, Li J. (2023). Positional cloning and characterization reveal the role of a miRNA precursor gene ZmLRT in the regulation of lateral root number and drought tolerance in maize. *Journal of Integrative Plant Biology*, 65, 772–790.

Zhao B, Liang R, Ge L, Li W, Xiao H, Lin H, Ruan K, Jin Y. (2007). Identification of drought-induced microRNAs in rice. *Biochemical and Biophysical Research Communications*, 354, 585–590.

Zhao J, Yuan S, Zhou M, Yuan N, Li Z, Hu Q, Bethea FG Jr, Liu H, Li S, Luo H. (2019). Transgenic creeping bentgrass overexpressing Osa-miR393a exhibits altered plant development and improved multiple stress tolerance. *Plant Biotechnology Journal*, 17, 233–251.

Zhou L, Liu Y, Liu Z, Kong D, Duan M, Luo L. (2010). Genome-wide identification and analysis of drought-responsive microRNAs in *Oryza sativa. Journal of Experimental Botany*, 61, 4157–4168.

9 Abiotic Stress-Related Functional Roles of microRNAs in Agronomically Important Cereal and Legume Crops

*Vidadala Rajendra**, *Jiban Shrestha, Mehidi Rehimi, and Shaikhul Islam*

9.1 INTRODUCTION

microRNAs are a family of very small RNAs that are typically 21–24 nucleotides long. They are also referred to as non-coding RNAs. They are important for controlling post-transcriptional gene expression (Lou et al., 2018). Primary, precursor, and mature miRNAs are the final products of transcription from DNA sequences, which produce the majority of miRNAs (O'Brien et al., 2018). Target messenger RNAs (mRNAs) attach to these molecules at complementary sequences, which typically causes translational repression or target destruction. microRNAs regulate post-transcriptional gene expression by either triggering the breakdown of specific mRNAs or blocking mRNA from being translated. miRNAs were first identified as significant controllers of numerous biological processes in both plants and mammals in the early 1990s (Lee et al., 1993). miRNAs are essential for a plant's growth, development, and differentiation. They are also crucial for the plant's ability to adjust to abiotic stressors that significantly impact plant growth and crop productivity. miRNA functional studies in economically significant plants have validated their importance in multiple biological and metabolic pathways (Gupta, 2015). Recent breakthroughs have demonstrated that miRNAs and their targets influence crop plant-specific agronomic parameters. According to Wang et al. (2012), Yang et al. (2019), and Zhang et al. (2013a), these results show that agricultural characteristics such as male sterility, crop output, plant height, panicle branching, tillering, grain size, and fruit quality—all associated with high yield and quality—are significantly improved by miRNAs.

By regulating genes related to various stress triggers, miRNAs help orchestrate a multitude of physiological and molecular adjustments that enhance plant resilience. This adaptive mechanism not only supports the survival of individual plants but also has implications for agricultural productivity and ecosystem stability. The practical applications of miRNAs include serving as potential targets for selective breeding programs aimed at enhancing crop resistance to adverse environmental conditions. miRNAs can be used in genetic engineering to improve crop plants' resistance to stress (Rani, 2021). The exact modification of miRNA genes is now possible thanks to advanced biotechnological technologies like CRISPR/Cas9 gene editing (Hussen et al., 2023). This enables the creation of transgenic crops that are more resilient to stress, develop at their optimal rate, and produce higher yields. This book chapter explores how miRNAs function to higher yields plants' responses to challenging environmental conditions, with a focus on cereal and legume crops that are significant to agronomy. The premise for this study arises from the urgent need to enhance crop resilience in light of the implications of climate change and increasing environmental pressures that threaten global food security. Understanding the miRNA-mediated regulatory networks

* *Corresponding Author*

DOI: 10.1201/9781003491378-9

provides a foundation for developing innovative strategies for abiotic stress mitigation. The discovery and synthesis of miRNAs, their mechanisms of action, and the specific miRNAs involved in the plant's response to various abiotic stresses are thoroughly covered in this chapter. It highlights the cross-talk between miRNA pathways and other molecular mechanisms, including hormonal signaling and reactive oxygen species (ROS) management, which are crucial for orchestrating a coordinated response to environmental challenges. The principal goal of this chapter is to clarify the diverse functional roles of miRNAs in different abiotic stresses in agronomically important cereals and legumes. By incorporating case studies and recent scientific discoveries, the chapter aims to provide a comprehensive picture of how miRNAs contribute to the intricate regulatory networks that control plant stress responses. Gaining this knowledge is essential for improving our capacity for sustainable agricultural practices and crop enhancement.

9.2 DISCOVERY AND BIOGENESIS OF miRNA

Plant fundamental functions and gene expression are significantly regulated by microRNAs (miRNAs). Ruvkun identified the first miRNA, lin-4, in Caenorhabditis elegans, which Ambros discovered in 1993 (Lee et al., 1993). microRNAs, or miRNAs, are small non-coding RNAs that have an average length of 22 nucleotides (Noopur et al., 2025a, b; Pooja et al., 2025; Rahul et al., 2024; Subhan et al., 2025; Tanvi et al., 2025). miRNAs are encoded by miRNA genes (Ha and Kim, 2014). By targeting protein-coding transcripts, miRNAs act as RNA-silencing guide molecules. Primary miRNA (pri-miRNAs) transcripts are created by RNA polymerase II from miRNA genes in the nucleus. Pri-miRNAs can range from hundreds to thousands of nucleotides in length. The pri-miRNAs are processed into stem–loop structures by an enzyme complex that includes Dicer-like I (DCL1), RNA-binding protein hyponastic leaves 1, C2H2 zinc finger protein serrate, and the cap-binding protein complex (Fang and Spector 2007). This processing results in precursor miRNAs (pre-miRNAs), with lengths of 60–70 nucleotides. Voinnet (2009) states that the pre-miRNAs are further processed in the nucleus to create a miRNA/miRNA duplex. The methyltransferase hua enhancer 1, which methylates the duplex to prevent uridylation and subsequent degradation, is crucial for the methylation of miRNAs throughout this process. Furthermore, to regulate pre-miRNA processing, DAWDLE (DDL) interacts with DCL1. While miRNA processing occurs in the nucleus, most mature miRNAs in plants perform their roles in the cytoplasm. Following export, the miRNA/miRNA duplex enters the cytoplasm via the transporter protein HST (HASTY). An RNA-induced silencing complex (RISC) is formed in this case by the combination of the miRNAs (Wang et al., 2019).

9.2.1 Modes of Action of miRNA

A single miRNA can affect the expression of numerous genes simultaneously since it can target hundreds of mRNAs. By controlling several genes and transcription factors (TFs), miRNAs regulate complex plant biological processes. The biogenesis of miRNAs involves three stages: (1) transcription of a pri-miRNA, typically carried out by RNA polymerase II; (2) partial processing of the pre-miRNA in the nucleus to produce a precursor miRNA; and (3) final maturation of the pre-miRNA in the cytoplasm to produce a functional miRNA (Beezhold et al., 2010). The degree of complementarity between each miRNA and its mRNA target determines the mechanism of action, which may involve cleavage or translational repression. Two primary mechanisms by which these plant miRNAs control target genes at the post-transcriptional level are translational repression and mRNA degradation. For the more common method of mRNA degradation, mature miRNAs are integrated into an RISC that also includes Argonaute (AGO) proteins, with AGO1 being the main protein involved in *Arabidopsis*. Within RISC, the miRNA guides it to partially or fully complementary target mRNAs (Franco-Zorrilla et al., 2007; Yan et al., 2012). The "seed region" of nucleotides 10–11 on the miRNA, where the miRNA and target mRNA complement each other almost perfectly, is where AGO1 cleaves the mRNA (Liu et al., 2014). This cleavage results in the breakdown of the target mRNA, preventing it from being translated into a protein. Numerous investigations have revealed that full base pairing between the mRNA and miRNA near the cleavage site is necessary for this cleavage. The second mechanism is translational repression, where the miRNA-programmed RISC binds to the mRNA without cleaving it. This binding prevents translation of the mRNA into protein without affecting mRNA levels. The ability of specific miRNAs to cleave or inhibit translation of their targets depends on how complementary they are to their binding sites (Ma and Hu, 2023).

9.2.2 Factors Influencing Production of miRNAs

Numerous factors and molecular mechanisms influence the production of miRNAs. These include TFs, methylation of CpGs, DNA copy number, and miRNA biogenesis. Among these, some important factors include transcriptional regulation, post-transcriptional processing, stability, and several others such as epigenetic modifications, transcriptional co-regulation, alternative splicing, RNA editing, regulatory small RNAs, nutrient availability, and hormonal regulation.

9.2.2.1 Transcriptional Regulation

This is the initial step in miRNA biogenesis. During abiotic stress, TFs such as DREB/C-repeat binding factor (CBF), AP2/EREBP, and MYB bind to specific cis-regulatory elements in the promoter regions of miRNA genes, thus activating or repressing their transcription (Zhang et al., 2014).

9.2.2.2 Post-Transcriptional Processing

Of miRNAs involves the cleavage of pri-miRNAs by the microprocessor complex. Under abiotic stress, the expression and activity of the components of the microprocessor complex, such as DCL1, can be modulated, leading to altered processing of pri-miRNAs and subsequent changes in miRNA levels (Michlewski and Cáceres, 2018)

9.2.2.3 miRNA Stability

This is crucial for their abundance and activity. Under abiotic stress, RNA-binding proteins, stress-responsive proteins, and stress-induced secondary structures can interact with miRNAs, influencing their stability (Khvorova et al., 2003).

9.2.2.4 Epigenetic Modifications

Including histone changes and DNA methylation, among others, this can also have an impact miRNA production under abiotic stress. These modifications can alter the accessibility of miRNA genes, thereby affecting the binding of TFs and other regulatory proteins (Yao et al., 2019).

9.2.2.5 Transcriptional Co-Regulation

This is another factor influencing miRNA production under abiotic stress. Stress-responsive TFs can simultaneously regulate both miRNA and target gene expression, ensuring coordinated gene expression networks involved in stress responses (Michlewski and Cáceres, 2018).

9.2.2.6 Alternative Splicing

Induced by environmental stresses, this may contribute to the synthesis of distinct isoforms of miRNA precursors. These alternative splicing events can generate miRNAs with altered sequences or structures, impacting their target specificity and regulatory functions (Shriram et al., 2016).

9.2.2.7 RNA Editing

This involves nucleotide modifications within RNA molecules, and can also affect miRNA production under abiotic stress. Stress conditions can induce RNA editing events within miRNA precursors, resulting in changes in miRNA sequence and stability (Sousa et al., 2019).

9.2.2.8 Regulatory Small RNAs

These, such as siRNAs and phasiRNAs, can cross-regulate miRNA production under abiotic stress. These small RNA molecules may compete for common factors involved in miRNA biogenesis or target mRNA degradation, influencing the abundance and activity of miRNAs (Wilson and Doudna, 2013).

9.2.2.9 Nutrient Availability

This contributes to miRNA production while plants are challenged with environmental alterations. Nutrient deficiencies or imbalances can indirectly affect miRNA production by influencing the expression of nutrient-responsive TFs or the availability of cofactors required for miRNA biogenesis (Chien et al., 2017).

9.2.2.10 Hormonal Regulation

This regulates miRNA production under abiotic stress. Plant hormones such as abscisic acid (ABA), ethylene, SA, and JA can modify the production of miRNA by changing the activity of TFs that direct the expression of miRNA genes or by modulating the expression of important components involved in miRNA biogenesis (Singh et al., 2021). miRNA production under abiotic stress is influenced by transcriptional and post-transcriptional regulation, stability, epigenetic modifications, RNA editing, regulatory small RNAs, nutrient availability, and hormonal regulation, contributing to intricate regulatory networks (Figure 9.1).

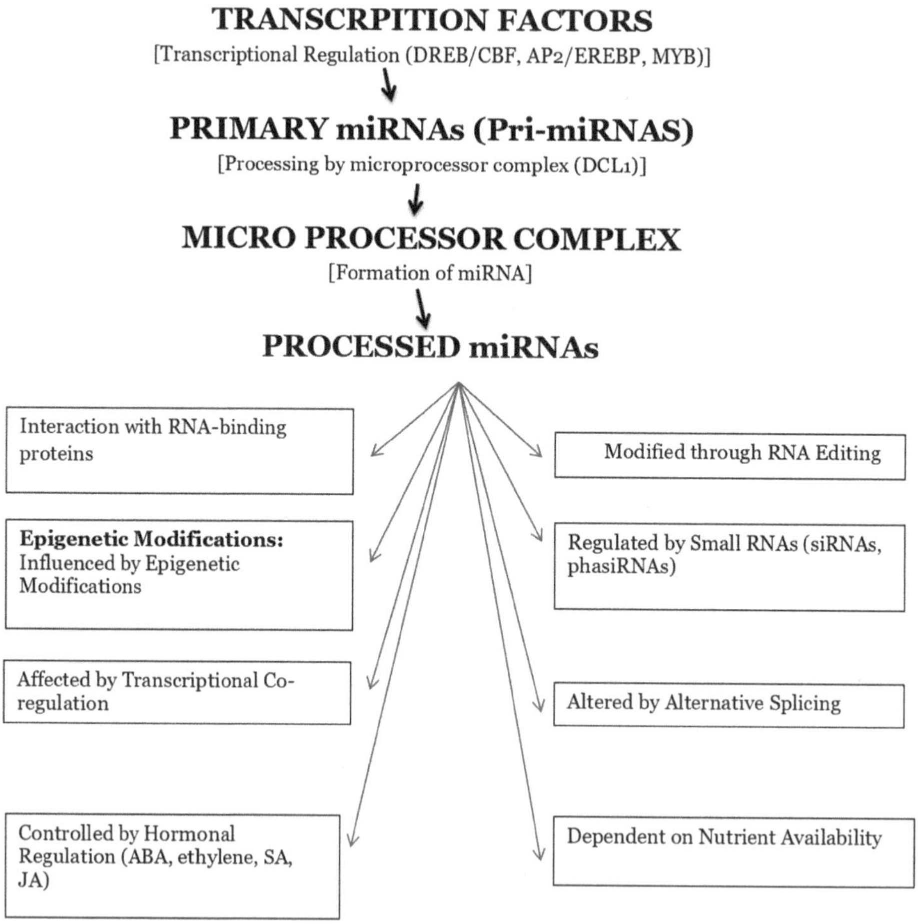

FIGURE 9.1 Diagram illustrating miRNA production under abiotic stress conditions.

This workflow elucidates the complex regulatory mechanisms at play, emphasizing the integrated nature of miRNA biogenesis and function under varying environmental conditions. The diagram demonstrates the regulatory checkpoints that collectively determine the dynamics of miRNA production in plants.

The sequential process and variables affecting the synthesis of miRNAs in response to environmental stress are depicted in Figure 9.1. The **transcriptional regulation** of pri-miRNAs is initiated by **transcription factors** such as AP2/EREBP, MYB, and DREB/CBF binding to cis-regulatory sites. These pri-miRNAs are then processed by the **microprocessor complex**, including DCL1, culminating in the formation of mature miRNAs. The pathway continues with the mature miRNAs interacting with **RNA-binding proteins**, which can affect miRNA stability and function. Concurrently, miRNA production is modulated by **epigenetic modifications** that influence gene accessibility, **transcriptional co-regulation** ensuring a coordinated stress response, and **alternative splicing**, which produces various miRNA isoforms with different regulatory capabilities. Further along the workflow, miRNAs may undergo **RNA editing** that alters their sequence, while **regulatory small RNAs** (such as siRNAs and phasiRNAs) may compete or cooperate within the miRNA regulatory networks. **Nutrient availability and hormonal regulation** by ABA, ethylene, SA, and JA also play crucial roles in modulating miRNA production during abiotic stress responses.

9.3 miRNAs AND STRESS-INDUCED SIGNALING PATHWAYS

In their native habitat, plants encounter several abiotic obstacles simultaneously. Consequently, complex signaling networks have emerged, allowing molecular-level communication among different stress response pathways. miRNAs play an essential role in facilitating this cross-talk by regulating gene expression in various stress signaling cascades. The expression of genes encoding TFs and proteins involved in the primary signaling pathways for abiotic stress, mediated by sucrose, calcium, ROS, and phytohormones, has been shown to be influenced by various miRNAs. This section will cover the current knowledge of how certain miRNAs interact with different stress signaling pathways and help to coordinate the plant's response to different environmental stressors.

9.3.1 Phytohormones Pathway

Through transcriptional regulation, phytohormones contribute to virtually every biological process in plants as well as their responses to stress. Key hormones involved in abiotic stress tolerance include ABA and ET. During stress, levels of certain miRNAs are altered to target stress-related genes or TFs via highly conserved pathways. The expression of these stress-responsive miRNAs is controlled by hormone-responsive elements in their promoters. Compelling evidence demonstrates cross-talk between miRNA and phytohormone pathways. In *Populus suaveolens*, freezing stress and exogenous GA, SA, and MeJA application modulated the activity of miR475b, its promoter, and its targets, highlighting regulation by multiple signals. Additionally, conserved miRNAs govern important developmental processes also regulated by phytohormones, providing a link between the two regulatory networks. Mutations affecting miRNA biogenesis, such as in hyl1 *Arabidopsis*, exhibit pleiotropic developmental defects along with reduced miRNA accumulation and hypersensitivity to auxin, cytokinin, and ABA. This establishes a functional connection, as these hormones depend on precise miRNA levels and targeting for normal signaling. Mobile tasiRNAs may also integrate hormonal and stress responses across cells as signaling molecules. The ability of miRNAs to trans-regulate genes and coordinate developmental pathways parallels phytohormone functions. Their interactions likely enable cross-talk mediating adaptive responses to environmental fluctuations through transcriptional and post-transcriptional gene regulation. Understanding these complex regulatory interactions is crucial for developing strategies that leverage miRNA pathways to improve stress tolerance in crops (Patel et al., 2019; Singh et al., 2021).

9.3.2 Calcium Pathway

Calcium serves as an essential auxiliary transmitter in transducing abiotic stress signals through discrete spatiotemporal surges in cytosolic calcium concentrations. Multiple calcium-binding proteins, such as calmodulins, calcium-regulated phosphatases, and calcium-dependent protein kinases (CDPKs), can be affected by these localized "calcium signatures." Several miRNAs play important roles in modulating calcium signaling pathways by targeting these calcium sensors at the post-transcriptional level. Increased soil arsenic levels and occurrences of water scarcity induce the inhibition of miR1432, leading to higher levels of its predicted calmodulin target in barley. In citrus, miR482 and novel miR049 differ in their targeting of a calcineurin-like phosphoesterase and CDPK6, respectively, under salt and drought conditions. The monocot-conserved miR444d family is anticipated to regulate calmodulin protein expression in response to powdery mildew or drought. During somatic embryogenesis, downregulation of miR1160, miR5638, miR1315, and miR5225 relieves repression of their CDPK and CBL targets, demonstrating miRNA-mediated calcium signaling. Calcineurin binding like 10, an important regulator of salt and drought tolerance pathways, is a conserved target of miR167 in apple and miRX7 in *Brassica rapa*. Under heavy metal and combined salt/thiourea stresses, miR1318 and miR1508 modulation of a calcium ATPase and CDPK alters the cellular response. In summary, abiotic stresses generate calcium signatures that are precisely controlled by miRNAs, which fine-tune the expression of key calcium sensors and effectors, integrating stress signals via the calcium signaling network in plant cells (Patel et al., 2019; Ranty et al., 2016).

9.3.3 Sugar Pathway

Sugars serve as a crucial regulator of plant signaling, integrating responses to various stresses through complex regulatory networks involving miRNAs. The sugar sensor SnRK1 kinase acts as a central hub, inducing the stress-responsive miR319, which targets growth repressors like TCP TFs. During nutrient scarcity, SnRK1 also modifies the levels of miRNAs 159b, 161, 775, and 824a. Sugars signal through hexokinases and the histone modifier HAC1, with HAC1 inhibition decreasing levels of SnRK1 subunits and impairing sugar sensitivity. Key sugars like glucose and sucrose influence gene expression and stress adaptation via interactions with ABA signaling components ABI4 and ABI5 during germination. Sucrose induces the stress-responsive miR398 and miR408 by freeing up copper reserves through the downregulation of copper proteins, linking sugars to post-transcriptional regulation by miRNAs. This illustrates how sugars mobilize carbon and energy reserves while fine-tuning the abiotic stress response through an intricate network of signaling pathways converging on miRNA expression and function at the crossroads of sugar, hormone, and small RNA signaling in plants (Keunen et al., 2013; Patel et al., 2019; Saddhe et al., 2021).

9.3.4 ROS Signaling Pathway

ROS are crucial signaling molecules that assist plants in balancing their defense and survival techniques while undergoing stress from environmental factors. However, improper regulation of excessive ROS buildup can lead to oxidative damage. Plants have developed intricate ROS signaling networks to induce specific miRNA expression profiles that help fine-tune stress-tolerance mechanisms. Superoxide, hydrogen peroxide, and hydroxyl radicals (ROS) are among the many reactive species generated by abiotic stresses such as dehydration, salinity, heavy metals, and abnormal temperature fluctuations. Elevated ROS levels can alter miRNA expression, hence influencing the synthesis of stress-responsive proteins. For example, arsenic toxicity induces ROS to upregulate miR408, miR528, and miR397b in rice. miR395, miR838, and miR854 target key sulfur assimilation enzymes involved in metal detoxification, such as ATP sulfurylase and sulfate transporters. ROS also activates the synthesis of proline and glutamate dehydrogenase through selective repression of

TFs by miRNAs. Modulation of miRNAs 164, 167, 390, and 159 affects auxin, ethylene, and ABA biosynthesis, influencing root architecture and defense responses. By balancing growth, development, and protective metabolite production through precise ROS-mediated spatiotemporal regulation of miRNAs and their target genes, plants can effectively manage oxidative damage resulting from various abiotic stresses. According to Huang et al. (2019) and Patel et al. (2019), this preserves cellular redox equilibrium and grants tolerance to unfavorable environmental conditions.

9.4 ROLE OF miRNAs IN DIFFERENT PLANT ABIOTIC STRESS

When plants are exposed to abiotic stress, miRNAs regulate gene expression, enhancing adaptability and survival. In important grains and legumes, they alter stress signaling pathways to improve resilience and stress tolerance. This section examines their critical role in plant stress responses.

9.4.1 Cold Stress

Economically important cereal crops and legumes are severely impacted by cold temperatures at critical developmental phases. The flowering stage is particularly sensitive in these crops, leading to reduced seed set due to pollen sterility under chilling conditions. Younger seedlings also struggle with stunted growth when faced with freezing early on. Prolonged cold further inhibits vegetative development in cereals, lowering biomass accumulation. Cold temperatures can also delay or prevent root nodule formation in legumes like soybean, restricting their nitrogen fixation capacity. Additional vulnerable growth windows include grain filling in maize and maturation in soybean, where cold may limit grain or pod development. The precise control of miRNA-mediated gene expression is crucial for enhancing resilience against cold-induced damage across various vulnerable developmental stages in plants. Cold tolerance mechanisms in plants rely on intricate cellular signal transduction pathways. The fluidity of the membrane varies at low temperatures, which in turn causes an increase in calcium levels and subsequent reactions. Plants commonly respond to low-temperature stress by activating the CBF-dependent signaling pathway (Guo et al., 2018). Within this context, many miRNAs actively respond to low-temperature stress. For example, miR398, which is often present in plants, inhibits Cu/Zn superoxide dismutases (SOD), which are required to reduce the generation of ROS. miR319 has been shown to positively enhance cold tolerance through the suppression of the rice OsPCF6 and OsTCP2 genes (Wang et al., 2014). In another investigation, researchers observed that overexpression of miR396b from *Poncirus trifoliata* enhanced *Citrus* lemon's resilience to cold stress. Compared to wild-type plants, the transgenic plants produced more polyamines and accumulated fewer ROS in cold environments. MiR156-expressing transgenic rice plants showed increased resilience to cold. Targeting the auxin receptor gene TIR1/AFB inhibits the auxin signaling pathway (Luo et al., 2022). Improved cold tolerance results from miR393 overexpression, which raises the expression of genes that respond to cold. The CBF-dependent pathway and ROS buildup are the usual mechanisms through which miR394, miR397, miR319, miR396, miR408, and miR402 are associated with cold stress. A variety of miRNAs that alter the ABA or IAA signaling pathway affect how plants respond to cold temperatures. A study by Schommer et al. (2008) indicates that miR319 targets the TF genes teosinte branched/cycloidea/PCF (TCP), which control several processes linked to plant development and growth, including cell proliferation. The elevated levels of miR156 promote cold stress survival and growth in rice, pine, and *Arabidopsis*. LEA proteins contribute to skeletal rigidity and persistence in low-water, non-metabolic states, particularly under cold stress.

9.4.2 Heat Stress

Elevated temperatures pose a substantial threat to the global food supply because they negatively impact the development and yield of vital crops such as legumes and cereals. Rice, wheat, and maize are particularly vulnerable to heat stress, especially during critical reproductive stages. In rice, flowering and anthesis are the most heat-sensitive periods, resulting in reduced pollen viability, lower fertilization rates, and decreased seed set under heat. Wheat exposed to high temperatures during anthesis exhibits heat sterility and floret abortion, leading to reduced grain numbers and yields. Maize experiences poor pollen development and grain filling at high temperatures, causing premature ripening and lower grain weights. Soybean, chickpea, and pea are also adversely affected during pod development and seed-fill. Soybean flowers and young pods often abort under heat, while chickpeas and peas have reduced pod set and fewer seeds per pod. High temperatures damage cell membranes, impact protein synthesis and cell division, and trigger oxidative stress. Heat shock-responsive TFs specifically target heat shock proteins and ROS-scavenging enzymes. They play a key protective role by refolding proteins that have been disrupted by heat shock. Heat stress activates alternative splicing mechanisms in maize that allow for splicing at higher temperatures but with reduced accuracy. In plants, miRNAs are essential for both heat stress memory and response. Heat treatment in maize raised the levels of NAC TFs and miR169 while lowering the expression of miR164, which was associated with enhanced SQUAMOSA promoter binding protein production. The AP2 TF, which is necessary for plant development in hot climates, is adversely affected by miR172. Wheat exhibited heat-induced miRNA regulation, with miR2020 regulating pentatricopeptide repeat proteins and miR9662 targeting mitochondrial transcription termination factor-like proteins. Forty-one miRNAs in *Brassica campestris*, including miR156, miR159, and miR171, also showed a preserved heat response. Identifying the target genes is critical to understanding the effects of miRNA in heat response. Conserved areas related to heat sensitivity, tolerance, and oxidative damage response were found using computational and sequencing investigations. While increased levels of miRNA395 and miRNA396 in *Arabidopsis* reduced ATP sulfurylase 1 and growth regulating factor 7 (GRF7), part of the heat stress response, overexpression of miR159 in wheat and TamiR165 in rice resulted in heat sensitivity. The heat stress tolerance of creeping bentgrass is influenced by changes in miRNA393a expression, whereas the heat stress tolerance of miRNA169 and decreased miRNA159 is enhanced. Cucumber miR9748 overexpression improved *Arabidopsis thaliana*'s ability to withstand high temperatures by modifying the phytohormone signaling system.

9.4.3 Drought

Drought reduces agricultural productivity, which is anticipated to worsen due to climate change in the coming years (Anil et al. 2023). Water scarcity hinders plant growth and restricts soil nutrient translocation. Drought stress, like other significant environmental constraints, increases the generation of free radicals, impacting antioxidant synthesis. Zhao et al. (2007) discovered the role of miR169 in rice during drought conditions and proposed that its regulation is controlled by CBF/DREBs (dehydration-responsive elements). A separate research team investigated the function of miR169 in wheat, specifically its ability to target the CCCAT-box TF and regulate drought stress. MiRNA156, one of the pioneering miRNAs discovered in plants, has been extensively associated with drought stress in many studies. Anthocyanins, a type of secondary metabolite, protect plants from ROS-mediated injuries. In *Arabidopsis*, rice, alfalfa, and poplar, the miR156/SPL pathway regulates the level of anthocyanin formation in response to drought-related stress. Plants have an adaptation mechanism that involves regulating anthocyanin accumulation in response to abiotic stress such as drought. Stomata are essential for the exchange of gases between plants and their surroundings (Meddya et al. 2023). In addition to being controlled by internal hormones, they open and close in response to external stimuli. These hormones, in turn, affect how well the plant responds to and endures drought stress. ABA is the primary hormone involved in regulating water loss, stomatal opening, and closure during drought stress. MiR393 exerts a positive influence on stomatal density and a negative influence on guard cell length. The phenotype exhibited by the amplification of miR393 levels was different from that of the deletion mutant. This could be attributed to the regulation of growth hormone response factor 5 (ARF5) and three genes associated with stomatal development: epidermal pattern factor 1 (EPF1), SPEECHLESS (SPCH), and MUTE, by miR393. Compared to the wild type, the strain overexpressing miR393 showed greater susceptibility to drought treatment, which led to a buildup of more malondialdehyde and hydrogen peroxide (H_2O_2). It also inhibited the synthesis of ABA in leaves. Additionally, studies show that miR393 regulates stomatal density and interacts with ABA to respond to drought stress in plants. Zhao et al. (2007) demonstrated that plants with increased expression of miR393a had an improved ability to withstand drought stress, which was linked to higher stomatal density and a denser epidermis. MiR393 regulates the expression of transport inhibitor response 1 (AsTIR1) and auxin signaling F-BOX 2 (AsAFB2). Plants require a hormone called monocrotaline lactone to survive drought stress. The external administration of monocrotaline lactone to tomatoes promotes the accumulation of miR156. Plants treated with monocrotaline lactone or miR156-OE exhibit increased ABA sensitivity and decreased stomatal conductance. MiR398c has been shown to negatively regulate soybean drought resistance. Increased expression of miR398c in *Arabidopsis* resulted in reduced expression of GmCSDla/b, GmCSD2a/b/c, and GmCCS. This led to a lower capacity to remove ROS, as well as an increase in relative electrolyte loss and stomatal opening. Consequently, there was a decrease in the germination rate, an increase in water loss from the leaves, a reduction in the capacity to survive in low-water environments, and a notable susceptibility to drought during seed germination and seedling growth. According to Li et al. (2015), auxins are vital phytohormones critical to plant growth and development, as well as the capacity to endure various abiotic stresses. Abiotic stress regulation is frequently mediated by TF-related regulators like miR156 and miR172. Similar to findings in *N. tabacum*, Li et al. (2013) identified three conserved miRNAs (miR156, miR159, and miR319) that are essential in the seedlings of inbred *Zea mays* lines under drought stress.

9.4.4 Salinity and Osmotic Stress

Soil salinity impacts approximately 6% of the Earth's area and 23% of land suitable for farming, resulting in significant economic losses due to crop stress and decreased yields. Salinity is crucial for plant growth, but when it exceeds a specific threshold, excessive soluble salts become toxic to plants. This toxicity affects the levels of many internal plant signaling molecules, including ABA, ethylene, gibberellin (GA), ROS, and nitric oxide (NO). These hormones can significantly influence the growth and development of plants, ultimately leading to lower agricultural output. Salt stress induces oxidative stress, causing substantial reductions in agricultural output through restricted yields, seed germination, and plant development (Chen et al., 2018). Soil salinity causes plants to grow more slowly, primarily due to osmotic stress, which is followed by ion toxicity (Gupta and Huang, 2014). Because crop plants cannot control the flow of sodium ions into their roots, they are particularly vulnerable to salt. Consequently, salt accumulates, and leaves begin to lose water due to osmosis (Munns, 2005). Osmotic stress triggers numerous bio-functional events, including impaired cell membrane function, disrupted nutritional balance, and diminished antioxidant and photosynthetic activities. Ion toxicity can result from osmotic stress, as the accumulation of Na+ ions can lead to decreased absorption of K+, which is essential for plant growth, and increased damage to various cellular components caused by ROS (Gupta and Huang, 2014). Plants have evolved a flexible system that enables them to respond swiftly to high salt levels at the cellular level to adapt to fluctuations in salinity. This involves regulating a group of genes that enhance the plant's resistance to salinity-induced stress (Wani et al., 2020). Plants have evolved a flexible system that enables them to respond swiftly to high salt levels at the cellular level to adapt to fluctuations in salinity. This involves regulating a group of genes that enhance the plant's resistance to salinity-induced stress (Wani et al., 2020). Plant responses to salinized environments encompass a wide range of mechanisms, including membrane trafficking, transcription, signal transduction, and protein synthesis. miRNAs play a crucial role in the complex regulatory network that governs how plants respond to salt stress. Comparative studies of various rice cultivars have revealed both novel and conserved miRNAs (miR393, miR172, miR167, and miR171), indicating a potential role for ethylene and auxin in salt tolerance. The activity of TFs (NAC, NFYA, TCP, HD-Zip III, and Myb) is regulated by a distinct set of miRNAs, including miR159, miR164, miR166, miR169, and miR319. These miRNAs regulate the expression of genes associated with abiotic stress responses. Wheat and maize exhibit different expressions of miR159 and miR319 in response to salt stress. Increased miR12477 in salt-tolerant rice targets L-ascorbate oxidase, which may mitigate oxidative stress. Salt-stressed tobacco shows alterations in miR156, miR482, miR6145, and miR6149, with miR156 extending the initial phase of resistance and miR482 targeting NBS-LRR proteins involved in stress response mechanisms. miR393 suppresses TIR1, reducing salinity tolerance; however, enhancing TIR1 expression in *Arabidopsis* improves adaptation to salinity. Transgenic creeping bentgrass with elevated rice miR528 exhibits improved salinity sensitivity. miR172a enhances salt tolerance in soybeans by reducing SSAC1 expression. Maize microarrays have demonstrated that miRNA regulation varies during salt stress, particularly with miR399 upregulating phosphate transporter genes and miR398b regulating Cu/Zn superoxide dismutase (CSD). High salinity inhibits plant proliferation and triggers significant changes in miRNA profiles, including miR168, miR359, miR165, miR156, miR158, and miR169. The results provide new insights into the regulatory mechanisms underlying plant responses to salinity.

9.4.5 Heavy Metal

Plants are seriously threatened by heavy metal stress, which interferes with physiological and metabolic functions and hinders crop development, growth, and output. The contamination of soil with heavy metals, often a consequence of anthropogenic activities, exacerbates this abiotic stress. In response to specific heavy metals, such as aluminum (Al), miRNA involvement in plant defense mechanisms becomes evident. In the case of *Medicago truncatula*, downregulation of miRNA160 during Al stress leads to downregulation of its two targets, ARF10 and ARF16, which are important for root cap growth. This ultimately reduces root length. Furthermore, treatment with Al_2O_3 nanoparticles enhances the expression of miRNA395, miRNA397, miRNA398, and miRNA399, which impacts the plant's ability to respond.

Furthermore, in *M. truncatula*, heavy metals such as cadmium (Cd), mercury (Hg), and aluminum (Al) cause the upregulation of certain miRNAs, including miRNA171, miRNA393, and miRNA529. On the other hand, amid such metallic stressors, miRNA166 and miRNA398 are downregulated. Tolerance to Hg is increased in *Arabidopsis* by the overexpression of a CSD2-miRNA398-resistant transcript. Moreover, heavy metal stress, such as that induced by arsenic (As), triggers the responsive behavior of multiple miRNAs. While miRNA397b, miRNA408, and miRNA528 show increased expression, miRNA390 and miRNA1316 exhibit decreased expression under As stress in indica rice roots.

The intricate regulatory role of miRNAs extends to heavy metal stress beyond individual metals. In *Brassica napus* exposed to cadmium, miRNA156a, miRNA167a/c, miRNA393, miRNA394a/b/c, and miRNA164b display differential expression, affecting various tissues. MiRNA160, which specifically targets ATP sulfurylases and sulfate transporters, is downregulated in response to Cd stress. Moreover, investigations on *Phaseolus vulgaris* have associated certain miRNAs, notably miRNA1515, miRNA1510/2110, miRNA1508, and miRNA1532, with manganese (Mn) stress. These miRNAs target specific genes involving CDPKs, heat shock proteins, nucleotide-binding site leucine-rich repeat resistance-like (NBS-LRR) proteins, and receptor kinases.

miRNAs mitigate the consequences of metallic stress by regulating gene expression, facilitating sulfur uptake, and aiding detoxification. Their significance in heavy metal stress responses demonstrates the complexity of plant defense systems and the potential for harnessing miRNA pathways to enhance crop tolerance.

9.4.6 Nutrient Deficiency

Plant growth and development depend on maintaining nutrient homeostasis, and new research has identified a number of miRNAs linked to the intake and movement of minerals in nutrients by plants. An often-limiting nutrient, phosphate, is essential for the production of nucleic acids and membrane lipids. Reduced UBC24 transcript levels result from MiR399's upregulation of a putative ubiquitin-conjugating enzyme (UBC24) in low phosphate environments. Phosphate-responsive genes are controlled by the phosphate starvation response 1 MYB TF, which is expressed when phosphate is scarce. Another miRNA, miR395, responds to sulfur deficiency by targeting ATP sulfatases and the sulfate transporter AST68. MiR398, which regulates copper homeostasis, is increased under copper-limiting conditions, influencing the allocation of copper to CSDs. In the context of nutrient stress, including nitrogen, phosphorus, and potassium deficiencies, miRNAs play a pivotal role in modulating plant responses. Nitrogen stress triggers miR167 to limit root growth and control nitrogen-metabolizing enzymes. MiR393, activated by nitrogen signaling during nitrification and uptake, influences root development. MiR399, a crucial element of the phosphorus starvation signaling pathway, is activated by phosphorus stress and controls the uptake and translocation of inorganic phosphorus. miRNA expression changes in response to potassium stress, with miR156, miR390, miR160, miR164, and miR393 being upregulated during potassium shortage. Beyond nitrogen, phosphorus, and potassium, other elements crucial for plant growth include magnesium (Mg), iron (Fe), sulfate (S), manganese (Mn), copper (Cu), and boron (B). miR395 targets genes involved in sulfate metabolism during sulfate restriction, while various miRNAs respond to manganese and boron stress. Copper homeostasis is regulated by miRNAs like miR397, miR398, miR408, and miR857, which target transcripts encoding Cu-containing proteins. miR398 also aids in detoxifying ROS during oxidative stress and copper shortage. Phosphate starvation response TFs trigger miR399 synthesis in shoots, which is then transported via phloem to roots to enhance phosphate uptake by cleaving specific PHO2 transcripts. miR395 targets ATP sulfurylase genes and the sulfur transporter SULTR2;1/AST68 during sulfate deficiency, aiding sulfate remobilization between leaves. Cu-responsive miRNAs like miR398 help conserve copper during copper deficiency and oxidative stress. Overall, miRNAs are key regulators in nutrient homeostasis, modulating gene expression and plant adaptation to various nutrient stresses, offering insights into the complex regulatory networks controlling plant responses to nutrient deficits.

9.4.7 UV-B Radiation

At large dosages, UV-B radiation can inhibit plant development and growth under abiotic stress. In contrast, it can cause plants to accumulate more secondary metabolites when used in small amounts. Plants respond to UV-B radiation by undergoing various physiological, biochemical, and molecular changes to protect themselves. The ability of plants to withstand UV-B radiation exposure is largely dependent on miRNAs. Numerous studies have identified distinct miRNAs that are regulated differently when exposed to UV-B rays. Zhou et al. (2007) found that when UV treatment was applied to *Arabidopsis*, 21 miRNA genes from 11 families were upregulated. Similarly, Jia et al. (2009) discovered that *Populus tremula* possesses 24 UV-B-sensitive miRNAs, of which 13 are upregulated and 11 are downregulated. Among these are miRNAs such as miR169, miR395, miR472, miR168, miR398, and miR408. Furthermore, it has been found that in both *Arabidopsis* and *P. tremula*, miRNAs such as miR156, miR160, miR165/166, miR167, and miR398 are responsive to UV-B stress. More recently, studies in other plants like grapes and chrysanthemum have also identified several miRNAs, such as miR156, miR482, miR530, and miR828, that are upregulated in response to UV-B radiation. Plants exposed to UV-B radiation stress benefit from the regulation of stress-responsive genes and the production of secondary metabolites, which is facilitated by these miRNAs.

9.5 ISOLATION, IDENTIFICATION, AND CHARACTERIZATION OF PLANT miRNAs

In the pursuit of studying plant miRNAs, the initial step involves isolating them from both control and stress-treated plant tissues. This process begins with the careful harvesting of plant tissues, such as leaves or roots, followed by immediate freezing in liquid nitrogen to preserve RNA integrity. Subsequently, the frozen tissues are ground into a fine powder, and total RNA is extracted using a commercial kit, ensuring lysis, removal of genomic DNA and proteins, and RNA purification on glass-fiber filters. The small RNA fraction, measuring less than 200 nt, is then separated from larger RNAs through either polyacrylamide gel electrophoresis or column chromatography using spin cartridges (Peng et al., 2014). The purified small RNAs are precipitated, washed, and resuspended in RNase-free water for downstream applications. Assessment of the quality, quantity, and size distribution of the isolated small RNAs on bioanalyzer chips ensures their high quality and integrity, making them suitable for miRNA discovery and expression analysis (Vennapusa et al., 2020; Vera-Hernández et al., 2019).

Once superior small RNAs are obtained, the focus shifts to miRNA identification. A small RNA library is constructed by ligating RNA adaptors to both the 3′ and 5′ ends of the small RNAs, enabling subsequent amplification and sequencing (Liu et al., 2014). The small RNAs in the 18–30 nt length range, which covers the typical size of plant miRNAs, are size-selected through polyacrylamide gel electrophoresis. The adaptor-ligated small RNAs are then amplified by PCR, and the resulting cDNA library is purified. Illumina sequencing technologies (Kulcheski et al., 2011), such as HiSeq or NovaSeq, are employed to sequence the small RNA library, generating millions of raw reads. Bioinformatic analysis follows, involving mapping reads to plant genomes/transcriptomes, removing non-miRNA entities, identifying sequences that match recognized phyto-miRNAs in miRBase with only two mismatches, and predicting hairpin structures and Dicer cleavage sites to identify novel miRNA candidates. Validation of novel miRNAs is achieved through stem–loop qRT-PCR, northern blotting, or in situ hybridization (Zhou et al., 2020).

Moving into the characterization phase, the identified miRNAs responsive to abiotic stress undergo expression analysis through qRT-PCR on small RNA isolated from stress-treated and control plant tissues. miRNA target genes are predicted using computational methods such as psRNATarget, which analyze complementarity between mRNA transcripts and miRNA sequences. Experimental validation of miRNA–target interactions is achieved through methods like 5′-RACE and degradome sequencing, which map cleavage sites in target transcripts guided by the miRNA (Zhang et al., 2014). Furthermore, the validation of the negative regulation of targets by miRNAs is supported by the qRT-PCR study of target gene expression under stress and normal conditions. Their roles in stress response and tolerance are further elucidated by phenotypic studies carried out in model plants through the overexpression or knockout of stress-related miRNAs. A comprehensive understanding of stress-responsive miRNA repertoires is provided by the genome-wide identification of novel and known miRNAs in plants, made feasible by integrating thorough bioinformatics analysis with next-generation sequencing (Addo-Quaye et al., 2008; German et al., 2008) (Figure 9.2).

This workflow diagram presents a comprehensive overview of the methodological steps involved in the discovery and functional analysis of miRNAs related to abiotic stress responses in major cereal and legume crops. Plant tissues are collected under specific stress conditions, such as drought, salt, heat, and cold, to initiate the process. Subsequently, RNA is extracted from these samples to isolate the small RNA fraction that includes miRNAs. Advanced sequencing technologies, particularly next-generation sequencing, are employed to sequence the extracted RNA, generating extensive datasets that are then analyzed to identify miRNAs exhibiting altered expression profiles under stress conditions. Once stress-responsive miRNAs are identified, the workflow progresses to the characterization of their target genes. Predictive computational tools are used to identify potential

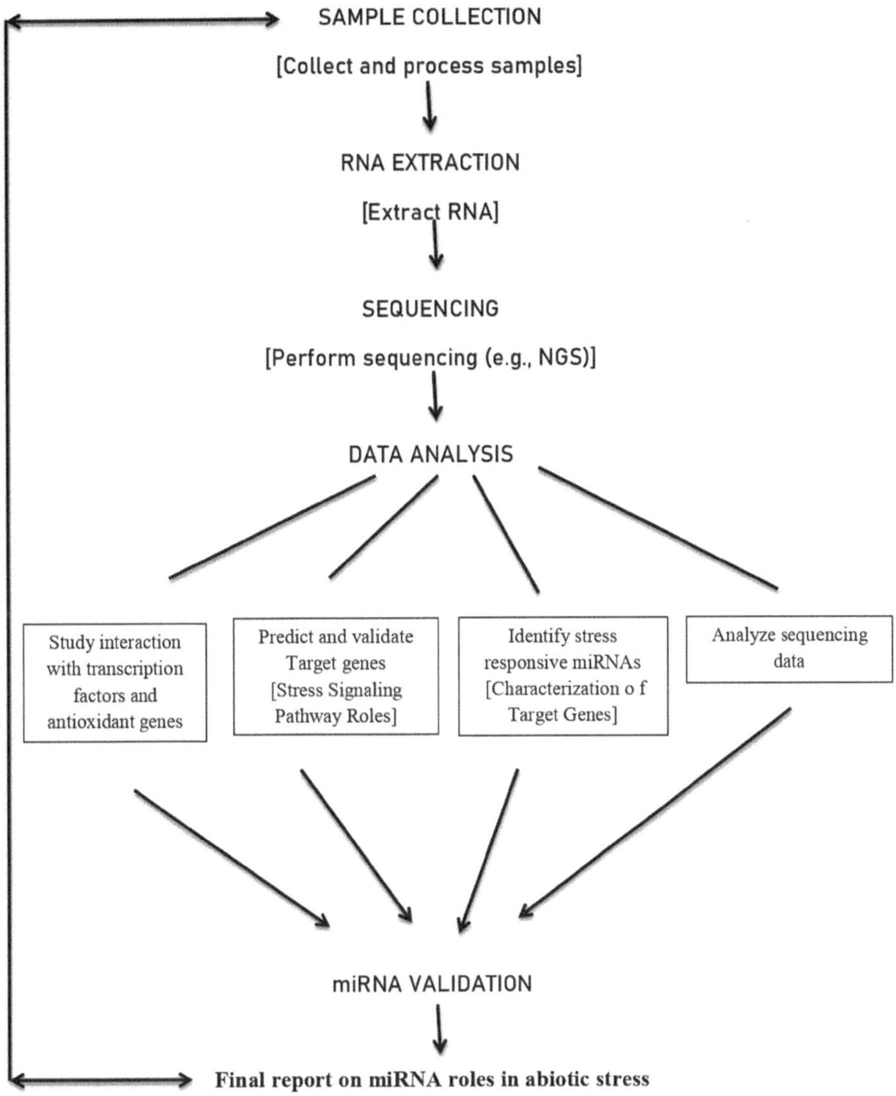

FIGURE 9.2 Illustration depicting miRNA discovery and functional analysis in response to abiotic stresses.

gene targets, which are then validated through experimental assays to confirm regulatory relationships. In parallel, the roles of these miRNAs in stress signaling pathways are elucidated, focusing on their interactions with TFs and antioxidant genes that govern plant stress responses. Feedback loops are crucial for iteratively improving and refining the miRNA profiles throughout the process. This workflow culminates in a thorough report that describes the functions of miRNAs in plant resistance to abiotic stress and provides valuable information for initiatives aimed at improving crop yield. For researchers studying plant stress biology, the diagram highlights the essential phases of miRNA analysis, from initial sample collection to final validation.

9.6 COMPARATIVE ANALYSIS OF miRNA REGULATORY NETWORKS

Understanding the complexity of miRNA regulatory networks across different plant groups provides critical insights into their evolutionary adaptations and functional mechanisms in stress responses. This section delves into the comparative analysis of miRNA regulatory networks, focusing on the differences and interactions between monocots and dicots, and exploring the coordination and cross-talk among miRNA families and other small RNA pathways.

9.6.1 Conserved and Unique miRNA Networks in Monocots vs Dicots

Both monocots and dicots share a set of conserved miRNAs that play fundamental roles in plant development and stress responses. For instance, miR156 and miR172, which regulate the transition from juvenile to adult phases in plants, are found in both groups, illustrating their evolutionary importance (Jung et al., 2011). Similarly, miR395, which is involved in sulfate assimilation and response to nutrient deficiency, shows conservation across monocots like rice and wheat, and dicots such as *Arabidopsis* and soybean, highlighting a common regulatory mechanism for nutrient management in plants.

Despite these conserved roles, unique miRNA networks often arise due to species-specific adaptations to their environments. In monocots, such as maize and rice, miR164 is vital for drought and salinity stress by targeting NAC TFs, which play crucial roles in mediating stress-related signaling pathways. In contrast, dicots like *Arabidopsis* utilize a distinct set of miRNAs, such as miR399, which is specifically upregulated under phosphate starvation and targets PHO2, a key regulator of phosphate homeostasis. This divergence exemplifies how miRNA networks have evolved to meet the specific physiological and environmental needs of different plant groups.

9.6.2 Coordination between miRNA Families and Cross-Talk with Other Small RNAs

MiRNA families often function in a coordinated manner, forming complex networks that enhance their regulatory capabilities. For example, in both monocots and dicots, miR159 and miR319 share targets among TCP TFs, coordinating developmental and stress responses by modulating growth and hormone signaling pathways. This coordination ensures that multiple aspects of plant physiology are harmonized in response to developmental cues and environmental stresses, optimizing the plant's overall fitness and adaptability.

MiRNAs also interact extensively with other small RNA pathways, such as siRNAs and piRNAs, to fine-tune gene expression. This cross-talk can significantly enhance the complexity of post-transcriptional regulation. In dicots, miRNAs may guide the RISC complex to target mRNAs also regulated by siRNAs, creating layers of regulation that can be crucial for responding to viral infections or severe stress conditions. Moreover, a recent study has demonstrated that miRNAs can regulate the expression of genes in monocots involved in siRNA manufacturing. This may impact RNA-directed DNA methylation pathways, which alter critical epigenetic states for stress memory and intergenerational inheritance.

9.7 FUNCTIONAL ROLES OF miRNA IN AGRONOMICALLY IMPORTANT CEREALS

miRNAs, intricate regulators of gene expression, were initially spotlighted in *Arabidopsis thaliana*, where the discovery of miR156 marked a groundbreaking revelation in developmental control by targeting SPL transcription factors. This early breakthrough paved the way for understanding miRNA functions in various plant species, revealing their roles in developmental milestones, including leaf morphogenesis, root development, and flowering time control (Navarro et al., 2006; Sun, 2012). As researchers delved deeper into the application of miRNAs in agronomically significant crops, the focus extended to abiotic stress responses, a critical factor in crop productivity. Crops like rice and maize emerged as pioneers in exploring miRNA-mediated stress regulation. For instance, in rice, miR319 was recognized as a key player in heat stress responses by targeting the OsTCP TF family (Zhou and Luo, 2013). Maize demonstrated how miR164 influences the activity of NAC TFs under water deficit conditions. These findings underscored the adaptability of miRNAs in addressing the challenges posed by environmental stresses. Recent advancements in technology have revolutionized the study of miRNAs. The high sensitivity and specificity of genome-wide detection of known and novel miRNAs is made possible by next-generation sequencing in conjunction with rigorous bioinformatics analysis (Budak et al., 2015). These state-of-the-art methods provide a thorough understanding of plant miRNA repertoires, especially in those exposed to abiotic stress.

For instance, by targeting the genes for copper SOD (Cu/ZnSOD), miR398 controls the wheat plant's reaction to oxidative stress (Jiang et al., 2018). Crucial legume soybeans use miR169 to target nuclear TF Y subunit A (NFYA) genes and modify drought stress responses (Luan et al., 2015). The discovery of these stress-responsive miRNAs in agriculturally significant crops opens avenues for developing strategies to enhance crop resilience in the face of changing environmental dynamics. Harnessing the potential of miRNAs for crop improvement requires a multifaceted approach. The prediction of miRNA target genes is facilitated by computational techniques like psRNATarget, which examines the complementarity between miRNA sequences and mRNA transcripts (Deng et al., 2021). A comprehensive understanding of miRNA–mRNA interactions at the genome-wide scale can be achieved through experimental validation techniques such as 5' RACE and degradome sequencing (Addo-Quaye et al., 2008; German et al., 2008). These methods enable a thorough characterization of miRNA-mediated stress regulation, especially when combined with functional genomics investigations that involve overexpressing or knocking down stress-related miRNAs.

The journey of miRNA research from model plants to agronomically important crops has unraveled the multifaceted functional roles of these small RNA molecules. From developmental milestones to stress adaptation, miRNAs emerge as versatile players in shaping the physiological responses of plants. The integration of modern technologies and computational tools not only refines our understanding of miRNA functions but also opens new avenues for innovative strategies in agriculture, crucial for ensuring global food security.

9.7.1 Functional Roles of miRNA in Rice

More than half of the world's population is fed by rice, or *Oryza sativa*, making it a critically important crop. However, adverse climatic factors impede rice cultivation and yields. Twenty miRNAs were identified in rice in 2004 by one of the earliest investigations using a cDNA library; the projected target genes for these miRNAs were involved in transcription, metabolism, disease resistance, transport, and other functions (Wang et al., 2004). Sunkar et al. (2005) conducted a follow-up study that found 35 rice miRNAs the following year, 14 of which were novel and anticipated to target genes involved in several physiological processes. There are now over 592 miRNA sequences identified in rice stored in miRBase version 21. Using data from next-generation sequencing, numerous studies have examined rice miRNAs and found conserved miRNAs that respond to abiotic stresses (Mittal et al., 2016; Sunkar et al., 2007; Zhang et al., 2007). osa-miR393 targets the auxin receptor genes OsTIR1 and OsAFB2 and is substantially expressed during droughts. Overexpression of this gene decreased tolerance to salinity and drought (Xia et al., 2012; Zhao et al., 2007; Zhou et al., 2010). Interestingly, osa-miR393 overexpression also led to beneficial high tillering and early flowering phenotypes (Jiao et al., 2010). According to Jeong et al. (2011), several cold-responsive miRNAs, including osa-miR167, osa-miR319, osa-miR812q, and osa-miR1425, are also crucial in controlling target genes in response to cold stress. The L-ascorbate oxidase gene is modulated by osa-miR397 in response to high temperatures, and its overexpression increased yield by enhancing panicle branching and grain size (Jeong et al., 2011; Yu et al. 2012; Zhang et al., 2013b).

9.7.2 Functional Roles of miRNA in Maize

Maize (*Z. mays* L.), a critical crop after rice, faces productivity challenges due to various abiotic stresses. miRNAs are key regulators in stress responses. A study identified 124 conserved and 68 novel miRNAs associated with drought tolerance in maize (Sheng et al., 2015). miR396, miR167, and miR169 target genes related to drought stress, while "Cu-miRNAs" like miR398 and miR408 regulate copper-related proteins (Kawashima et al., 2011). miR395 controls sulfur assimilation by regulating sulfur transport genes (Lauter et al., 2005). Other miRNAs, including miR156, miR160, and miR169, exhibit differential expression throughout maize growth (Djami-Tchatchou et al., 2017). Additionally, miRNAs respond to low nitrate, phosphate, and salt concentrations, enhancing maize productivity and stress tolerance by regulating stress-related genes (Ding et al., 2011; Qin et al., 2011; Zhang, et al. 2012).

9.7.3 Functional Roles of miRNA in Wheat

Triticum aestivum L., or wheat, is a staple grain grown worldwide and a vital source of calories and nutrients for both humans and animals. However, abiotic factors such as drought, salinity, high temperatures, and UV radiation frequently lead to significant losses in wheat development and yields. miRNAs have been shown in numerous studies to play important regulatory roles in wheat's tolerance and response to various abiotic stresses. Using deep sequencing, one of the first investigations identified 58 wheat miRNAs from 43 conserved and novel miRNA families (Zhang et al., 2009). Under UV-B (280–320 nm) stress, several of these miRNAs, including miR156, miR159, miR164, miR167a, miR171, and miR395, exhibited distinct expression patterns. It has also been demonstrated that miR398 regulates the genes encoding Cu–Zn superoxide dismutase (CSD), which is involved in wheat's resistance to various stresses such as salinity, heavy metals, UV radiation, and drought (Inal et al., 2014). Wheat has been shown to express miR169g, miR171a, and miR393 in drought-prone conditions. Of these, miR169 has been shown to regulate TFs known as CBF/DREB, which promote drought tolerance (Inal et al., 2014). Similarly, Sunkar and Zhu (2004) observed the differential regulation of miRNAs, including miR408, miR169, miR396, miR172, miR393, miR397, and miR165/166, under cold stress in *Triticum turgidum* ssp. *durum*. More recently, wheat sequencing revealed the presence of abiotic stress response miRNAs such as Ta-miR1122, Ta-miR1117, Ta-miR1113, and Ta-miR113 (Li et al., 2015). The aforementioned underscores the significant regulatory functions of miRNAs in controlling gene expression and enhancing wheat's ability to adapt to various environmental stresses through molecular mechanisms such as response pathways and stress signaling. Wheat cultivars with increased resistance to abiotic stresses may be developed with a deeper understanding of the roles played by these stress-related miRNAs.

9.7.4 Functional Roles of miRNA in Barley

Barley (*Hordeum vulgare* L.), a grain crop of substantial economic importance, is grown worldwide. In addition to its applications in human nutrition and animal husbandry, barley serves as a model plant belonging to the Triticeae family (Curaba et al., 2012; Sreenivasulu et al., 2008). A total of 56 conserved miRNAs, including miR156/157, miR159, miR319, miR160, miR172, miR393, miR396, miR408, and others, were first identified in barley using deep sequencing. Similarly, Shuzuo et al. (2012) used short RNA and mRNA degradome analysis to describe 84 conserved miRNAs and 7 novel miRNAs. More recently, Deng et al. (2015) discovered novel salt-responsive miRNAs that enhance salinity tolerance in barley, along with their target genes. Additionally, Shuzuo et al. (2012) identified 133 new and 126 conserved barley miRNAs through Solexa sequencing, including miR-n026, miR-n028, and miR-n029, which have been proposed to be involved in responses to salt stress and drought.

9.7.5 Functional Roles of miRNA in Sorghum

Sorghum (*Sorghum bicolor* L. Moench), an important grain crop, is renowned for its resilience to environmental challenges, including flooding and drought. It can be utilized for various purposes, including as animal feed, food, and as a raw material for products like biofuels, alcohol, and starch (Mutegi et al., 2010). Sorghum miRNAs and their regulation of gene expression have been extensively studied (Katiyar et al., 2012). To date, over 446 sorghum miRNAs are cataloged in miRBase. Previous research identified 17 new miRNAs from 11 families that regulate genes involved in processes such as the cell cycle, metabolism, protein degradation, RNA processing, and stress responses (Du et al., 2010). An additional investigation discovered 13 new miRNAs, 29 conserved miRNA families, and predicted targets that included genes related to the ubiquitin system, SNARE proteins, SAM decarboxylase, receptor kinases, and oxidoreductases (Zhang et al., 2011). Individual miRNAs targeted distinct mRNAs involved in different processes. For example, miR398 targets genes related to ROS detoxification, while miR393 targets an auxin receptor, indicating spatial specificity in their regulatory roles. The miR169 family has been linked to sorghum's resistance to heat stress and drought, suggesting it plays a role in the plant's resilience to abiotic stresses (El Sanousi et al., 2016; Hamza et al., 2016). Remaining functional roles of miRNA in minor agronomically important cereals are included in Table 9.1.

9.7.6 Functional Roles of miRNA in Soybean

Soybean (*Glycine max* L.) is an economically important crop cultivated worldwide for its high protein and oil content. Early studies on soybean miRNAs identified 20 conserved families in 2008. Currently, over 639 soybean miRNA sequences have been documented in miRBase v21. Several miRNAs regulate seed development by targeting TFs and other genes. For example, 26 miRNAs target MYB, ARF, NAC, GRF, and TCP TFs, as well as F-box proteins and G-proteins, revealing regulatory networks in seed formation. Like other crops, soybean faces various abiotic stresses that impact productivity, especially drought. Studies have examined how salinity, alkalinity, and drought affect soybean miRNA regulation (Kulcheski et al., 2011). miRNA396 and miRNA172 are expressed in response to drought in soybean (Zhang et al., 2016), and miR396 interacts with GRFs to govern plant development, growth, and stress tolerance. Several miRNAs, including members of the miR167, miR319, and miR393 families, were found to be similarly regulated under diverse stresses such as drought, salinity, and alkalinity. This suggests that they target the negative regulators of stress-tolerance pathways. According to Sahito et al. (2017), miR172 targets nodule number control 1 and is stimulated by salt stress, while the expression level of soybean NNC1 (Nodule Number Control 1) influences its response to salt stress. Furthermore, miR395 was found to be upregulated in response to salt stress, likely targeting genes involved in sulfate starvation response pathways that sustain energy supply. Under various abiotic stresses, soybean miRNAs and their predicted target genes exhibit different expression patterns. In wild soybean, key miRNAs also modulate the response to aluminum stress (Qiao-Ying et al., 2012).

9.7.7 Functional Roles of miRNA in Cowpea

The cowpea (*Vigna unguiculata* L.) is a crucial crop in tropical and subtropical regions of Africa, Asia, and South America. Research has examined the genetic factors, mechanisms, and miRNAs involved in cowpea's drought tolerance. Early studies identified 47 miRNAs from 13 families computationally. These miRNAs were predicted to target over 30 genes related to development, growth, metabolism, and stress response (Lu and Yang, 2010). Several cowpea miRNAs, such as miR156, miR160, miR162, miR164, miR166, miR159, miR167, miR169, miR171, miR319, miR390, miR393, miR396, miR403, and miR482, have since been placed in miRBase (Barrera-Figueroa et al., 2011). A few miRNAs are essential for drought tolerance. For instance, by triggering TFs related to nuclear factor Y, miR169 improves tolerance. Under water stress, Vun-miR156b and Vun-miR156f regulate multi-cystatin, controlling protein degradation rates during drought. This helps maintain optimal cellular protein levels for plant survival under water deficit conditions. Some additional miRNAs are listed below (miRNA name—its target—its target description) in Table 9.1 (Gul et al., 2017).

9.7.8 Functional Roles of miRNA in Groundnut

Groundnuts, or peanuts (*Arachis hypogaea* L.), rank among the top five oilseed crops grown worldwide. Early research on peanut miRNAs utilized high-throughput sequencing, first identifying 14 novel miRNA families (miRn1 to miRn14), 13 of which were peanut-specific, along with 75 conserved miRNA families (miR156 to miR894) in 2010 (Zhao et al., 2010).

Auxin signaling, transport, metabolism, and stress responses were all regulated by additional miRNA targets. In 2011, Chi et al. (2011) identified 126 conserved miRNAs from 33 families and 25 novel peanut miRNAs. Gene ontology and KEGG pathway analysis linked 34 miRNA families to metabolic processes involving fatty acids, carbohydrates, signaling, and responses to biotic and abiotic stresses. The miR156, miR159, miR171, and miR14 families mostly targeted genes involved in amino acid, lipid, and fatty acid metabolism. Newly identified miRNAs targeted genes for chlorophyll binding, heat shock, membrane proteins, and nutrient transport. More recently, research has demonstrated the key roles of miRNAs in peanut responses to the damaging bacterial wilt pathogen *Ralstonia solanacearum*, with numerous miRNAs differentially expressed upon infection (Zhao et al., 2015). These miRNAs were anticipated to target proteins linked to pathogenesis, TFs, receptors, and defense genes for aquaporins.

9.7.9 Functional Roles of miRNA in Common Bean

The common bean or *Phaseolus vulgaris* L., also referred to as the kidney bean, navy bean, or string bean, is a significant legume crop in terms of both economics and nutrition. miRNAs have regulatory functions in how common beans react to abiotic stresses. Differential expression of miR156, miR159, miR160, miR166, miR167, miR168, miR169, miR171, miR172, miR393, miR396, and miR398 has been observed in response to heat, salt, and drought stresses (Arenas-Huertero et al., 2009; Peláez et al., 2012; Valdés-López et al., 2010; Wu et al., 2017). During drought stress, miR156 and miR167 were elevated, while miR159 and miR169 were downregulated. Common beans are vulnerable to a range of biotic stresses. Infection by the bean anthracnose pathogen, *Colletotrichum lindemuthianum*, elicits the differential expression of miR482 and miR2118. Specifically, miR482 targets NBS-LRR, which are implicated in plant defense (Naya et al., 2014). Additionally, miR1507, miR1510, and miR2118 were upregulated upon *Pseudomonas syringae* pv. *phaseolicola* pathogenesis, while miR159, miR160, miR166, and miR396 were downregulated (Islam et al., 2017). The functional aspects of miRNA and their roles in biotic stress responses in agronomically important cereal and legume crops (Begum, 2022; Djami-Tchatchou et al., 2017; Mantri et al., 2013; Mittal et al., 2023; Samynathan et al. 2023) are shown in Table 9.1.

9.8 EMERGING OMICS APPROACHES FOR miRNA DISCOVERY AND FUNCTIONAL ANALYSIS

The integration of advanced omics technologies has significantly enhanced our understanding of miRNAs' roles in plant tolerance. These technologies, encompassing genomics, transcriptomics, and proteomics, provide a multi-layered approach to studying plant–environment interactions. Here's a detailed exploration of how each of these omics approaches contributes to miRNA research, particularly in the context of stress adaptation in plants.

9.8.1 Genomics and High-Throughput Sequencing

Genomics has been foundational in identifying miRNA genes across various plant genomes. High-throughput sequencing methods, such as next-generation sequencing (NGS), have allowed the rapid and cost-effective sequencing of complete plant genomes. This comprehensive genomic data

TABLE 9.1
Functional Aspects of miRNA and Their Roles in Biotic Stress Response in Agronomically Important Cereal and Legume Crops

Crop	miRNA	Functions
Rice	miR159, miR169, miR171, miR319, miR395, miR474, miR845, miR851, miR854, miR896, miR901, miR903, miR1026 and miR1125, miR444b.1, miR529	Drought response
	miR393	Drought response High tillering and early flowering
	miR397	Heat stress response and adaptation
	miR820, mi166, miR172a	Response to salt
	miR169g, miR171a and miR393	Induced in response to dehydration
	miR169, miR397, miR528, miR827, miR1425, miR319a.2 and miR408-5p	H_2O_2-responsive
	miR156k, miR166m, miR166k, miR167a/b/c, miR168b, miR169e, miR169f, miR169h, miR171a, miR1884b, miR319a/b, miR444, miR535, miR1320, miR1435, miR1850, miR1868, and miR1876	Cold stress
	miR319a/b	Cold tolerance
	miR528, and miR827	H_2O_2-oxidative stress
	miR2118	Photoperiod-sensitive male sterility
	miR168	Flower and yield improvement
	OsmiR535	Leaf senescence alteration
Wheat	miR1867, miR896, miR398, miR528, miR474, miR1450, miR396, miR1881, miR894, and miR156	Drought-resistant wild emmer
	miR396a, miR444c.1,172a, miR393, miR167d, miR167c) and tasiRNA	Cold stress
	miR156, miR159, miR160, miR166, miR168, miR169, miR393, miR395, miR1118, miR1120/1/2, miR827, and miR9772	Heat stress
	miR156, miR159, miR164, miR171, miR395	UV-B stress
Maize	miR164 and miR160	Regulation of ABA metabolism
	miR162, miR156, miR164, miR167, miR168, miR395, and miR474	Salinity stress
	miR169	Drought stress response
	miR529, miR396, miR171, miR165, miR166	UV-B stress
Barley	miR156d, miR399b/d	Drought stress response
	miR166a/b/c	Different stress
	miR5049c/f	Salt stress
	miR-n026, miR-n028, and miR-n029	Drought and stress
	miR166a, miR166′b, miR166c	Vascular differentiation, development of leaf and root formation
	miR397a	Plant and flowering development
	miR164	Lateral root and shoot development
Sorghum	miR529, miR398,miR156a, miR156b, miR156c-f, miR156g-h, miR529, novelsbi-miR-119, novel-sbi-miR-383, novelsbi-miR-329, and a novel-sbi-miR-350	Response to drought stress
Soybean	miR156, miR159, miR164, miR169, 319, 396, 398, 408	Response to drought stress
	miR156, miR319, miR166, miR171, miR172a, miR398	Salt stress

(Continued)

TABLE 9.1 (*Continued*)
Functional Aspects of miRNA and Their Roles in Biotic Stress Response in Agronomically Important Cereal and Legume Crops

Crop	miRNA	Functions
Cowpea	miR166, miR160	Drought stress
	miR159, miR167	Enhanced drought tolerance
	miR169, miR319, miR390, miR393, miR396, miR403	Metabolic pathways associated with drought stress
	miR156b/f	Protein degradation/drought stress
Groundnut	miR156, miR172, miR397, miR398, miR408 miR393, miR408, miR1511, miR1507	Heat and drought stress
Common bean	miR159, miR167, miR393, miR169, miR172, miR395, miR396, miR398	Responsive to salt stress
	miR2118, pvu-miR393, pvu-miR1514a, pvu-miR2119, pvu-miR159.2, pvu-miRS1	Responsive to drought stress
	pvu-miR159.2, pvu-miR393, pvu-miR2118, pvu-miR2119, pvu-miR1514a, pvu-miRS1	Responsive to cold stress

Source: This above information was gathered from the literary works of Samynathan et al. (2023), Djami-Tchatchou et al. (2017), Yasmin Begum (2022) and Mittal et al. (2023).

enables researchers to locate miRNA genes and predict their potential regulatory roles based on their sequences and conservation across species.

9.8.2 Whole-Genome Sequencing

Whole-genome sequencing enables investigators to pinpoint both conserved and entirely novel miRNAs across multiple plant species, providing insights into the evolutionary dynamics of miRNA regulation.

9.8.3 Small RNA Sequencing

Specifically tailored to capture small RNAs, this technique provides detailed profiles of miRNA expression levels and patterns, helping to clarify their functional roles under various stress conditions.

9.8.4 Transcriptomics

Transcriptomics refers to the decoding of all RNA molecules from a plant, including mRNAs and non-coding RNAs such as miRNAs. RNA sequencing provides a snapshot of gene regulation shifts under stressful conditions while identifying miRNA targets by examining the patterns of gene expression as miRNA levels fluctuate.

9.8.5 Differential Expression Analysis

By comparing the transcriptomes of stressed and unstressed plants, researchers can identify which miRNAs are differentially expressed following exposure to specific abiotic stresses.

9.8.6 Integrative Network Analysis

Combining transcriptomic data with miRNA profiles allows the construction of regulatory networks that detail the interactions between miRNAs and their target mRNAs, providing a comprehensive understanding of how miRNAs modulate stress response pathways.

9.8.7 Proteomics

Proteomics complements genomic and transcriptomic data by analyzing the protein products that function within the cell. Changes in miRNA levels affect the production of specific proteins, particularly those involved in stress responses.

9.8.8 Mass Spectrometry

Advanced mass spectrometry techniques enable precise quantification and identification of stress-responsive proteins, linking changes in miRNA levels to tangible shifts in protein expression *profiles.*

9.8.9 Protein Interaction Networks

By mapping the interactions between proteins, researchers can infer the downstream effects of miRNA-mediated regulation, elucidating how miRNAs impact complex biological pathways and stress responses.

9.8.10 Bioinformatics and Systems Biology

The vast datasets generated by genomic, transcriptomic, and proteomic studies require sophisticated bioinformatics tools for analysis and interpretation (Johri et al., 2022). Systems biology approaches integrate these multiple layers of data to model and simulate the complex biological processes influenced by miRNAs.

9.8.11 Computational Modeling

Bioinformatics tools can predict miRNA targets and simulate the impacts of miRNA regulation on metabolic and signaling pathways, offering insights into the broader implications of miRNA activity under stress conditions (Verma et al., 2023).

9.8.12 CRISPR/Cas9 and Gene Editing

Utilizing omics data, researchers can design CRISPR/Cas9 strategies to edit miRNA genes or their regulatory elements, precisely enhancing or reducing their expression to improve stress tolerance. These gene-editing experiments are guided by comprehensive omics analyses, ensuring targeted and effective modifications.

9.9 FUTURE APPLICATIONS OF miRNA IN CROP IMPROVEMENT

9.9.1 Pyramiding Favorable Alleles via miRNA–Target Networks

Identifying miRNA-regulated gene networks involved in stress signaling helps breeders introgress favorable alleles controlling stress response pathways. Marker-assisted selection allows tracking stress-responsive miRNA and target gene loci during pedigree breeding. Favorable alleles associated with stress tolerance are introgressed simultaneously. This pyramids additive and epistatic effects for enhanced stress resilience. Network-based breeding leverages natural gene regulatory mechanisms mediated by miRNAs to develop climate-resilient varieties with optimized stress adaptation.

9.9.2 Tissue-Specific Modulation of Stress Response

Understanding spatial patterns of miRNA expression aids the precision engineering of stress response. Tissue-specific promoters help restrict modified miRNA levels to cells directly impacted by stress. For example, inducing a dehydration-responsive miRNA in leaves but not roots prevents trade-offs between stress defense and normal growth. Cell-type targeting also avoids undesirable systemic effects. This precision modification of stress signaling optimizes stress protection versus growth maintenance.

9.9.3 Stable Inheritance of Stress Traits via miRNAs

miRNAs offer an advantage as heritable carriers of stress tolerance. In contrast to transient expression from transcriptional fusions, engineered miRNA loci stably maintain modified stress–gene networks. Orthologous miRNA genes transfer reliably across sexual generations, allowing fixation of stress-adaptive phenotypes. Epigenetic inheritance of miRNA-directed changes in gene expression also facilitates development of perennial crops with fixed stress resilience. Together, the heritability and mitotic stability of engineered miRNA loci aid in the development of climate-resilient varieties.

9.9.4 Rapid Field Screening via Stress-Responsive miRNA Profiling

High-throughput miRNA profiling enables developing stress indices for field-based screening. Stress-induced changes in specific miRNA profiles correlate with tolerance and serve as 'molecular barcodes' of stress resilience. Profiling large germplasm collections under field conditions allows for ranking genotypes based on their miRNA stress indices. This rapidly identifies promising sources of tolerance to evaluate in breeding. Combined with phenotyping, the approach helps eliminate sensitive accessions early. Selected genotypes can then be subjected to detailed omics analyses to discover tolerance mechanisms and genes/miRNAs for crop improvement.

9.9.5 Epigenetic Priming of Stress Defenses

Stress-responsive miRNAs play important roles in epigenetically priming stress defense pathways. They fine-tune the expression of signaling molecules and TFs that contribute to the stress response. Engineering crops to modulate specific miRNAs associated with priming could allow for faster defenses upon subsequent stress exposure via altered chromatin states and DNA methylation patterns. This epigenetic priming would mitigate yield losses from rapid stress imposition.

9.9.6 Stage-Specific miRNA-Mediated Stress Protection

Careful profiling of miRNA expression during critical crop developmental stages like flowering and grain filling can reveal miRNAs that protect these phases. Engineering such stage-specific miRNAs may help shield vital growth processes even under stress. For example, modulating miRNAs that are upregulated only during flowering prevents stress-induced floral abortion and maintains productivity.

9.9.7 Resource Reallocation for Stress Tolerance

Certain miRNAs mediate the reallocation of resources from growth to stress defense pathways under stress. Overexpressing these miRNAs could help crops prioritize tolerance over vegetative growth when resources are limited. This ensures survival and protects harvests, allowing crops to withstand variable climatic conditions with minimal yield penalties.

9.9.8 miRNA-Based Bio-Stress Protectants

Synthetic constructs that deliver stress-related miRNA triggers into plant soils show potential as bio-pesticides or bio-fertilizers. Induced in root systems, these could epigenetically prime above-ground stress defenses via grafting or signaling without transgenesis. This provides a novel, non-transgenic route to enhance crop resilience to abiotic and biotic stress.

9.9.9 miRNAs as Molecular Markers

Stress-responsive miRNA profiles serve as molecular markers in breeding programs, enabling the identification and selection of crop varieties with innate stress tolerance through high-throughput screening.

9.9.10 Elucidating miRNA Target Modules

Detailed analysis of miRNA–target interactions under stress conditions informs the development of crop varieties with optimized stress responses, enhancing traits such as hormone balance and oxidative stress management.

9.9.11 miRNA-Guided Nutrient Efficiency

Unveiling miRNA interactions with nutrient signaling can guide the breeding of nutrient-efficient crops that thrive in low-input agriculture. Profiling miRNA expression under varying nutrient levels pinpoints crucial post-transcriptional regulators. Targeting miRNAs that restrict nutrient uptake and utilization during scarcity holds promise for crops excelling in sub-optimal fertilization scenarios.

9.9.12 miRNA-Mediated Grafting for Perennial Improvement

miRNAs offer potential for engineering stress-tolerant rootstocks and scions in perennial crops through grafting. Profiling root and shoot miRNAs identifies genotypes with differential expression of stress-responsive miRNAs. Engineering these tissue-specific miRNAs into model rootstocks and scions allows for the evaluation of stress resilience. Successful models can then be leveraged as frameworks for grafting-based stress resistance in commercial orchards.

9.9.13 Field-based Stress Profiling and Breeding

High-throughput miRNA profiling facilitates the development of stress indices for rapid field-level screening of stress responses in large germplasm collections. Correlating these indices with visual stress scores aids in identifying tolerant accessions for breeding programs. Selected accessions can then undergo detailed omics analyses to elucidate tolerance mechanisms and identify genes and miRNAs for varietal development and crop improvement.

9.9.14 Coordinated Stress Response via miRNA Circuits

Synthetic genetic circuits regulated by a single engineered miRNA hold promise for dynamic and optimized stress responses by modulating multiple stress-responsive genes. Elucidating circuit design principles that integrate abiotic and biotic signals with endogenous developmental pathways is crucial. Successful prototypes could provide a blueprint for multigene engineering, conferring robust and coordinated stress-tolerance traits.

9.9.15 Broad-Spectrum Protection with Viral Vectors

Novel plant viral vectors carrying stress-related miRNA payloads show potential for conferring resistance to multiple stresses through foliar application. This method provides a means to epigenetically prime above-ground defenses without resorting to transgenesis. Evaluating the ability of vector-delivered miRNA triggers to protect crops against major diseases and abiotic threats could establish them as tools for developing broad-spectrum stress resilience in important crops.

9.10 CHALLENGES IN HARNESSING miRNA POTENTIAL FOR STRESS TOLERANCE

A major issue is identifying the miRNAs responsible for stress responses, as we understand how they assist in stress defense. Despite the increase in high-throughput sequencing technologies that allow for the detection of miRNAs, functional characterization can be time-consuming because of the complexity involved. Moreover, miRNAs modulate gene expression by degrading or repressing the translation of specific mRNAs. However, interpreting the intricate relationships involving miRNAs, their target genes, and their interactions with stress conditions remains a significant challenge; therefore, comprehensive computational and experimental methods must be adopted. Furthermore, effectively delivery of delivering miRNA or genetically modifying the target organs or cells poses a considerable challenge. Developing successful biotechnology platforms for selective transportation and optimal expression in response to stress factors is one of the key obstacles.

Additionally, miRNAs may have multiple targets and exhibit non-specificity at off-target sites, which can lead to undesired consequences. To enhance the positive impacts and mitigate the negative effects of miRNA-based therapies, deviations from targets and resilience to stress must be taken into consideration.

9.11 FUTURE RESEARCH DIRECTIONS AND ADVANCEMENTS HARNESSING miRNA POTENTIAL FOR STRESS TOLERANCE

Instead of solely focusing on stress-responsive miRNAs, researchers can explore the untapped potential of non-canonical or non-conserved miRNAs. By investigating these lesser-known miRNAs, we may uncover unique stress-tolerance mechanisms and identify novel targets for manipulation. Embracing synthetic biology principles can revolutionize the use of miRNAs for stress tolerance. Researchers can engineer synthetic miRNAs or design novel regulatory components to fine-tune miRNA expression and target specificity. A comprehensive understanding of miRNA-mediated stress responses may be achieved by merging multi-omics research, which encompasses genomics, transcriptomics, proteomics, metabolomics, and epigenomics. Through the integration of multiple data layers, scientists may decipher intricate regulatory networks, pinpoint miRNA target genes, and gain insight into post-transcriptional and post-translational regulation. Innovative forms of stress endurance can be discovered by examining the interactions and cross-talk among various types of non-coding RNAs, including circular RNAs and long non-coding RNAs. New avenues for intervention may emerge by understanding whether miRNAs cooperate with or compete against other non-coding RNAs in stressful situations.

While genome editing technologies have primarily focused on modifying miRNA target sites, there is potential to explore editing techniques to directly modify miRNA sequences themselves. Precise editing of miRNAs can enhance their stability, specificity, or target range, ultimately fine-tuning stress-responsive pathways. Developing comprehensive computational models that integrate miRNA-mediated gene regulation, signaling pathways, and physiological responses can provide predictive insights into stress tolerance. These models can guide the design of optimized miRNA-based strategies and identify potential combinatorial approaches with other stress-tolerance mechanisms. In addition to traditional delivery methods, exploring non-conventional delivery systems, such as extracellular vesicles or engineered nanoparticles, can enhance the efficiency and specificity of miRNA delivery to target tissues. These innovative delivery platforms offer opportunities for precise and controlled release of stress-responsive miRNAs. Furthermore, understanding how miRNAs function in concert with environmental cues can lead to the development of context-specific stress-tolerance strategies. Implementing high-throughput screening platforms, such as CRISPR-based genetic screens or small molecule libraries, can identify novel regulators or modulators of miRNA activity. These approaches can uncover small molecules or genetic factors that enhance or suppress the activity of stress-responsive miRNAs, thereby providing new avenues for stress-tolerance enhancement.

9.12 CONCLUSION

This chapter emphasizes the critical role of miRNAs in regulating cereal and legume crop responses to abiotic stimuli such as oxidative stress, salt, and drought. Despite their small size, miRNAs significantly impact plant stress regulatory networks by influencing mRNA stability, translation, and gene expression at both the transcriptional and post-transcriptional levels. Specific stress responses are regulated by key miRNAs such as miR169 (drought), miR393 (salinity), and miR398 (oxidative stress), which modify physiological and biochemical pathways to enhance plant resilience.

Our understanding of miRNAs has advanced due to technical developments in omics and bioinformatics, which have facilitated the identification and characterization of both novel and existing miRNAs, as well as their target genes. miRNAs regulate genes involved in signaling pathways, antioxidant systems, hormone production, and free radical defenses, including ethylene formation.

When miRNA research is integrated with traditional and modern breeding techniques, the development of stress-tolerant agricultural cultivars becomes a feasible goal. Enhancing nutritional security in the face of environmental challenges, growing populations, and climate change requires collaboration. To increase agricultural productivity and sustainability amid changing climate conditions, innovative solutions must incorporate ongoing research and the development of advanced genetic technologies, including CRISPR/Cas9-mediated genome editing.

REFERENCES

Addo-Quaye, C., Eshoo, T. W., Bartel, D. P., & Axtell, M. J. (2008). Endogenous siRNA and miRNA targets identified by sequencing of the Arabidopsis degradome. *Current Biology*, *18*(9), 758–762. https://doi.org/10.1016/j.cub.2008.04.042

Anil, P., Dharmendra, P., Sachidanand, S., Yuriy, E., Snezhan, B., Dilfuza, J., & Chinmayi, J. (2023). Beneficial effects of biochar application on mitigating the drought and salinity stress implications on plants. *Plant Science Today*, *10*(2), 188–193. http://doi.org/10.14719/pst.2591

Arenas-Huertero, C., Pérez, B., Rabanal, F., Blanco-Melo, D., Rosa, C., Estrada-Navarrete, G., Sánchez, F., Covarrubias, A., & Reyes, J. (2009). Conserved and novel miRNAs in the legume *Phaseolus vulgaris* in response to stress. *Plant Molecular Biology*, *70*(4), 385–401. https://doi.org/10.1007/s11103-009-9480-3

Barrera-Figueroa, B. E., Gao, L., Diop, N. N., Wu, Z., Ehlers, J. D., & Roberts, P. A. (2011). Identification and comparative analysis of drought-associated microRNAs in two cowpea genotypes. *BMC Plant Biology*, *11*(1), 127. https://doi.org/10.1186/1471-2229-11-127

Beezhold, K. J., Castranova, V., & Chen, F. (2010). Microprocessor of microRNAs: regulation and potential for therapeutic intervention. *Molecular Cancer*, *9*(1), 1–9. https://doi.org/10.1186/1476-4598-9-1

Begum, Y. (2022). Regulatory role of microRNAs (miRNAs) in the recent development of abiotic stress tolerance of plants. *Gene*, *821*, 146283. https://doi.org/10.1016/j.gene.2022.146283

Budak, H., Khan, Z., & Kantar, M. (2015). History and current status of wheat miRNAs using next-generation sequencing and their roles in development and stress. *Briefings in Functional Genomics*, *14*(3), 189–198. https://doi.org/10.1093/bfgp/elu021

Chen, Y. E., Mao, J. J., Sun, L. Q., Huang, B., Ding, C. B., Gu, Y., Liao, J. Q., Hu, C., Zhang, Z. W., & Yuan, S. (2018). Exogenous melatonin enhances salt stress tolerance in maize seedlings by improving antioxidant and photosynthetic capacity. *Physiologia Plantarum*, *164*(3), 349–363. https://doi.org/10.1111/ppl.12707

Chi, X., Yang, Q., Chen, X., Wang, J., & Pan, L. (2011). Identification and characterization of microRNAs from peanut (*Arachis hypogaea* L.) by high-throughput sequencing. *PLoS One*, *6*(12), e27530. https://doi.org/10.1371/journal.pone.0027530

Chien, P., Chiang, C., Wang, Z., & Chiou, T. (2017). microRNA-mediated signaling and regulation of nutrient transport and utilization. *Current Opinion in Plant Biology*, *39*, 73–79. https://doi.org/10.1016/j.pbi.2017.06.007

Curaba, J., Spriggs, A., Taylor, J., Li, Z., & Helliwell, C. (2012). MiRNA regulation in the early development of barley seed. *BMC Plant Biology*, *12*, 120. https://doi.org/10.1186/1471-2229-12-120

Deng, P., Wang, L., Cui, L., Feng, K., Liu, F., & Du, X. (2015). Global identification of microRNAs and their targets in barley under salinity stress. *PLoS One*, *10*(10), e0137990. https://doi.org/10.1371/journal.pone.0137990

Deng, Y., Qin, Y., Yang, P., Du, J., Kuang, Z., Zhao, Y., Wang, Y., Li, D., Wei, J., Guo, X., & Yang, X. (2021). Comprehensive annotation and functional exploration of microRNAs in lettuce. *Frontiers in Plant Science*, *12*, 781836. https://doi.org/10.3389/fpls.2021.781836

Ding, X., Li, X., & Xiong, L. (2011). Evaluation of near-isogenic lines for drought resistance QTL and fine mapping of a locus affecting flag leaf width, spikelet number, and root volume in rice. *Theoretical and Applied Genetics*, *123*(5), 815–826. https://doi.org/10.1007/s00122-011-1629-1

Djami-Tchatchou, A. T., Sanan-Mishra, N., Ntushelo, K., & Dubery, I. A. (2017). Functional roles of microRNAs in agronomically important plants—potential as targets for crop improvement and protection. *Frontiers in Plant Science*, *8*, 378. https://doi.org/10.3389/fpls.2017.00378

Du, J. F., Wu, Y. J., Fang, X. F., Xia, C. J., & Liang, Z. (2010). Prediction of sorghum miRNAs and their targets with computational methods. *Chinese Science Bulletin*, *55*(14), 1263–1270. https://doi.org/10.1007/s11434-010-0035-4

El Sanousi, R. S., Sanan-Mishra, N., Abdelmula, A. A., Mohammed, I. A., Seif, M. G., & Hamza, N. B. (2016). Differential expression of miRNAs in *Sorghum bicolor* under drought and salt stress. *American Journal of Plant Sciences*, *7*(06), 870–878. https://doi.org/10.4236/ajps.2016.76082

Fang, Y., & Spector, D. L. (2007). Identification of nuclear dicing bodies containing proteins for microRNA biogenesis in living Arabidopsis plants. *Current Biology*, *17*(9), 818–823. https://doi.org/10.1016/j.cub.2007.04.005

Franco-Zorrilla, J. M., Valli, A., Todesco, M., Mateos, I., Puga, M. I., Rubio-Somoza, I., Leyva, A., Weigel, D., García, J. A., & Paz-Ares, J. (2007). Target mimicry provides a new mechanism for regulation of microRNA activity. *Nature Genetics*, *39*(8), 1033–1037. https://doi.org/10.1038/ng2079

German, M. A., Pillay, M., Jeong, D. H., Hetawal, A., Luo, S., Janardhanan, P., Kannan, V., Rymarquis, L. A., Nobuta, K., German, R., Paoli, E., Lu, C., Schroth, G., Meyers, B. C., & Green, P. J. (2008). Global identification of microRNA-target RNA pairs by parallel analysis of RNA ends. *Nature Biotechnology*, *26*(8), 941–946. https://doi.org/10.1038/nbt1417

Gul, Z., Barozai, M. Y. K., & Din, M. (2017). In-silico based identification and functional analyses of miRNAs and their targets in cowpea (*Vigna unguiculata* L.). *AIMS Genetics*, *4*(2), 138–165. https://doi.org/10.3934/genet.2017.2.138

Guo, X., Liu, D., & Chong, K. (2018). Cold signalling in plants: insights into mechanisms and regulation. *Journal of Integrative Plant Biology*, *60*(9), 745–756. https://doi.org/10.1111/jipb.12706

Gupta, B., & Huang, B. (2014). Mechanism of salinity tolerance in plants: physiological, biochemical, and molecular characterization. *International Journal of Genomics*, 2014, 701596. https://doi.org/10.1155/2014/701596

Gupta, P. K. (2015). microRNAs and target mimics for crop improvement. *Current Science*, *108*(9), 1624–1633.

Ha, M., & Kim, V. N. (2014). Regulation of microRNA biogenesis. *Nature Reviews Molecular Cell Biology*, *15*(8), 509–524. https://doi.org/10.1038/nrm3838

Hamza, N. B., Sharma, N., Tripathi, A., & Sanan-Mishra, N. (2016). microRNA expression profiles in response to drought stress in *Sorghum bicolor*. *Gene Expression Patterns*, *20*(2), 88–98. https://doi.org/10.1016/j.gep.2016.01.001

Huang, H., Ullah, F., Zhou, D., Yi, M., & Zhao, Y. (2019). Mechanisms of ROS regulation of plant development and stress responses. *Frontiers in Plant Science*, *10*, 800. https://doi.org/10.3389/fpls.2019.00800

Hussen, B. M., Rasul, M. F., Abdullah, S. R., Hidayat, H. J., Faraj, G. S., Ali, F. A., Salihi, A., Baniahmad, A., Ghafouri-Fard, S., & Rahman, M. (2023). Targeting miRNA by CRISPR/Cas in cancer: advantages and challenges. *Military Medical Research*, *10*, 32. https://doi.org/10.1186/s40779-023-00468-6

Inal, B., Türktaş, M., Eren, H., Ilhan, E., & Okay, S. (2014). Genome-wide fungal stress responsive miRNA expression in wheat. *Planta, 240*(6), 1287–1298. https://doi.org/10.1007/s00425-014-2153-8

Islam, W., Islam, S., Qasim, M., & Wang, L. (2017). Host-pathogen interactions modulated by small RNAs. *RNA Biology, 14*(7), 891–904. https://doi.org/10.1080/15476286.2017.1318009

Jeong, D. H., Park, S., Zhai, J., Gurazada, S. G., Paoli, E., & Meyers, B. C. (2011). Massive analysis of rice small RNAs: mechanistic implications of regulated microRNAs and variants for differential target RNA cleavage. *Plant Cell, 23*(12), 4185–4207. https://doi.org/10.1105/tpc.111.089045

Jia, X., Ren, L., Chen, Q. J., Li, R., & Tang, G. (2009). UV-B-responsive microRNAs in *Populus tremula. Journal of Plant Physiology, 166*(18), 2046–2057. https://doi.org/10.1016/j.jplph.2009.06.008

Jiang, W., Yang, L., He, Y., Zhang, H., Li, W., Chen, H., Ma, D., & Yin, J. (2018). Genome-wide identification and transcriptional expression analysis of superoxide dismutase (SOD) family in wheat (*Triticum aestivum*). *PeerJ, 7*, e8062. https://doi.org/10.7717/peerj.8062

Jiao, Y., Wang, Y., Xue, D., Wang, J., Yan, M., & Liu, G. (2010). Regulation of *OsSPL14* by *OsmiR156* defines ideal plant architecture in rice. *Nature Genetics, 42*(6), 541–544. https://doi.org/10.1038/ng.591

Johri, P., Rajput, M. S., Trivedi, M., & Singh, A. (2022). Artificial intelligence and deep learning: current impact and perspectives in life sciences and medicine. *Advances in Bioresearch, 13*(1), 184–196. https://doi.org/10.15515/abr.0976-4585.13.1.184196

Jung, J. H., Seo, P. J., Kang, S. K., & Park, C. M. (2011). MiR172 signals are incorporated into the miR156 signaling pathway at the SPL3/4/5 genes in Arabidopsis developmental transitions. *Plant Molecular Biology, 76*(1–2), 35–45. https://doi.org/10.1007/s11103-011-9759-z

Katiyar, A., Smita, S., Chinnusamy, V., Pandey, D. M., & Bansal, K. C. (2012). Identification of miRNAs in sorghum by using bioinformatics approach. *Plant Signaling & Behavior, 7*(2), 246–259. https://doi.org/10.4161/psb.18914

Kawashima, C. G., Yoshimoto, N., Nakashita, M. A., Tsuchiya, Y. N., Saito, K., & Takahashi, H. (2011). Sulphur starvation induces the expression of microRNA395 and one of its target genes but in different cell types. *Plant Journal, 57*(2), 313–321. https://doi.org/10.1111/j.1365-313X.2008.03690.x

Keunen, E., Peshev, D., Vangronsveld, J., Ende, W., & Cuypers, A. (2013). Plant sugars are crucial players in the oxidative challenge during abiotic stress: extending the traditional concept. *Plant, Cell & Environment, 36*(7), 1242–1255. https://doi.org/10.1111/pce.12067

Khvorova, A., Reynolds, A., & Jayasena, S. D. (2003). Functional siRNAs and miRNAs exhibit strand bias. *Cell, 115*(2), 209–216. https://doi.org/10.1016/S0092-8674(03)00801-8

Kulcheski, F. R., Oliveira, L. F., Molina, L. G., Almerao, M. P., Rodrigues, F. A., Marcolino, J., Barbosa, J. F., Stolf, R., Nepomuceno, A. L., & Marcelino-Guimaraes, F. C. (2011). Identification of novel soybean microRNAs involved in abiotic and biotic stresses. *BMC Genomics, 12*(1), 307. https://doi.org/10.1186/1471-2164-12-307

Lauter, N., Kampani, A., Carlson, S., Goebel, M., & Moose, S. P. (2005). microRNA172 down-regulates *glossy15* to promote vegetative phase change in maize. *Proceedings of the National Academy of Sciences, 102*(26), 9412–9417. https://doi.org/10.1073/pnas.0503927102

Lee, R. C., Feinbaum, R. L., & Ambros, V. (1993). The *C. elegans* heterochronic gene *lin-4* encodes small RNAs with antisense complementarity to *lin-14*. *Cell, 75*(5), 843–854. https://doi.org/10.1016/0092-8674(93)90529-y

Li, J. S., Fu, F. L., Ming, A., Zhou, S. F., She, Y. H., & Li, W. C. (2013). Differential expression of microRNAs in response to drought stress in maize. *Journal of Integrative Agriculture, 12*(8), 1414–1422. https://doi.org/10.1016/S2095-3119(13)60311-1

Li, X. P., Zeng, Q. F., Zhang, G. S., & Zhao, J. (2015). *GmMIR160A*, a class of soybean microRNA gene, negatively regulates progress of leaf senescence. *Guihaia, 35*(1), 84–91. https://www.cabidigital library.org/doi/full/10.5555/20153213915

Liu, B., Li, J., & Cairns, M. J. (2014). Identifying miRNAs, targets and functions. *Briefings in Bioinformatics, 15*(1), 1–19. https://doi.org/10.1093/bib/bbs075

Lou, S., Sun, T., Li, H., & Hu, Z. (2018). Mechanisms of microRNA-mediated gene regulation in unicellular model alga *Chlamydomonas reinhardtii. Biotechnology for Biofuels, 11*, 118. https://doi.org/10.1186/s13068-018-1249-y

Lu, Y., & Yang, X. (2010). Computational identification of novel microRNAs and their targets in *Vigna unguiculata. International Journal of Genomics*, 2010, 128297. https://doi.org/10.1155/2010/128297

Luan, M., Xu, M., Lu, Y., Zhang, L., Fan, Y., & Wang, L. (2015). Expression of zma-miR169 miRNAs and their target ZmNF-YA genes in response to abiotic stress in maize leaves. *Gene, 555*(1), 178–185. https://doi.org/10.1016/j.gene.2014.11.001

Luo, P., Di, D., Wu, L., Yang, J., Lu, Y., & Shi, W. (2022). microRNAs are involved in regulating plant development and stress response through fine-tuning of TIR1/AFB-dependent auxin signaling. *International Journal of Molecular Sciences*, *23*(1), 510. https://doi.org/10.3390/ijms23010510

Ma, Z., & Hu, L. (2023). microRNA: a dynamic player from signalling to abiotic tolerance in plants. *International Journal of Molecular Sciences*, *24*(11), 11364. https://doi.org/10.3390/ijms241411364

Mantri, N., Basker, N., Ford, R., Pang, E., & Pardeshi, V. (2013). The role of micro-ribonucleic acids in legumes with a focus on abiotic stress response. *The Plant Genome*, *6*(1), 0013. https://doi.org/10.3835/plantgenome2013.07.0022

Meddya, S., Shweta, M., Deepranjan, S., Rakesh, S., Rahul, D., Sachidanand, S., Gosangi, A., Arun, K. K., Ajit, K. S., & Thiyagarajan, T. (2023). Plant stomata: an unrealized possibility in plant defense against invading pathogens and stress tolerance. *Plants*, *12*(19), 3380. https://doi.org/10.3390/plants12193380

Michlewski, G., & Cáceres, J. (2018). Post-transcriptional control of miRNA biogenesis. *RNA*, *25*(1), 1–16. https://doi.org/10.1261/rna.068692.118

Mittal, D., Sharma, N., Sharma, V., Sopory, S. K., & Sanan-Mishra, N. (2016). Role of microRNAs in rice plant under salt stress. *Annals of Applied Biology*, *168*(2), 2–18. https://doi.org/10.1111/aab.12241

Mittal, M., Dhingra, A., Dawar, P., Payton, P., & Rock, C. D. (2023). The role of microRNAs in responses to drought and heat stress in peanut (*Arachis hypogaea*). *The Plant Genome*, *16*(2), e20350. https://doi.org/10.1002/tpg2.20350

Munns, R. (2005). Genes and salt tolerance: bringing them together. *New Phytologist*, *167*(3), 645–663. https://doi.org/10.1111/j.1469-8137.2005.01487.x

Mutegi, E., Fabrice, S., Moses, M., Ben, K., Bernard, R., & Caroline, M. (2010). Ecogeographical distribution of wild, weedy and cultivated *Sorghum bicolor* (L.) Moench in Kenya: implications for conservation and crop-to-wild gene flow. *Genetic Resources and Crop Evolution*, *57*(2), 243–253. https://doi.org/10.1007/s10722-009-9466-7

Navarro, L., Dunoyer, P., Jay, F., Arnold, B., Dharmasiri, N., & Estelle, M. (2006). A plant miRNA contributes to antibacterial resistance by repressing auxin signaling. *Science*, *312*(5772), 436–439. https://doi.org/10.1126/science.1126088

Naya, L., Paul, S., Valdés-López, O., Mendoza-Soto, A., Nova-Franco, B., Sosa-Valencia, G., Reyes, J., & Hernández, G. (2014). Regulation of copper homeostasis and biotic interactions by microRNA 398b in common bean. *PLoS One*, *9*(1), e84416. https://doi.org/10.1371/journal.pone.0084416

Noopur, K., Pragati, K., Adarsh, K. S., Sachidanand, S., Tanvi, J., & Megha, B. (2025a). Role of non coding RNA in effector trigger immunity. In Dutta, R., Kapadia, C., Danish, S., & Singh, S. (Eds), *Non-Coding RNA in Plants Modulation and Stress Responses Plant Biology, Sustainability and Climate Change*. Academic Press, New York.

Noopur, K., Pragati, K., Megha, B., Arpan, M., Tanvi, J., Shrisendu, M., Adarsh, K. S., & Sachidanand, S. (2025b). Role of non-coding RNA in plant architecture. In Dutta, R., Kapadia, C., Danish, S., & Singh, S. (Eds), *Non-Coding RNA in Plants Modulation and Stress Responses Plant Biology, Sustainability and Climate Change*. Academic Press, New York.

O'Brien, J., Hayder, H., Zayed, Y., & Peng, C. (2018). Overview of microRNA biogenesis, mechanisms of actions, and circulation. *Frontiers in Endocrinology*, *9*, 388354. https://doi.org/10.3389/fendo.2018.00402

Patel, P., Yadav, K., Ganapathi, T. R., & Penna, S. (2019). Plant miRNAome: cross talk in abiotic stressful times. In Rajpal, V., Sehgal, D., Kumar, A., & Raina, S. (Eds), *Genetic Enhancement of Crops for Tolerance to Abiotic Stress: Mechanisms and Approaches*, Vol. I, pp. 25–52. Springer, Cham. https://doi.org/10.1007/978-3-319-91956-0_2

Peláez, P., Trejo, M., Iñiguez, L., Estrada-Navarrete, G., Covarrubias, A., & Reyes, J. (2012). Identification and characterization of microRNAs in *Phaseolus vulgaris* by high-throughput sequencing. *BMC Genomics*, *13*(1), 83–83. https://doi.org/10.1186/1471-2164-13-83

Peng, J., Xia, Z., Chen, L., Shi, M., Pu, J., Guo, J., & Fan, Z. (2014). Rapid and efficient isolation of high-quality small RNAs from recalcitrant plant species rich in polyphenols and polysaccharides. *PloS One*, *9*(6), e95687. https://doi.org/10.1371/journal.pone.0095687

Pooja, G., Megha, B., Arpan, M., Chintan, K., Rahul, D., Subhan, D., & Sachidanand, S. (2025). Introduction to long noncoding RNA. In Dutta, R., Kapadia, C., Danish, S., & Singh, S. (Eds), *Non-Coding RNA in Plants Modulation and Stress Responses Plant Biology, Sustainability and Climate Change*. Vol. 1, pp. 1–12. Academic Press, New York.

Qiao-Ying, Z., Yang, C. Y., Ma, Q. B., Li, X. P., Dong, W. W., & Nian, H. (2012). Identification of wild soybean miRNAs and their target genes responsive to aluminum stress. *BMC Plant Biology*, *12*(1), 182. https://doi.org/10.1186/1471-2229-12-182

Qin, Y., Duan, Z., Xia, X., & Yin, W. (2011). Expression profiles of precursor and mature microRNAs under dehydration and high salinity shock in *Populus euphratica*. *Plant Cell Reports*, *30*(10), 1893–1907. https://doi.org/10.1007/s00299-011-1096-9

Rahul, D., Chintan, K., Subhan, D., & Sachidanand, S. (2024). *Non-Coding RNA in Plants: Modulation and Stress Responses*. Academic Press, New York.

Rani, J. (2021). microRNA utilization as a potential tool for stress tolerance in plants. In Ghimire, B. K. (Ed), *Botany-Recent Advances and Applications*. IntechOpen, New York. https://doi.org/10.5772/intechopen.96352

Ranty, B., Aldon, D., Cotelle, V., Galaud, J. P., Thuleau, P., & Mazars, C. (2016). Calcium sensors as key hubs in plant responses to biotic and abiotic stresses. *Frontiers in Plant Science*, *7*, 327. https://doi.org/10.3389/fpls.2016.00327

Saddhe, A., Manuka, R., & Penna, S. (2021). Plant sugars: homeostasis and transport under abiotic stress in plants. *Physiologia Plantarum*, *171*(4), 739–755. https://doi.org/10.1111/ppl.13283

Sahito, Z. A., Wang, L. X., Sun, Z. X., Yan, Q. Q., Zhang, X. K., & Jiang, Q. (2017). The miR172c-NNC1 module modulates root plastic development in response to salt in soybean. *BMC Plant Biology*, *17*(1), 229. https://doi.org/10.1186/s12870-017-1161-9

Samynathan, R., Venkidasamy, B., Shanmugam, A., Ramalingam, S., & Thiruvengadam, M. (2023). Functional role of microRNA in the regulation of biotic and abiotic stress in agronomic plants. *Frontiers in Genetics*, *14*, 1272446. https://doi.org/10.3389/fgene.2023.1272446

Schommer, C., Palatnik, J. F., Aggarwal, P., Chételat, A., Cubas, P., Farmer, E. E., Nath, U., & Weigel, D. (2008). Control of jasmonate biosynthesis and senescence by miR319 targets. *PLoS Biology*, *6*(9), e230. https://doi.org/10.1371/journal.pbio.0060230

Sheng, L., Chai, W., Gong, X., Zhou, L., Cai, R., & Li, X. (2015). Identification and characterization of novel maize miRNAs involved in different genetic backgrounds. *International Journal of Biological Sciences*, *11*(7), 781–793. https://doi.org/10.7150/ijbs.11619

Shriram, V., Kumar, V., Devarumath, R. M., Khare, T. S., & Wani, S. H. (2016). microRNAs as potential targets for abiotic stress tolerance in plants. *Frontiers in Plant Science*, *7*, 817. https://doi.org/10.3389/fpls.2016.00817

Shuzuo, L., Nie, X., Le, W., Du, X., Biradar, S. S., & Jia, X. (2012). Identification and characterization of microRNAs from barley (*Hordeum vulgare* L.) by high-throughput sequencing. *International Journal of Molecular Sciences*, *13*(3), 2973–2984. https://doi.org/10.3390/ijms13032973

Singh, P., Dutta, P., & Chakrabarty, D. (2021). miRNAs play critical roles in response to abiotic stress by modulating cross-talk of phytohormone signaling. *Plant Cell Reports*, *40*(8), 1617–1630. https://doi.org/10.1007/s00299-021-02736-y

Sousa, M., Gjorgjieva, M., Dolicka, D., Sobolewski, C., & Foti, M. (2019). Deciphering miRNAs' action through miRNA editing. *International Journal of Molecular Sciences*, *20*(24), 6249. https://doi.org/10.3390/ijms20246249

Sreenivasulu, N., Graner, A., & Wobus, U. (2008). Barley genomics: an overview. *International Journal of Plant Genomics*, 2008, 486258. https://doi.org/10.1155/2008/486258

Subhan, D., Uzma, Y., Niaz, A., Khurram, S., Rahul, D., & Sachidanand, S. (2025). Fertilizer-responsive non-coding RNA and their significance. In Dutta, R., Kapadia, C., Danish, S., & Singh, S. (Eds), *Non-Coding RNA in Plants Modulation and Stress Responses Plant Biology, Sustainability and Climate Change*. Academic Press, New York.

Sun, G. (2012). microRNAs and their diverse functions in plants. *Plant Molecular Biology*, *80*(1), 17–36. https://doi.org/10.1007/s11103-011-9817-6

Sunkar, R., & Zhu, J. K. (2004). Novel and stress-regulated microRNAs and other small RNAs from Arabidopsis. *Plant Cell*, *16*(8), 2001–2019. https://doi.org/10.1105/tpc.104.022830

Sunkar, R., Chinnusamy, V., Zhu, J. H., & Zhu, J. K. (2007). Small RNAs as big players in plant abiotic stress responses and nutrient deprivation. *Trends in Plant Science*, *12*(7), 301–309. https://doi.org/10.1016/j.tplants.2007.05.001

Sunkar, R., Girke, T., Jain, P. K., & Zhu, J. K. (2005). Cloning and characterization of microRNA from rice. *Plant Cell*, *17*(5), 1397–1411. https://doi.org/10.1105/tpc.105.031682

Tanvi, J., Arpita, S., Praneta, Y., Palak, S., & Sachidanand, S. (2025). Epigenetic modulation of plant developmental pathways by non-coding RNA. In Dutta, R., Kapadia, C., Danish, S., & Singh, S. (Eds), *Non-coding RNA in Plants Modulation and Stress Responses Plant Biology, Sustainability and Climate Change*. Academic Press, New York.

Valdés-López, O., Yang, S., Aparicio-Fabre, R., Graham, P. H., Reyes, J. L., Vance, C. P., & Hernández, G. (2010). microRNA expression profile in common bean (*Phaseolus vulgaris*) under nutrient deficiency stresses and manganese toxicity. *The New Phytologist*, *187*(3), 805–818. https://doi.org/10.1111/j.1469-8137.2010.03320.x

Vennapusa, A. R., Somayanda, I. M., & Doherty, C. J. (2020). A universal method for high-quality RNA extraction from plant tissues rich in starch, proteins and fiber. *Scientific Reports*, *10*(1), 16887. https://doi.org/10.1038/s41598-020-73958-5

Vera-Hernández, P. F., Folter, S., & Rosas-Cárdenas, F. F. (2019). Isolation and detection methods of plant miRNAs. In de Folter, S. (Ed), *Plant microRNAs. Methods in Molecular Biology*, Vol. 1932, pp. 109–120. Humana Press, New York. https://doi.org/10.1007/978-1-4939-9042-9_8

Verma, A., Kumar, R., Shukla, S., Singh, S., Johri, P., & Singh, M.S. (2023). Genome wide association studies: an outlook. *Journal of Innovation in Applied Research*, *1*(6), 56–64. https://doi.org/10.51323/JIAR.2023/6/1/56-62

Voinnet, O. (2009). Origin, biogenesis, and activity of plant microRNAs. *Cell*, *136*(4), 669–687. https://doi.org/10.1016/j.cell.2009.01.046

Wang, J. F., Zhou, H., Chen, Y. Q., Luo, Q. J., & Qu, L. H. (2004). Identification of 20 microRNAs from *Oryza sativa*. *Nucleic Acids Research*, *32*(5), 1688–1695. https://doi.org/10.1093/nar/gkh332

Wang, J., Mei, J., & Ren, G. (2019). Plant microRNAs: biogenesis, homeostasis, and degradation. *Frontiers in Plant Science*, *10*, 360. https://doi.org/10.3389/fpls.2019.00360

Wang, S. T., Sun, X. L., Hoshino, Y., Yu, Y., Jia, B., Sun, Z. W., Sun, M. Z., Duan, X. B., & Zhu, Y. M. (2014). microRNA319 positively regulates cold tolerance by targeting OsPCF6 and OsTCP21 in rice (*Oryza sativa L.*). *PLoS One*, *9*(3), e91357. https://doi.org/10.1371/journal.pone.0091357

Wang, S., Wu, K., Yuan, Q., Liu, X., Liu, Z., Lin, X., Zeng, R., Zhu, H., Dong, G., & Qian, Q. (2012). Control of grain size, shape and quality by *OsSPL16* in rice. *Nature Genetics*, *44*(8), 950–954. https://doi.org/10.1038/ng.2326

Wani, S. H., Kumar, V., Khare, T., Guddimalli, R., Parveda, M., Solymosi, K., Suprasanna, P., & Kishor, P. B. K. (2020). Engineering salinity tolerance in plants: progress and prospects. *Planta*, *251*(4), 1–29. https://doi.org/10.1007/s00425-020-03366-6

Wilson, R. C., & Doudna, J. A. (2013). Molecular mechanisms of RNA interference. *Annual Review of Biophysics*, *42*(1), 217–239. https://doi.org/10.1146/annurev-biophys-083012-130404

Wu, J., Wang, L., & Wang, S. (2017). microRNAs associated with drought response in the pulse crop common bean (*Phaseolus vulgaris* L.). *Gene*, *628*, 78–86. https://doi.org/10.1016/j.gene.2017.07.038

Xia, K., Wang, R., Ou, X., Fang, Z., Tian, C., Duan, J., Wang, Y., & Zhang, M. (2012). *OsTIR1* and *OsAFB2* downregulation via *OsmiR393* overexpression leads to more tillers, early flowering and less tolerance to salt and drought in rice. *PLoS One*, *7*(1), e30039. https://doi.org/10.1371/journal.pone.0030039

Yan, J., Gu, Y., Jia, X., Kang, W., Pan, S., Tang, X., Chen, X., & Tang, G. (2012). Effective small RNA destruction by the expression of a short tandem target mimic in *Arabidopsis*. *The Plant Cell*, *24*(1), 415–427. https://doi.org/10.1105/tpc.111.094144

Yang, T., Wang, Y., Teotia, S., Wang, Z., Shi, C., Sun, H., Gu, Y., Zhang, Z., & Tang, G. (2019). The interaction between miR160 and miR165/166 in the control of leaf development and drought tolerance in *Arabidopsis*. *Scientific Reports*, *9*(1), 2832. https://doi.org/10.1038/s41598-019-39397-7

Yao, Q., Chen, Y., & Zhou, X. (2019). The roles of microRNAs in epigenetic regulation. *Current Opinion in Chemical Biology*, *51*, 11–17. https://doi.org/10.1016/j.cbpa.2019.01.024

Yu, X., Wang, H., Lu, Y., Ruiter, M., Cariaso, M., & Prins, M. (2012). Identification of conserved and novel microRNAs that are responsive to heat stress in *Brassica rapa*. *Journal of Experimental Botany*, *63*(2), 1025–1038. https://doi.org/10.1093/jxb/err337

Zhang, B. H., Wang, Q. L., Wang, K. B., Pan, X. P., Liu, F., & Guo, T. L. (2007). Identification of cotton microRNAs and their targets. *Gene*, *397*(1–2), 26–37. https://doi.org/10.1016/j.gene.2007.03.020

Zhang, B., Wang, Q., & Pan, X. (2011). microRNAs and their regulatory roles in animals and plants. *Journal of Cellular Physiology*, *226*(2), 279–289. https://doi.org/10.1002/jcp.22309

Zhang, C., Ding, Z., Wu, K., Yang, L., Li, Y., Yang, Z., Shi, S., Liu, X., Zhao, S., Yang, Z., Wang, Y., Zheng, L., Wei, J., Du, Z., Zhang, A., Miao, H., Li, Y., Wu, Z., & Wu, J. (2016). Suppression of jasmonic acid-mediated defense by viral-inducible microRNA319 facilitates virus infection in rice. *Molecular Plant*, *9*(9), 1302–1314. https://doi.org/10.1016/j.molp.2016.06.014

Zhang, L. W., Song, J. B., Shu, X. X., Zhang, Y., & Yang, Z. M. (2013a). miR395 is involved in detoxification of cadmium in Brassica napus. *Journal of Hazardous Materials*, *250*, 204–211. https://doi.org/10.1016/j.jhazmat.2013.01.070

Zhang, L., Chia, J. M., Kumari, S., Stein, J. C., Liu, Z., Narechania, A., & Ware, D. (2009). A genome-wide characterization of microRNA genes in maize. *PLoS Genetics*, *5*(11), e1000716. https://doi.org/10.1371/journal.pgen.1000716

Zhang, Y. C., Yu, Y., Wang, C. Y., Li, Z. Y., Liu, Q., Xu, J., & Chen, X. M. (2013b). Overexpression of microRNA *OsmiR397* improves rice yield by increasing grain size and promoting panicle branching. *Nature Biotechnology*, *31*(9), 848–852. https://doi.org/10.1038/nbt.2646

Zhang, Y., Zhu, X., Chen, X., Song, C., Zou, Z., Wang, Y., & Fang, W. (2014). Identification and characterization of cold-responsive microRNAs in tea plant (*Camellia sinensis*) and their targets using high-throughput sequencing and degradome analysis. *BMC Plant Biology*, *14*, 271. https://doi.org/10.1186/s12870-014-0271-x

Zhang, Z., Lin, H., Shen, Y., Gao, J., Xiang, K., Liu, L., & Zhang, M. (2012). Cloning and characterization of miRNAs from maize seedling roots under low phosphorus stress. *Molecular Biology Reports*, *39*(8), 8137–8146. https://doi.org/10.1007/s11033-012-1661-5

Zhao, B., Liang, R., Ge, L., Li, W., Xiao, H., Lin, H., & Tang, G. (2007). Identification of drought-induced microRNAs in rice. *Biochemical and Biophysical Research Communications*, *354*(2), 585–590. https://doi.org/10.1016/j.bbrc.2007.01.022

Zhao, C. Z., Xia, H., Frazier, T. P., Yao, Y. Y., Bi, Y. P., & Li, A. Q. (2010). Deep sequencing identifies novel and conserved microRNAs in peanuts (*Arachis hypogaea* L.). *BMC Plant Biology*, *10*, 3. https://doi.org/10.1186/1471-2229-10-3

Zhao, C., Xia, H., Cao, T., Yang, Y., Zhao, S., Hou, L., & Li, C. (2015). Small RNA and degradome deep sequencing reveals peanut microRNA roles in response to pathogen infection. *Plant Molecular Biology Reporter*, *33*(5), 1013–1029. https://doi.org/10.1007/s11105-014-0806-1

Zhou, L., Liu, Y., Liu, Z., Kong, D., Duan, M., & Luo, L. (2010). Genome-wide identification and analysis of drought-responsive microRNAs in *Oryza sativa*. *Journal of Experimental Botany*, *61*(15), 4157–4168. https://doi.org/10.1093/jxb/erq237

Zhou, M., & Luo, H. (2013). microRNA-mediated gene regulation: potential applications for plant genetic engineering. *Plant Molecular Biology*, *83*(1–2), 59–75. https://doi.org/10.1007/s11103-013-0089-1

Zhou, X., Khare, T., & Kumar, V. (2020). Recent trends and advances in identification and functional characterization of plant miRNAs. *Acta Physiologiae Plantarum*, *42*(2), 25. https://doi.org/10.1007/s11738-020-3013-8

Zhou, X., Wang, G., & Zhang, W. (2007). UV-B responsive microRNA genes in *Arabidopsis thaliana*. *Molecular Systems Biology*, *3*(1), 103. https://doi.org/10.1038/msb4100143

10 Role of microRNA for Abiotic Stress Tolerance in Tuber and Fruit Crops

Anil A. Hake, Pranita P. Thakur*, Pravin V. Jadhav, Gopika Mote, Prerona Boruah, Mala Parab, and Ramesh Palakurthi*

10.1 INTRODUCTION

In today's climate change scenarios, abiotic stresses like drought, salinity, high temperatures, flooding, and nutritional deficits affect crops more frequently. Crop productivity is constrained by these stresses. Recent developments in genetics, physiology, molecular biology, and biochemistry have substantially enhanced our knowledge of how crops respond to various stresses and the causes of varietal tolerance variations. The role of microRNA (miRNA) in fruit and tuber crops that can withstand abiotic stress will be elucidated in detail in this chapter. According to a definition from physics, stress is the force applied per unit area of a material that causes strain and changes its dimensions. In biological terms, stress can be a biotic or abiotic factor that may produce an adverse effect on individual organisms, populations, or communities. In agriculture, stress is a phenomenon that limits crop productivity or destroys biomass (Grime, 1979). Scientists have distinguished plant stresses into two main types: biotic and abiotic. Abiotic stress relies on interactions between organisms and their physical surroundings, whereas biotic stresses arise from interactions between living organisms. The potentially harmful impacts of salinity, flooding, drought, metal toxicity, nutrient shortage, and extremes in temperature are all considered forms of abiotic stress (Anil et al., 2023). Over vast regions, abiotic stress consistently restricts crop choices and agricultural output, and extreme occurrences lead to crop failures. Abiotic stresses have a negative impact on national economies, food security, and the standard of living for individual farmers and their families (Tungare et al., 2022).

In this chapter, we elaborate on the role of miRNA in tuber and fruit crops. "Tuber crops" refer to those plants that produce tubers, which are enlarged, fleshy, underground stems or root structures that serve as storage organs for the plant. These tubers are rich in carbohydrates and serve as an important food source for humans and animals. Some common tuber crops include potatoes, sweet potatoes, yams, cassava, and taro. In addition to their nutritional value, tuber crops are cultivated on a large scale for commercial purposes, used in various culinary dishes, processed into products like flour or starch, and serve as feed for livestock. Fruit crops, on the other hand, are cultivated primarily for their edible fruits, which are typically sweet or sour in flavor and are consumed fresh or processed into various food products. They play a vital role in agriculture, providing important sources of nutrition, income, and diversity in diets worldwide. Some common fruit crops include apples, oranges, bananas, grapes, strawberries, and mangoes.

* *Corresponding Author*

DOI: 10.1201/9781003491378-10

10.2 miRNA: BRIEF INTRODUCTION AND BIOGENESIS

miRNAs are naturally occurring small non-coding RNAs, typically 19–25 nucleotides in length (mostly 22 nucleotides) (Noopur et al., 2025a; Noopur et al., 2025b; Pooja et al., 2025; Rahul et al., 2024; Subhan et al., 2025). They can be found in certain viruses, animals, and plants. The first miRNA, named lin-4, was discovered in *Caenorhabditis elegans* by Lee et al. (1993). However, in plants, particularly in *Arabidopsis thaliana,* the first evidence of miRNAs was reported by Reinhart et al. (2002). miRNAs typically regulate post-transcriptional gene expression through RNA silencing. By regulating several target genes, they play multiple important roles in controlling a wide range of biological processes, such as cell differentiation, proliferation, and metabolism. By attaching to the 3′-UTR, 5′-UTR, promoter, and coding regions of their target mRNAs, miRNAs control biological processes by causing the destruction or suppression of the mRNA. It has been found that each miRNA functions as a post-transcriptional modulator for many genes (Meenakshi et al., 2025).

miRNA biogenesis is a multi-step, intricate process that takes place inside the cell. Although many of the miRNA genes reside in clusters of two to seven genes with short intervening regions, they are often expressed independently. Their co-transcriptional expression, as suggested by experimental data, indicates that they are regulated by shared regulatory regions. The first step of miRNA production is the transcription of miRNA genes by RNA polymerase II or III, which produces primary miRNA transcripts. Further processing of primary miRNAs by Drosha, a primary nuclease (ribonuclease III), initiates miRNA processing in the nucleus, resulting in a short hairpin loop RNA known as precursor miRNA (pre-miRNA). Furthermore, exportin-5, a protein involved in the export of pre-miRNA, transports it from the nucleus to the cytoplasm. In the cytoplasm, Dicer, an RNase III endonuclease, processes the pre-miRNA to produce a short, functional double-stranded RNA molecule (19–25 nucleotides). One of the strands of the resulting double-stranded RNA is recognized as the mature miRNA, which is subsequently loaded onto the Argonaute protein to create an RNA-induced silencing complex responsible for silencing target RNAs. The other strand, known as the passenger or miRNA, is usually destroyed. The RNA-induced silencing complex is directed by the mature miRNA to complementary sequences found in target messenger RNAs (mRNAs), and the process of miRNA binding to its target mRNA results in translational repression or mRNA degradation, which controls the expression of specific genes. Figure 10.1 depicts an overview of miRNA biogenesis.

10.3 METHODS TO IDENTIFY miRNA

The fields of traditional biology and computational biology have both greatly benefited from the discovery of miRNAs. The direct cloning method was a commonly employed technique in conventional biology to identify plant miRNAs by following the discovery of the highly conserved stem–loop hairpin structure of miRNAs. Bioinformatics techniques were developed to distinguish miRNAs from known miRNAs of other plant species in order to detect miRNAs in novel plant species. High-throughput sequencing of small RNA has been widely utilized to identify functional miRNAs in plants, as it has become more affordable. To meet the needs of scientists conducting miRNA research, a growing variety of computational algorithms and web-based resources are being developed, the majority of which are related to computational approaches that include the prediction of miRNA genes and their targets. An increasing number of miRNAs have been identified in various taxa, including nematodes, algae, mammals, insects, viruses, and plants, leading to rapid development of techniques for identifying plant miRNAs (http://www.mirbase.org). The Mirbase database (http://www.mirbase.org) contains miRNA data for more than 75 plant species. For example, information about 604 miRNAs in rice (*O. sativa*) is provided within the plant category. Through complementary base pairing, miRNAs may completely or partially target mRNAs. According to Peterson et al. (2014), there are four groups of binding patterns that miRNAs are known to bind to in the 3′-UTRs of target transcripts: seed match, conservation, free energy, and site accessibility properties, which are described in the following sections.

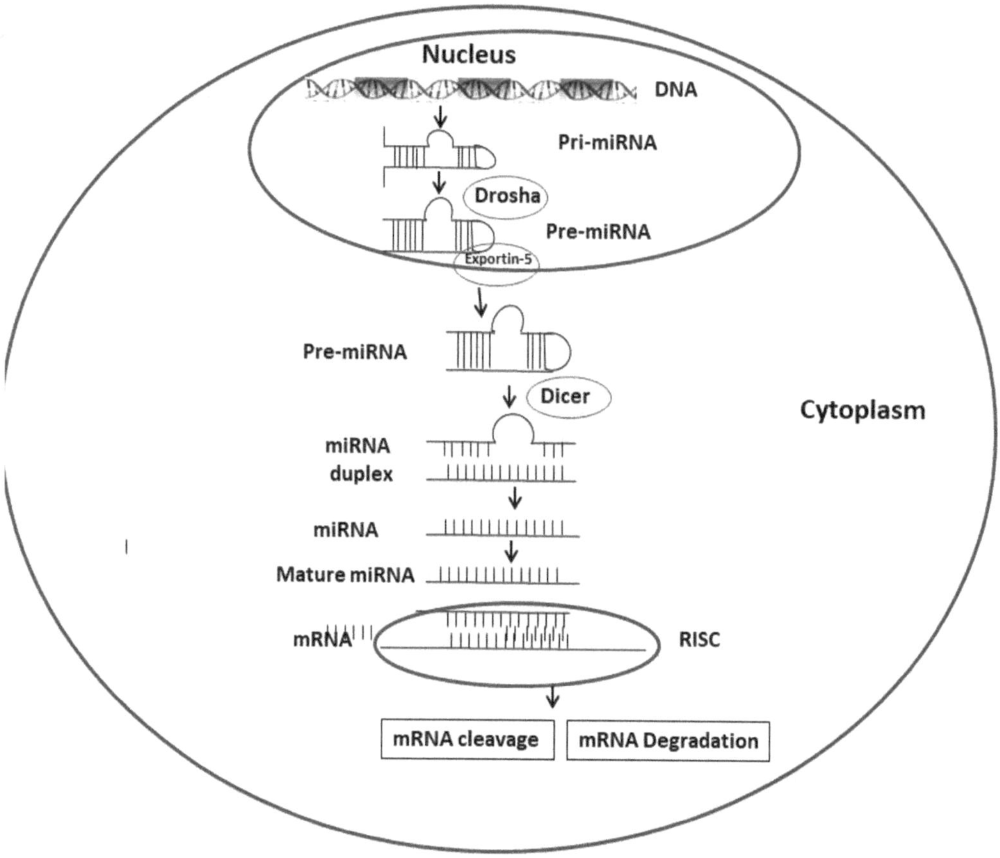

FIGURE 10.1 microRNA biogenesis process.

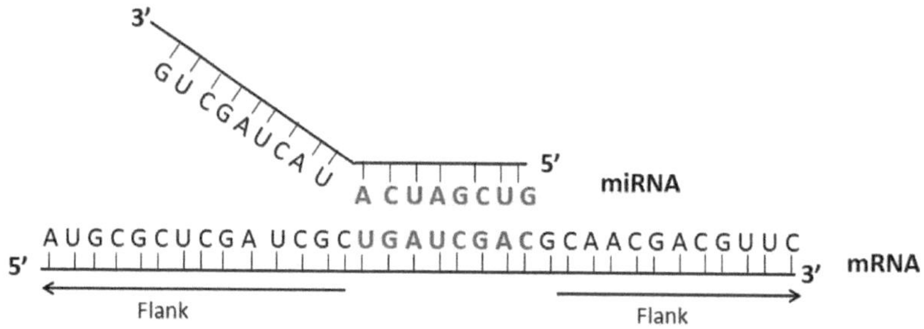

FIGURE 10.2 An overview of how miRNA interacts with target mRNA.

10.3.1 Seed Match

The seed sequence of a miRNA typically includes the first 2–8 nucleotides, starting at the 5′ end and counting toward the 3′ end (Figure 10.2). A Watson–Crick (WC) model formed between a miRNA and its target based on the seed sequence is referred to as a seed match. When adenosine (A) pairs with the complementary sequence uracil (U), and guanine (G) pairs with the complementary sequence cytosine (C), a miRNA and mRNA nucleotide form a WC match. There are no alignment gaps in the WC matching for a perfect seed match between the miRNA and the mRNA target. Depending on the methodology, different types of seed matches may also be considered. These are listed here: (1) 6mer: A perfect WC match spanning six nucleotides between the mRNA and miRNA seed. (2) 7mer-m8: A perfect WC match from the miRNA seed's nucleotides 2–8. (3) 7mer-A1: An A opposite miRNA nucleotide 1 and a perfect WC match from nucleotides 2–7 of the miRNA seed. (4) 8mer: An A opposite miRNA nucleotide 1 and a perfect WC match from nucleotides 2–8 of the miRNA seed.

The position of the miRNA is indicated in blue. The miRNA positions 2–8 nucleotides are referred to as the seed sequence. The mRNA sequences on each side of the region that matches the miRNA seed sequence are referred to as the flank.

10.3.2 Conservation

In general language, conservation means the preservation of living and non-living things. Here, the conservation concept related to genomics studies refers to the maintenance of a nucleotide sequence across species. The miRNA seed region generally exhibits higher conservation than the non-seed region. The regions in the 3′ UTR, 5′ UTR, and miRNA, or any combination of these three, are the focus of conservation analyses. The 3′ compensatory sites are conserved pairing sites at the 3′ end of miRNA that can compensate for mismatches in the seed in a small percentage of miRNA–mRNA target interactions. Conservation analysis provides evidence that a predicted miRNA target is functional when predicting miRNA targets in 3′ UTRs. Furthermore, the genomic regions flanking the miRNA gene and miRNA target genes are increasingly important for conservation studies. For example, conservation analysis has been used to examine the promoter regions of miRNAs and their target genes, as well as the co-localization of independently transcribed miRNAs and surrounding protein-coding genes. Thus, conservation plays a broad role in miRNA target prediction, and the analysis of conserved components can be utilized in various ways.

10.3.3 Free Energy

A biological system's stability can be gauged using free energy, also known as Gibbs free energy. A putative target mRNA is thought to have a higher probability of being the actual target of a miRNA if the miRNA's binding to it is predicted to be stable. Since it is challenging to measure free energy precisely, the change in free energy throughout a reaction is typically considered (ΔG). Reactions with a negative ΔG produce more stable systems because they have less energy available to react in the future. The overall ΔG can be used as a measure of binding strength, and regions of high and low free energy can be predicted (Figure 10.3) by anticipating how the miRNA and its putative target hybridize.

A hairpin loop is displayed, where the stem represents a region of low free energy (a negative ΔG) and the loop represents an area of high free energy (a positive ΔG).

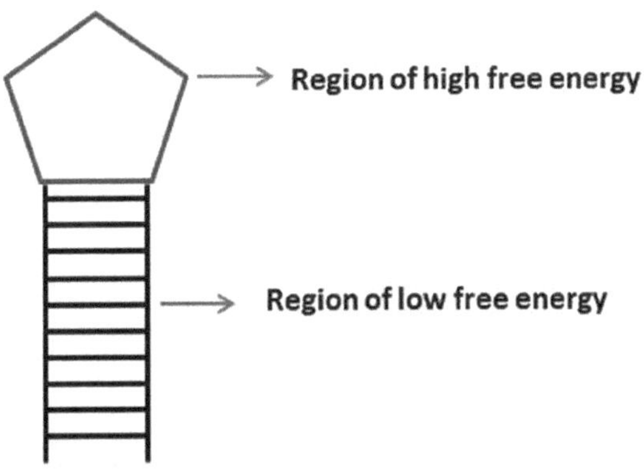

FIGURE 10.3 Overview of the projected RNA hybridization structure's free energy analysis.

10.3.4 Site Accessibility

The ease with which a miRNA can find and hybridize with an mRNA target is measured by its site accessibility. The secondary structure that mRNA adopts after transcription may obstruct miRNA's ability to attach to a target site. A miRNA first attaches to a brief, accessible region of the mRNA in the two-step process of miRNA–mRNA hybridization. Once the miRNA has fully bound to a target, the mRNA secondary structure begins to unfold. Thus, the estimated energy needed to make a site accessible to a miRNA can be used to determine the probability that an mRNA is the target of a miRNA.

Additional variables are introduced to predict the efficacy of miRNA–mRNA target interactions as new discoveries are made in their characterization. With each new development in the characterization of miRNA–mRNA target interactions, more features are included to forecast the target's efficacy. The number of target sites within a 3′ UTR is known as target-site abundance. A and U nucleotide concentrations that border the matching seed region of the miRNA are referred to as local AU content. The term "GU wobble" describes the seed match situation where a G pairing is allowed with a U instead of a C. Base pair matching with miRNA nucleotides 12–17 is referred to as 3′ compensatory pairing. The anticipated duplex's calculated free energy is known as seed pairing stability. The target site's location within the mRNA is examined through position contribution analysis. Machine-learning methods construct a model of miRNA targets from training data and use the model to predict miRNAs. Because machine-learning techniques can be trained to estimate the predictive potential of each feature on both positive and negative datasets, it is likely that they will incorporate more features in their predictions.

10.4 TOOLS/SOFTWARE USED FOR IDENTIFICATION miRNA TARGET

To identify putative miRNA targets, several computational techniques were developed relatively quickly, including miRanda (John et al., 2004), mirSVR (Betel et al., 2010), PicTar (Krek et al., 2005), TargetScan (Lewis et al., 2003), TargetScanS (Lewis et al., 2005), RNA22 (Miranda et al., 2006), PITA (Kertesz et al., 2007), RNAhybird (Rehmsmeier et al., 2004), and DIANA-microT (Maragkakis et al., 2009).

10.5 ROLE OF miRNA IN ABIOTIC STRESS TOLERANCE IN TUBER AND FRUIT CROPS

A primary response mechanism of plants to many abiotic stressors is the activation of the miRNA network. The influence of miRNAs on DNA in different plant species has been associated with oxidative stress response mechanisms (Pagano et al., 2021). Just as diverse as the reactions of miRNAs to different environmental stimuli are the responses of these miRNAs to plant growth and development. MiRNAs are critical for plants to respond to harsh environments because they regulate the expression of related target genes.

10.5.1 miRNAs Response to Water Stress

An inadequate amount of water in soil for plants leads to drought conditions, while an excess of water leads to flood stress. Both extreme conditions can cause tremendous damage to plants, often resulting in plant death.

The majority of the world's climate is characterized by regular and frequent droughts, which significantly reduce crop yield. Plants, however, have their own defensive mechanisms to withstand unfavorable weather. miRNAs reprogram gene expression as one of these mechanisms. The miRNAs govern the expression of genes involved in drought response pathways, such as those related to stomatal regulation, water uptake and transport, osmotic adjustment, and antioxidant defense.

Specific miRNAs have been found to target transcription factors, such as MYB, NAC, and bZIP, which subsequently control the expression of downstream stress-responsive genes. In apple, Niu et al. (2019) identified two miRNAs, namely miR156p and miRn-249, which exhibited drought tolerance under laboratory conditions using mannitol. A total of 531, 471, 535, and 487 mature miRNAs were identified by Eldem et al. (2012) from peach (*Prunus persica* L.) leaf control, leaf stress, root control, and root stress samples, respectively. They also found that the expression levels of these miRNAs changed during drought stress. Eight miRNAs (Novel_8, Novel_9, Novel_105, miR156d-3p, miR160a-5p, miR162a-3p, miR172b-3p, and miR398a-5p) were examined for expression in potatoes by Öztürk et al. (2021). According to their research, these miRNAs and their targets are essential for preventing heat stress and drought at different post-transcriptional phases. Therefore, plants that excessively overexpress miRNAs may be more resilient to abiotic stressors. Drought lowers the transcription level of mdm-miR171i, which negatively impacts *MsSCL26.1* (*SCARECROW-LIKE PROTEINS26.1*) in wild apple (*Malus sieversii*) trees, particularly in the roots (Wang et al., 2020). According to Li et al. (2020), *miR166f* may have a favorable regulatory effect on the drought tolerance of mulberry (*Morus multicaulis*). According to Shen et al. (2022), the mdm-miR160-MdARF17-MdHYL1 miRNA-transcription factor complex functions as a regulatory network and is essential for the survival of apple trees under drought stress.

Similarly, in tuber and fruit crops, miRNAs play a critical role in controlling systems for enduring floods. They regulate ethylene signaling-related genes, which are essential for the plant's response to flooding stress. Aerenchyma development, adventitious root production, and elongation growth under flooded conditions are just a few of the biological activities in which ethylene is essential. The miRNAs also target genes implicated in anaerobic metabolism, including those that encode enzymes for fermentation pathways, which help plants adapt to flooding-induced oxygen deprivation.

10.5.2 miRNAs Response to Temperature Stress in Plants

miRNAs play essential roles in regulating gene expression and several biological processes in plants under severe temperature conditions, much as they do in response to other abiotic stresses. Plants undergo intricate molecular reactions to adapt and survive when exposed to extreme temperature stress, whether it be heat or cold. By regulating the expression of target genes essential to stress tolerance mechanisms, miRNAs contribute to these responses. Below we describe how miRNAs respond to extreme temperature stress in plants:

Under extreme temperature conditions, certain miRNAs are up- or downregulated, which modifies the target genes they affect, leading to changes in gene expression. These target genes may encode signaling molecules, stress tolerance-related proteins, or transcription factors. MiRNAs can also be integrated into plant stress signaling networks. For instance, miRNAs can interfere with metabolic processes and target components of stress signaling cascades to alter genes encoding enzymes involved in energy metabolism, antioxidant defense systems, and the plant's overall response. Stress from extreme temperatures can cause epigenetic modifications in plants, such as histone modifications and DNA methylation, which can influence patterns of gene expression. By directly affecting the expression of genes involved in epigenetic alterations or by targeting genes encoding epigenetic regulators, miRNAs may contribute to these epigenetic changes. MiRNAs play a variety of roles in plants' responses to high temperatures, controlling gene expression, signaling pathways, metabolic processes, and epigenetic alterations that help plants adapt to harsh climatic conditions and survive. It has been determined how some miRNAs react to cold stress. For example, homeostasis is regulated by the miRNAs stuABF8011, stu-novel-miR5125, and StuAAO36826 via the abscisic acid metabolic pathways. A crucial enzyme, gibberellin-3-beta-dioxygenase (GA3ox1), involved in the production of gibberellin, is negatively controlled by stu-novel-miR10881 in tuber crops. Some plants overexpress miR1320, which provides them with tolerance to cold. Furthermore, the AP2/ERF transcription factor *OsERF096*, a *miR1320* target, co-regulates cold tolerance in several fruit crops by inhibiting the cold signaling pathway mediated by jasmonic acid (Sun et al., 2022).

Concurrently, several significant regulatory miRNA target modules have been discovered in fruit crops to counteract heat stress and mitigate its negative effects. These include miR169-SBP, miR159-MYB, miR156-SBP/SPL, miR172-AP2/ERF, miR164-NAC, miR166-HD zip, miR396-GRF, and miR5381-SAC2. The SPL gene family is co-targeted by miR535 and miR156, and the miR156-SPL module regulates heat stress in bananas. These miRNAs in bananas showed downregulation in response to high-temperature stress (Stief et al., 2014; Zhu et al., 2019). *MiR156* was the first miRNA discovered to be highly induced in plants at elevated temperatures. MiR393-SlTIR1 and miR160-SlARF10/16 regulate the auxin signaling cascade, and there is a substantial association between miR398b-SlCSD1 and reactive oxygen species (ROS)-mediated cell division and proliferation.

Following rice, wheat, and maize in the world's food crop rankings, potatoes rank fourth. It has been shown that, in potatoes, miRNAs are connected to growth, development, metabolism, regulation, transcription factors, and physiological activities. Fruit crop-specific miRNAs such as miR824, miR858, miR856, and miR857 have been identified. Among these, miR824 targets the MYB12 gene (*At2g47460*, R2R3-MYB family), miR858 targets the AGL16 gene (*At3g57230*, MADSbox), and miR856 targets an unidentified transcript that codes for the efflux protein ZAT1. The transcript of the laccase family gene LAC7 (*At3g09220*), which is expected to have multicopper oxidase activity, was identified as the target of miR857.

MiRNAs play a role in both responding to and mitigating the negative effects of high-temperature stress due to their ability to control the abundance of transcription factors and proteins linked to the metabolism of ROS and multiple mitogen-activated protein kinase signaling pathways (Ahmed et al., 2019; Zhang et al., 2009). Transcription factors such as MYB, BZIP, Bzip21, homeobox-leucine zipper protein, and HAFA9 are regulated by miRNAs including stu-miR1446, stu-miR2090, stu-miR3951, stu-miR5150, stu-miR5205, and stu-miR6438. According to Din et al. (2014), the new miRNAs stu-miR1446, stu-miR2090, stu-miR3951, stu-miR5150, stu-miR5205, stu-miR5449, and stu-miR6438 also enhance potato resistance to biotic, xenobiotic, and abiotic stresses.

One unpredictable weather aspect during the potato growing season is freezing damage to potatoes caused by low temperatures. These low temperatures impede the growth and spread of the potato business. Several miRNAs, including stu-novel-miR5125, stu-miR167b-3p, stu-miR171b-3p, stu-miR827–5p, and stu-novel-miR43095, have been reported in relation to low-temperature stress. These findings offer valuable insights into the role of miRNA during low-temperature stress and can aid in the future selection of potato varieties resistant to such conditions. Yu et al. (2020) examined the regulatory function of 26 miRNAs and their 14 target genes in the sweet potato's response to stress from high temperatures and chilling. These miRNAs target genes include IbMAA, IbAP2, IbARF10, IbARF8, IbATHB, and others. Some of these miRNAs are IbmiR156, IbmiR160, IbmiR164, IbmiR166, IbmiR172, IbmiR390, IbmiR395, and IbmiR397. According to a study on grapevines conducted by Sun et al. (2015), a variety of grapevine miRNAs are cold-inducible and play an important role in cold tolerance. They identified 67 unique miRNAs in addition to 163 known miRNAs; of these, 44 miRNAs were expressed differently in *Vitis vinifera* during the cold response. The results indicate that miRNAs may regulate several transcription factors, such as AP2, SBP, MYB, bHLH, GRAS, and bZIP, in grapevines subjected to cold stress. MiR535 and miR156 may regulate heat stress in bananas by targeting the SPL gene family (Zhu et al., 2019). During the profiling of temperature-responsive miRNAs in bananas, 26 new unique miRNAs and 113 recognized miRNAs were discovered. Among the identified miRNAs in bananas, 42 were found to be expressed differently in response to heat and cold stresses (Zhu et al., 2019).

According to Zhu et al. (2019), degradome sequencing identified 60 different genes controlled by known miRNAs, of which 15 were determined to be temperature-responsive miRNAs responsible for half of the target genes. The findings imply that the *miR535* and *miR156* families cooperate to control the expression of SPL genes in bananas and may have had a common origin during evolution (Zhu et al., 2019). Zhu et al. (2019) also found that the discovery of miRNA-triggered phased secondary siRNAs in bananas resulted in the production of niR393-TIR1/AFB phasiRNA, suggesting enrichment specific to cold stress.

In orange (*Poncirus trifoliata*), the miRNA Ptr-miR396b was identified as a target of the ACO gene (1-aminocyclopropane-1-carboxylic acid oxidase) in response to cold stress conditions (Zhang et al., 2016). The expression of the ACC oxidase gene is regulated by miRNA396b, which also modulates ethylene-polyamine homeostasis, thus conferring cold tolerance. Phospholipase A and Phospholipase D are associated with plant temperature stress-responsive processes. In mango, phospholipase A and phospholipase D are targeted by MmiR78769 and MmiR101928, respectively (Moh et al., 2021).

10.5.3 miRNAs Response to Salt Stress in Plants

Through oxidative stress, osmotic stress, and homeostasis (ion exchange), plants respond to salt stress in a complex manner. miRNAs play a significant role in regulating gene expression and various biological processes during salt stress. Plants utilize intricate molecular systems to counteract the negative effects of salt stress and preserve cellular homeostasis when exposed to high salinity levels in the soil. By regulating the expression of target genes associated with resistance to salt stress, miRNAs contribute to these responses. Salt stress disrupts the balance of ions in plant cells, leading to the accumulation of toxic ions such as sodium (Na+) and chloride (Cl-).

The miRNAs are involved in the regulation of ion transporters and channels that facilitate ion uptake, compartmentalization, and efflux. For example, certain miRNAs target genes encoding ion transporters like HKT high-affinity K+ transporter (HKT) and Na+/H+ exchanger, thereby modulating ion fluxes and maintaining ion balance under salt stress conditions. Additionally, salt stress induces osmotic stress in plants due to reduced water availability caused by increased soil salinity. miRNAs play a crucial role in regulating the expression of genes responsible for osmotic stress responses. This includes the regulation of genes related to the production of osmoprotectants such as compatible solutes (e.g., proline and glycine betaine) and proteins (aquaporins) involved in water uptake and retention. The miRNAs help plants cope with osmotic stress induced by salt accumulation by regulating the expression of specific genes. ROS produced by salt stress in plant cells cause oxidative damage to biomolecules. The miRNAs regulate the expression of genes encoding antioxidant enzymes (e.g., peroxidase, catalase, and superoxide dismutase) and antioxidant metabolites (e.g., glutathione and ascorbate) involved in ROS scavenging.

Plant growth, development, yield, and quality are all significantly impacted by salt stress. In recent years, it has been reported that various plant species, such as cotton, apple, switchgrass, Jerusalem artichoke, alfalfa, tamarisk trees, and cereal crops, have been found to regulate miRNA as a means of protecting themselves from the harmful effects of salts. This mechanism of salt tolerance has been observed in multiple plant species, excluding the model plant *Arabidopsis* (Cheng et al., 2021). Limited research has been conducted on the involvement of miRNAs in salt stress tolerance in fruit and tuber crops. Among a wide range of distantly related plant species, the identified miRNA precursors demonstrated strong sequence conservation and constituted stable stem–loop structures. Under salinity stress, the differential expression patterns of seven selected guava miRNAs (pgu-miR156f-5p, pgu-miR160c-5p, pgu-miR162-3p, pgu-miR164b-5p, pgu-miR166t, pgu-miR167a-5p, and pgu-miR390b-5p) were recorded. The most affected miRNAs were pgu-miR162-3p, pgu-miR166t, and pgu-miR166t.

Salt stress-related genes are regulated by miRNA, which helps plants adapt to salt (Xu et al., 2021). Numerous cellular pathways linked to salt stress tolerance are influenced by WRKY transcription factors (Jiang et al., 2017). In apples, Ma et al. (2021) revealed that overexpression of MdWRKY100 increased salt tolerance due to the upregulation of the miR156/SPL module. Salinity stress impacted the physiological and morphological traits of potatoes. One study found 110 novel miRNAs in response to alkali stress in potato; the miRNAs novel-mo64-5p and miR4243-x play roles in this response by negatively regulating genes for sucrose phosphate synthase and shikimate o-hydroxy-cinnamoyl transferase, respectively (Kang et al., 2021). Recently, using high-throughput transcriptome sequencing, hundreds of miRNAs have been identified in response to salt sensitivity in the sweet potato (*Ipomoea batatas*) cultivar "Xushu 32", proving that miRNAs play a crucial role in acclimatizing to salt stress (Yang et al., 2020). A comparative study of salt-responsive miRNAs in two cultivars of sweet potato demonstrates the contribution of miRNAs to salt tolerance

(Yang et al., 2022). Since certain miRNAs control responses to salt stress, it is possible to create salt-tolerant plants by genetically modifying the expression of specific miRNAs (Gao et al., 2022).

10.5.4 miRNAs Response to Heavy Metal Stress in Plants

Very limited data is available on the role of miRNA in heavy metal tolerance in tuber and fruit crops, as research on this topic is in its nascent phase. Heavy metal stress adversely impacts the yield and growth of plants. Heavy metals include essential metals like iron, copper, manganese, and zinc, as well as non-essential metals like lead, arsenic, aluminum, mercury, and cadmium. Under conditions of heavy metal stress, plants employ a variety of intricate gene regulation mechanisms. miRNAs play an important role in the post-transcriptional regulation of gene expression. Numerous biological processes, including protein folding and assembly, metal chelation, hormone signaling, heavy metal absorption and transport, and ROS scavenging, are regulated by metal-regulated miRNAs and their target genes (Ding et al., 2020). It is currently believed that miRNA master controls regulate metal absorption, transportation, detoxification, and phytohormone signaling (Zhou et al., 2020). Du et al. (2021) identified novel miRNAs for copper stress in mulberry (*Morus alba*) leaves through high-throughput transcriptome sequencing. The identified miRNAs consist of 65 known and 78 novel mature miRNAs. Among the identified miRNAs, 40 were expressed differentially under copper stress. They discovered that under copper stress, miR408-3p was downregulated. Similarly, in *Arabidopsis thaliana*, miR408-3p was downregulated under copper treatment, suggesting that miR408-3p may be related to the copper regulatory response in other fruit and tuber crops. In radish, miRNAs are significantly expressed under chromium stress. Chromium-responsive miRNAs target several genes, including different transcription factor families like SPLs, MYBs, ERFs, and bZIPs (Liu et al., 2015).

10.6 CONCLUSION

Research in this area has identified specific miRNAs and their target genes associated with abiotic stress tolerance in various tuber and fruit crops. Researchers aim to develop crop types that are more tolerant to these abiotic stresses by better understanding the regulatory activities of miRNAs, which would enhance resilience and productivity in harsh environmental conditions. Furthermore, genetic engineering or breeding techniques that modify miRNA expression have the potential to improve the resistance of these significant crop species to water stress. Additionally, the manipulation of miRNA expression through genetic engineering or breeding strategies holds promise for enhancing water stress tolerance in these important crop species.

REFERENCES

Ahmed, W., Xia, Y., Zhang, H., Li, R., Bai, G., Siddique, K.H., & Guo, P. 2019. Identification of conserved and novel miRNAs responsive to heat stress in flowering Chinese cabbage using high-throughput sequencing. *Scientific Reports*, 9(1), 14922.

Anil, P., Dharmendra, P., Sachidanand, S., Yuriy, E., Snezhan, B., Dilfuza, J., & Chinmayi, J. 2023. Beneficial effects of biochar application on mitigating the drought and salinity stress implications on plants. *Plant Science Today*, 10, 188–193. https://doi.org/10.14719/pst.2591.

Betel, D., Koppal, A., Agius, P., Sander, C., & Leslie, C. 2010. Comprehensive modeling of microRNA targets predicts functional non-conserved and non-canonical sites. *Genome Biology*, 11, 1–14.

Cheng, X., He, Q., Tang, S., Wang, H., Zhang, X., Lv, M., & Liu, J. 2021. The miR172/IDS1 signaling module confers salt tolerance through maintaining ROS homeostasis in cereal crops. *New Phytologist*, 230(3), 1017–1033.

Din, M., Barozai, M.Y.K., & Baloch, I.A. 2014. Identification and functional analysis of new conserved microRNAs and their targets in potato (*Solanum tuberosum* L.). *Turkish Journal of Botany*, 38, 1199–1213.

Ding, Y., Ding, L., Xia, Y., Wang, F., & Zhu, C. 2020. Emerging roles of microRNAs in plant heavy metal tolerance and homeostasis. *Journal of Agricultural and Food Chemistry*, 68(7), pp.1958–1965.

Du, Q., Guo, P., & Shi, Y. 2021. Genome-wide identification of copper stress-regulated and novel micrornas in mulberry leaf. *Biochemical Genetics*, 59, 589–603.

Eldem, V., Çelikkol Akçay, U., Ozhuner, E., Bakır, Y., Uranbey, S., & Unver, T. 2012. Genome-wide identification of miRNAs responsive to drought in peach (*Prunus persica*) by high-throughput deep sequencing. *PLoS One*, 7(12), e50298.

Gao, Z., Ma, C., Zheng, C. et al. 2022. Advances in the regulation of plant salt-stress tolerance by miRNA. *Molecular Biology Reports*, 49, 5041–5055.

Grime, J. P. 1979. Plant strategies and vegetation processes. Chichester: Wiley and Sons: 222.

Jiang, J., Ma, S., Ye, N., Jiang, M., Cao, J., & Zhang, J. 2017. WRKY transcription factors in plant responses to stresses. *Journal of Integrative Plant Biology*, 59, 86–101.

John, B., Enright, A.J., Aravin, A., Tuschl, T., Sander, C., & Marks, D.S. 2004. Human microRNA targets. *PLoS Biology*, 2(11), e363.

Kang, Y., Yang, X., Liu, Y., Shi, M., Zhang, W., Fan, Y., Yao, Y., Zhang, J., & Qin, S. 2021. Integration of mRNA and miRNA analysis reveals the molecular mechanism of potato (*Solanum tuberosum* L.) response to alkali stress. *International Journal of Biological Macromolecules*, 182, 938–949.

Kertesz, M., Iovino, N., Unnerstall, U., Gaul, U., & Segal, E. 2007. The role of site accessibility in microRNA target recognition. *Nature Genetics*, 39(10), 1278–1284.

Krek, A., Grün, D., Poy, M.N., Wolf, R., Rosenberg, L., Epstein, E.J., MacMenamin, P., Da Piedade, I., Gunsalus, K.C., Stoffel, M., & Rajewsky, N. 2005. Combinatorial microRNA target predictions. *Nature Genetics*, 37(5), 495–500.

Lee, R. C., Feinbaum, R. L., & Ambros, V. 1993. The C-elegans heterochronic gene lin-4 encodes small RNAs with antisense complementarity to lin-14. *Cell*, 75(5), 843–854.

Lewis, B.P., Burge, C.B., & Bartel, D.P. 2005. Conserved seed pairing, often flanked by adenosines, indicates that thousands of human genes are microRNA targets. *Cell*, 120(1), 15–20.

Lewis, B.P., Shih, I.H., Jones-Rhoades, M.W., Bartel, D.P., & Burge, C.B. 2003. Prediction of mammalian microRNA targets. *Cell*, 115(7), 787–798.

Li, X., Chen, P., Xie, Y., Yan, Y., Wang, L., Dang, H., Zhang, J., Xu, L., Ma, F., & Guan, Q. 2020. Apple SERRATE negatively mediates drought resistance by regulating MdMYB88 and MdMYB124 and microRNA biogenesis. *Horticulture Research*, 7, 98.

Liu, W., Xu, L., Wang, Y. et al. 2015.Transcriptome-wide analysis of chromium-stress responsive microRNAs to explore miRNA-mediated regulatory networks in radish (*Raphanus sativus* L.). *Scientific Reports*, 5, 14024.

Ma, Y., Xue, H., Zhang, F., Jiang, Q., Yang, S., Yue, P., Wang, F., Zhang, Y., Li, L., He, P., & Zhang, Z. 2021. The miR156/SPL module regulates apple salt stress tolerance by activating MdWRKY100 expression. *Plant Biotechnology Journal*, 19(2), 311–323.

Maragkakis, M., Alexiou, P., Papadopoulos, G.L., Reczko, M., Dalamagas, T., Giannopoulos, G., Goumas, G., Koukis, E., Kourtis, K., Simossis, V.A., & Sethupathy, P. 2009. Accurate microRNA target prediction correlates with protein repression levels. *BMC Bioinformatics*, 10, 1–10.

Meenakshi, S., Rahul, D., Sachidanand, S., Anita, G.M., & Rajpal, S. 2025. Plant microRNAs: An overview. *microRNA Advances and Application in Plant Biology Plant Biology, Sustainability and Climate Change*, 1, 1–33.

Miranda, K.C., Huynh, T., Tay, Y., Ang, Y.S., Tam, W.L., Thomson, A.M., Lim, B., & Rigoutsos, I. 2006. A pattern-based method for the identification of microRNA binding sites and their corresponding heteroduplexes. *Cell*, 126(6), 1203–1217.

Moh, N.M.M., Zhang, P., Chen, Y., & Chen, M. 2021. Computational identification of miRNAs and temperature-responsive LncRNAs from mango (*Mangifera indica* L.). *Frontier in Genetics*, 12, 607248.

Niu, C., Li, H., Jiang, L., Yan, M., Li, C., Geng, D., Xie, Y., Yan, Y., Shen, X., Chen, P., & Dong, J. 2019. Genome-wide identification of drought-responsive microRNAs in two sets of Malus from interspecific hybrid progenies. *Horticulture Research*, 6, 75.

Noopur, K., Pragati, K., Adarsh, K.S., Sachidanand, S., Tanvi, J., & Megha, B. 2025a. Role of non coding RNA in effector trigger immunity. In Dutta, R., Kapadia, C., Danish, S., & Singh, S. (Eds), *Non-Coding RNA in Plants Modulation and Stress Responses Plant Biology, Sustainability and Climate Change*. Academic Press, New York.

Noopur, K., Pragati, K., Megha, B., Arpan, M., Tanvi, J., Shrisendu, M., Adarsh, K.S., & Sachidanand, S. 2025b. Role of non-coding RNA in plant architecture. In Dutta, R., Kapadia, C., Danish, S., & Singh, S. (Eds), *Non-Coding RNA in Plants Modulation and Stress Responses Plant Biology, Sustainability and Climate Change*. Academic Press, New York.

Öztürk Gökçe, Z.N., Aksoy, E., Bakhsh, A., Demirel, U., Çalışkan, S., & Çalışkan, M.E. 2021. Combined drought and heat stresses trigger different sets of miRNAs in contrasting potato cultivars. *Functional & Integrative Genomics*, 21(3), 489–502.

Pagano, L., Rossi, R., Paesano, L., Marmiroli, N., & Marmiroli, M. 2021. miRNA regulation and stress adaptation in plants. *Environmental and Experimental Botany*, 184, 104369.

Peterson, S.M., Thompson, J.A., Ufkin, M.L., Sathyanarayana, P., Liaw, L., & Congdon, C.B. 2014. Common features of microRNA target prediction tools. *Frontiers in Genetics*, 5, 76191.

Pooja, G., Megha, B., Arpan, M., Chintan, K., Rahul, D., Subhan, D., & Sachidanand, S. 2025. Introduction to long noncoding RNA. In Dutta, R., Kapadia, C., Danish, S., & Singh, S. (Eds), *Non-Coding RNA in Plants Modulation and Stress Responses Plant Biology, Sustainability and Climate Change*. Academic Press, New York.

Rahul, D., Chintan, K., Subhan, D., & Sachidanand, S. 2024. *Non-Coding RNA in Plants: Modulation and Stress Responses*. Academic Press, New York.

Reinhart, B. .J., Weinstein, E. G., Rhoades, M. W., Bartel, B. & Bartel, D. P. 2002. MicroRNAs in plants. *Genes and Development*, 16, 1616–1626.

Rehmsmeier, M., Steffen, P., Höchsmann, M., & Giegerich, R. 2004. Fast and effective prediction of microRNA/target duplexes. *RNA*, 10(10), 1507–1517.

Shen, X.X., He, J.Q., Ping, Y.K., Guo, J.X., Hou, N., Cao, F.G., Li, X.W., Geng, D.L., Wang, S.C., Chen, P.X., et al. 2022. The positive feedback regulatory loop of miR160-Auxin Response Factor 17-HYPONASTIC LEAVES 1 mediates drought tolerance in apple trees. *Plant Physiology*, 188, 1686–1708.

Stief, A., Altmann, S., Hoffmann, K., Pant, B.D., Scheible, W.R., & Bäurle, I. 2014. Arabidopsis miR156 regulates tolerance to recurring environmental stress through SPL transcription factors. *The Plant Cell*, 26(4), 1792–1807.

Subhan, D., Uzma, Y., Niaz, A., Khurram, S., Rahul, D., & Sachidanand, S. 2025. Fertilizer-responsive non-coding RNA and their significance. In Dutta, R., Kapadia, C., Danish, S., & Singh, S. (Eds), *Non-Coding RNA in Plants Modulation and Stress Responses Plant Biology, Sustainability and Climate Change*. Academic Press, New York.

Sun, M., Shen, Y., Chen, Y., Wang, Y., Cai, X., Yang, J., & Sun, X. 2022. Osa-miR1320 targets the ERF transcription factor OsERF096 to regulate cold tolerance via JA-mediated signaling. *Plant Physiology*, 189(4), 2500–2516.

Sun, X., Fan, G., Su, L., Wang, W., Liang, Z., & Li, S. 2015. Identification of cold-inducible microRNAs in grapevine. *Frontier in Plant Sciences*, 6, 595.

Tungare, K., Shahu, R., Zambare, V., Agarwal, P., Jobby, R., Nisar, N., Alabdallah, N.M., Al-Saeed, F.A., Johri, P., Singh, S., Saeed, M., & Jha, P. 2022. Toxicity mitigation of textile dye reactive blue 4 by hairy roots of *Helianthus annuus* and testing its effect in *in vivo* model systems, *BioMed Research International*, 1958939, 13. https://doi.org/10.1155/2022/1958939

Wang, Y.T., Feng, C., Zhai, Z.F., Peng, X., Wang, Y.Y., Sun, Y.T., Li, J., Shen, X.S., Xiao, Y.Q., Zhu, S.J., et al. 2020. The apple microR171i-scarecrow-like proteins 26.1 module enhances drought stress tolerance by integrating ascorbic acid metabolism. *Plant Physiology*, 184, 194–211.

Xu, T., Zhang, L., Yang, Z., Wei, Y., & Dong, T. 2021. Identification and functional characterization of plant miRNA under salt stress shed light on the salinity resistance improvement through miRNA manipulation in crops. *Frontier in Plant Sciences*, 12, 972.

Yang, Z., Dong, T., Dai, X., Wei, Y., Fang, Y., Zhang, L., Zhu, M., Nawaz, G., Cao, Q., & Xu, T. 2022. Comparative analysis of salt responsive microRNAs in two sweetpotato [*Ipomoea batatas* (L.) Lam.] cultivars with different salt stress resistance. *Frontier in Plant Sciences* 13, 879819.

Yang, Z., Zhu, P., Kang, H., Liu, L., Cao, Q., Sun, J,, Dong, T., Zhu, M., Li, Z., & Xu, T. 2020. High-throughput deep sequencing reveals the important role that microRNAs play in the salt response in sweet potato (*Ipomoea batatas* L.). *BMC Genomics*, 21, 164.

Yu, J., Su, D., Yang, D., Dong, T., Tang, Z., Li, H., Han, Y., Li, Z., & Zhang, B. 2020. Chilling and heat stress-induced physiological changes and microRNA-related mechanism in sweetpotato (*Ipomoea batatas* L.). *Frontier in Plant Sciences*, 11, 687.

Zhang, W., Luo, Y., Gong, X., Zeng, W., & Li, S. 2009. Computational identification of 48 potato microRNAs and their targets. *Computational Biology and Chemistry*, 33(1), 84–93

Zhang, X., Wang, W., Wang, M., Zhang, H.-Y., & Liu, J.-H. 2016. The mir396b of *Poncirus trifoliata* functions in cold tolerance by regulating ACC oxidase gene expression and modulating ethylene–polyamine homeostasis. *Plant Cell Physiology*, 57, 1865–1878.

Zhou, X., Khare, T., & Kumar, V. 2020. Recent trends and advances in identification and functional characterization of plant miRNAs. *Acta Physiologiae Plantarum*, 42, 1–21.

Zhu, H., Zhang, Y., Tang, R., Qu, H., Duan, X., & Jiang, Y. 2019. Banana sRNAome and degradome identify microRNAs functioning in differential responses to temperature stress. *BMC Genomics*, 20, 1–16.

11 Abiotic Stress-Related Functional Roles of microRNAs in Agronomically Important Biofuel and Beverage Crops

*Reshmi Das**, *Srishti Jagiasi, Riya Bhiwandkar, Navami Pai, and Pratima Gurung*

11.1 INTRODUCTION

Plants are complex stationary organisms that face numerous challenges in their environment as they struggle for survival. These challenges are collectively termed stress. Stress in plants arises from external conditions that detrimentally impact their growth, development, or productivity (Verma et al., 2013). The most common type of stress is abiotic stress, which is caused by unfavorable conditions stemming from non-living environmental factors. Abiotic stress encompasses a broad range of physical and chemical factors, including extreme temperatures, drought, flooding, elevated salinity, heavy metals, oxidative stress, nutrient deficiencies, and ultraviolet radiation. Abiotic stress is primarily responsible for reducing crop yields by more than 50%, thereby impacting the economy (Rodriguez et al., 2005). Abiotic stress triggers a series of alterations, including changes in morphology, physiology, biochemistry, and molecular composition, all of which impact plant growth and productivity (Tungare et al., 2022; Wang et al., 2001). Plants possess sophisticated mechanisms to perceive stress, which triggers molecular changes in cells. This signal transduction activates or modulates different messenger proteins that relay the signal at the target site. Ultimately, this process alters the expression of stress-responsive genes through activation or repression. Changes in gene expression initiate the production of stress-responsive proteins and enzymes, thus helping the plant overcome stress. Responses generated to counter stress include alterations in growth and development, changes in metabolism, desaturation of membrane lipids, induction of molecular chaperones, scavenging of reactive oxygen species (ROS), conservation of water, accumulation of osmolytes, and activation of defense mechanisms to combat various cellular damage (Zhang et al., 2023; He et al., 2018).

The involvement of microRNAs (miRNAs) in adaptive responses to environmental stresses has been confirmed by numerous molecular biology research studies (Sanan-Mishra et al., 2009; Bej and Basak, 2014; Djami et al., 2015). This small class of non-coding RNA acts as a conserved post-transcriptional regulator of gene expression required for various metabolic processes during abiotic stress (Noopur et al., 2025a,b; Pooja et al., 2025; Rahul et al., 2024; Subhan et al., 2025). The activation of miRNA expression under stress conditions contributes to strengthening the plant's stress signaling system, thereby enhancing its capacity to endure adverse environmental conditions (Song et al., 2018). This nucleic acid regulator plays a significant role in crop improvement and helps enhance plant productivity (Djami et al., 2017).

* *Corresponding Author*

DOI: 10.1201/9781003491378-11

Beverage crops encompass a large number of crops cultivated for their liquid products. Beverages are liquid products ranging from tea, coffee, fruit juices, and smoothies to the refined tastes of vintage alcoholic drinks such as wine, vodka, beer, and spirits. These drinks hold deep significance in human civilization, serving as symbols of hospitality and celebration. Some examples of beverage crops are *Camellia sinensis, Coffea arabica, Oryza sativa, Saccharum officinarum, Zea mays, Hordeum vulgare,* and various fruit crops. The beverage industry provides economic opportunities for small farmers, artisanal producers, and large-scale businesses. Notably, the tea, coffee, and alcohol industries generate billions of dollars in revenue annually. Biofuel crops are key players in the transition from non-renewable energy to renewable sources of energy. Biofuel is a green alternative to conventional fossil fuels, resulting in decreased emissions of greenhouse gases. Biofuels include bioethanol, biogas, and biodiesel, which are produced using crops such as sugarcane, switchgrass, sugar beet, corn, sorghum, soybean, and canola. Beverage crops and biofuel crops are potential drivers of economic growth through job creation, expansion of market opportunities, trade, and investment across diverse regions of the nation. The cultivation and utilization of these crops have garnered significant attention, but these crops need to manage stress conditions. Stress management in crops is crucial to mitigate undesirable environmental conditions, ensuring enhanced crop performance, optimal growth, and productivity.

This chapter provides a brief description of the miRNA-mediated regulatory mechanism of stress responses in beverage and biofuel plants and its impact on crop yield, biomass production, and quality, emphasizing their potential applications for crop enhancement.

11.2 BIOGENESIS OF miRNA

miRNA genes reside within intergenic sequences or intronic regions of DNA. Subsequently, these genes undergo transcription by RNA polymerase II within the nucleus to generate primary miRNA molecules. The primary miRNA molecule exhibits a hairpin-like structure spanning nucleotide lengths of 70–100 bases, necessitating its processing. The conversion of primary miRNA to pre-miRNA requires a micro-processing machinery comprising various enzymes. Dicer like-1 protein is one of the enzymes present in this machinery that cuts the primary miRNA and brings the miRNAs together to form an miRNA–miRNA* duplex. The miRNA–miRNA* duplex is then methylated at the 3′ ribose end. Methylation protects the miRNA from degradation by 3′ exonucleases. The miRNA–miRNA* duplex is transported into the cytoplasm via the HASTY protein, which recognizes the target gene. Following export, a helicase unwinds the miRNA–miRNA* duplex, liberating the mature miRNA into the cytoplasm. Subsequently, the mature miRNA associates with an RNA-induced silencing complex, culminating in the formation of an miRNA-induced silencing complex (miRISC) assembly (Dang et al., 2014; Ramkumar et al., 2023). The core of the miRISC comprises the binding of single-stranded mature miRNA to proteins from the Argonaute (AGO) family (Lekha and Priya, 2023). At the post-transcriptional level, the miRISC can regulate the expression of target genes through two distinct mechanisms: (1) endonucleolytic cleavage of target mRNA and (2) translational repression. Therefore, miRNA acts as a master modulator of gene expression in plants by regulating responses to environmental conditions (Krishnatreya et al., 2021).

11.3 BEVERAGE CROPS

Beverage crops are cultivated to produce various kinds of drinks catering to diverse tastes. These crops are grown for their specific properties, such as enticing aroma, strong taste, or natural sweetness. Drinks can be broadly categorized into two types: non-alcoholic drinks and alcoholic drinks, leading to the division of beverage crops into two categories: non-alcoholic beverage crops and alcoholic beverage crops.

11.3.1 Non-Alcoholic Beverage Crops

Non-alcoholic beverage crops are primarily cultivated for the production of drinks such as tea, coffee, smoothies, and fruit juices. The main crops cultivated for producing these beverages include *Camellia sinensis, Coffea arabica, Coffea canephora,* and various fruit crops such as *Mangifera indica, Prunus persica, Psidium guajava,* and citrus fruits. Tea and coffee are the world's most consumed beverages, and their cultivation relies on precise climatic conditions to yield high-quality crops. Significant productivity and yield losses are observed during abiotic stress due to drought, temperature fluctuations, and salinity (Anil et al. 2023; Jabborova et al. 2023; Verma et al. 2021). Therefore, resistance to abiotic stress is necessary. During such stressful times, miRNA is known to regulate gene expression and mediate the plant's response to stress (Meenakshi et al., 2025).

11.3.2 Stress Responsiveness in Non-Alcoholic Beverage Plants Due to miRNA

Research indicates that miR156, miR166a, and miR398 play pivotal roles in conferring drought stress resistance to *Camellia sinensis* tea plants (Guo, 2017). miR156 acts as a negative regulator of transcription factors within the squamosa promoter-binding protein-like (SPL) family. Studies conducted on alfalfa have demonstrated that during drought stress, miR156 promotes the development of thicker, denser, and more adventitious roots. Additionally, miR156 reduces water loss from stems and diminishes stomatal conductance on the adaxial surface of leaves, thereby aiding in water conservation and enhancing drought tolerance. Furthermore, miR156 upregulates the expression of a proline biosynthesis gene, pyrroline-5-carboxylate synthetase, leading to the accumulation of osmoprotectants like proline. This accumulation helps absorb more water from the soil and maintain cellular stability and pressure during periods of drought. miR156 is also responsible for the upregulation of glutathione synthase (GSH), resulting in the production of antioxidants to combat oxidants and free radicals generated during stress. Upregulation of protein phosphatase 2C mediates ABA signaling, which regulates drought tolerance in plants (Arshad et al., 2017). The miR156-SPL-DFR module present in tea is responsible for catechin biosynthesis from dihydroflavonols (Zhao et al., 2020). Therefore, miR156 plays a crucial role in the accumulation of secondary metabolites, including catechins, which constitute approximately 70% of the polyphenols found in tea (Liu et al., 2015). miR166a is a highly conserved drought-responsive miRNA present in *Camellia sinensis.* HD-ZIP III genes are the main targets of miR166s. These genes code for transcription factors that regulate the expression of stress-associated genes. miR166 causes the cleavage of HD-ZIP III family members in response to drought stress (Tian et al., 2022). During drought stress, the levels of miR398 in the plant increase. miR398 regulates an enzyme known as cytochrome c oxidase subunit V, a component of the electron transport chain (Jones and Bartel, 2004; Sunkar and Jian, 2004). In *Prunus persica* peach plants, downregulation of miR160, miR165/166, and miR167 has been observed in both root and leaf tissues under water deficit conditions (Eldem et al., 2012). Both miR160 and miR167 are known to target transcripts of the auxin response factors gene, contributing to the plant's response to drought tolerance.

Increased temperature causes changes in the expression of conserved miRNAs present in coffee plants, such as miR160 and miR175, which alter the expression of their target genes (Kellen and Antonio, 2023). Studies on cotton revealed that miR160 represses the expression of ARF10, ARF16, and ARF17 by cleaving their respective mRNA and mediating auxin signaling during high temperatures (Ding et al., 2017). Repression of ARF10, ARF16, and ARF17 caused high seed germination in *Arabidopsis* plants during heat stress (Lin et al., 2018). Another miRNA present in coffee, miR393, is known to target genes that encode transport inhibitors, transcription factors, DNA binding proteins, and a GRR1-like protein during chitin, cold, salt stress, and water stress (Dezulian et al., 2005). During heat stress in *Coffea arabica*, car-miR397b targets the gene responsible for encoding laccase 2, resulting in the accumulation of monolignols in the leaves (Del Terra et al., 2013).

In mango trees, 10 miRNAs have been identified that target six ethylene-responsive genes, including ethylene-insensitive proteins and ethylene-responsive transcription factors, during cold stress. Thus, these miRNAs hold promise for conferring tolerance to cold stress (Nann et al., 2021). Similarly, in *Camellia sinensis*, cold-tolerant tea plants exhibited upregulation of 31 miRNAs and downregulation of 43 miRNAs under cold conditions (Zhang et al., 2014). One of the miRNAs present in tea plants, miR858, may modulate the phenylpropanoid pathway by cleaving at the eighth and ninth bases of the binding region of *MYB114* transcription factor genes during environmental stresses like cold and drought. This may lead to an accumulation of flavonoid content and an increase in anthocyanin content during cold stress (Krishnatreya et al., 2021).

Salinity stress in guava leaves led to the downregulation of miR164 and miR166 (Sharma et al.,2020). Since miR166 targets a type of HD-ZIP gene, its downregulation will cause the expression of the Athb-12 gene. Athb-12 is responsible for suppressing the Na^+-sensitive phenotype and regulates the expression of the PM2RA gene, which encodes for sodium ion transport through the plasma membrane. Therefore, the downregulation of miR166 potentially improves salt tolerance in plants (Ding et al., 2008). The downregulation of miR164 leads to the expression of the NAC gene, which also contributes to the adaptation of plants to salt stress (Sharma et al., 2020).

11.3.3 Alcoholic Beverage Crops

Alcoholic beverages are crafted from a diverse array of crops and then subjected to fermentation for the production of alcohol and distillation. Major alcoholic beverage crops include *Hordeum vulgare, Vitis vinifera, Oryza sativa, Zea mays, Saccharum officinarum, Solanum tuberosum*, and *Triticum aestivum*, which are used for the production of beer, wine, sake, spirits, vodka, whiskey, bourbon, etc. Abiotic stress can hamper the productivity, yield, and quality of crops. Therefore, these crops must use regulatory mechanisms such as miRNA to increase plant tolerance against stress.

11.3.4 Stress Responsiveness in Alcoholic Beverage Plants Due to miRNA

During drought stress in *Triticum aestivum* (wheat), miR159 and miR169 are downregulated in the roots, whereas miR159 is upregulated in the leaves. miR159 targets MYB transcription factors such as MYB33 and MYB101 transcripts, which are responsible for the accumulation of the plant hormone abscisic acid during drought stress. miR169 targets a gene family of nuclear transcription factor (NFY) transcription factors. The downregulation of miR169 observed in wheat leaves causes the overexpression of NFYA. In *Arabidopsis*, overexpression of NFYA leads to reduced water loss from leaf parts, making the plant more tolerant to drought stress (Li et al., 2008). However, there is an upregulation of miR169 in the roots of wheat plants. An upregulation of the miR395 family, which contains 34 miRNAs, is observed during drought stress in the leaves of the wheat plant. miR395 regulates genes involved in the assimilation and distribution of sulfate by targeting ATP-sulfurylase and sulfate transporters. It is speculated that miR395 contributes to drought stress tolerance in plants, as its levels significantly increase in leaves during drought (Akdogan et al., 2015). Studies on the *Solanum tuberosum* potato plant revealed that miR811, miR814, miR835, and miR4398 play important roles in the post-transcriptional regulation of drought-related genes. These miRNAs target genes such as MYB transcription factors, hydroxyproline-rich glycoprotein aquaporins, and WRKY transcription factors. The increased expression of stu-miR396 and the downregulation of stu-miR156a and stu-miR157 shortly after drought treatment indicate that these miRNAs are early responders to drought conditions (Aksoy et al., 2015).

During cold stress in sugarcane, miR319 mRNA was found to be upregulated. miR319 is the most conserved miRNA (Yang et al., 2013). It has been observed to regulate TCP6 and GAMyb transcription factor families by causing efficient cleavage in both genes (Thiebaut et al., 2011). Research conducted on transgenic rice plants subjected to cold stress unveiled the overexpression of miR319. In these plants, the leaves exhibited heightened levels of osa-miR319, leading to an increase in longitudinal small veins, likely resulting in an expanded leaf blade width. Consequently, miR319 is implicated in enhancing cold tolerance (Yang et al., 2013).

During heat stress in rice (*Oryza sativa*) plants, the expression of miR160, miR169, and miR1884 differed in shoot and root tissues. During heat stress, there was upregulation of miR160e, miR160f, miR396a, miR396e, miR164a, miR166k, and miR166g, along with downregulation of miR5054, miR408, and miR5788. The upregulation of miR169r-5p caused the downregulation of its target gene LOC_Os12g42400. High expression of miR169r-5p during the flowering stage increased heat tolerance in rice plants. osa-miR172 might play a significant role in protecting the rice panicle during heat stress. Moreover, osa-miR414 potentially inhibits the expression of flowering genes by downregulating LOC_Os05g51830, thereby delaying the heading of rice (Peng et al., 2020).

During high salinity stress in maize, *Zea mays*, the miR474 and miR395 families are observed to be upregulated. miR474 and miR395 likely target the gene expression of NADP-dependent malic enzyme (NADP-ME). The NADP-ME enzyme is involved in the conversion of L-malate to pyruvate through oxidative decarboxylation. Pyruvate is then oxidized in the Krebs cycle. As the miRNA targets NADP-ME, it will cause downregulation of the NADP-ME gene, hampering the conversion of L-malate. osa-miR528-like miRNA is also consistently regulated under salt stress conditions. miR528 targets the superoxide dismutase gene, which scavenges ROS that are typically produced in excess during salt stress. Therefore, scavenging ROS is another strategy plants use to increase tolerance against this stress (Ding et al., 2008)

Amid heavy metal stress in rice, miR604 plays a crucial role in regulating cadmium tolerance. This miRNA targets the expression of the wall-associated kinase gene, which is pivotal in enhancing plant tolerance to heavy metals such as Al, Cu, Cd, and Zn. Another miRNA, Osa-miR602, is downregulated during Cd stress in rice plants. Osa-miR602 targets the gene encoding the xyloglucan endotransglucosylase/hydrolases (XET) enzyme. XET, an enzyme facilitating cell wall loosening reactions, acts by cleaving xyloglucan chains endolytically and subsequently rejoining them to other xyloglucan molecules or H_2O. The downregulation of Osa-miR602 by Cd in leaves may imply the potential upregulation of XET activity. However, the process by which heavy metals cause cell wall stiffening and limit cell elongation is still poorly understood (Si et al., 2009).

11.4 ABIOTIC STRESS ENCOUNTERED BY BIOFUEL CROPS

Biofuels are chemicals resulting from the chemical modification of biomass or obtained through biological processes. To meet the ever-increasing demand for energy from burning fossil fuels, which contributes to surging environmental concerns like global warming, biofuels can be considered a sustainable alternative to satisfy global energy needs (Rodionova et al., 2017). Bioethanol, biodiesel, and biogas can be derived from plant biomass and account for second-generation biofuels (Dragone et al., 2010).

Bioalcohols are produced from crops rich in starch and sugars, as sugar serves as a substrate for bioalcohol production. Starch can be hydrolyzed into simpler sugars, which can then be utilized as a sugar source. Bioalcohols are yielded from carbohydrate fermentation performed by microorganisms, followed by distillation and dehydration to produce fuels that serve as a significant energy source. Among bioalcohols, ethanol is the most frequently used, as it has lower emissions of carbon monoxide, nitric oxide, and volatile organic compounds due to the presence of oxygen as a chemical component, which in turn reduces the temperature required for combustion. Bioethanol can be derived from sugarcane and sugar beet, and starch from corn, wood, and bagasse (Connor and Hernandez, 2009).

Biodiesel serves as a renewable alternative to petroleum diesel. Producing biodiesel involves two distinct steps: the first step consists of extracting oil from plant sources. Major sources include oilseed plants like *Jatropha curcas*, which is the most common crop for producing substantial biodiesel due to its rapid growth and increased seed production. Other sources include rapeseeds, soybeans, and sunflower oil. The second step involves the formation of methyl esters by breaking down long-chain fat molecules into shorter chains, accompanied by esterification with alcohols (methanol or ethanol) in a process known as transesterification (Brown and Mark, 2009; Rodionova et al., 2017).

Using biodegradable matter or energy-rich crops as substrates for anaerobic digestion by anaerobic microorganisms results in methane, while byproducts can be utilized as biofertilizers; the former can be used as biogas. Predominantly, maize and sunflower are the primary plant sources, while grass and Sudan grass are also widely used crops (Redman, 2008).

Dwindling translation caused by downregulation of protein-encoding genes by miRNA is referred to as gene regulation mediated by miRNA. Modification of miRNA could be implemented to increase plant biomass yield and tackle several abiotic stress-related effects by modulating the activity of target genes (Naqvi et al., 2012; Zhang and Qinglian, 2015).

11.4.1 Role of Certain miRNAs in Biofuel Crops

A few miRs (mature miRNAs) that are evolutionarily conserved within plant species that produce biofuels and their activity in response to abiotic stress are discussed as follows.

A. miR156 has been reported to boost biomass levels in important biofuel crops. The target transcript for cleavage by miR156 is the transcription factor protein, SPL, which positively impacts the size of plant organs. miR156 influences the transition from immature to mature cell walls and is abundantly found in the early phases of plant growth and development, suggesting its elevated expression in juvenile plants and a decline upon reaching maturity. The importance of plant biomass, conducive to increased biofuel production, is achieved by the suppressed overexpression of miR156, which impedes flowering—a desirable characteristic for enhancing biomass content. In a study, it was observed that with the overexpression of miR156, flowering is inhibited in switchgrass. Conversely, if the threshold is exceeded, it could have a negative impact, such as reduced growth, ultimately affecting overall biomass content. miR156 activity is augmented under stress due to elevated temperatures, drought, and decreased nitrogen by altering the expression of SPL in response to abiotic stress, as noted in plant species like sugarcane, switchgrass, poplar, maize, and Jatropha. For these underlying reasons, miR156 expression could be exploited in various ways to achieve beneficial traits in biofuel crops (Fu et al., 2012; Jennifer et al., 2014).

B. miR160 and miR167 are found to regulate the activity of loci coding for conserved sequences known as auxin response factors 10 and 16. They govern root cap formation and the demarcation of lateral root cells. Among various stresses, including heat, ultraviolet radiation, and pathogen responses, miR160 activity is observed to be stimulated, which is beneficial for biofuel production. The disintegration of miR160 and ARF 16 diminishes lateral root formation and induces alterations in cell and root conformation (Khraiwesh et al., 2012; Yang et al., 2014).

C. miR164: The activity of the sequence transcribing NAC and MYB transcription factors is associated with miR164 and corresponds to plant metabolism and metamorphosis (Wei et al., 2009). miR164 enhances biomass turnover as it facilitates lateral rooting under various stress conditions. Additionally, miR164 suppresses the encoding of transcription factor which is ethylene insensitive that encompasses transcripts which bring about senescence of leaves. Furthermore, downregulation of miR164 promotes cell death induced by aging and senescence through the ORE transcription factor. Leaf senescence channels nutrients to developing organs, enhancing plant fitness as it delays senescence. For instance, in response to abiotic stress such as drought and salinity, miR164 enhances water uptake in sugarcane and its expression is upregulated in maize, respectively. Biofuel plant species containing miR164 as an evolutionarily conserved site include switchgrass, sorghum, maize, poplar, and cassava (Kim et al., 2014; Li et al., 2013; Liu et al., 2010).

D. There is a close resemblance between miR166 and miR167, leading to their clustering together. Transcription factors, namely the ARF family, along with the protein-coding gene Argonaute 10, are the key targets for miR167 and miR166, respectively. They induce alterations in metabolism, morphogenesis, and development in plants. For instance, the effective roles of miR166 and miR167 are evident in maize during the developmental period. Under abiotic stress such as salinity, miR166 and miR167 function like a switch in switchgrass. The former is downregulated while the latter is upregulated. This upregulation results from miR167 targeting glycosyltransferase, which accounts for reduced recalcitrance, thereby accelerating biofuel generation (Jennifer et al., 2014; Ji et al., 2011; Sun et al., 2012; Xie et al., 2010; Xie et al., 2014).

E. NFY A promotes an increase in cell size and number in *Brachypodium* through positive regulation of miR169 when the crop is subjected to drought stress via the plant stress pathway. In addition, amid externally present abscisic acid and elevated temperature, miR169 is downregulated and upregulated, respectively. This phenomenon could be applied to increase biomass productivity, which could be a desirable attribute for biofuel generation (Bertolini et al., 2013; Li et al., 2010; Zhao et al., 2013).

11.4.2 miRNA Regulation with Regard to Drought, Temperature, and Salinity in Biofuel Producing Crops: Highlighting Effect on Sugarcane Yield: The Principal Producer of Biofuel

Sugarcane is a significant source of sugar as well as a fuel to facilitate sustainable energy. Studies on the metabolic pathways associated with lignin and the breakdown components comprising cellulose are gathering interest. Sugarcane has gained the spotlight as an effective feedstock for cellulosic bioethanol (Swapna and Kumar, 2017). A complicated regulatory mechanism for economically significant features, such as sugar content and resilience to biotic and abiotic stressors, is posed by a high degree of polyploidy. Research on sugarcane miRNAs is made more problematic by the crop's inherent complexity and lack of a well-sequenced full genome. Additionally, the relatively unstable nature of the miRNA precursors renders it challenging to locate them in EST databases (Gentile et al., 2015). The most analogous diploid relative of sugarcane is sorghum, and comparative genetic research has revealed parallels between the two (Zanca et al., 2010).

One key characteristic of crops used to produce biofuels, in this regard, is the amount of biomass sugarcane produces, and miRNAs play a crucial role in this process. The sugarcane miRNA-mediated regulation associated with drought has garnered enormous attention and is under tremendous scientific investigation. The expression pattern of miRNAs can differ depending on the genetic makeup, type of drought treatment, tissue, species, and developmental conditions. miRNAs function as moderators for crop molecular dynamics. Plant miRNAs instantaneously splice or silence the translation of their target transcripts by virtually appropriate base-pair complementarity. Numerous species might possess identical target gene sequences with conserved miRNAs belonging to equivalent biological lineages. Because the miRNA species in various plants share a substantial amount of parallelism, methods for predicting and validating the target genes that these miRNAs regulate have been established (Gentile et al., 2015).

Drought stress has varied impacts on the four phases of sugarcane growth: germination, tillering, grand growth, and maturity. Water shortages particularly affect the tillering and grand growth stages, which account for 80% of sugarcane yield during these two phases (Bamber et al., 2005; Gascho and Shih, 1983; Ramesh, 2000; Singh and Rao, 1987). From the commencement of planting until the mature cane is harvested, the sugarcane crop cycle typically lasts between 12 and 18 months. However, given that sugarcane is subject to year-round climate fluctuations, farmers are unable to predict or avoid drought periods (Gentile et al., 2015). The drying time scale has multiple benefits:

it reduces soil compaction after harvest, conserves water, thereby lowering irrigation costs, and may increase sucrose content in the soil. Since growth has a greater impact on sucrose concentration than photosynthesis, the carbon dioxide uptake by leaves and stalks could instead be directed toward sucrose accumulation in the culm (Inman, 2013; Robertson and Donaldson, 1998; Singels et al., 2000).

Elevated ssp-miR166 expression is detected during drought stress. Target transcripts that encode transcription factors associated with the homeobox-leucine zipper cause a reduction in internode length when a transcription factor is excessively expressed. One of the most prominent phenotypes in sugarcane plants subjected to drought is a truncated stalk. The transcription factor responsible for the shortening of internodes is stimulated by the decreased ssp-miR166 (Bamber et al., 2005; Silva et al., 2008).

The sugarcane gene that produces a protein significantly resembling members of the SCL–GRAS domain protein family is the intended target of miR171. This protein's *Arabidopsis* analogs cause branching and are major miR171 targets. In transgenic mutant plants, upregulation of miR171 leads to decreased shoot branching. When sugarcane is stressed by drought, shoot branching diminishes, endangering plant viability and lowering crop output.

Reduced ssp-miR171 concentrations and, on the contrary, increased activity of the SCL transcription factor may indicate a sugarcane response to mitigate the detrimental effects of drought on tillering (Bamber et al., 2005; Gentile et al., 2015; Kapur et al., 2011; Silva et al., 2008).

A sugarcane transcript that closely resembles the VNI2 protein is a potential target of miR160. The transcription factor vascular related NAC-domain7, a key inducer of xylem development, is inhibited by this protein. To the best of our understanding, no studies have explored how sugarcane's xylem differentiates in response to drought; rather, poplar trees exhibit this response to drought. Consequently, VIN2 levels might decline as a result of ssp-miR160 induction, unleashing VND7's activity to aid in xylem differentiation. This may enhance the sugarcane plants' capacity to translocate water to different parts of the plant (Arend and Jorg, 2007; Yamaguchi et al., 2010).

A hallmark of agroecological practices that optimizes plant productivity under drought stress is the prolonged retardation of leaf senescence. The decline in agricultural produce is attributed to leaf senescence, which occurs in sugarcane plants exposed to drought conditions. Drought triggers the expression of ssp-miR399, which targets a protein responsible for leaf senescence. Elevated levels of this miRNA may help sustain a green leaf phenotype, enabling sugarcane plants to continue photosynthesis under prolonged stress conditions (Bamber et al., 2005; Borrell et al., 2000; Rivero et al., 2007; Swapna and Kumar, 2017; Lopes et al., 2011).

Sugarcane encounters oxidative stress due to drought, which increases the levels of H_2O_2 and lipid peroxidation. The target sequence for miR169 is glutathione S-transferase (GST), which is downregulated under drought conditions. The detoxification of toxins produced under stress is aided by enzymes, and transgenic plants that overexpress GSTs exhibit a higher degree of resistance to stress. The decline in ssp-miR169 could increase the concentration of GST, potentially mitigating the detrimental effects of ROS (George et al., 2010; Li et al., 2010). When transcription factors are significantly overexpressed, plants can modify their microtranscriptome in various ways to cope with different phases and magnitudes of stress induced by water scarcity. These transcription factors include NAC domain, homeobox-leucine zipper, Nuclear Factor YA, GRAS/SCL, APETALA2, and bZIP. Functional assays and mutants with altered expression levels of certain miRNAs are necessary due to the complexity of expression patterns. In transgenic plants, miRNA precursors can be overexpressed or suppressed to generate mutants, which would be invaluable in determining how miRNAs function in response to drought stress (Swapna and Kumar, 2017). One of the primary abiotic factors negatively affecting plant yield is that tropical and sub-tropical crops cannot thrive well in cold conditions, which retards their growth. Although it is primarily grown in the sub-tropics, where it frequently experiences cold stress, sugarcane (*Saccharum* spp.) is a tropical plant. By preventing metabolic processes that lead to osmotic, oxidative, and other stressors, low temperatures restrict the growth and development of plants (Yang et al., 2017). C-repeat-binding factors (CBFs) are regulated by the promoter of C-repeat-binding factor expression 1 (ICE1) during low-temperature stress. CBFs bind to dehydration-sensitive regions in the promoter region to upregulate the expression of downstream cold-responsive (COR) genes, collectively referred to as the ICE-CBF-COR pathway. While lower temperatures enable sugarcane to accumulate sucrose, they also initiate bud browning, which hinders growth. Stresses involving salinity and aridity trigger an increase in miR156, indicating that miR156 plays a role in sugarcane's response to abiotic stress (Yang et al., 2017). A higher expression of miR167 suggests that the auxin signaling pathway and the developmental response to cold stress are regulated by miR167 and the trans-acting short-interfering RNA-auxin response factor (tasiRNA-ARF). To fine-tune their target transcripts under non-stress conditions, minimal amounts of miRNA are sufficient. Upregulated miRNAs under stress can decrease ARF-mediated auxin response gene expression by reducing ARF and TIR transcript levels. This attenuates the plant's progression and ultimately enhances plant stress tolerance (Tian et al., 2022).

Salt tolerance is attributed to the control of the plasma membrane, the shape of the cell wall, osmotic coordination, functional genes, transcription factors, and the activation of several cell signaling pathways (Gupta et al., 2023). Elevated concentrations of salt sense stimuli and generate intracellular mediators such as calcium, phosphatidylinositol, and ROS. This leads to an influx of calcium ions into the cell, thereby raising their concentration, which ultimately drives signaling pathways through protein phosphorylation. To confer tolerance against abiotic stresses, phosphorylated proteins modulate several transcription factors, including MYC/MYB, ABF, CBF/DREB, and bZIP (Song et al., 2018; Zhang et al., 2023). The *Arabidopsis* vacuolar pyrophosphatase gene enhances sugarcane's ability to withstand salinity and drought when transformed by Agrobacterium. The luminous energy captured by chlorophyll during photosynthesis positively influences tolerance to salinity due to increased gaseous exchange. The majority of the identified mRNA targets are primarily transcription factors involved in plant development, including the AGO1 protein, GAMyB (miR159XVI), HAP12 (miR169III), and growth-regulating factor (GRF) (miR396II). Targets of miR159 have been shown to include the transcription factor GAMyB. When exposed to mild to severe salt stress, roots exhibit increased levels of GAMyB, and over time, the expression levels of GAMyB in shoots remain relatively stable. Nuclear Factor Y, also known as transcription factor HAP (NF-Y), is also involved in this process. The miRNA concentrations in roots were negatively correlated with the expression of miR169III, showing downregulation at 1 hour and upregulation in shoots, followed by a rise at 6 hours and a reduction at 24 hours under salt stress conditions. A category of transcription factors known as GRFs is involved in the proliferation and maturation of plants. Based on expression analysis, miR396II likely regulates GRF1. After 1 hour of salt stress, GRF mRNA levels began to decline, whereas miR396II levels remained unchanged (Bertolini et al., 2013). One critical step is the function of miR168 in regulating AGO1 mRNA through feedback. The tissues of plants cultivated in mildly salinized environments showed a strong induction of AGO1. When root and leaf tissues were subjected to severe salt treatment, AGO1 expression levels were found to be lower than miR168II levels at the same salt concentration.

Other plants' miR397 targets were identified using laccases. Ascorbate oxidase and several other enzymes belong to a broader class of enzymes known as multi-copper enzymes (Vaucheret et al., 2004; Li et al., 2012).

11.5 CONCLUSION

A plethora of investigations has delineated the critical role of miRNA in plant genomes. Gaining insights into their function and how they interact with target genes can help agricultural crops develop tolerance to stress, which is vital for long-term food security and sustainable agriculture in a changing environment. The most effective approach to enhance the agronomic features of crops is to create transgenic plants that express the desired miRNA. Beverage and biofuel crops can benefit from the application of computational analysis and experimental approaches to assess gene expression targeting (Ramkumar et al., 2023). A breakthrough in small RNA research and biofuels will be achieved through deep sequencing alongside additional technological innovations that facilitate genome analysis with larger data sets, allowing for experimental validation. By fully utilizing miRNA regulation, researchers can create crops that yield higher produce suitable for beverage crops and possess greater biomass content, ideal for biofuel production. Studies on transgenic technology have primarily examined how stress conditions affect seed germination and growth. However, the lack of comprehensive miRNA profiles in many agronomically significant crops makes the creation of complete miRNA profiling under various abiotic stresses necessary. The success of a crop, its health, and the revenue it generates all depend on seed germination and seedling establishment, enabling the cultivation of desirable crops even on compromised lands. Therefore, there is compelling evidence to support the application of innovative miRNA-based biotechnology to increase crop tolerance to a variety of environmental challenges, as demonstrated in previous transgenic investigations involving the overexpression of specific miRNAs (Trumbo et al., 2015). To fully utilize miRNAs for robust and sustainable agriculture in the face of escalating environmental stresses, it will be necessary to address current research gaps and challenges, as well as to embrace future technologies and interdisciplinary approaches (Ramkumar et al., 2023).

11.5.1 Questions for Further Investigation of the Scope of miRNA

Does miRNA regulate the expression of all transcripts involved in several metabolic pathways? Is a miRNA binding site located on every gene that directs a specific function? Is there a bias in selecting a target to be altered by miRNA, or does it occur arbitrarily? Does the complexity of the genome or ploidy number influence the activity of miRNA? What is the scope of applying comparative genomics to the regulatory function of miRNA in crop yield enhancement? (Swapna and Kumar, 2017)

BIBLIOGRAPHY

Akdogan, G., Tufekci, E. D., Serkan, U., and Turgay, U. (2015). miRNA-based drought regulation in wheat. *Functional & Integrative Genomics* 16(3): 221–233.

Aksoy, E., Ufuk, D., Ozturk, Z. N., Sevgi, C., and Mehmet, E. C. (2015). Recent advances in potato genomics, transcriptomics, and transgenics under drought and heat stresses: a review. *Turkish Journal of Botany* 39: 920–940.

Anil, P., Dharmendra, P., Sachidanand, S., Yuriy, E., Snezhan, B., Dilfuza, J., and Chinmayi, J. (2023). Beneficial effects of biochar application on mitigating the drought and salinity stress implications on plants. *Plant Science Today* 10(2): 188–193. http://doi.org/10.14719/pst.2591

Arend, M., and Jorg, F. (2007). Seasonal change in the drought response of wood cell development in poplar. *Tree Physiology* 27(7): 985–992.

Arshad, M., Biruk, A. F., Lisa, A., Banyar, A., and Abdelali, H. (2017). microRNA156 improves drought stress tolerance in alfalfa (*Medicago Sativa*) by silencing SPL13. *Plant Science* 258(5): 122–136.

Bej, S., & Basak, J. (2014). MicroRNAs: the potential biomarkers in plant stress response. American Journal of Plant Sciences, 05(05), 748–759. https://doi.org/10.4236/ajps.2014.55089

Bertolini, E., Wim, V., David, S. H., Luca, G., Viviana, P., Dirk, I., Mario, E. P., and Erica, M. (2013). Addressing the role of microRNAs in reprogramming leaf growth during drought stress in *Brachypodium distachyon*. *Molecular Plant* 6(2): 423–443.

Borrell, A. K., Graeme, L. H., and Robert, G. H. (2000). Does maintaining green leaf area in sorghum improve yield under drought? II. Dry matter production and yield. *Crop Science* 40(4): 1037–1048.

Brown, R. C., and Mark, M. W. (2009). *Biomass Conversion to Fuels and Electric Power*. Iowa State University, Ames, IA.

Carnavale, B. M., Sabrina, R., Clicia, G., Flavia, T., Cristian, A. R., Laurent, F., Adriana, S. H., and Paulo, C.G.F. (2013). High-throughput sequencing of small RNA transcriptome reveals salt stress regulated microRNAs in sugarcane. *PLoS One* 8(3): e59423.

Cia, M. C., Guimarães, A. C. R., Medici, L. O., Chabregas, S. M., and Ricardo, A. A. (2009). Antioxidant responses to water deficit by drought-tolerant and-sensitive sugarcane varieties. *Annals of Applied Biology* 161(3): 313–324.

Connor, D. J., and Hernandez, C. G. (2009). *Crops for Biofuel: Current Status and Prospects for the Future.* The University of Melbourne, Spain.

Dang, T. H. Y., Mark, Z., and Bhave, M. (2014). *Abiotic Stress Response in Barley and the Emergent Roles of microRNAs.* Nova Science Publishers, Inc., Hauppage, NY. https://hdl.handle.net/10536/DRO/DU:30122356.

Dezulian, T., Michae, L. R., Javier, F. P., Detlef, W., and Daniel, H. H. (2005). Identification of plant microRNA homologs. *Bioinformatics* 22(3): 359–360.

Del Terra, L., Lonzarich, V., Asquini, E., Navarini, L., Graziosi, G., Liverani, F. S., & Pallavicini, A. (2013). Functional characterization of three Coffea arabica L. monoterpene synthases: Insights into the enzymatic machinery of coffee aroma. Phytochemistry, 89, 6–14. https://doi.org/10.1016/j.phytochem.2013.01.005

Ding, D., Lifang, Z., Hang, W., Zhijie, L., Zuxin, Z., and Yonglian, Z. (2008). Differential expression of MiRNAs in response to salt stress in maize roots. *Annals of Botany* 103(1): 29–38.

Ding, Y., Yizan, M., Nian, L., Jiao, X., Qin, H., Yaoyao, L., and Yuanlong, W. (2017). microRNAs involved in auxin signalling modulate male sterility under high-temperature stress in cotton (*Gossypium Hirsutum*). *The Plant Journal* 91(6): 977–994.

Djami, T., Arnaud, T., and Ian, A. D. (2015). Lipopolysaccharide perception leads to dynamic alterations in the microtranscriptome of *Arabidopsis thaliana* cells and leaf tissues. *BMC Plant Biology* 15(1): 79.

Djami, T., Arnaud, T., Neeti, S. M., Khayalethu, N., and Ian, A. D. (2017). Functional roles of microRNAs in agronomically important plants—potential as targets for crop improvement and protection. *Frontiers in Plant Science* 8(3): 378.

Dragone, G., Bruno, D. F., Antonio, A. V., and Jose, A. T. (2010). *Third Generation Biofuels from Microalgae.* Kocaeli University, Turkey.

Eldem, V., Akçay, U. Ç., Ozhuner, E., Bakır, Y., Uranbey, S., & Unver, T. (2012). Genome-Wide Identification of miRNAs Responsive to Drought in Peach (Prunus persica) by High-Throughput Deep Sequencing. PLoS ONE, 7(12), e50298. https://doi.org/10.1371/journal.pone.0050298

Fu, C., Ramanjulu, S., Chuanen, Z., Hui, S. J., Yi, Z., Jessica, M., and Jennifer, W. (2012). Overexpression of miR156 in switchgrass (*Panicum virgatum* L.) results in various morphological alterations and leads to improved biomass production. *Plant Biotechnology Journal* 10(4): 443–452.

Gascho, G. J., and Shih, S. F. (1983). Sugarcane. *Crop Water Relations* 1983: 445–479.

Gentile, A., Lara, I. D., Raphael, S. M., Thais, H. F., and Marcelo, M. (2015). microRNAs and drought responses in sugarcane. *Frontiers in Plant Science* 6: 58.

Gentile, A., Thais, H. F., Raphael, S. M.., Lara, I. D., Andrea, A. H., Monalisa, S. C., Glaucia, M. S., et al. (2013). Effects of drought on the microtranscriptome of field-grown sugarcane plants. *Planta* 237: 783–798.

George, S., Gayatri, V., and Ajay, P. (2010). A chloroplast-localized and auxin-induced glutathione S-transferase from phreatophyte *Prosopis juliflora* confer drought tolerance on tobacco. *Journal of Plant Physiology* 167(4): 311–318.

Guo, Y., Zhao, S., Zhu, C., Chang, X., Yue, C., Wang, Z., Lin, Y., & Lai, Z. (2017). Identification of drought-responsive miRNAs and physiological characterization of tea plant (Camellia sinensis L.) under drought stress. BMC Plant Biology, 17(1). https://doi.org/10.1186/s12870-017-1172-6

Gupta, S., Gupta, K., Nehra, C., Gaur, R.K., and Yadav, D. (2023). Biotechnological intervention for sugarcane improvement under salinity. *Sugar Tech* 25(1): 15–31.

He, M., Cheng, Q. H., and Nai Z. D. (2018). Abiotic stresses: general defenses of land plants and chances for engineering multistress tolerance. *Frontiers in Plant Science* 9(12): 1771.

Inman, B. G. (2013). Sugarcane yields and yield limiting processes. *Sugarcane: Physiology, Biochemistry, and Functional Biology* 2013: 579–600.

Inman-Bamber, N. G., and David M. Smith. (2005). Water relations in sugarcane and response to water deficits. *Field Crops Research* 92(23): 185–202.

Jabborova, D., Nigora, Z., Yuriy, E., Abdujalil, N., Anand, D., Khurshid, S., Zafarjon, J., Sachidanand, S., and Rahul, D. (2023). Growth of spinach as influenced by biochar and bacillus endophyticus IGPEB 33 in drought condition. *Pakistan Journal of Botany* 55(SI): 1–7. https://www.researchgate.net/publication/368753511_GROWTH_OF_SPINACH_AS_INFLUENCED_BY_BIOCHAR_AND_BACILLUS_ENDOPHYTICUS_IGPEB_33_IN_DROUGHT_CONDITION.

Ji, L., Xigang, L., Jun, Y., Wenming, W., Rae, E. Y., Yun, J. K.., Thanh, T. D. et al. (2011). ARGONAUTE10 and ARGONAUTE1 regulate the termination of floral stem cells through two microRNAs in Arabidopsis. *PLoS Genetics* 7(3): e1001358.

Jones, R. M. W., and Bartel, D. P. (2004). Computational identification of plant microRNAs and their targets, including a stress-induced miRNA. *Molecular Cell* 14(6): 787–799.

Kapur, R., Duttamajumder S. K, and Krishna, K. R. (2011). A breeder's perspective on the tiller dynamics in sugarcane. *Current Science* 100: 183–189.

Kellen, R., and Antonio, C. (2023). Small RNAs: promising molecules to tackle climate change impacts in coffee production. *Plants* 12(20): 3531.

Khraiwesh, B., Jian, K. Z., and Jianhua, Z. (2012). Role of miRNAs and siRNAs in biotic and abiotic stress responses of plants. *Biochimica et Biophysica Acta (BBA)-Gene Regulatory Mechanisms* 1819(2): 137–148.

Kim, H. J., Sung, H. H., You, W. K., Hwan, L., Ji, H. J., Bong, K. P., Timilsina, R. et al. (2014). Gene regulatory cascade of senescence-associated NAC transcription factors activated by ethylene-insensitive2-mediated leaf senescence signalling in Arabidopsis. *Journal of Experimental Botany* 65(14): 4023–4036.

Krishnatreya, D. B., Niraj, A., Sarvajeet, S. G., and Tirthankar, B. (2021). Understanding the role of MiRNAs for improvement of tea quality and stress tolerance. *Journal of Biotechnology* 328(2): 34–46.

Lekha, B., and Priya, S. (2023). *microRNA-Induced Silencing Complex Assembly and microRNA Turnover*. CRC Press, Boca Raton, FL.

Li, W. X., Youko, O., Jianhua, Z., Xin, J. H., Jian, M. W., Kei, I., Xiao, Y. L., Xinping, C., Hailing, J., and Jian, K. Z. (2008). The *Arabidopsis* NFYA5 transcription factor is regulated transcriptionally and posttranscriptionally to promote drought resistance. *The Plant Cell* 20(8): 2238–2251.

Li, W., Xiao, C., Zhaolu, M., Xiahe, H., Qi, X., Heng, W., Hailing, J., Dabing, Z., and Wanqi, L. (2012). Transcriptional regulation of Arabidopsis *MIR168a* and *ARGONAUTE1* homeostasis in abscisic acid and abiotic stress responses. *Plant Physiology* 158(3): 1279–1292.

Li, Y., Yaru, F., Lusha, J., Changai, W., and Chengchao, Z. (2010). Characterization and expression analysis of the Arabidopsis mir169 family. *Plant Science* 178(3): 271–280.

Li, Z., Jinying, P., Xing, W., and Hongwei, G. (2013). Ethylene-insensitive3 is a senescence-associated gene that accelerates age-dependent leaf senescence by directly repressing miR164 transcription in Arabidopsis. *The Plant Cell* 25(9): 3311–3328.

Lin, J., Kuo, C., Yang, I., Tsai, W., Shen, Y., Lin, C., Liang, Y., Li, Y., Kuo, Y., King, Y., Lai, H., & Jeng, S. (2018). MicroRNA160 modulates plant development and heat shock protein gene expression to mediate heat tolerance in arabidopsis. Frontiers in Plant Science, 9. https://doi.org/10.3389/fpls.2018.00068

Liu, L., Yong, Z., Mark, W. S., Xianghua, L., and Yongjun, L. (2010). Identification and application of a rice senescence-associated promoter. *Plant Physiology* 153(3): 1239–1249.

Liu, M., Heng, T., Jian, H. W., Ren, R. C., Run, X. W., Xiao, H. Q., Qiang, X., and Xue, H. C. (2015). Relationship between gene expression and the accumulation of catechin during spring and autumn in tea plants (*Camellia Sinensis* L.). *Horticulture Research* 2(1): 15011.

Liu, Y., Dan, L., Shengquan, Z., Liping, Z., Jie, G., Yanhong, L., Jiamin, C. et al. (2022). Integrated analysis of microarray, small RNA, and degradome datasets uncovers the role of microRNAs in temperature-sensitive genic male sterility in wheat. *International Journal of Molecular Sciences* 23(15): 8057.

Lopes, M. S., Jose, L. A., Philippus, D. R., Van, H., and Christine, H. F. (2011). Enhancing drought tolerance in C_4 crops. *Journal of Experimental Botany* 62(9): 3135–3153.

Lorenzo, D. T., Valentina, L., Elisa, A., Luciano, N., Graziosi, G., Liverani, F. S., and Alberto, P. (2013). Functional characterization of three *Coffea arabica* L. Monoterpene synthases: insights into the enzymatic machinery of coffee aroma. *Phytochemistry* 89(5): 6–14.

Meenakshi, S., Rahul, D., Sachidanand, S., Anita, G. M., and Rajpal, S. (2025). Plant microRNAs: An overview. In Dutta, R., Kapadia, C., Danish, S., and Singh, S. (Eds), *microRNA Advances and Application in Plant Biology Plant Biology, Sustainability and Climate Change*. Academic Press, New York.

Nann, P. Z., Yujie, C., and Ming, C. (2021). Computational identification of MiRNAs and temperature-responsive LncRNAs from mango (*Mangifera Indica* L.). *Frontiers in Genetics* 12(6): 607248.

Naqvi, A. R., Maryam, S., Shirin, H., and Nirupam, R. (2012). Biogenesis, functions and fate of plant microRNAs. *Journal of Cellular Physiology* 227(9): 3163–3168.

Neeti, S. M., Vikash, K., Sudhir, K. S., and Sunil, K. M. (2009). Cloning and validation of novel miRNA from basmati rice indicates cross talk between abiotic and biotic stresses. *Molecular Genetics and Genomics* 282(5): 463–474.

Noopur, K., Pragati, K., Adarsh, K. S., Sachidanand, S., Tanvi, J., and Megha, B. (2025a). Role of non coding RNA in effector trigger immunity. In Dutta, R., Kapadia, C., Danish, S., and Singh, S. (Eds), *Non-Coding RNA in Plants Modulation and Stress Responses Plant Biology, Sustainability and Climate Change*. Academic Press, New York.

Noopur, K., Pragati, K., Megha, B., Arpan, M., Tanvi, J., Shrisendu, M., Adarsh, K. S., and Sachidanand, S. (2025b). Role of non-coding RNA in plant architecture. In Dutta, R., Kapadia, C., Danish, S., and Singh, S. (Eds), *Non-Coding RNA in Plants Modulation and Stress Responses Plant Biology, Sustainability and Climate Change*. Academic Press, New York.

Peng, Y., Zhang, X., Liu, Y., & Chen, X. (2020). Exploring Heat-Response Mechanisms of MicroRNAs based on microarray data of rice post-Meiosis panicle. International Journal of Genomics, 2020, 1–13. https://doi.org/10.1155/2020/7582612

Pooja, G., Megha, B., Arpan, M., Chintan, K., Rahul, D., Subhan, D., and Sachidanand, S. (2025). Introduction to long noncoding RNA. In Dutta, R., Kapadia, C., Danish, S., and Singh, S. (Eds), *Non-Coding RNA in Plants Modulation and Stress Responses Plant Biology, Sustainability and Climate Change*. Vol. 1, pp. 1–12. Academic Press, New York.

Rahul, D., Chintan, K., Subhan, D., and Sachidanand, S. (2024). *Non-Coding RNA in Plants: Modulation and Stress Responses*. Academic Press, New York.

Ramesh, P. (2000). Sugarcane Breeding Institute, Coimbatore, India effect of different levels of drought during the formative phase on growth parameters and its relationship with dry matter accumulation in sugarcane. *Journal of Agronomy and Crop Science* 185(2): 83–89.

Ramkumar, S., Baskar, V., Ashokraj, S., Sathishkumar, R., and Muthu, T. (2023). Functional role of microRNA in the regulation of biotic and abiotic stress in agronomic plants. *Frontiers in Genetics* 14(10): 1272446.

Redman, G. (2008). The Andersons Centre. "Assessment of on-farm AD in the UK". National Non-Food Crops Centre: 06–09.

Rivero, R. M., Mikiko, K., Amira, G., Hitoshi, S., Ron, M., Shimon, G., and Eduardo, B. (2007). Delayed leaf senescence induces extreme drought tolerance in a flowering plant. *Proceedings of the National Academy of Sciences of the United States of America* 104(49): 19631–19636.

Robertson, M. J., and Donaldson, R. A. (1998). Changes in the components of cane and sucrose yield in response to drying-off of sugarcane before harvest. *Field Crops Research* 55: 201–208.

Rodionova, M. V., Poudyal, R.S., Tiwari, I., Voloshin, R. A., Sergei, K. Z., Hong, G. N., Bolatkhan, K. Z., Barry, D. B., Harvey, J. M. H., and Suleyman, I. A. (2017). Biofuel production: challenges and opportunities. *International Journal of Hydrogen Energy* 42(12): 8450–8461.

Rodriguez, M., Canales, E., and Borras, H. O. (2005). Molecular aspects of abiotic stress in plants. *Biotecnologia Aplicada* 22: 110.

Sanan-Mishra, N., Kumar, V., Sopory, S. K., & Mukherjee, S. K. (2009). Cloning and validation of novel miRNA from basmati rice indicates cross talk between abiotic and biotic stresses. Molecular Genetics and Genomics, 282(5). https://doi.org/10.1007/s00438-009-0478-y

Sharma, A., Luis, M. R. M., Francisco, I. S. C., Paula, R. R. P., Cynthia, K., Esmeralda, B., Ana, K., Aashish, S., and Sujay, P. (2020). Identification of microRNAs and their expression in leaf tissues of guava (*Psidium Guajava* L.) under salinity stress. *Agronomy* 10(12): 1920.

Si, Q. H., Jie, P., Cheng, X. Q., and Zhi, M. Y. (2009). Heavy metal-regulated new microRNAs from rice. *Journal of Inorganic Biochemistry* 103(2): 282–287.

Silva, M. A., Rogerio, A. B. S., Marcos, G. A. L., and Mario, P. C. (2008). Agronomic performance of sugarcane families in response to water stress. *Bragantia* 67: 655–661.

Singels, A., Kennedy, A. J., and Bezuidenhout, C. N. (2000). The effect of water stress on sugarcane biomass accumulation and partitioning. *Proceedings of the South African Sugar Technologists Association* 74: 169.

Singh, S., and Rao, P. N. G. (1987). Varietal differences in growth characteristics in sugar cane. *The Journal of Agricultural Science* 108(1): 245–247.

Song, S., Yi, X., Dongmei, H., Muhammad, A. A., Jingyang, L., Wei, H., Zhiqiang, J., et al. (2018). Identification and characterization of miRNA169 family members in banana (*Musa Acuminata* L.) that respond to *fusarium oxysporum f.* sp. *cubense* infection in banana cultivars. *PeerJ* 6(12): e6209.

Subhan, D., Uzma, Y., Niaz, A., Khurram, S., Rahul, D., and Sachidanand, S. (2025). Fertilizer-responsive non-coding RNA and their significance. In Dutta, R., Kapadia, C., Danish, S., and Singh, S. (Eds), *Non-Coding RNA in Plants Modulation and Stress Responses Plant Biology, Sustainability and Climate Change*. Academic Press, New York.

Sun, G. C., Neal, S. J., Peng, X., and Baohong, Z. (2012). microRNA expression analysis in the cellulosic biofuel crop switchgrass (*Panicum virgatum*) under abiotic stress. *PLoS One* 7(3): e32017.

Sunkar, R., and Jian, K. Z. (2004). Novel and stress-regulated microRNAs and other small RNAs from Arabidopsis. *The Plant Cell* 16(8): 2001–2019.

Swapna, M., and Kumar, S. (2017). microRNAs and their regulatory role in sugarcane. *Frontiers in Plant Science* 8: 997.

Thiebaut, F., Cristian, A. R., Karla, L. A., Clicia, G., Giselli, C. D., Caren, R. C. L., Janice, D. A. E., Adriana, S. H., and Paulo, C. G. F. (2011). Regulation of MiR319 during cold stress in sugarcane. *Plant, Cell & Environment* 35(3): 502–512.

Tian, C., Chengzhe, Z., Chen, Z., Lan, C., Biying, S., Yuling, L., Zhongxiong, L., and Yuqiong, G. (2022). Genome-wide investigation of the MiR166 family provides new insights into its involvement in the drought stress responses of tea plants (*Camellia Sinensis* (L.) O. Kuntze). *Forests* 13(4): 628.

Trumbo, J. L., Baohong, Z., and Charles, N. S. (2015). Manipulating micro RNA s for improved biomass and biofuels from plant feedstocks. *Plant Biotechnology Journal* 13(3): 337–354.

Tungare, K., Shahu, R., Zambare, V., Agarwal, P., Jobby, R., Nisar, N., Alabdallah, N. M., Al-Saeed, F. A., Johri, P., Singh, S., Saeed, M., and Jha, P. (2022). Toxicity mitigation of textile dye reactive blue 4 by hairy roots of *helianthus annuus* and testing its effect in *in vivo* model systems. *BioMed Research International* 13: 1958939. https://doi.org/10.1155/2022/1958939

Vahap, E., Ufuk, C. A., Esma, O., Yakup, B., Serkan, U., and Turgay, U. (2012). Genome-wide identification of MiRNAs responsive to drought in peach (prunus persica) by high-throughput deep sequencing. *PLoS One* 7(12): e50298.

Vaucheret, H., Franck, V., Patrice, C., and David, P. B. (2004). The action of ARGONAUTE1 in the miRNA pathway and its regulation by the miRNA pathway are crucial for plant development. *Genes & Development* 18(10): 1187–1197.

Verma, N., Sao, P., Srivastava, A., and Singh, S. (2021). Physiological and molecular responses to drought, submergence and excessive watering in plants. In Husen, A. (Ed), *Harsh Environment and Plant Resilience* (pp. 305–321). Springer International Publishing, Cham.

Verma, S., Shadab, N., and Praveen, K. V. (2013). Biotic and abiotic stress signaling in plants. *Stress Signaling in Plants: Genomics and Proteomics Perspective* 1: 25–49.

Wang, W., Basia, J. V., Oded, S., and Arie, A. (2001). Biotechnology of plant osmotic stress tolerance physiological and molecular considerations. *International Journal of Molecular Sciences* 24(13): 10915.

Wei, B., Tao, C., Rongzhi, Z., Aili, L., Naxin, H., Shan, L., Yong, Q. et al. (2009). Novel microRNAs uncovered by deep sequencing of small RNA transcriptomes in bread wheat (*Triticum aestivum* L.) and *Brachypodium distachyon* (L.) Beauv. *Functional & Integrative Genomics* 9: 499–511.

Xie, F., Charles, N. S., Faten, A. T., Qiuling, H., Huawei, L., and Baohong, Z. (2014). High throughput deep sequencing shows that micro RNA s play important roles in switchgrass responses to drought and salinity stress. *Plant Biotechnology Journal* 12(3): 354–366.

Xie, F., Taylor, P. F., and Baohong, Z. (2010). Identification and characterization of microRNAs and their targets in the bioenergy plant switchgrass (*Panicum virgatum*). *Planta* 232: 417–434.

Yamaguchi, M., Misato, O., Nobutaka, M., Minoru, K., Masaru, O. T., Hiroo, F., and Taku, D. (2010). VND-INTERACTING2, a NAC domain transcription factor, negatively regulates xylem vessel formation in Arabidopsis. *The Plant Cell* 22(4): 1249–1263.

Yang, C., Dayong, L., Donghai, M., Xue, L., Chengjun, J., Xiaobing, L., Xianfeng, Z., Zhukuan, C., Caiyan, C., and Lihuang, Z. (2013). Overexpression of microRNA319, impacts leaf morphogenesis and leads to enhanced cold tolerance in rice (*Oryza Sativa* L.). *Plant, Cell & Environment* 36(12): 2207–2218.

Yang, Y., Xu, Z., Yachun, S., Jiake, Z., Zhoutao, W., Liping, X., and Youxiong, Q. (2017). miRNA alteration is an important mechanism in sugarcane response to low-temperature environment. *BMC Genomics* 18(1): 1–18.

Yang, Z. B., Xiaoyu, G., Chunmei, H., Feng, Z., Rong, W., Walter, J. H., and Zhaojun, D. (2014). TAA1-regulated local auxin biosynthesis in the root-apex transition zone mediates the aluminum-induced inhibition of root growth in Arabidopsis. *The Plant Cell* 26(7): 2889–2904.

Zanca, A. S., Vicentini, R., Ortiz-Morea, F. A., Del Bem, L. E. V., da Silva, M. J., Vincentz, M., and Nogueira, F. T. S. (2010). Identification and expression analysis of microRNAs and targets in the biofuel crop sugarcane. *BMC Plant Biology* 10(1): 1–13.

Zhang, B., and Qinglian, W. (2015). microRNA based biotechnology for plant improvement. *Journal of Cellular Physiology* 230(1): 1–15.

Zhang, H., Jinshan, Z., Jun, Y., Feng, G., Yanfei, M., Guiliang, T., Jose, R. B., and Jian, K. Z. (2017). Short tandem target mimic rice lines uncover functions of MiRNAs in regulating important agronomic traits. *Proceedings of the National Academy of Sciences of the United States of America* 114(20): 5277–5282.

Zhang, Y., Jing, X., Ruofan, L., Yanrui, G., Yufei, L., and Ruili, L. (2023). Plants' response to abiotic stress: mechanisms and strategies. *International Journal of Molecular Sciences* 24(13): 10915.

Zhang, Y., Xujun, Z., Xuan, C., Changnian, S., Zhongwei, Z., Yuhua, W., Mingle, W., Wanping, F., and Xinghui, L. (2014). Identification and characterization of cold-responsive microRNAs in tea plant (*Camellia Sinensis*) and their targets using high-throughput sequencing and degradome analysis. *BMC Plant Biology* 14(1): 271.

Zhao, J., Penghui, L., Tao, X., and Xiaochun, W. (2020). Exploring plant metabolic genomics: chemical diversity, metabolic complexity in the biosynthesis and transport of specialized metabolites with the tea plant as a model. *Critical Reviews in Biotechnology* 40(5): 667–688.

Zhao, Y., Zhenhua, X., Qiaocheng, M., Cheng, Z., Wenxue, L., Yunbi, X., and Chuanxiao, X. (2013). Combined small RNA and degradome sequencing reveals novel miRNAs and their targets in response to low nitrate availability in maize. *Annals of Botany* 112(3): 633–642.

12 Modulating Plant Responses
Epigenetic Perspectives on Climate Variability

Heeram Hashmi, Parul Johri, Manish Singh Rajput, and Mala Trivedi*

12.1 INTRODUCTION

Plant epigenetic regulation in response to stress is of immense significance in agriculture (Kooke et al., 2019). Plants are at the center of organismic interactions, always interacting—either cooperatively or competitively—with bacteria, nematodes, oomycetes, insects, fungi, and other plants. They have evolved complex regulatory mechanisms to modify the molecular processes required for forming advantageous linkages, as well as for defense against herbivores and pathogens. In addition to the primary defense provided by the canonical plant immune system, pathogens have begun to target epigenetic mechanisms as a potential regulatory strategy. It is essential to understand the subtle differences and mechanisms between the two types of epigenetic inheritance (Li et al., 2022). It is well known that epigenetic alterations significantly enhance the refunctioning and expression of genes within the plant transcriptome during stress responses. Modifications to the chromatin structure, such as "DM, histone modifications (HMs), and non-coding RNA molecules," facilitate these alterations (Noopur et al., 2025; Pooja et al., 2025; Rahul et al., 2024; Zhao et al., 2023).

Plant functions including growth, development, reproduction, and disease resistance depend on epigenetic controls, which also help plants become more resilient to environmental stresses like salt and temperature. It takes time to modify epigenetic-related phenomena to improve agricultural production, crop quality, and yield. These modifications promote sustainable agriculture. Epigenetic mechanisms also control important agronomic aspects in crops. These epigenetic pathways have a direct impact on agricultural productivity, yield, and quality (Kakoulidou et al., 2021).

Understanding the mechanisms underlying plant epigenetic regulation in stress tolerance can transform crop breeding and advance genetic engineering, aiding in the development of stress-tolerant crops and resulting in sustainable and improved food production. The precise meaning and scope of the phrase will probably remain open for discussion. Numerous epigenetic mechanisms have been identified, including sumoylation, ubiquitylation, acetylation, phosphorylation, and methylation. As research continues, other epigenetic mechanisms and factors are likely to come to light. Although epigenetic processes are normal and necessary for many biological functions, they can have serious negative impacts on behavior and health when they go awry (Omony, 2022).

12.2 UNDERSTANDING CLIMATE VARIABILITY AND ITS IMPACT ON PLANT BIOLOGY

Temperature and other environmental factors are often linked to the timing of phenological events like flowering. Thus, it is expected that shifting environments will result in changes to life cycles, and numerous plant species have shown evidence of these changes. Crop plant diseases will

* *Corresponding Author*

DOI: 10.1201/9781003491378-12

presumably rise as a result of climatic alterations. First, over the past few decades, globalization and trade have increased crop pathogen movement between continents, raising the possibility of disease transmission to disease-free areas. In new geographic locations, plant species or cultivars that have not coevolved with the introduced pathogen are likely to see increased pathogen prevalence and disease outbreaks (Guarino et al., 2022; Parveen et al., 2023).

The influence of climate change on plant diseases is intricate and challenging to foresee due to its relationship with various aspects of plants and the environment. These factors encompass the distribution and abundance of taxa (such as geographical range and niche preferences), the fitness and virulence of pathogens, interactions with the environment, evolutionary processes of plant–microorganism interactions, characteristics of hosts and vectors, and prevailing environmental conditions. Climate warming can notably affect the dynamics of pathogen populations, including overwintering and survival rates, population growth, and the number of generations in species with multiple life cycles. For instance, warmer temperatures can reduce the latency period of *Hemileia vastatrix*, the coffee leaf rust pathogen, thereby fostering epidemics in Central America. Moreover, higher temperatures can shorten the pathogen's incubation period, thereby increasing its prevalence throughout the growing season (Kumari et al., 2022).

Stress can influence flowering time in part due to epigenetic changes. These epigenetic modifications occur at various levels within the nucleosome via reversible biochemical reactions such as DNA methylation (DM) and histone tail modifications (Chandana et al., 2022). DM occurs when a methyl group is covalently added to cytosine in the DNA backbone, whereas HMs occur when specific lysine or arginine residues within the amino acid terminal tail of histones are post-translationally modified, either by acetylation or ADP-ribosylation (Papolu et al., 2022). Elevated average winter temperatures have exacerbated the infection rates of the American chestnut fungal pathogen *Phytophthora* spp., leading to severe tree mortality incidents across North America. Ultimately, a 30-year study found a correlation between higher rates of *Phacidium infestans* (snow blight) infection in pine trees and early snowmelt.

The prevalence of pathogens in natural and agricultural ecosystems can be significantly impacted by changes in global temperatures, which raises the danger of encountering new pests and diseases. Numerous fungal soil-borne plant pathogens are expected to become more prevalent as a consequence of global warming, which will have a major impact on primary productivity. Increased warmth has the potential to give rise to more virulent and better-adapted pathogen strains.

Climatic changes in humidity likely affect future plant disease outbreaks because variations in soil moisture and relative humidity are two primary factors influencing the abundance and infectivity of plant pathogens. High humidity is necessary for the spores of many fungal diseases to germinate and infect their host plants. In general, pathogens infecting aerial plant tissues become more virulent when humidity levels are high (Vaucheret & Voinnet, 2024).

The geographical spread of diseases due to climate change can be influenced by alterations in plant–pathogen interactions driven by drought across different levels of aridity. For example, while more favorable climate conditions may promote disease transmission, the reduced presence of alternate hosts at higher elevations can decrease the likelihood of infection. Conversely, drought conditions and increased tree mortality in arid regions may accelerate the decline of pine blister rust at lower elevations (Anil et al, 2023). Additionally, drought can lead to the emergence of new pathogens that are resilient to extreme weather and capable of exploiting changes in plant physiology triggered by stress responses. Understanding these dynamics is crucial for anticipating how climate change will reshape disease patterns and impact plant health across varied environmental conditions. It underscores the need for adaptive strategies that consider both the direct effects of climate on pathogens and the complex interplay with host plants and environmental factors. As the climate continues to evolve, monitoring and managing these interactions will be essential for mitigating the broader impacts on ecosystem health and agricultural productivity.

12.3 FUNCTION OF DNA METHYLATION IN MODIFYING CROP BEHAVIOR

DM plays a critical role in maintaining the stability of the plant genome by regulating the mobility of transposable elements (TEs). These elements are methylated across all sequence contexts in the genomes of all plant species, ranging from mosses to angiosperms. DM involves the addition of a methyl group to the C5 position of cytosine, resulting in the formation of 5-methylcytosine. Unlike mammals, where CG methylation predominates and non-CG methylation is restricted to specific tissues like embryonic stem cells, higher plants exhibit both CG and non-CG methylation. Non-CG methylation occurs in CHG (symmetric) and CHH (asymmetric) contexts, where H represents adenine, cytosine, or thymine. This broad pattern of DM in plants reflects its diverse regulatory roles in genome stability and gene expression, highlighting its significance across plant evolution and adaptation. Understanding these methylation dynamics is crucial for unraveling how plants respond to environmental cues and maintain genomic integrity amidst diverse ecological challenges (Manavella et al., 2023).

DM is an epigenetic modification essential for regulating various biological processes in plants, such as gene expression and genome stability. Understanding its role often involves studying mutants and using pharmacological interventions; however, the widespread effects across the genome can complicate linking methylation changes directly to specific phenotypes. To address this challenge, tools that enable precise manipulation of DM at specific loci have become invaluable. These tools not only help elucidate the precise functions of methylation in different biological processes but also facilitate the creation of novel epigenetic variants known as epialleles. Recent advancements include techniques that employ programmable DNA-binding proteins to recruit methylation machinery directly to target sites, greatly enhancing researchers' ability to manipulate and study DM dynamics in plants. This review provides a comprehensive exploration of DM mechanisms in plants, highlighting recent innovations in methodologies for its targeted manipulation and their implications for advancing our understanding of plant epigenetics (Halder et al., 2023).

Recent advancements and extensive ongoing research efforts have propelled us into the epigenetic era, succeeding the genetic era marked by the sequencing of the human genome. Epigenetics, literally meaning "beyond genetics," investigates heritable changes in gene expression that occur independently of alterations in the primary DNA sequence. These epigenetic modifications are essential for mammalian development and cellular proliferation, yet they can be disrupted in mature mammals due to stochastic factors or environmental influences. Epigenetic changes refer to inheritable variations in gene expression that are not encoded in the DNA sequence itself. Over the past decade, significant strides have been made in understanding these modifications during normal development and in disease conditions such as cancer. However, the epigenetic landscape has proven increasingly intricate, encompassing diverse mechanisms such as DM, the histone code, non-coding RNA regulation, nucleosome positioning, and the interplay with DNA sequence (Kapazoglou et al., 2021).

DM, facilitated by enzymes known as DNA methyltransferases (DNMTs), is particularly pivotal in the silencing of gene transcription through epigenetic mechanisms. These insights underscore the multifaceted nature of epigenetic regulation, highlighting its critical role in shaping gene expression patterns and influencing cellular function across different biological contexts. As research continues to unravel the complexities of epigenetic processes, novel therapeutic strategies and diagnostic tools may emerge, offering new avenues for understanding and treating various diseases linked to epigenetic dysregulation. The following are features of DM:

1. It contributes to the long-term silencing of genes.
2. It contributes to the repression of repetitive elements.
3. It contributes to the inactivation of the X chromosome.
4. It is used in the creation and maintenance of genes with imprints.
5. It suppresses the expression of the viral genome and other harmful elements that have gradually merged into the host's genome (Stability, 2020).

During the cell cycle, DNMTs catalyze the process at different stages:

1. DNMTI-Maintenance methylase in mammals.
2. DNMT 2.
3. DNMT 3b as well as 3a.
4. DNMT 3L.

DM plays a significant role in modulating various plant responses, including environmental adaptation and defense mechanisms. Here's an overview of its role:

1. **Gene Expression Regulation**: Methylation at promoter regions can directly influence the expression of genes by affecting the binding of transcription factors that alter chromatin structure, leading to either gene silencing or activation. Thus, DM acts as an epigenetic mark that regulates the expression of genes involved in stress responses, growth, and development (Khan et al., 2021).
2. **Environmental Stress Response**: Plants face numerous challenges in their environment, including drought, salinity, temperature fluctuations, and attacks from pathogens. These adversities can prompt dynamic changes in DM patterns within plants. These alterations in methylation have the potential to significantly impact the activation or suppression of stress-responsive genes, enabling plants to effectively adjust and thrive in unfavorable conditions. This flexibility in gene expression regulation through methylation changes underscores plants' remarkable ability to adapt and endure in diverse and challenging environments.
3. **Developmental Processes**: DM is involved in regulating various growth-related mechanisms in plants such as seed germination, embryogenesis, flowering, and senescence. Alterations in DM patterns during development impact the expression of genes related to developmental transitions and tissue differentiation.
4. **Transgenerational Inheritance**: DM can be heritable and can persist across multiple generations in plants. Environmental stimuli experienced by parent plants can induce changes in DM patterns, which can be transmitted to offspring and influence their phenotypic traits and stress responses (Chao et al., 2022). This phenomenon, known as epigenetic inheritance, provides a mechanism for crops to adapt to altered climatic conditions and environments over generations.
5. **Defense against Pathogens**: DM plays a crucial role in the defense of various plants against several pathogens. Pathogen infection can trigger changes in methylation patterns, leading to expression profiles associated with defense responses. DM-mediated silencing of TEs and defense-related genes can enhance plant resistance to pathogens.
6. **Epigenetic Regulation of TEs**: DM also plays a significant role in silencing TEs in plant genomes, thereby maintaining genome stability. TEs are mobile genetic elements that can cause genomic rearrangements and mutations. DM at TE sequences prevents their activation and transposition, thereby safeguarding genome integrity.

Overall, DM is a dynamic epigenetic mechanism that plays diverse roles in modulating plant responses to environmental cues, developmental processes, and defense against various environment-related stresses.

12.4 HISTONE MODIFICATIONS AND THEIR INFLUENCE ON PLANT STRESS RESPONSES

HMs are another crucial aspect of epigenetic regulation in plants, influencing various cellular processes, including stress responses. Around histones, the genetic material DNA is folded in a wrap fashion to generate the model called nucleosomes, which is the basic unit of chromatin.

Post-translational modifications of histone proteins such as ubiquitination, sumoylation, phosphorylation, methylation, and acetylation can alter chromatin structure and gene expression. Here's how HMs influence plant stress responses:

1. **Acetylation**: Acetylation generally correlates with gene regulation by relaxing the structure of chromatin, making the DNA more accessible to transcription factors and RNA polymerase. During stress responses, acetylation of histones at stress-responsive gene loci can enhance their transcription, leading to the activation of stress-related pathways. Conversely, deacetylation, catalyzed by histone deacetylases, can repress gene expression (Deng et al., 2022).
2. **Methylation**: Histone methylation can occur on different lysine and arginine residues and can create various effects on gene expression that depend on the specific lysine or arginine residue being modified and the degree of methylation. For example, trimethylation of histone H3 lysine 4 (H3K4me3) is linked with gene activation, while trimethylation of histone H3 lysine 27 (H3K27me3) is linked with gene repression. In response to stress, alterations in histone methylation profiles can modulate the expression of genes that contribute to stress in plants.
3. **Phosphorylation**: Histone phosphorylation is involved in regulating various cellular processes related to stress responses. Stress signaling pathways can lead to the phosphorylation of histones, which can alter chromatin structure and gene expression. For example, phosphorylation of histone H2AX is linked with DNA damage response and repair mechanisms under stress conditions.
4. **Ubiquitination and Sumoylation:** Histone ubiquitination and sumoylation involve the covalent attachment of ubiquitin or small ubiquitin-like modifier proteins to histone proteins, respectively. These modifications influence chromatin structure and gene expression. They are also implicated in plant stress responses, although their specific roles are less well-characterized compared to other HMs (Mladenov et al., 2021).
5. **Cross-Talk between Histone Modifications:** The extensive cross-talk between various HMs allows for intricate gene expression regulation in response to environmental challenges. For example, histone acetylation can promote the recruitment of histone methyltransferases or demethylases, thereby influencing histone methylation patterns and vice versa.

By modulating chromatin structure and accessibility to DNA, HMs alter the expression of genes responsible for causing and spreading the harmful and unaltered functioning of various plants, enabling them to adapt to changing environmental conditions. HMs can change chromatin structure by altering the binding nature of DNA to transcription factors and enzymes such as RNA polymerase. For example, acetylation of histones typically relaxes chromatin, making it more accessible for gene transcription. This increased accessibility allows stress-responsive genes to be activated more readily, promoting appropriate responses to environmental challenges. Certain HMs serve as binding sites for proteins involved in transcriptional regulation. The recruitment of regulatory proteins helps to modulate the expression of genes that induce stress in plants. HMs can confer epigenetic memory, allowing plants to "remember" past environmental exposures and mount more efficient responses to subsequent encounters with similar stresses. This memory is often mediated by the establishment and maintenance of specific HM patterns at stress-responsive gene loci, enabling rapid and precise activation of stress-related pathways (Ashapkin et al., 2020).

HMs are dynamically regulated in response to environmental cues. Stress signaling pathways can trigger changes in HM profiles, leading to rapid modifications in gene expression profiles. For example, exposure to drought stress may induce histone acetylation or methylation at the promoters of drought-responsive genes, enhancing their transcriptional activity and promoting stress tolerance. There can be extensive cross-talk among different HMs, allowing for intricate gene expression regulation in response to external environmental stimuli. For instance, histone acetylation can promote the recruitment of histone methyltransferases or demethylases, influencing histone methylation patterns and vice versa. This cross-talk enables plants to integrate multiple signals and fine-tune their responses to specific stress conditions. HMs can be heritable across cell divisions and sometimes even across generations, contributing to the transmission of stress memory in plants. Through mechanisms such as HM maintenance or replication-coupled deposition, plants can inherit specific HM patterns from parental cells or organisms, allowing for the inheritance of stress-induced traits (Agius et al., 2023).

Overall, HMs serve as critical epigenetic regulators of gene expression in plants, mediating their responses to environmental stresses by modulating chromatin structure and transcriptional activity at stress-responsive loci (Chao et al., 2022). This dynamic and finely tuned regulatory mechanism enables plants to adapt and thrive in diverse environments.

From an epigenetic perspective, the mechanism of HM in plant responses to environmental stimuli, including climate variability, involves intricate processes that regulate gene expression through chromatin remodeling. Here's a detailed breakdown of the mechanism:

1. **Sensing Environmental Signals:** Plants have sophisticated mechanisms to sense environmental cues such as temperature fluctuations, water availability, light intensity, and nutrient levels. These signals are perceived by receptors and signaling pathways, triggering cellular responses to adapt to changing conditions (Deng et al., 2022).
2. **Activation of HM Machinery:** Environmental signals activate specific signaling pathways that ultimately lead to changes in the activity of HM enzymes. These enzymes include histone acetyltransferases, some histone methyltransferases, histone demethylases, and histone deacetylases. Their activity is modulated in response to environmental cues, resulting in alterations in HM patterns.
3. **Chromatin Remodeling:** HMs alter the structure of chromatin, which consists of DNA wrapped around histone proteins. For example, acetylation of histones generally relaxes chromatin structure, making the DNA more accessible for transcriptional machinery. Conversely, methylation of histones can have activating or repressive effects on gene expression, depending on the specific histone residue and the degree of methylation (Mladenov et al., 2021).
4. **Regulation of Gene Expression:** HMs influence the interaction of DNA with transcription factors and RNA polymerase, thereby modulating gene expression. In response to environmental stimuli, specific genes involved in stress responses, such as those encoding stress-related transcription factors, osmoprotectants, and detoxification enzymes, may undergo changes in HMs, leading to their upregulation or downregulation (Yang, 2023).
5. **Epigenetic Memory:** HMs can confer epigenetic memory, allowing plants to "remember" past environmental exposures and mount more efficient responses upon subsequent encounters with similar stresses. This memory is often mediated by the establishment and maintenance of specific HM patterns at stress-responsive gene loci, enabling rapid and precise activation of stress-related pathways.
6. **Transgenerational Inheritance:** Some HMs can be heritable across cell divisions and even transmitted to offspring, contributing to the transmission of stress memory in plants. Through mechanisms such as HM maintenance or replication-coupled deposition, plants can inherit specific HM patterns from parental cells, allowing for the inheritance of stress-induced traits across generations (Jeynes-cupper & Catoni, 2023).

Overall, the mechanism of HM in plant responses to environmental stimuli involves dynamic changes in chromatin structure and gene expression mediated by the activity of HM enzymes. These epigenetic processes play a critical role in enabling plants to adapt and survive in fluctuating environmental conditions, including those resulting from climate variability.

12.5 siRNAs: REGULATORS OF EPIGENETIC CHANGES IN PLANTS

Small interfering RNAs (siRNAs) are significant regulators of epigenetic changes in plants, playing a crucial role in gene silencing and chromatin modification. Here's how siRNAs function as regulators of epigenetic changes in plants. siRNAs are short RNA molecules, typically 21–24 nucleotides, generated from dsRNA precursors by the Dicer-like enzyme. In plants, siRNAs can be derived from endogenous sources, such as TEs, repetitive sequences, or natural antisense transcripts, as well as from exogenous sources, such as viral RNA (Ashapkin et al., 2020). TEs are specialized genetic elements that can disrupt gene function and genome stability. siRNAs play a crucial role in silencing TEs by guiding the RNA-induced silencing complex (RISC) to complementary sequences within TE transcripts. Once bound to the target RNA, RISC cleaves the RNA molecule, leading to its degradation and preventing TE mobilization. siRNAs can also regulate gene expression by guiding DM and HM to specific genomic loci. siRNAs derived from repetitive sequences or other endogenous sources can target complementary sequences in the genome, leading to the recruitment of DNMTs and the establishment of DM at these loci. This DM can result in transcriptional gene silencing, effectively repressing gene expression. siRNAs can also mediate changes in chromatin structure through HMs. For example, siRNAs associated with heterochromatic regions can guide histone methyltransferases to these regions, leading to the deposition of histone marks that further reinforce transcriptional silencing. siRNAs can contribute to the establishment of epigenetic memory and the transmission of gene silencing across generations (Agius et al., 2023). Through mechanisms such as RNA-directed methylation and maintenance of HMs, siRNAs can maintain stable epigenetic states that persist through cell divisions and even across generations, ensuring the long-term stability of gene silencing and genome integrity. siRNAs serve as crucial regulators of the epigenetic modification process in plants, orchestrating gene silencing to maintain genome stability, regulating gene expression, and defending against invasive genetic elements such as TEs and viruses. Their role in epigenetic regulation highlights their importance in shaping plant growth, adaptation, and response to environmental stresses. The detailed mechanisms are as follows:

1. **siRNA Generation:** siRNAs are generated from dsRNA precursors, often originating from various sources, including TEs, repetitive sequences, or viral RNA.
2. **Recognition and Targeting:** These siRNAs recognize and bind to complementary sequences within the genome. They can target specific genomic loci, including promoter regions, TEs, or other regulatory sequences.
3. **RISC Complex Formation:** The siRNAs, along with other proteins, form the RISC (Jeynes-cupper & Catoni, 2023). This complex is guided by the siRNA to the targeted genomic loci.
4. **Gene Silencing:** The siRNA-loaded RISC complex can induce transcriptional gene silencing by directing DNMTs to methylate cytosine residues in the targeted genomic loci. This process, known as RNA-directed methylation, leads to heterochromatin formation and transcriptional repression. Alternatively, siRNAs can induce post-transcriptional gene silencing by instructing the RISC complex to cleave complementary mRNA transcripts, resulting in their degradation and inhibition of protein synthesis.
5. **Chromatin Remodeling:** Epigenetic changes induced by siRNAs can alter chromatin structure and organization. This includes changes in HMs, such as methylation or acetylation, which further contribute to transcriptional regulation and gene silencing (Papolu et al., 2022).
6. **Epigenetic Inheritance:** The epigenetic changes induced by siRNAs can be heritable across cell divisions, maintaining gene silencing states over time. This ensures the stability of transcriptional repression and the long-term regulation of gene expression patterns.

This higher-level mechanism illustrates how siRNAs serve as key regulators of epigenetic changes in plants, orchestrating transcriptional and post-transcriptional regulation through targeted HM and chromatin remodeling processes.

12.6 CHALLENGES AND FUTURISTIC APPROACHES IN EPIGENETIC RESEARCH FOR CLIMATE ADAPTATION

Epigenetics holds immense promise in understanding how organisms adapt to changing environments, including climate shifts. One challenge is to decipher the complex interplay between epigenetic modifications and siRNA regulation in response to climate stressors. This involves understanding different mechanisms of the epigenetic process and siRNA activity under varying environmental conditions. Identifying siRNAs that play crucial roles in climate adaptation is essential (Chandana et al., 2022). High-throughput sequencing technologies combined with bioinformatics approaches can help identify siRNAs that are differentially expressed in response to environmental cues (Verma et al., 2023). Integrating epigenetic profiling data can provide insights into the regulatory mechanisms controlling these siRNAs. Futuristic approaches may involve using CRISPR-based epigenome editing tools to manipulate epigenetic marks associated with climate-responsive siRNAs. This could enable researchers to engineer crops with enhanced stress tolerance by modulating the expression of specific siRNAs through targeted epigenome modifications. Integrating epigenomic data can provide a better understanding of the molecular mechanisms underlying climate adaptation mediated by siRNAs. Advanced computational methods, such as machine learning algorithms, can be employed to analyze multi-omics datasets and identify key regulatory networks involved in siRNA-mediated stress responses (Kumari et al., 2022). Environmental factors can influence epigenetic modifications and siRNA function, a phenomenon known as transgenerational epigenetic change. Studying the transgenerational effects of climate stress on siRNA-mediated adaptation requires innovative experimental designs and long-term monitoring of plant populations exposed to varying environmental conditions. Synthetic biology techniques can be used to engineer synthetic siRNAs with enhanced stability and efficacy under adverse climatic conditions. Rational design and optimization of synthetic siRNAs can enable precise manipulation of gene expression networks to confer climate resilience in crops and other organisms. As with any emerging technology, there are ethical and regulatory considerations associated with the epigenetic manipulation of siRNAs for climate adaptation. Addressing concerns related to environmental safety, intellectual property rights, and equitable access to technology is essential for the responsible development and deployment of epigenetic interventions in agriculture and conservation efforts (Ray et al., 2022).

By addressing these challenges and leveraging futuristic approaches, epigenctic research on siRNA-mediated climate adaptation can contribute to the development of resilient crops and ecosystems capable of withstanding the challenges posed by climate change (Rymen et al., 2020).

Certainly, focusing specifically on plant response modulation through epigenetic mechanisms involving siRNA for climate adaptation poses its own set of challenges and futuristic approaches:

12.6.1 Challenges

1. **Complexity of Epigenetic Regulation:** Plant responses to climate change involve intricate regulatory networks governed by epigenetic modifications and siRNA-mediated gene regulation. Understanding these networks and deciphering their specific roles in climate adaptation poses a significant challenge (Guarino et al., 2022).
2. **Environmental Variability:** Climate change introduces unpredictable environmental variability, including temperature fluctuations, water availability, and biotic stresses, which can influence epigenetic modifications and siRNA expression patterns in plants. Capturing the dynamic nature of these responses under varying environmental conditions is challenging.

3. **Species-Specific Responses:** Different plant species exhibit unique responses to climate stressors, mediated by distinct epigenetic mechanisms and siRNA pathways. Investigating these species-specific responses and identifying conserved regulatory elements across diverse plant taxa require comprehensive comparative genomics and functional studies (Holoch et al., 2015).
4. **Heritability and Stability:** Ensuring the heritability and stability of epigenetic modifications and siRNA-mediated traits across generations presents a challenge. Translating induced epigenetic changes into stable and heritable phenotypes that endure under changing environmental conditions remains a major hurdle.
5. **Delivery and Targeting:** Efficient delivery of epigenetic modifiers or siRNA-based therapeutics to target tissues in plants poses technical challenges. Developing systems that enable precise targeting and uptake of epigenetic modulators or siRNAs into plant cells while minimizing off-target effects is essential (Omony, 2022).

12.6.2 Futuristic Approaches

1. **Genome Editing Technologies:** Advancements in genome editing technologies, such as CRISPR/Cas-based epigenome editing, offer futuristic approaches for precisely modulating plant responses to climate change. Targeted manipulation of epigenetic marks or siRNA pathways associated with stress tolerance genes could enhance resilience to environmental challenges (Elvira-matelot et al., 2016).
2. **Epigenome Engineering:** Engineering synthetic epigenetic regulators or siRNAs with enhanced stability and specificity holds promise for fine-tuning plant responses to climate stressors. Designing customized epigenetic modulators that can dynamically regulate gene expression in response to changing environmental cues represents a futuristic approach (Kakoulidou et al., 2021).
3. **Systems Biology and Multi-Omics Integration:** Integrating systems biology approaches with multi-omics data analysis can provide holistic insights into the regulatory networks governing plant responses to climate change (Guo et al., 2016). Applying advanced computational models to unravel complex interactions between epigenetic modifications, siRNA pathways, and environmental factors is crucial for predictive modeling of plant adaptation.
4. **Epitranscriptomics:** Exploring the role of RNA modifications (epitranscriptomics) in modulating siRNA biogenesis and function offers promising avenues for understanding plant responses to climate change. Characterizing RNA modifications and their impact on siRNA-mediated gene regulation under stress conditions could reveal novel mechanisms of adaptation (Zhai et al., 2008).
5. **Environmental Epigenetics and Ecological Engineering:** Integrating environmental epigenetics with ecological engineering enables the design of resilient ecosystems capable of mitigating the impacts of climate change. Utilizing epigenetic mechanisms to promote stress tolerance in key plant species and engineering diverse plant communities for ecosystem resilience represents a futuristic strategy for sustainable adaptation.

Addressing these challenges and implementing futuristic approaches in epigenetic research to modulate plant responses to climate change through siRNA-mediated mechanisms holds immense potential for enhancing agricultural productivity, preserving biodiversity, and ensuring food security in a changing climate (Akhter et al., 2021).

12.7 CONCLUSION

In conclusion, the modulation of plant responses to climate variability through epigenetic mechanisms represents a dynamic and promising field of research with significant implications for agriculture, ecology, and environmental sustainability. Epigenetic processes, which include

DM, histone modification, and sRNA-regulated gene regulation, play crucial roles in mediating plant reactions to climate variability. Understanding the intricate interplay between these epigenetic processes and environmental cues is essential for deciphering the molecular basis of plant adaptation (Chen et al., 2018). Despite recent advances, challenges such as the complexity of regulatory networks, species-specific responses, and the heritability of induced epigenetic changes remain significant hurdles in the field. Addressing these challenges requires interdisciplinary approaches that integrate genomics, bioinformatics, and functional genomics techniques. Future research directions encompass innovative approaches such as genome editing technologies, epigenome engineering, and systems biology integration to manipulate plant responses to climate variability. Synthetic biology tools, including synthetic epigenetic regulators and designer siRNAs, offer promising avenues for tailoring plant traits to withstand environmental stresses. Integration of multi-omics data, including genomics, epigenomics, transcriptomics, and metabolomics, provides comprehensive insights into the molecular mechanisms underlying plant adaptation to climate variability (Zhao et al., 2023). Advanced computational methods and predictive modeling techniques facilitate the elucidation of complex regulatory networks governing stress responses. Leveraging environmental epigenetics in ecological engineering strategies enables the design of resilient ecosystems capable of mitigating the impacts of climate change. Promoting biodiversity conservation and ecosystem stability through epigenetically mediated plant adaptations contributes to sustainable agriculture and ecosystem resilience (Guyot et al., 2022).

In summary, epigenetic perspectives on modulating plant responses to climate variability offer novel insights into the molecular basis of adaptation and resilience in natural and agricultural systems (Li et al., 2022). By addressing current challenges and embracing futuristic approaches, epigenetic research holds immense potential for enhancing crop productivity, preserving biodiversity, and fostering environmental sustainability in the face of ongoing climate change. Epigenetic modifications and small RNA-mediated pathways are integral components of plants' adaptive strategies. By tapping into these natural mechanisms, we can unlock the potential to develop crops and ecosystems resilient to the challenges posed by climate variability (Gangwar et al., 2023).

Success in this endeavor hinges on collaborative efforts across disciplines. Integrating expertise from genomics, epigenetics, bioinformatics, agronomy, and ecology is crucial for enhancing our understanding of plant–environment interactions and translating research findings into practical solutions. Futuristic approaches, including genome editing, synthetic biology, and multi-omics integration, offer unprecedented opportunities to manipulate plant responses to climate variability with precision and efficacy (Kooke et al., 2019). Embracing innovation and leveraging cutting-edge technologies will drive progress in epigenetic research. Epigenetic insights not only hold promise for enhancing agricultural productivity and food security but also for conserving biodiversity and restoring degraded ecosystems. By promoting resilient plant traits and fostering ecosystem stability, we can address the dual challenges of climate change and environmental degradation. As we embark on this journey, it is necessary to evaluate the legal, social, and regulatory implications of the epigenetic phenomenon in agriculture and conservation. Ensuring equitable access to emerging technologies and safeguarding environmental integrity are central tenets of responsible innovation. In essence, modulating plant responses to climate variability through epigenetic perspectives represents a forward-thinking approach to tackling one of the most pressing challenges of our time. By embracing nature's adaptive toolkit and pushing the boundaries of scientific inquiry, we can pave the way for a more resilient and sustainable future for generations to come (Ashra et al., 2020).

In conclusion, exploring epigenetic mechanisms to modulate plant responses to climate variability presents a transformative avenue in agricultural and ecological sciences. As we navigate the complexities of a changing climate, understanding how plants perceive, interpret, and respond to environmental cues at the epigenetic level is paramount.

REFERENCES

Agius, D. R., Kapazoglou, A., Avramidou, E., Baranek, M., Carneros, E., Caro, E., Castiglione, S., Cicatelli, A., Radanovic, A., Ebejer, J., Gackowski, D., Guarino, F., Hidve, N., Hoenicka, H., Gulya, A., Johannes, F., Karalija, E., Lieberman-lazarovich, M., & Martinelli, F. (2023). Exploring the crop epigenome: a comparison of DNA methylation pro fi ling techniques. *Frontiers in Plant Science*, *14*, 1181039. https://doi.org/10.3389/fpls.2023.1181039

Akhter, Z., Bi, Z., Ali, K., Sun, C., Fiaz, S., Haider, F. U., & Bai, J. (2021). In response to abiotic stress, DNA methylation confers epigenetic changes in plants. *Plants*, *10*, 1–21.

Anil, P., Dharmendra, P., Sachidanand, S., Yuriy, E., Snezhan, B., Dilfuza, J., & Chinmayi, J. (2023). Beneficial effects of biochar application on mitigating the drought and salinity stress implications on plants. *Plant Science Today*, *10*(2), 188–193. http://doi.org/10.14719/pst.2591

Ashapkin, V. V, Kutueva, L. I., Aleksandrushkina, N. I., & Vanyushin, B. F. (2020). Epigenetic mechanisms of plant adaptation to biotic and abiotic stresses. *International Journal of Molecular Sciences*, *21*(20), 1–33.

Ashra, M., Zarrabi, A., Hushmandi, K., Hashemi, F., Moghadam, E. R., Raei, M., Kalantari, M., Tavakol, S., Mohammadinejad, R., Naja, M., Tay, F. R., & Makvandi, P. (2020). Progress in natural compounds/siRNA Co-delivery employing nanovehicles for cancer therapy. *ACS Combinatorial Science*, *22*(12), 669–700. https://doi.org/10.1021/acscombsci.0c00099

Chandana, B. S., Mahto, R. K., Singh, R. K., & Ford, R. (2022). Epigenomics as potential tools for enhancing magnitude of breeding approaches for developing climate resilient chickpea. *Frontiers in Genetics*, *13*, 1–25. https://doi.org/10.3389/fgene.2022.900253

Chao, H., Hu, Y., Zhao, L., Xin, S., Ni, Q., & Zhang, P. (2022). Biogenesis, functions, interactions, and resources of non-coding RNAs in plants. *International Journal of Molecular Science*, *13*, 1–25.

Chen, W., Zhang, X., Fan, Y., Li, B., Ryabov, E., Shi, N., & Zhao, M. (2018). A genetic network for systemic RNA silencing in plants. *Plant Physiology*, *176*, 2700–2719. https://doi.org/10.1104/pp.17.01828

Deng, Z., Ma, L., Zhang, P., & Zhu, H. (2022). Small RNAs participate in plant – virus interaction and their application in plant viral defense. *International Journal of Molecular Science*, *23*, 696.

Elvira-matelot, E., Shamandi, N., Comella, P., & Zytnicki, M. (2016). Arabidopsis RNASE THREE LIKE2 modulates the expression of protein-coding genes via 24-nucleotide small interfering RNA-di33rected DNA Methylation. *Plant Cell*, *28*, 406–425. https://doi.org/10.1105/tpc.15.00540

Gangwar, P., Johri, P., Dutta, R., Singh, S., & Trivedi, M. (2023). Overview of phytopharmaceuticals and biotechnology of herbal plants. *Phytopharmaceuticals and Biotechnology of Herbal Plants*, *1*, 1–17

Guarino, F., Cicatelli, A., Castiglione, S., Agius, D. R., Orhun, G. E., Fragkostefanakis, S., Leclercq, J., & Dobránszki, J. (2022). An epigenetic alphabet of crop adaptation to climate change. *Frontiers in Genetics*, *13*, 1–18. https://doi.org/10.3389/fgene.2022.818727

Guo, Q., Liu, Q., Smith, N. A., Liang, G., & Wang, M.-B. (2016). Bentham science impact factor: 2.43 RNA silencing in plants: mechanisms, technologies and applications in horticultural crops. *Current Genomics*, *17*(6), 476–489. https://doi.org/10.2174/1389202917666160520103117

Guyot, V., Rajeswaran, R., Chu, H. C., Karthikeyan, C., Laboureau, N., Galzi, S., Mukwa, L. F. T., Krupovic, M., Kumar, P. L., & Pooggin, M. M. (2022). A newly emerging alphasatellite affects banana bunchy top virus replication, transcription, siRNA production and transmission by aphids. *PLoS Pathogens*, *18*(4), e1010448. https://doi.org/10.1371/journal.ppat.1010448

Halder, K., Chaudhuri, A., Abdin, M. Z., & Datta, A. (2023). Tweaking the small non-coding RNAs to improve desirable traits in plant. *International Journal of Molecular Science*, *1*, 1–26.

Holoch, D., Moazed, D., & Avenue, L. (2015). RNA-mediated epigenetic regulation of gene expression. *Nature Reviews Genetics*, *16*(2), 71–84. https://doi.org/10.1038/nrg3863

Jeynes-cupper, K., & Catoni, M. (2023). Long distance signalling and epigenetic changes in crop grafting. *Frontiers in Plant Science*, *14*, 1121704. https://doi.org/10.3389/fpls.2023.1121704

Kakoulidou, I., Avramidou, E. V, Bar, M., Brunel-muguet, S., Farrona, S., Johannes, F., Kaiserli, E., Lieberman-lazarovich, M., & Martinelli, F. (2021). Epigenetics for crop improvement in times of global change. *Biology*, *10*(8), 766.

Kapazoglou, A., Tani, E., Avramidou, E. V, Abraham, E. M., Gerakari, M., Megariti, S., Doupis, G., & Doulis, A. G. (2021). Epigenetic changes and transcriptional reprogramming upon woody plant grafting for crop sustainability in a changing environment. *Frontiers in Plant Science*, *11*, 1–16. https://doi.org/10.3389/fpls.2020.613004

Khan, P., Siddiqui, J. A., Lakshmanan, I., Ganti, A. K., Salgia, R., Jain, M., Batra, S. K., & Nasser, M. W. (2021). RNA-based therapies: a cog in the wheel of lung cancer defense. *Molecular Cancer*, *20*, 1–24.

Kooke, R., Morgado, L., Becker, F., Eekelen, H. Van, Hazarika, R., Zheng, Q., Vos, R. C. H. De, Johannes, F., & Keurentjes, J. J. B. (2019). Epigenetic mapping of the Arabidopsis metabolome reveals mediators of the epigenotype-phenotype map. *Genome Research, 29*, 96–106. https://doi.org/10.1101/gr.232371.117

Kumari, P., Khan, S., Wani, I. A., Gupta, R., & Verma, S. (2022). Unravelling the role of epigenetic modifications in development and reproduction of angiosperms: a critical appraisal. *Frontiers in Genetics, 13*(May), 1–23. https://doi.org/10.3389/fgene.2022.819941

Li, C., Gent, J. I., Xu, H., Fu, H., Russell, S. D., & Sundaresan, V. (2022). Resetting of the 24-nt siRNA landscape in rice zygotes. *Genome Research, 35*(6), 309–323. https://www.genome.org/cgi/doi/10.1101/gr.275981.121

Manavella, P. A., Herz, M. A. G., Kornblihtt, A. R., Sorenson, R., Sieburth, L. E., Nakaminami, K., Seki, M., Ding, Y., Sun, Q., Kang, H., Ariel, F. D., Crespi, M., Giudicatti, A. J., Cai, Q., Agrobiotecnología, I. De, Biología, C. De, & Bioquímica, F. De. (2023). Beyond transcription: compelling open questions in plant RNA biology. *The Plant Cell, 35*(6), 1626–1653. https://doi.org/10.1093/plcell/koac346

Mladenov, V., Fotopoulos, V., Kaiserli, E., Karalija, E., Maury, S., & Gašparovi, M. (2021). Deciphering the epigenetic alphabet involved in transgenerational stress memory in crops. *International Journal of Molecular Sciences, 22*(13), 7118.

Noopur, K., Pragati, K., Megha, B., Arpan, M., Tanvi, J., Shrisendu, M., Adarsh, K. S., Sachidanand, S. (2025). Role of non-coding RNA in plant architecture. In Dutta, R., Kapadia, C., Danish, S., & Singh, S. (Eds), *Non-Coding RNA in Plants Modulation and Stress Responses Plant Biology, Sustainability and Climate Change*. Academic Press, New York.

Omony, J. (2022). DNA methylation analysis in plants: review of computational tools and future perspectives. *Briefings in Bioinformatics, 21*(3), 906–918. https://doi.org/10.1093/bib/bbz039

Papolu, P. K., Ramakrishnan, M., Mullasseri, S., Kalendar, R., Wei, Q., Zou, L. H., Ahmad, Z., Vinod, K. K., & Yang, P. (2022). Retrotransposons: how the continuous evolutionary front shapes plant genomes for response to heat stress. *Frontiers in Plant Science, 13*, 1–23. https://doi.org/10.3389/fpls.2022.1064847

Parveen, R., Khare, N., Singh, S., & Johri, P. (2023). An in silico outlook for the detection and surveillance of evolving and persistent plant pathogens. In Singh, U. B. et al. (Eds), *Detection, Diagnosis and Management of Soil-borne Phytopathogens*, pp. 17–40. Springer, Singapore.

Pooja, G., Megha, B., Arpan, M., Chintan, K., Rahul, D., Subhan, D., & Sachidanand, S. (2025). Introduction to long noncoding RNA. In Dutta, R., Kapadia, C., Danish, S., & Singh, S. (Eds), *Non-Coding RNA in Plants Modulation and Stress Responses Plant Biology, Sustainability and Climate Change*. Vol. 1, pp. 1–12. Academic Press, New York.

Rahul, D., Chintan, K., Subhan, D., & Sachidanand, S. (2024). *Non-Coding RNA in Plants: Modulation and Stress Responses*. Academic Press, New York.

Ray, P., Sahu, D., Aminedi, R., & Chandran, D. (2022). Concepts and considerations for enhancing RNAi efficiency in phytopathogenic fungi for RNAi-based crop protection using nanocarrier-mediated dsRNA delivery systems. *Frontiers in Fungal Biology, 3*(September), 1–25. https://doi.org/10.3389/ffunb.2022.977502

Rymen, B., Ferrafiat, L., & Blevins, T. (2020). Non-coding RNA polymerases that silence transposable elements and reprogram gene expression in plants. *Transcription, 11*(3–4), 172–191. https://doi.org/10.1080/21541264.2020.1825906

Stability, G. (2020). HHS public access. *Chemical Reviews, 118*(8), 4365–4403. https://doi.org/10.1021/acs.chemrev.7b00487

Vaucheret, H., & Voinnet, O. (2024). The plant siRNA landscape. *The Plant Cell, 36*(2), 246–275. https://doi.org/10.1093/plcell/koad253

Verma, A., Kumar, R., Shukla, S., Singh, S., Johri, P. & Singh, M.S. (2023). Genome wide association studies: an outlook. *Journal of Innovation in Applied Research, 6*(1), 56–64. https://doi.org/10.51323/JIAR.2023/6/1/56-62

Yang, J. (2023). Drug conjugates for targeting regulatory T cells in the tumor microenvironment: guided missiles for cancer treatment. *Experimental & Molecular Medicine, 55*(9), 1996–2004. https://doi.org/10.1038/s12276-023-01080-3

Zhai, J., Liu, J., Liu, B., Li, P., Meyers, B. C., Chen, X., & Cao, X. (2008). Small RNA-directed epigenetic natural variation in *Arabidopsis thaliana*. *PLoS Genetics, 4*(4), e1000056. https://doi.org/10.1371/journal.pgen.1000056

Zhao, Y., Liu, G., Yang, F., Liang, Y., Gao, Q., Xiang, C., Li, X., & Yang, R. (2023). Multilayered regulation of secondary metabolism in medicinal plants. *Molecular Horticulture, 3*, 1–24. https://doi.org/10.1186/s43897-023-00059-y

13 Emerging Role of miRNA in Genome Stability and Editing

*Stuti Maurya and Sayali Mukherjee**

13.1 INTRODUCTION

The realm of plant biology is an intricate and ever-changing system that relies on a highly advanced molecular network to control the coordination of reactions to environmental signals. Among the essential regulators that play a significant role in shaping plant growth, development, and responses to these signals are microRNAs (miRNAs). Since the identification of the initial miRNA, lin-4, in *Caenorhabditis elegans* in 1993, the investigation of miRNAs has transformed our understanding of gene regulation and molecular biology. It has helped us to understand their essential functions in shaping various biological processes across different organisms (Xu et al., 2019). The process of miRNA formation begins with the transcription of a primary miRNA (pri-miRNA) transcript by RNA polymerase II, which usually consists of several hundred nucleotides. The initial transcript adopts a distinct hairpin conformation, which is identified and modified by Dicer-like (DCL) proteins to produce precursor miRNAs (pre-miRNAs) that are around 70 nucleotides in length. Subsequently, the pre-miRNAs are transported from the nucleus to the cytoplasm, where they undergo additional processing by DCLs to produce mature miRNAs, which are usually 19–25 nucleotides long. The significance of miRNAs in coordinating reactions to environmental stresses and regulating plant adaptability to evolving ecological circumstances has been increasingly recognised (Wani et al., 2020).

Plants encounter various obstacles in their native environments, such as lack of water, high salt levels, harsh temperatures, and attacks from pathogens, which can significantly impact agricultural output and food supply security. Plants have developed intricate mechanisms to detect and respond to these environmental signals, allowing them to adjust and thrive in different ecological habitats. The regulatory capacities of miRNAs are pivotal in these adaptive responses, enabling plants to change gene expression precisely and synchronously. Through the selective degradation or translational suppression of particular messenger RNAs (mRNAs), miRNAs can regulate the expression of genes involved in stress signalling, osmotic regulation, and antioxidant defence. This mechanism strengthens a plant's ability to withstand environmental challenges, increasing its resilience. Furthermore, miRNAs also play a role in controlling developmental processes that impact plant adaptability to environmental conditions, such as root growth, timing of blooming, and reproductive success. The functional importance of miRNAs in stress responses and ecological adaptation has generated considerable interest in utilising their regulatory capacity to enhance crop resilience and yield. Researchers aim to genetically modify crops by controlling the production of stress-responsive miRNAs to create crops with an enhanced ability to withstand environmental challenges, thereby increasing agricultural productivity and sustainability (Patil et al., 2021).

The emergence of gene editing tools, particularly CRISPR/Cas9, has expanded the possibilities of precise genetic modification, providing exceptional prospects for incorporating miRNA-mediated pathways into strategies for enhancing crops. This chapter seeks to provide a detailed analysis of the emerging field of miRNA-mediated genome stability and editing. It will examine the potential collaboration between miRNAs and gene editing technologies to enhance the ability of crops to withstand climate-related difficulties. This narrative aims to contribute to the

* *Corresponding Author*

DOI: 10.1201/9781003491378-13

discussion on sustainable agriculture and the urgent need for innovative solutions to ensure global food security in the face of climate change. It explores the complexities of miRNA biogenesis, processing pathways, and their connection to gene editing tools. Researchers are using experimental studies, computational modelling, and bioinformatics analysis to understand the intricate process of miRNA-mediated gene regulation and its potential impact on crop enhancement (Abbas et al., 2022). Through our expanding comprehension of miRNA functionality and control, we may utilise the potential of miRNAs to manipulate crops with improved durability, productivity, and nutritional value. This approach enables us to tackle the challenges of global food security and sustainable agriculture in the 21st century with innovative solutions.

13.2 miRNA BIOSYNTHESIS IN PLANTS

miRNAs are a class of small RNA molecules that play a crucial role in regulating gene expression in plants after transcription. miRNAs do not code for proteins but function as post-transcriptional gene expression regulators. The biogenesis of miRNAs is a complex and tightly regulated process that involves several phases and a range of protein components. The process of miRNA biogenesis begins with the transcription of specific non-coding RNA sequences, known as MIR genes, by RNA polymerase II in the nucleus. The MIR genes generate pri-miRNAs transcripts, which have a similar structure to protein-coding mRNAs. The regulation of MIR genes at the transcriptional level is crucial in defining the quantity and specificity of miRNA expression. Various transcription factors and epigenetic changes influence this regulation, ensuring the precise management of miRNA production. This allows plants to fine-tune their gene expression patterns in response to developmental signals and environmental cues.

After transcription, pri-miRNAs undergo complex processing stages to produce mature, functional miRNAs. Pri-miRNA processing occurs within the nucleus and involves the participation of multiple protein complexes. DCL1 is the prominent enzyme processing pri-miRNAs, working alongside the RNA-binding protein HYL1 and the zinc finger protein SE. Collectively, these proteins enzymatically cleave pri-miRNAs to produce pre-miRNAs characterised by their hairpin-shaped structures. The accuracy of this processing step is essential for generating functional miRNAs with appropriate target specificity. Recent research indicates that the formation of the RNA-induced silencing complex (RISC), which is crucial for gene silencing mediated by miRNA, may occur in the nucleus with the aid of a protein known as EXPO1. The localisation of RISC assembly components within the nucleus emphasises the spatial control of miRNA production and the intricate nature of the process.

After processing in the nucleus, pre-miRNAs are transported to the cytoplasm for further maturation. HASTY proteins facilitate the nuclear export of pre-miRNAs, playing a role similar to that of exportin-5 in mammals. The accurate intracellular positioning of miRNA biogenesis components ensures spatial and temporal control of miRNA synthesis, enabling adaptable responses to developmental and environmental stimuli. DCL1 performs additional processing of pre-miRNAs in the cytoplasm, resulting in the production of mature miRNAs. These mature miRNAs are characterised by their distinctive double-stranded complexes and specific features at the 3′ end. The stability of mature miRNAs is governed by proteins such as Hua enhancer1, which methylates the 3′ ends of miRNAs, protecting them against exonuclease degradation. Once mature, miRNAs are incorporated into the RISC complex, the active complex responsible for gene silencing mediated by miRNA. The formation of the RISC complex is a carefully controlled process involving multiple auxiliary proteins and cofactors; the structure and behaviour of the RISC complex impact the effectiveness and accuracy of gene control through miRNA. Mature miRNAs within the RISC complex facilitate the recognition and binding of target mRNAs through the complementarity of sequences.

The interaction between miRNAs and target mRNAs can result in several outcomes, such as the enzymatic cleavage of the mRNA, the prevention of translation initiation, and the removal of the poly(A) tail from the target RNA. miRNAs provide a wide range of regulatory capabilities

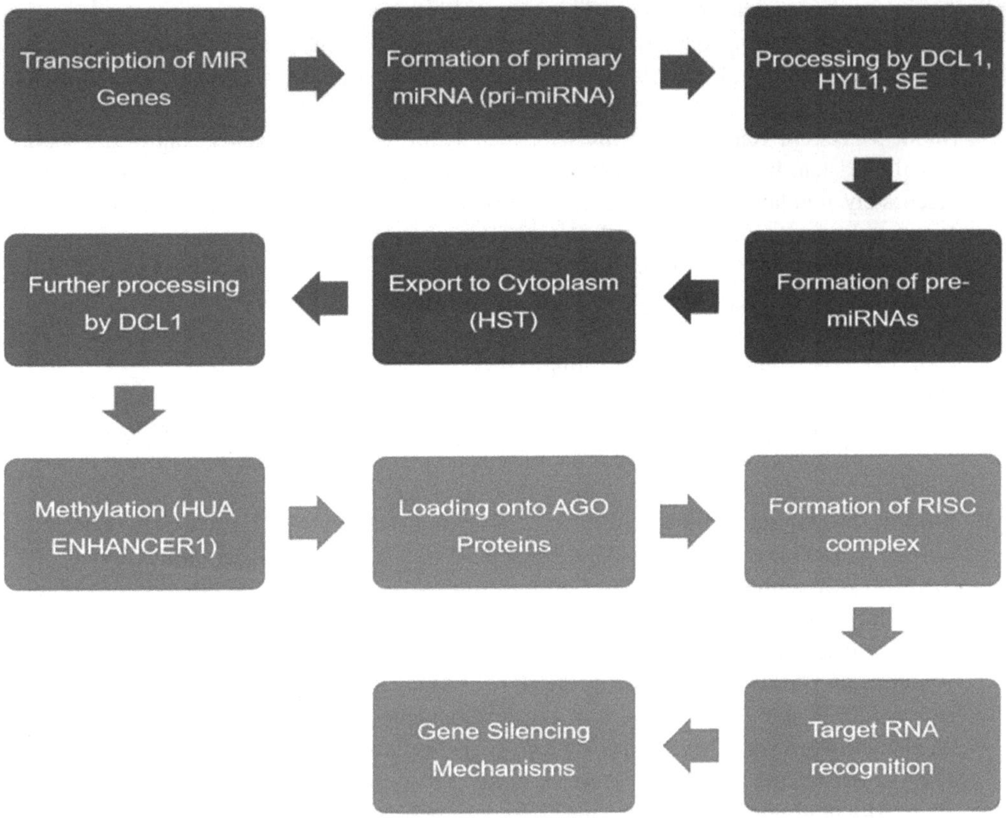

FIGURE 13.1 The significant steps in miRNA biosynthesis.

that allow them to precisely adjust gene expression networks involved in various physiological processes, such as plant development, growth, and responses to environmental stimuli. Besides their involvement in gene silencing, miRNAs have also been linked to additional regulatory processes that enhance their functional range. For instance, miRNAs can inhibit translation initiation, induce deadenylation, and destabilise target mRNAs, exerting diverse regulatory effects on gene expression. miRNAs have regulatory impacts beyond specific gene targets and affect broader cellular processes and pathways. Furthermore, miRNAs have demonstrated their involvement in RNA interference (RNAi) mechanisms, where they cooperate with other small RNA molecules, such as small interfering RNAs (siRNAs), to enhance gene silencing effects and regulate gene expression at both the transcriptional and post-transcriptional levels. Incorporating miRNA-mediated gene regulation into RNAi pathways introduces additional complexity to the regulatory networks that govern plant physiology. Figure 13.1 shows the significant steps in miRNA biosynthesis. Understanding the complexities of miRNA production and regulation is crucial for deciphering the regulatory mechanisms underlying plant development, growth, and environmental responses.

13.3 ROLE OF miRNA IN PLANT DEVELOPMENT

miRNAs are small, non-coding RNA molecules that play a significant role in plant development by regulating multiple processes essential for growth, shape formation, and the plant's ability to respond to environmental cues (Noopur et al., 2025; Pooja et al., 2025; Rahul et al., 2024). These small RNA molecules delicately restrict gene expression to enable the proper progression of developmental programmes, from shoot meristem control to root organ development, and facilitate adjustments to fluctuating environmental conditions. This extensive examination delves deeper into the various roles of miRNAs, including their specific functions in influencing plant structure, orchestrating changes in development, and facilitating adjustments to fluctuating environmental conditions.

- **Shoot Meristem Regulation**: miRNAs, such as miR394, significantly control the shoot apical meristem by targeting genes involved in the STM-WUS-CLV signalling pathway. Accurately regulating miRNA levels is essential for preserving shoot apical meristem functionality and facilitating the development of aboveground structures.
- **Leaf Development**: miRNAs, including miR160, miR165/166, miR319, and miR396, influence leaf development by regulating the differentiation of leaf primordia, thereby affecting leaf size and structure. They also restrict genes linked to auxin response, transcription factors, and growth-regulating proteins.
- **Vascular Development**: Vascular development is influenced by specific miRNAs, such as miR165/166 and miR397, which are crucial for differentiating xylem and phloem tissues. They regulate vascular cell differentiation, impacting both plant development and the formation of vascular bundles.
- **Flower Development**: Flower development is controlled by miRNAs, such as miR156, miR157, miR172, and miR164. These miRNAs play specific roles in various aspects of the process, ranging from the onset of flowering to the production of floral organs. They selectively target particular transcription factors, influencing the timing and characteristics of floral structures.
- **Root Organ Development**: Specific miRNAs, namely miR160, miR164, and miR165/166, regulate the formation of the root cap, the development of lateral roots, and the production of adventitious roots. They control stem cell differentiation and determine the direction of growth, influencing the overall structure of the root system.
- **Plant Morphological Structure and Yield**: miRNAs, such as miR156 and miR172, affect the physical structure of plants, specifically altering characteristics like the formation of side shoots and the overall height of crops. Their regulatory function is vital for optimising crop productivity and overall yield.
- **Response to Stress**: The dynamic regulation of gene expression through miRNAs under abiotic stress conditions facilitates plant adaptability to environmental challenges. miRNAs, such as miR393 and miR397, respond to low temperatures, drought, salt stress, and viral infections.
- **Integration with Hormonal Pathways**: miRNAs integrate with hormone pathways to precisely regulate developmental processes, controlling specific gene expression patterns in tissues. This complements the broader effects of hormones on plant growth and development.
- **Fine-Tuning Gene Expression**: miRNAs function as fine-tuners of gene expression, exerting control after transcription by selectively targeting mRNAs for degradation or translational suppression. This level of precise control enables the delicate adjustment of gene activity that is essential for developmental processes.
- **Developmental Timing**: miRNAs control shoot meristem regulation and influence the timing of developmental events by regulating the transition from vegetative to reproductive growth. For instance, miR156 and miR172 specifically target genes that regulate flowering time, ensuring that plants' reproductive growth is synchronised with environmental signals.
- **Regulation of Signalling Pathways**: miRNAs also play a role in controlling different signalling pathways that are crucial for plant growth and development. These pathways are influenced by hormones such as auxins, cytokinins, gibberellins, and abscisic acid. miRNAs regulate developmental responses by selectively influencing components of various pathways, hence modulating signalling outputs.

TABLE 13.1
The Roles of miRNAs in Different Aspects of Plant Development

Developmental Aspect	Key miRNAs	Target Genes/Processes Regulated	References
Shoot Meristem	miR394	STM-WUS-CLV signalling pathway, SAM-specific gene (WUS)	Xue et al. (2020)
Leaf Development	miR160, miR165/166, miR319, miR396	Auxin response factors (ARFs), HD-ZIP III genes, TCP transcription factors, MYB transcription factor	Choudhary et al. (2021)
Vascular Development	miR165/166, miR397, miR857	HD-ZIP III genes, ATHB15, LACCASE7 (involved in xylem formation)	Milhinhos et al. (2020)
Flower Development	miR156, miR157, miR172, miR164, miR165/166, miR396, miR397	SPL proteins, AP2-like genes, CUC transcription factors, FT gene, MADS-box genes, ARF6/8, pistil and stamen development	Dong et al. (2022)
Root Organ Development	miR160, miR164, miR165/166	ARF10, stem cell differentiation, lateral root growth	Gautam et al. (2021)
Morphological Structure	miR156, miR172, miR319, miR397, miR1432	SPL genes, AP2-like genes, TCP transcription factors, yield-related traits, grain weight, tillering, height	Tao and Yu (2023)
Response to Stress	miR393, miR397, miR160, miR164, miR165/166	Stress-related genes, auxin response, plant biotic stress response, viral infections, abiotic stress response	Verma and Sarkar (2021)
Integration with Hormones	miR156, miR172, miR159, miR319, miR160, miR393	Interaction with SPL and AP2-like genes, fine regulation of phase transition, prevention of over-activation of miR156	Dong et al. (2022)

- **Maintenance of Developmental Homeostasis**: miRNAs preserve developmental homeostasis by inhibiting the abnormal expression of developmental regulators. Dysregulation of miRNA function can result in developmental abnormalities, underscoring their significance in maintaining the proper progression of plant growth and morphogenesis.
- **Epigenetic Regulation**: miRNAs can impact gene expression through epigenetic processes, including DNA methylation and histone modifications. By targeting specific elements within the epigenetic machinery or transcription factors responsible for epigenetic control, miRNAs can influence the structure of chromatin and the accessibility of genes, consequently affecting numerous developmental processes.
- **Interplay with Other Regulatory Elements**: miRNAs do not function in isolation but interact with other regulatory elements to fine-tune developmental processes. By collaborating with various regulatory elements, miRNAs can precisely control gene expression, leading to optimal plant growth and development (Table 13.1).

13.4 ROLE OF miRNA IN MANAGING STRESS

Understanding the complex role of miRNAs in regulating plant stress responses is crucial for improving agricultural output and ensuring food security in the face of growing environmental challenges. miRNAs are small RNA molecules that do not code for proteins but play a critical role in controlling the expression of genes. They coordinate a sophisticated molecular response system that allows plants to adjust and thrive in stressful situations. This article delves into the various roles of miRNAs in regulating stress and discusses their potential use in developing crops more resistant to stress.

One of the critical roles of miRNAs is regulating gene expression by selectively targeting specific mRNAs for degradation or translational suppression. In plant stress responses, miRNAs control the expression of genes involved in several physiological processes, including growth, development, and adaptation to stress. By fine-tuning gene expression levels, miRNAs enable plants to effectively allocate resources and prioritise survival strategies in challenging environmental conditions.

The regulatory function of miRNAs has exceptional specificity, which is influenced by factors such as plant species, developmental stages, tissue types, and stress stimuli. Various types of stress elicit unique miRNA expression patterns, indicating the intricate regulatory mechanisms designed to respond to specific environmental signals. Gaining insight into the context-specific functions of miRNAs is crucial for understanding the complexities of plant stress responses and developing precise therapies to improve stress tolerance.

Recent breakthroughs in high-throughput sequencing and bioinformatics techniques have significantly transformed the discovery and characterisation of stress-responsive miRNAs in plants (Bellato et al., 2019). Through a thorough analysis of miRNA expression patterns in different stress conditions, scientists can identify crucial regulators that play a role in stress perception, signal transduction, and adaptive responses. These stress-responsive miRNAs serve as useful molecular indicators for evaluating the ability of plants to withstand stress and for identifying specific genes to target in efforts for genetic enhancement.

Various miRNAs, such as miR156, miR159, miR169, and miR319, significantly regulate plant growth, development, and stress resistance. These miRNAs impact several developmental processes, including leaf formation, regulation of flowering time, and alteration of root structure, which ultimately determines the overall characteristics of plants. Furthermore, they play a role in integrating environmental signals into developmental processes, allowing plants to adapt their growth patterns to shifting ecological circumstances.

Plants frequently face the challenge of being exposed to multiple stressors simultaneously or one after another in their natural environments. This requires intricate regulatory mechanisms to coordinate their adaptive responses. miRNAs that can regulate reactions to different types of stress are essential components of the plant's ability to withstand and recover from adverse conditions. By selectively targeting genes involved in overlapping stress pathways, these miRNAs that respond to multiple stresses enable plants to initiate coordinated and complementary responses specifically adapted to the current environmental conditions.

miRNAs directly impact mRNA targets and play a role in epigenetic control by affecting DNA methylation, histone modifications, and chromatin remodelling processes. Epigenetic modifications function as flexible regulatory indicators that adjust the accessibility of genomic DNA and finely regulate miRNA expression in response to stress signals. miRNAs contribute to the long-term adaptability and memory formation necessary for plants to cope with stress by reprogramming gene expression patterns through epigenetic mechanisms.

miRNAs coordinate stress responses at the molecular level by interacting with hormone signalling pathways and transcriptional regulatory networks. They engage with crucial elements of hormone signalling cascades, such as auxin, abscisic acid, and jasmonic acid, to regulate stress-related physiological processes, including stomatal closure, osmotic control, and defence signalling. In addition, miRNAs specifically target transcription factors involved in the production of genes responsive to stress, adjusting the transcriptional environment to enhance stress adaptation mechanisms.

Utilising miRNAs' regulatory capabilities has excellent potential for creating crops that can tolerate stress and exhibit increased resilience. By understanding the intricate role of miRNAs in regulating plant stress responses, scientists can develop precise strategies to improve crop productivity and ensure global food security.

13.5 miRNA AND GENOME STABILITY IN PLANTS

miRNAs are small, non-coding RNA molecules essential for maintaining genomic stability and precise gene regulation in plants. They are involved in various biological processes, including development, differentiation, and response to environmental stresses such as DNA damage. The primary function of miRNAs is to regulate gene expression by binding to the 3′ untranslated region of target mRNAs, leading to their degradation or inhibition of translation. By doing so, miRNAs help preserve the integrity of the genome, which is crucial for the survival and adaptation of plants. In this chapter, we will delve deeper into the various roles of miRNAs in preserving genome integrity, specifically focusing on their involvement in DNA damage response (DDR), genome editing (GE), and the regulation of miRNA stability. The DDR pathway is a critical mechanism for cells to repair DNA damage and maintain genomic stability. It involves a complex network of signalling pathways and DNA repair mechanisms that work together to detect, signal, and repair DNA damage. Recent studies have shown that miRNAs play a crucial role in regulating the DDR pathway by targeting various genes involved in DNA repair, checkpoint activation, and apoptosis. Furthermore, miRNAs have emerged as promising tools for GE, as they can precisely target and manipulate specific genes and gene networks. They can also improve the efficiency and specificity of GE technologies, such as CRISPR/Cas9. In addition, miRNA stability is another critical factor that affects their function and regulation. Recent studies have shown that various factors, including RNA-binding proteins, stress, and environmental cues, can regulate miRNA stability and turnover, thereby affecting their function and gene regulatory activities. Overall, miRNAs are essential contributors to maintaining genomic stability in plants, and their roles in DDR, GE, and miRNA regulation make them potential targets for future biotechnological applications in crop improvement and environmental stress tolerance.

13.5.1 DNA Damage Response (DDR) and miRNAs

Plants are continuously exposed to various environmental stressors that can potentially damage their genetic material. This damage can threaten the stability of their DNA and lead to mutations, which can adversely affect the growth and development of the plant. However, plants have evolved advanced mechanisms called DDR to quickly identify and repair DNA damage. Recent research has shown that miRNAs play a crucial role in the DDR cascade by regulating the expression of genes vital for DNA damage repair and signalling. miRNAs help precisely adjust the plant's reaction to DNA damage, preserving the genome's stability. Furthermore, miRNAs play a significant role in regulating the balance between DNA repair and cell cycle progression. This ensures that the plant can quickly repair DNA damage while preventing abnormal cell division, which can lead to mutations. The precise control exerted by miRNAs is crucial in reducing the accumulation of mutations and genomic abnormalities, safeguarding the stability of the plant's genome. This is particularly important in the face of adverse environmental conditions, which can cause significant damage to the plant's DNA. By regulating DDR-related proteins, miRNAs help the plant effectively cope with environmental stressors, ensuring its survival.

13.5.2 Genome Editing (GE) and miRNAs

The field of plant genetics has undergone a significant transformation with the advent of GE tools, particularly the revolutionary CRISPR-Cas9 system, which enables precise modifications to the genome. In addition to their role in naturally maintaining the genome, miRNAs, have emerged as potential targets for GE, opening up new avenues to enhance desired traits and improve crop stress tolerance. Recent studies have demonstrated the feasibility of using CRISPR-Cas9 to manipulate miRNA genes, modulating their expression and function to achieve specific desirable traits (Mandal et al., 2021). By targeting particular miRNAs involved in stress response pathways, researchers can improve plant resilience to environmental stressors such as drought, salinity, and infections (Kumar et al., 2021). Furthermore, manipulating miRNA-mediated regulatory networks can enhance crop yield, improve nutrient uptake efficiency, and augment agronomic performance. However, further research is needed to gain a deeper understanding of miRNA-mediated gene regulation and to refine GE techniques for the precise manipulation of miRNA expression in plants.

13.5.3 miRNA Stability Control and Genome Stability

Maintaining the stability of miRNAs is a crucial aspect of preserving precise gene control and genomic stability in plants. miRNAs are small non-coding RNA molecules that regulate gene expression by binding to mRNA, either inhibiting its translation or promoting its degradation. Several mechanisms regulate miRNA stability, including post-transcriptional changes, interactions with RNA-binding proteins, and ribonucleoprotein destruction. Disruption of these pathways can result in abnormal miRNA production and reduced genome integrity, highlighting the significance of the mechanisms that regulate miRNA stability. Recent studies have provided insight into the dynamic nature of miRNA stability control, revealing complex feedback loops and interactions with other regulatory mechanisms. For instance, specific RNA-binding proteins can bind to miRNAs and protect them from ribonuclease degradation. Conversely, the binding of other proteins can destabilise miRNAs and enhance their degradation. These interactions can be influenced by various factors, including environmental cues such as changes in temperature and nutrient availability. Environmental cues can impact miRNA stability, which in turn can affect patterns of gene expression and the resulting characteristics of an organism. For example, exposure to high temperatures can increase miRNA degradation, leading to changes in the expression of heat-responsive genes. Similarly, nutrient deprivation can alter miRNA stability, leading to changes in the expression of nutrient-responsive genes. Understanding the interaction between miRNA stability regulation and genome stability is crucial for deciphering the molecular mechanisms that drive plant adaptability to fluctuating environmental conditions. By unravelling these mechanisms, researchers can gain insight into how plants respond to environmental stress and how they can be engineered to better tolerate changing conditions.

13.5.4 Role of miRNAs in Stress Response and Genome Stability

Plants face various environmental stressors such as drought, heat, cold, salinity, and pathogens that can significantly impact their growth and survival. In response to these stressors, plants have evolved complex mechanisms to detect and respond to the changing environment. One such mechanism is the involvement of miRNAs in regulating gene expression. miRNAs are small, non-coding RNAs that play a crucial role in modulating plant stress responses. They can precisely regulate gene expression to help plants withstand unfavourable conditions. Stress-responsive miRNAs regulate genes implicated in stress detection, signal transduction, and stress tolerance pathways. These miRNAs assist plants in adapting to the changing environment by modulating the expression of stress-related genes. By doing so, they maintain genome stability under challenging conditions and facilitate communication between several stress response pathways. Furthermore, miRNAs are involved in optimising a plant's survival strategies by prioritising the distribution of resources. They can regulate various processes that assist in coordinating intricate physiological and metabolic adaptations, which support a plant's ability to withstand environmental stress. This ultimately helps maintain genomic integrity and promotes agricultural sustainability. In summary, miRNAs are crucial in coordinating plant stress responses by precisely regulating gene expression in response to environmental signals. Their involvement in stress response pathways helps plants adapt to unfavourable conditions and maintain genomic stability, ultimately promoting agricultural sustainability.

13.5.5 Impact of Environmental Factors on miRNA-Mediated Genome Stability

Environmental conditions significantly influence the production and activity of miRNA in plants (Meenakshi et al., 2025). This, in turn, affects how plants respond to environmental stress and can impact the stability of their genomes. Abiotic stressors like drought, salinity, and extreme temperatures can alter the expression patterns of miRNA, leading to adaptive changes in the regulation of genes and stress response pathways (Anil et al. 2023; Tiwari and Rajam 2022; Verma and Sarkar 2021). In the case of biotic stressors such as pests and pathogens, miRNA can activate defence mechanisms, enhancing plant immunity and maintaining genome stability. Furthermore, environmental contaminants and UV radiation can cause DNA damage, which triggers miRNA-mediated DDR pathways to reduce genomic instability. Gaining a deeper understanding of the influence of environmental conditions on the maintenance of genomic stability through miRNA is crucial, especially in the face of climate change. By analysing the molecular pathways that govern the responses of miRNA to environmental stress, scientists can pinpoint specific genetic targets for enhancement. This knowledge can help develop more robust crop varieties that can thrive even in demanding conditions, ensuring food security and reducing the adverse effects of climate change on agricultural output.

13.5.6 miRNA-Mediated Regulation of Transposable Elements and Genome Stability

Transposable elements (TEs) are DNA sequences that can move and replicate within a genome. Although they have driven evolutionary changes over millions of years, they pose a substantial risk to the genome's stability as they can induce mutagenesis by inserting themselves into the genome and causing rearrangements. This can result in a wide range of genetic disorders and diseases, including cancer. Fortunately, plants have evolved a sophisticated system to control the activity of TEs and maintain genome stability. miRNAs are small, non-coding RNAs that play a crucial role in regulating the activity of TEs by repressing their production through post-transcriptional gene silencing. These miRNAs can target and degrade TE transcripts or inhibit their translation, thereby preventing the expression of TEs. The regulation of TE activity is vital for plant growth and development. Disruption of TE activity can lead to genome instability and reduced plant fitness, underscoring the significance of miRNA-mediated TE regulation in preserving genome stability. Recent research has revealed the presence of tiny RNAs produced from TEs, such as miRNAs, which are involved in pathways that silence TEs (Liu and Zhao, 2023). TE-derived miRNAs contribute to the regulation of TE activity and are involved in defending the genome against intrusive DNA elements. However, the production of miRNAs derived from TEs and their effects on genome stability are not yet fully understood. Further research is needed to explore the exact mechanisms involved in producing these miRNAs and their role in maintaining genome stability.

13.5.7 Role of miRNAs in Epigenetic Regulation and Genome Stability

miRNAs are a class of small RNA molecules that play a significant role in the regulation of gene expression. Recent studies have shown that miRNAs are essential in regulating epigenetics. Epigenetics refers to the study of heritable changes in gene expression that do not involve changes to the underlying DNA sequence. MiRNAs selectively target genes that encode epigenetic modifiers, including DNA methyltransferases and histone-modifying enzymes. By doing so, they regulate the structure and stability of chromatin, the material that makes up chromosomes. As a result, miRNAs impact patterns of gene expression and the stability of the genome. The dysregulation of miRNA-mediated epigenetic pathways can result in abnormal gene expression and genomic instability. This highlights the significance of miRNAs in preserving the integrity of the epigenome. Moreover, miRNAs have been linked to transgenerational epigenetic inheritance, in which the environmental stimuli encountered by parent plants affect the epigenetic characteristics and stress resilience of offspring plants. By transmitting epigenetic information from one generation to another, miRNAs play a role in adapting plant populations to dynamic environmental conditions. Studying the function of miRNAs in controlling gene expression and maintaining the integrity of the genome will provide valuable insights into how plants adapt to their environment. This understanding may enhance agricultural productivity and make plants more resilient to environmental changes, ultimately benefiting the economy, food security, and the environment.

13.6 miRNA EDITING IN PLANTS

13.6.1 Functional Analysis of miRNAs

miRNAs are small, non-coding RNA molecules that play a crucial role in regulating plant gene expression at the post-transcriptional level. These molecules usually comprise 21–24 nucleotides and participate in various biological activities, including development, growth, response to environmental stimuli, and adaptation to stress. By interacting with target mRNAs, miRNAs exert their regulatory effects, resulting in either mRNA degradation or translational repression. Understanding the functional significance of miRNAs is vital for a comprehensive understanding of plant biology. It can also help enhance crop productivity and ensure long-term agricultural sustainability. Therefore, functional analysis of miRNAs is an essential research area that allows us to study the intricate mechanisms underlying gene expression regulation in plants and identify potential targets for crop improvement.

13.6.2 Manipulation of miRNA Expression

miRNAs are small, non-coding RNAs that play a crucial role in regulating gene expression in plants. Researchers often manipulate their expression levels in various ways to better understand the functions of miRNAs. There are two common approaches for manipulating miRNA expression: overexpression and knockdown. Overexpression involves introducing exogenous miRNA molecules or miRNA-encoding genes into plant cells, which increases their abundance. This method can be used to study the effects of increased miRNA levels on gene expression and plant development. On the other hand, knockdown techniques aim to decrease miRNA levels. This can be accomplished using various methods, such as antisense oligonucleotides, RNAi, or CRISPR/Cas9-mediated gene editing. These methods can be used to study the effects of reduced miRNA levels on gene expression and plant development. Overall, manipulating miRNA expression levels is a powerful tool for understanding the biological functions of miRNAs in plants and can provide valuable insights into gene regulation and plant development.

13.6.3 Validation of miRNA–Target Interactions

The validation of the relationships between miRNAs and their targets is of utmost importance for a better understanding of the regulatory networks controlled by miRNAs. Various experimental approaches are commonly used to achieve this, such as 5′ RACE (Rapid Amplification of cDNA Ends) and degradome sequencing. 5′ RACE is a technique that enables the detection of cleaved mRNA products that arise from miRNA-mediated target cleavage. This method provides valuable information about the precise binding sites and regulatory impacts of miRNAs on their targets. By identifying the cleaved mRNA products, researchers can determine the exact location of the miRNA–target interaction. This information can further help us understand the functional roles of miRNAs in various biological processes. On the other hand, degradome sequencing is a highly efficient method that helps identify miRNA cleavage sites in target transcripts. This method enables a systematic analysis of miRNA–target interactions, providing a global view of miRNA-mediated gene regulation. By identifying the sites of miRNA-mediated target cleavage, researchers can gain insights into the regulatory circuitry of miRNAs and their targets. This information can further aid in developing therapeutic strategies to treat miRNA-related diseases.

13.6.4 Small RNA Inactivating Techniques

Controlling miRNA activity is a vital aspect of gene regulation in plants. While established methodologies for regulating miRNA expression exist, there is a growing need for more advanced techniques to manipulate miRNA activity at the post-transcriptional level. Recently, several innovative approaches have been developed to effectively isolate miRNAs and prevent them from interacting with their natural targets. One such method is miRNA target mimic (TM). TM involves the introduction of synthetic RNA molecules into plant cells that contain sequences imitating the binding site of miRNA. These mimetic molecules act as decoys, effectively isolating miRNAs and inhibiting their interaction with their natural targets. Another method is the miRNA sponge, designed to have several tandem binding sites for a single miRNA. These constructs act as competitive inhibitors by binding firmly to the miRNA, effectively isolating it and preventing interaction with its natural targets. The short tandem target mimic is another method that employs short RNA sequences containing repeated miRNA binding sites to effectively suppress miRNA activity. This method provides a powerful way to prevent the cleavage and suppression of miRNA-mediated targets. Artificial miRNAs are synthetic molecules that mimic the function of miRNAs. They are specifically intended to silence or inhibit the expression of particular genes. Researchers can achieve accurate and customisable gene regulation by utilising artificial miRNA constructs targeting genes of specific interest (Singh and Das, 2021). Finally, synthetic trans-acting siRNAs are produced from synthetic precursor transcripts. The synthetic siRNAs can selectively silence target genes by following a specified sequence, making them a versatile tool for regulating gene expression in plants. These advanced techniques provide researchers with precise control over gene expression, enabling them to study gene function in greater detail.

13.6.5 CRISPR/Cas9-Mediated miRNA Editing

CRISPR/Cas9 technology has revolutionised the field of plant genetics by enabling precise and targeted modifications to the genome. This novel tool is derived from the bacterial adaptive immune system. It utilises the Cas9 endonuclease along with a single-guide RNA to accurately edit DNA sequences at specific genomic loci. While initially designed for GE, scientists have repurposed CRISPR/Cas9 to target plant miRNAs. This breakthrough has enormous potential for exploring new approaches to analyse and control miRNA expression and functionality. With this technology, researchers can gain unprecedented insights into the mechanisms underlying miRNA function and regulation, paving the way for the development of more effective strategies for crop improvement and disease management.

13.6.6 CRISPR/Cas9-Mediated miRNA Editing and Modulation of miRNA Expression

CRISPR/Cas9-mediated miRNA editing is a highly effective and versatile method that offers numerous benefits for the functional study and modulation of miRNA expression and function in plants. One of the key advantages of this technique is its ability to accurately target specific miRNAs within the plant genome. This precision is achieved by designing single-guide RNAs complementary to the particular miRNA sequences or their genomic loci, allowing for the precise delivery of the Cas9 endonuclease to the desired target sites. This level of accuracy ensures that the targeted modifications are made exclusively in the designated areas, reducing unintended consequences and improving the reliability of experimental outcomes.

Moreover, CRISPR/Cas9 is highly efficient in generating plant miRNA knockout mutants. It allows researchers to impair miRNA synthesis or function by deliberately introducing specific mutations in the genomic regions that encode miRNAs or their precursor sequences. This facilitates the creation of loss-of-function mutants for particular miRNAs or entire miRNA families, enabling researchers to explore their roles in plant development, physiology, and stress responses. Using CRISPR/Cas9-mediated miRNA knockout mutagenesis enhances research efficiency and expedites the discovery of crucial regulatory circuits governed by miRNAs.

In addition, CRISPR/Cas9 can precisely edit miRNA target locations within gene transcripts and alter the miRNA sequences. This innovative method enables researchers to finely control gene expression levels through miRNA regulation with unparalleled accuracy. By manipulating miRNA target sites through the introduction of precise mutations or alterations, researchers can adjust the regulatory connections between miRNAs and their target mRNAs. This facilitates a detailed exploration of the intricate mechanisms of miRNA-mediated gene regulation in plants. This specific editing capability provides valuable insights into the functional importance of individual miRNA–target interactions and their impact on plant development, stress responses, and other biological processes. Overall, utilising CRISPR/Cas9-mediated miRNA editing offers various benefits for precise and efficient functional studies and modulation of miRNA expression and function in plants.

13.7 APPLICATIONS OF miRNA EDITING

miRNAs are small, non-coding RNA molecules that regulate gene expression in various biological processes. They participate in numerous biological processes, such as development, stress responses, and genomic stability, making them a topic of considerable research interest in recent years. The importance of miRNAs in maintaining genome stability and facilitating plant genetic modifications has led to a growing interest in their potential applications. In this context, there are various ways in which miRNA editing can be applied to achieve these goals. To name a few, miRNA editing can regulate gene expression, manipulate plant traits, and improve plant resistance to environmental stressors.

13.7.1 Targeted Gene Regulation

Gene regulation is a crucial process that involves turning genes on or off in response to various signals. When it comes to targeted gene regulation, the focus is on controlling the activity of specific genes precisely and deliberately. This is achieved using several techniques, such as small molecules or proteins that directly interact with the gene's regulatory elements. The ultimate goal of targeted gene regulation is to provide an effective method to control gene expression in plants, particularly in pathways that maintain genome stability. MiRNA editing is one of the most effective techniques for achieving targeted gene regulation. MiRNAs are short RNA molecules that play a vital role in regulating gene expression by binding to specific mRNAs and reducing their translation into proteins. Researchers can manipulate the sequence or structure of miRNAs to control their target specificity and affinity accurately. By doing so, they can regulate the expression of genes involved in DNA repair, recombination, and other elements of genome maintenance. This level of control allows for the manipulation of critical regulatory networks, which helps us understand the molecular mechanisms that govern genome stability. The ability to regulate gene expression precisely using targeted gene regulation has significant implications for plant genetics. It opens up possibilities for developing therapeutic approaches to treat plant genetic abnormalities. By better understanding the molecular mechanisms that govern genome stability, we can create more effective treatments for various plant diseases and disorders. In addition, targeted gene regulation can help us understand how plants respond to different environmental stresses, ultimately leading to more resilient and sustainable crop production.

13.7.2 Enhancing Stress Tolerance

Plant growth and production are crucial for ensuring global food security and agricultural sustainability. However, environmental stresses such as drought, salinity, and extreme temperatures can significantly impact crop yields by causing genomic instability and reducing plant resilience. Researchers have explored various strategies to enhance plant stress tolerance to address this challenge. One promising approach is editing miRNA, which involves manipulating the expression of stress-responsive miRNAs to reprogram the molecular responses of plants to unfavourable environments. By editing miRNA, researchers can alter the expression of stress-related genes, which play a crucial role in regulating plant responses to environmental stress. This technique holds great potential for producing robust crop varieties better adapted to challenging growing conditions. By enhancing the plant's resilience to environmental challenges, we can ensure sustainable food production and agricultural practices in the face of climate change. In summary, miRNA editing represents an innovative and effective strategy for improving plant stress tolerance and crop yields. By reprogramming plant molecular responses to environmental stresses, we can create more resilient crop varieties better equipped to withstand the challenges posed by a changing climate.

13.7.3 Genome Editing

GE is an advanced technique that enables scientists to make precise modifications to the DNA sequence of an organism. This technology has revolutionised the field of genetics, providing researchers with unprecedented control over genetic information. GE can be performed by adding, deleting, or altering specific genes. It has numerous applications in various areas, including agriculture. One of the most significant advancements in plant biotechnology has been the introduction of CRISPR-Cas GE technology. This innovative approach has transformed the field by providing unparalleled accuracy and effectiveness in making specific changes to the genome. Moreover, the utilisation of modified miRNAs in miRNA-guided GE has further amplified the adaptability and precision of this technique by exploiting the specific targeting characteristics of these altered miRNAs. The modified miRNAs are designed to act as guide RNAs for CRISPR-Cas nucleases, which can accurately direct DNA cleavage to specific genomic locations. By doing so, researchers can precisely modify genes or regulatory regions that significantly maintain genome stability. This novel approach facilitates the simultaneous modification of several genes in the genome, allowing for improved crop types with enhanced characteristics while minimising unintended effects on non-target genes. Overall, GE technology holds enormous potential for transforming agriculture by enabling the development of crops with enhanced characteristics such as better yields, increased disease resistance, and improved nutritional content. The use of CRISPR-Cas GE technology and modified miRNAs has significantly enhanced the accuracy and effectiveness of this technology. It is expected to drive further advancements in the field of plant biotechnology.

13.7.4 Epigenetic Regulation

Epigenetic regulation is a complex process that involves modifications to the structure of DNA without altering the underlying genetic code. These modifications, known as epigenetic marks, play a critical role in controlling gene expression and can be influenced by various factors such as environmental conditions, lifestyle choices, and ageing. Epigenetic miRNAs are essential to this process and have pivotal functions in regulating epigenetic alterations, including DNA methylation and histone modifications. These alterations are vital for preserving the genome's stability and gene expression patterns. Scientists can manipulate miRNAs involved in epigenetic regulation to control chromatin states and gene expression profiles. Ultimately, this process impacts the pathways responsible for maintaining genomic stability. The fascinating process of epigenetic reprogramming provides crucial insights into how miRNAs and epigenetic modifiers work together to preserve genome stability. This understanding opens up prospects for modifying plant traits associated with stress responses, developmental processes, and disease resistance. By manipulating the epigenetic modifications and miRNAs, scientists can develop new strategies to enhance plant growth, improve crop yields, and increase disease resistance. Epigenetic regulation is a rapidly growing field of research, and scientists are optimistic about the potential of this process to develop new ways to combat various diseases and improve quality of life.

13.7.5 Functional Genomics

Functional genomics is a vast and complex field of study that focuses on understanding the functions and interactions of genes within an organism. It involves using various techniques and technologies to analyse genes' structure, expression, and regulation, and their impact on cellular processes and overall organismal function. To study the functions of genes in more detail, researchers have developed a technique called miRNA editing, which is used to analyse the roles of miRNAs in genome stability pathways and other biological processes. This innovative technique involves creating modified miRNAs with different target specificities or production levels, which enables researchers to gain insight into the functions of individual miRNAs in DNA repair, recombination, and other elements of genome maintenance. By simplifying the organised examination of miRNA activity, miRNA editing allows for the discovery of new regulators of genome stability and offers significant knowledge about plant biology. Additionally, miRNA editing provides potential avenues for conducting functional genomics research to elucidate the genetic pathways that drive plant development, stress responses, and adaptability to environmental stimuli. This groundbreaking technique has enormous potential for advancing our understanding of the complex relationship between genes and their functions. It will undoubtedly play a critical role in shaping the future of functional genomics research.

13.7.6 Crop Improvement

The manipulation of miRNA editing has emerged as a promising approach to enhance various agricultural characteristics, including productivity, resistance to stress, and nutritional value. In recent years, researchers have been exploring the potential of miRNA editing to improve crop performance under different environmental conditions. By manipulating miRNAs associated with critical agricultural properties such as drought tolerance, disease resistance, and nutrient uptake, they can create more robust agricultural cultivars that can withstand environmental pressures and reduce crop output losses. This method effectively ensures food security, promotes sustainability, and tackles global food supply and distribution issues. Moreover, miRNA editing can improve crops' nutritional quality, longevity, and post-harvest attributes. For example, by editing miRNAs linked to the biosynthesis of specific nutrients, researchers can enhance the nutritional value of crops and reduce the risk of malnutrition. Additionally, miRNA editing can enhance crops' resistance to pests and diseases, reducing the need for harmful pesticides and herbicides. Furthermore, it can improve the post-harvest attributes of crops and extend their shelf life. All in all, miRNA editing holds great promise for the future of agriculture, providing a powerful tool for creating more resilient, nutritious, and sustainable crops. By leveraging the potential of miRNA editing, researchers can develop new solutions to address the agricultural sector's challenges, including climate change, food security, and global sustainability.

13.7.7 Disease Resistance

Plants are constantly threatened by microbial diseases and herbivorous pests, which can negatively impact crop yields and quality. One of the reasons why plants struggle to defend themselves against these threats is the instability of their genome. However, recent research has shown that modulating immune signalling pathways using miRNA editing can be a promising approach to bolster crop disease resistance. By selectively targeting miRNAs involved in plant defence systems, researchers can enhance the plant's inherent immunity and provide sustainable alternatives to chemical pesticides, which can have harmful environmental consequences. This method minimises the adverse effects of pest control and fosters ecological equilibrium in agricultural settings. As a result, crops become more resilient to various infections and pests, leading to higher yields and better-quality produce.

13.8 CHALLENGES

The study of miRNAs in plants has become an increasingly captivating field of research, particularly regarding their roles in maintaining genome integrity and facilitating genetic modifications. However, this topic presents both challenges and opportunities for scientists and biotechnologists. This analysis delves into the various facets of this emerging area, highlighting significant obstacles and potential avenues for progress.

Firstly, identifying and characterising miRNAs has proven to be a significant challenge despite recent developments in sequencing technologies. Although these technologies have facilitated the identification of numerous miRNAs, a considerable proportion remains uncharacterised, especially those that play specific roles in genome maintenance. The lack of knowledge in this area limits our understanding of the complex regulatory networks that control plant genomes. To overcome this hurdle, we need to increase our efforts to utilise advanced sequencing techniques and bioinformatics tools to better understand the intricate nature of the plant miRNA ecosystem. Combining high-throughput sequencing data with functional genomics methods, including CRISPR-based screens, can help identify the most crucial miRNA candidates for further study. Additionally, it will be necessary to devise experimental approaches to functionally validate uncharacterised miRNAs to elucidate their roles in preserving genomic integrity.

Secondly, understanding the precise functions of specific miRNAs in genomic stability and editing pathways is challenging. Although certain miRNAs have clearly defined targets and functions, many demonstrate pleiotropic effects or interact with several targets. To comprehensively understand the impact of miRNAs on plant genome stability, it is essential to incorporate a wide range of omics data, including transcriptomics, proteomics, and metabolomics, into functional annotation efforts. Utilising sophisticated bioinformatics tools that incorporate network biology and pathway analysis can assist in predicting and verifying the functional consequences of miRNA-mediated regulation.

Thirdly, comprehending the complex regulation of DNA repair pathways by miRNAs in plants is crucial for uncovering the mechanisms that maintain genomic integrity. MiRNAs likely regulate gene expression in DNA repair processes, including homologous recombination, non-homologous end joining, and base excision repair. To understand the unique regulatory roles of miRNAs in DNA repair, it is crucial to conduct comprehensive functional investigations that utilise genetic and biochemical methods. Employing CRISPR-based screens to target essential elements of DNA repair pathways alongside miRNA profiling can offer valuable insights into the dynamic interaction between miRNAs and DNA repair mechanisms.

Fourthly, incorporating miRNA-mediated regulation with CRISPR-based GE presents both obstacles and opportunities. Utilising miRNAs to fine-tune the expression of CRISPR components can enhance the accuracy and effectiveness of GE in plants. Moreover, the modulation of repair outcomes of DNA breaks caused by CRISPR through miRNA regulation holds significant promise. Researchers must investigate methodologies for developing synthetic miRNA mimics that precisely target and control CRISPR components. To improve the effectiveness of these techniques, it is essential to comprehend the spatial and temporal patterns of miRNA expression during GE.

Fifthly, ensuring the specificity of miRNA-mediated regulation is vital to prevent off-target effects, especially in GE. Unforeseen consequences can undermine the precision and safety of genetic alteration in plants. An exhaustive examination of the factors that affect miRNA–target interactions is necessary. By utilising advanced computational algorithms to predict off-target effects and validating them through experiments, it is possible to enhance the specificity of miRNA-based strategies by guiding their design. Incorporating bioinformatics methods that consider the secondary structure of target mRNA can help reduce unwanted outcomes.

Lastly, transporting miRNAs to specific plant tissues or cell types poses a significant technical challenge. Delivery mechanisms that efficiently target miRNAs to specific genomic loci or cellular compartments are crucial for accurately manipulating genome stability and editing activities.

Nanotechnology-driven delivery technologies, such as nanoparticles or viral vectors, promise to deliver miRNAs precisely to specific targets. To successfully employ miRNA-mediated GE, it is essential to optimise these delivery platforms for plant systems and ensure their safety and effectiveness.

13.9 CONCLUSION AND FUTURE PROSPECTIVES

The increasing significance of miRNAs in genome stability, GE, and stress responses in plants is a rapidly expanding field of study that has the potential to revolutionise plant biology and agriculture. This complex topic has yielded numerous critical findings and insights, shifting our understanding of how miRNAs coordinate crucial molecular processes for plant development, stress responses, and environmental adaptation. Maintaining cells' structural and functional integrity and accurately inheriting genetic material necessitates genomic stability. The intricate network of biological systems that safeguard the genome includes miRNAs, which regulate gene expression in DNA repair, recombination, and replication. miRNAs target gene transcripts that encode proteins involved in the DDR to regulate the biological response precisely. They can coordinate genome repair mechanisms and restore integrity. Regulatory interactions between miRNAs affect the expression of DDR genes, enabling plants to resist genotoxic stimuli and maintain genomic stability in environmental challenges. GE methods, particularly CRISPR-Cas9, have revolutionised plant genetic modification. However, concerns about unwanted effects on non-targeted areas necessitate optimising GE technologies to improve efficacy and precision. MiRNAs can enhance GE accuracy and efficiency. Identifying miRNAs that target critical genes involved in GE offers new ways to modify the editing process precisely, reduce unwanted consequences, and boost specific genetic modifications. Scientists may regulate these miRNAs to manage the editing process, providing new avenues for crop improvement and plant genetic engineering. In addition to preserving and modifying genome stability, miRNAs help plants respond to drought, salt, extreme temperatures, and disease invasions. MiRNAs regulate gene expression networks to coordinate stress-responsive pathways, assisting plants in adapting to difficult conditions. By understanding the roles of specific miRNAs in stress tolerance, scientists can identify genetic targets to create crops that can survive environmental stressors. Using miRNA-mediated stress tolerance pathways could reduce the detrimental effects of climate change on agricultural productivity and provide food security in a fast-changing world. MiRNAs regulate chromatin structure and DNA methylation patterns to control epigenetics and stress responses. MiRNAs selectively repress genes involved in epigenetic changes, which affect epigenetic states and gene expression in plants. The complex connection between miRNA-mediated gene regulation and epigenetic modifications reveals new molecular processes that drive plant development and allow for environmental response. By understanding these complex regulatory networks, scientists can better comprehend plant biology and develop new methods for crop improvement and sustainable agriculture. Studying miRNAs in genomic stability, editing, stress responses, and epigenetic control has significant implications for agricultural biotechnology and crop development. Scientists can use miRNAs to improve plant traits, agricultural productivity, and food security in a changing climate. Using miRNA-based methodologies alongside traditional breeding and GE technology could speed agricultural improvement and address global concerns, including population growth, climate change, and environmental degradation. Several significant research and development areas could improve our understanding of miRNA-mediated activities in plants and benefit agriculture and food security. To understand complex miRNA-mediated regulatory networks and uncover genetic modification targets, miRNAs and their target genes must be functionally characterised. The advancements in high-throughput sequencing, bioinformatics, and GE will make miRNA identification and manipulation for plant biotechnology easier. Fundamental research and practical science must be linked to apply miRNA-related findings to agriculture, food security, and environmental sustainability. To ensure safety, efficacy, and public approval, miRNA-based technologies in agriculture must be regulated and ethically guided. Researchers can address regulatory and ethical issues to build public trust in miRNA-based agricultural enhancement and genetic engineering. miRNAs' growing role in genome stability, modification, and stress response offers significant research opportunities in plant biology and biotechnology. We can use miRNAs' regulatory abilities to find new ways to boost crop resilience, productivity, and sustainability. This will shape agriculture and ensure food security for future generations.

REFERENCES

Abbas, A., A. N. Shah, M. Tanveer, W. Ahmed, A. A. Shah, S. Fiaz, M. M. Waqas, and S. Ullah. 2022. miRNA fine tuning for crop improvement: using advance computational models and biotechnological tools. *Molecular Biology Reports*, 49(6), 188–193. https://doi.org/10.1007/s11033-022-07231-5.

Anil, P., P. Dharmendra, S. Sachidanand, E. Yuriy, B. Snezhan, J. Dilfuza, and J. Chinmayi. 2023. Beneficial effects of biochar application on mitigating the drought and salinity stress implications on plants. *Plant Science Today*, 10(2), 188–193. http://doi.org/10.14719/pst.2591.

Bellato, M., D. De Marchi, C. Gualtieri, E. Sauta, P. Magni, A. Macovei, and L. Pasotti. 2019. A bioinformatics approach to explore microRNAs as tools to bridge pathways between plants and animals. Is DNA damage response (DDR) a potential target process? *Frontiers in Plant Science*, 10, 1535. https://doi.org/10.3389/fpls.2019.01535.

Choudhary, A., A. Kumar, H. Kaur, and N. Kaur. 2021. miRNA: the taskmaster of plant world. *Biologia*, 76(5), 1551–1567. https://doi.org/10.1007/s11756-021-00720-1.

Dong, Q., B. Hu, and C. Zhang. 2022. microRNAs and their roles in plant development. *Frontiers in Plant Science*, 13, 824240. https://doi.org/10.3389/fpls.2022.824240.

Gautam, V., P. Gupta, P. Salvi, A. Sharma, D. Kumar, and A. Dwivedi. 2021. miRNA mediated signaling involved in Arabidopsis thaliana root development. *Signaling and Communication in Plants*, 393, 93–113. https://doi.org/10.1007/978-3-030-84985-6_7.

Kumar, A., R. Kaur, and M. V. Rajam. 2021. Regulatory role of microRNAs in plants under challenging environmental conditions with special focus on drought and salinity. In: Husen, A. (Ed), *Harsh Environment and Plant Resilience*. Springer, Cham, pp. 121–140. https://doi.org/10.1007/978-3-030-65912-7_6.

Liu, B., and M. Zhao. 2023. How transposable elements are recognized and epigenetically silenced in plants? *Current Opinion in Plant Biology*, 75, 102428. https://doi.org/10.1016/j.pbi.2023.102428.

Mandal, K., P. Boro, and S. Chattopadhyay. 2021. microRNA based gene regulation: a potential way for crop improvements. *Plant Gene*, 27, 100312. https://doi.org/10.1016/j.plgene.2021.100312.

Meenakshi, S., D. Rahul, S. Sachidanand, G. M. Anita, and S. Rajpal. 2025. Plant microRNAs: an overview. In: Dutta, R., Kapadia, C., Danish, S., and Singh, S. (Eds), *microRNA Advances and Application in Plant Biology Plant Biology, Sustainability and Climate Change*. Academic Press, New York.

Milhinhos, A., S. Lopes, and C. Miguel. 2020. microRNA-mediated regulation of plant vascular development and secondary growth. In: Miguel, C., Dalmay, T., and Chaves, I. (Eds), *Concepts and Strategies in Plant Sciences*. Springer, Cham, pp. 143–168. doi:https://doi.org/10.1007/978-3-030-35772-6_8.

Noopur, K., K. Pragati, K. S. Adarsh, S. Sachidanand, J. Tanvi, and B. Megha. 2025. Role of non coding RNA in effector trigger immunity. In: Dutta, R., Kapadia, C., Danish, S., and Singh, S. (Eds), *Non-Coding RNA in Plants Modulation and Stress Responses Plant Biology, Sustainability and Climate Change*. Academic Press, New York.

Patil, S., S. Joshi, M. Jamla, X. Zhou, M. J. Taherzadeh, P. Suprasanna, and V. Kumar. 2021. microRNA-mediated bioengineering for climate-resilience in crops. *Bioengineered* 12(2), 10430–10456. https://doi.org/10.1080/21655979.2021.1997244.

Pooja, G., B. Megha, M. Arpan, K. Chintan, D. Rahul, D. Subhan, and S. Sachidanand. 2025. Introduction to long noncoding RNA. In Dutta, R., Kapadia, C., Danish, S., and Singh, S. (Eds), *Non-Coding RNA in Plants Modulation and Stress Responses Plant Biology, Sustainability and Climate Change*. Vol. 1, pp. 1–12. Academic Press, New York.

Rahul, D., K. Chintan, D. Subhan, and S. Sachidanand. 2024. *Non-Coding RNA in Plants: Modulation and Stress Responses*. Academic Press, New York.

Singh, A., and S. Das. 2021. Slicing messengers by artificial designs: artificial microRNA induced gene silencing in polyploid plants for functional genomics and trait modification. *Concepts and Strategies in Plant Sciences*, 6, 77–129. https://doi.org/10.1007/978-3-030-64994-4_6.

Tao, J., and Y. Yu. 2023. Plant small RNA in food crops. In: Guleria, P., Kumar, V., and Mo, B. (Eds), *Chapter 4- Small RNAs as Emerging Regulators of Agricultural Traits of Food Crops*. Academic Press, New York. https://www.sciencedirect.com/science/article/pii/B9780323917223000051.

Tiwari, R., and M. V. Rajam. 2022. RNA- and miRNA-interference to enhance abiotic stress tolerance in plants. *Journal of Plant Biochemistry and Biotechnology*, 31(4), 689–704. https://doi.org/10.1007/s13562-022-00770-9.

Verma, S., and A. K Sarkar. 2021. MiRNA-mediated regulation of biotic and abiotic stress responses in plants. *Agricultural Biotechnology: Latest Research and Trends*, 20, 463–492. https://doi.org/10.1007/978-981-16-2339-4_20.

Wani, S. H., V. Kumar, T. Khare, P. Tripathi, T. Shah, C. Ramakrishna, S. Aglawe, and S. K. Mangrauthia. 2020. MiRNA applications for engineering abiotic stress tolerance in plants. *Biologia*, 75(7), 1063–1081. https://doi.org/10.2478/s11756-019-00397-7.

Xu, J., Q.-M. Hou, T. Khare, S. K. Verma, and V. Kumar. 2019. Exploring miRNAs for developing climate-resilient crops: a perspective review. *Science of the Total Environment*, 653, 91–104. https://doi.org/10.1016/j.scitotenv.2018.10.340.

Xue, Z., L. Liu, and C. Zhang. 2020. Regulation of shoot apical meristem and axillary meristem development in plants. *International Journal of Molecular Sciences*, 21(8), 2917. https://doi.org/10.3390/ijms21082917.

14 microRNA-Mediated Bioengineering for Climate Resilience in Crops

Priyamvada Amol Arte, Khadija Turabi, Mustansir Bhori, Kanchanlata Tungare, Renitta Jobby*, and Pranavi Mishra*

14.1 INTRODUCTION

Climate encompasses alterations in temperature and atmospheric conditions. The main reason for climate change is said to be the emission of greenhouse gases generated from processes like the burning of fossil fuels, leaks from industrial activities, and consequent chemical reactions. The weather patterns in a particular area can greatly affect different industries, potentially causing droughts, water shortages, wildfires, rising sea levels, floods, melting polar ice, severe storms, and declines in biodiversity (Anil et al. 2023; Verma et al. 2021). These factors all have adverse effects on plant life due to shifts in temperature, rainfall, and levels of CO_2 and N_2. This can lead to reduced crop yields, poorer quality produce, and the extinction of numerous plant species. Extreme temperature changes significantly affect pollination, as specific species germinate pollen at certain temperatures only. Variation in temperature can lead to denaturation of the pollen, thus decreasing the growth rate (Hatfield & Prueger, 2015). This has created a severe problem for humanity because a decline in plant yield immediately causes food scarcity. Abiotic stresses, such as increases in salinity, temperature changes, and droughts, affect the morphology, physiology, and biochemistry of plants (Hussain et al., 2018). The abiotic stress caused by changes in climate is said to result in around a 50% reduction in yield across the world (Singhal et al., 2016). Consequently, multiple strategies have been implemented to lessen the effects on the plant realm. Recent research has focused on developing plants that are more resilient to climate change, with one approach being the utilization of microRNAs (miRNA).

miRNAs are a type of non-coding RNA molecule, typically measuring between 21 and 25 nucleotides long, that play a role in controlling gene expression (Noopur et al., 2025, 2025b; Pooja et al., 2025). For most miRNA, the primary miRNA is transcribed from the DNA sequence, which is then converted into precursor miRNA and lastly develops into mature miRNA. The 3′ untranslated region (3′ UTR), 5′ UTR, promoter region, and coding sequence are commonly identified as sites susceptible to miRNA action (Meenakshi et al., 2025). When miRNA binds to these sites, it initiates an interaction that can modify gene expression and, in some cases, regulate processes such as gene translation and transcription (O'Brien et al., 2018). To regulate gene expression, miRNA employs two protein complexes called Dicer and RNA-induced silencing complex. Dicer not only facilitates the formation of RNA-induced silencing complex but also contributes to gene silencing, consequently influencing RNA interference-mediated gene expression (Hammond, 2005).

In plants, miRNAs regulate meristem development, leaf polarity, flowering patterns, and more. Plant development is influenced by miRNAs, which control the expression of target genes post-transcriptionally, either through direct mechanisms or by hindering the process of translation (Dong et al., 2022). Mutations affecting miRNA transcription or processing can have diverse impacts on both gene expression and plant growth. For instance, miR172 in Arabidopsis plays a role in

* *Corresponding Author*

DOI: 10.1201/9781003491378-14

regulating flowering time, miR159 in Arabidopsis is linked to various developmental abnormalities, and miRNA165/6 contribute to determining the types of xylem cells (Allen et al., 2007; Aukerman & Sakai, 2003; Carlsbecker et al., 2010). Research indicates that modifying the activity of a specific miRNA in plants can enhance their tolerance to different stressors. For instance, boosting the levels of miR408 in Arabidopsis has been associated with increased resistance to salinity, cold, and oxidative stress (Ma et al., 2015). All the evidence suggests that miRNA manipulation can be an effective way to develop plants resilient to abiotic stresses caused by climate change. Over the last 10 years, extensive studies have focused on primary agricultural crops such as rice, corn, wheat, and cotton, examining how different stress factors influence their miRNA composition and subsequent phenotypic responses through miRNA investigations (Chaudhary et al., 2021).

This chapter highlights strategies such as genetic modification techniques and genome editing, miRNA target mimicry, topical delivery of miRNA via nanocarriers, miRNA cross-kingdom transfer and epigenetic memory, as well as the developmental challenges and benefits of miRNA bioengineering.

14.2 miRNA-MEDIATED TECHNIQUES FOR THE DEVELOPMENT OF CLIMATE-RESISTANT PLANTS

MiRNAs facilitate gene expression by binding to target transcripts in a complementary manner, leading to either mRNA cleavage or the suppression of protein translation (Ayan et al., 2023). Mature miRNAs are the active forms of miRNA molecules, which arise from a precursor stem–loop structure. This structure is derived from the primary miRNA, transcribed by RNA polymerase II from the MIR gene, representing an intermediate stage in miRNA formation (O'Brien et al., 2018). In plants, miRNAs play pivotal roles in regulating gene expression after transcription, targeting specific mRNA molecules for either degradation or translational inhibition. This regulatory activity enables miRNAs to finely adjust gene expression, impacting various biological processes such as development, growth, and responses to stress. Given their involvement in nearly all facets of plant life, there is a growing emphasis on understanding the intricate mechanisms of miRNA-mediated gene regulation, particularly in response to stress and climate change. Developing crops resilient to elevated CO_2 levels, temperature shifts, floods, and droughts, which significantly impede crop yield and growth, is essential to tackle the challenges posed by climate change (Lamaoui et al., 2018). Heat stress, for instance, disrupts protein function, leading to excessive membrane fluidity and metabolic imbalances. Plant cells, however, possess complex molecular mechanisms and signaling pathways, often mediated by miRNAs, to detect and respond to temperature shifts. As such, miRNAs emerge as primary targets for engineering stress-tolerant plant varieties (Zuo et al., 2021). Targeting miRNAs presents a promising strategy to engineer plant varieties capable of thriving under high environmental stress. Through the application of genetic engineering and contemporary biotechnological methodologies like genomics, proteomics, and transcriptomics, we can address these challenges and cultivate crops resilient to stress while minimizing yield loss (Joshi et al., 2024). Nevertheless, miRNAs have emerged as potent molecules, acting as expression markers under environmental stress conditions, thus underscoring their significance in the quest for resilient crop varieties. By combining insights into miRNA-triggered stress responses with genetic manipulation methods, we can lay the groundwork for environmentally friendly farming amidst shifting climate conditions (Figure 14.1, Table 14.1).

14.2.1 Genetic Modification Techniques

Genetic modification (GM) is a powerful tool in modern biotechnology that holds immense promise for developing climate-resistant plants capable of withstanding the challenges posed by shifting environmental conditions. Having been extensively researched since the 1990s, GM was popularized by the successful creation of the genetically modified and fortified rice variety called "golden rice",

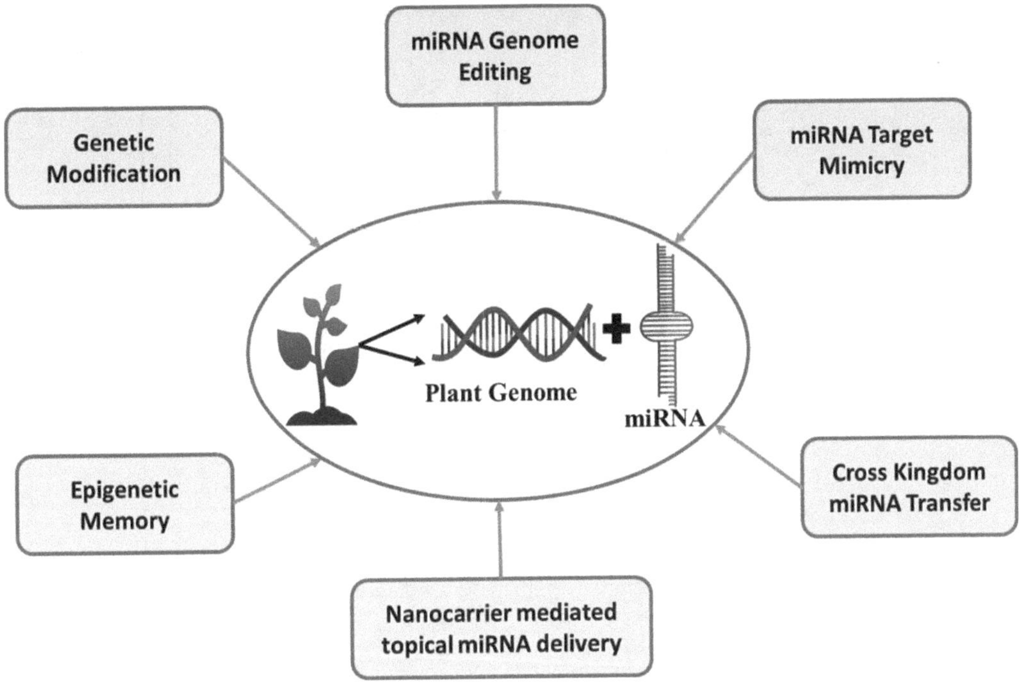

FIGURE 14.1 Biotechnology tools for miRNA-targeted development of climate-resistant plants.

cultivated as a potent source of Vitamin A in impoverished countries. This development opened a wide range of opportunities to selectively improve plant varieties through GM and motivated modern breeding methods to identify, select, and analyze individual organisms with genetically enhanced traits that can help other species adopt desirable traits (Kobayashi et al., 2023). Therefore, the potential of GM can be expanded to create climate-resilient crops by focusing on miRNAs. These molecules interact with particular sequences of target mRNA's 3′UTR, leading to the inhibition of translation. This process is crucial for aligning miRNAs with their respective signaling pathways, facilitating the development of various stress-resistant crops (Huntzinger & Izaurralde, 2011; Ipsaro & Joshua-Tor, 2015). miRNA binding sites are detected in mRNA regions containing the 5' UTR with coding sequences within the promoter region. This is because the interaction of miRNAs with promoter regions induces transcription (O'Brien et al., 2018; Riolo et al., 2020). This technology enables the creation of various genetic and epigenetic modifications, encompassing both random and targeted alterations of DNA sequences at specific genomic sites. Various techniques have emerged to manipulate the genetic makeup of organisms, each offering unique strategies for introducing desired traits. Among these techniques are transgenesis, cisgenesis, and intragenesis, which differ in their approaches to modifying the genetic material of organisms. Moreover, through the application of recombinant DNA technology, researchers have the capability to integrate sizable foreign DNA segments into the host genome, whether at predetermined or random sites. This mechanism facilitates modifications in epigenetic indicators, allowing for the development of innovative genetic permutations while refining the regulation and function of miRNAs implicated in plants' resilience to environmental stress (Rai et al., 2023). When engineering plants using the GM approach, several criteria must be considered. Firstly, genetic uniformity across the population is crucial, particularly if these miRNAs differ from those found in other breeds of the same species. Additionally, genetically modified seeds must exhibit stability over successive generations. Given the lengthy development process of plant species, typically spanning 12 years for full development, evaluation, and eventual release of a new variety, meticulous attention to these criteria is paramount to ensure the successful integration and sustained efficacy of engineered miRNAs (Health, 2004).

TABLE 14.1
Summary of Climate-Resistant Transgenic Plants Developed Using miRNA-Targeted Approaches

miRNA-Mediated Tools	miRNA	Target Plant	Stress	Outcome	References
miRNA genome editing	MIR169a knockout by CRISPR/Cas9 system	*Arabidopsis thaliana*	Drought resistance	Deletion and replacement of MIR169a improved drought tolerance	Zhao et al. (2016)
miRNA target mimicry	miRNA165/166 single mutant	*Arabidopsis thaliana*	Drought tolerance	Superior drought tolerance due to reduced stomatal conductance and transpiration rates	Yang et al. (2019)
	miR319	*Arabidopsis thaliana, Oryza sativa*	Salt tolerance	Upregulation of ethylene synthesis in C3 plants	Liu et al. (2019)
Nanocarrier-mediated topical delivery of miRNA	miR166a (ds-MIR166a/SPc nanocarrier)		Environmental stress resistance	Penetration of the root cortex and systematic transportation of ds-miRNA/SPc complex through the vascular tissue in seedlings of *Arabidopsis* and maize	Yang et al. (2022)
Cross-kingdom miRNA transfer	miR168a	*Homo sapiens*	Disease resistance	Inhibition of mammals' low-density lipoprotein receptor adapter protein 1 (LDLRAP1)	Zhang et al. (2012)
	MiR159 miR166	*Gossypium herbaceum*	Pathogenic stress	Binds to virulence-related proteins HiC-15 (isotrichodermin C-15 hydroxylase) and Clp-1 (Ca2+-dependent cysteine protease)	Zhang et al. (2016)
Epigenetic memory	AtHUB2	*Gossypium herbaceum*	Drought tolerance	Monoubiquities H2B	Chen et al. (2019)
	TaHKT2;1	*Triticum aestivum* L.	Salt tolerance	Downregulating the expression of genes associated with potassium transporters in wheat (*Triticum aestivum* L.)	Kumar et al. (2017)
	TaHKT2;3	*Triticum aestivum* L.	Salt tolerance	Manipulation of genes responsible for ion transfer	Kumar et al. (2017)

Furthermore, while implementing these techniques, it is imperative to consider the diverse spectrum of qualitative and quantitative changes that may arise. Throughout the development process, unintended changes may occur due to specific characteristics of the genetic modification (GM) process or other biotechnological techniques employed. For example, processes involving the cultivation of plant cells in a laboratory setting, the creation and application of plant protoplasts, and the ability to grow new plants from individual cells or tissue cultures all carry the risk of unintended changes. Therefore, careful consideration and rigorous assessment of these potential changes are essential to ensure the safety and efficacy of genetically modified organisms. Despite advancements in protocols and GM systems adopted by researchers, plant GM remains highly regulated, following biosafety and environmental risk assessment protocols established by international bodies such as the Food and Agriculture Organization (FAO), World Health Organization (WHO), and Organization for Economic Co-operation and Development (OECD) (Eckerstorfer et al., 2019).

14.2.1.1 Transgenesis

Transgenesis is a genetic engineering technique that involves the insertion of genes, known as transgenes, from unrelated plants or non-plant organisms into the target plant's DNA using recombinant DNA technology. These transgenes can confer traits that do not naturally occur in the target species. The purpose of miRNA-targeted transgenesis is to imbue plants with beneficial traits that enhance their resilience against drastic environmental changes and the abiotic stresses associated with them (Grover et al., 2013). Prior studies on transgenesis primarily focus on evaluating how the incorporation of a foreign gene into a plant's genome impacts the expression of miRNA in transgenic varieties of the plant under study. When sorghum (*Sorghum bicolor*) was genetically modified with the mtlD transgene, which encodes mannitol-1-phosphate dehydrogenase from E. coli, there was a notable enhancement in the plant's ability to withstand water scarcity and NaCl stress. Several studies have successfully adopted this approach to develop transgenic tobacco (*Nicotiana tabacum*), rice (*Oryza indica*), *Arabidopsis* (*Arabidopsis thaliana*), and peanut (*Arachis hypogaea* L.) crops to improve their growth in drought conditions (Khan et al., 2019; Maheswari et al., 2010). Yu et al. achieved comparable results by utilizing bacterial cold shock proteins, such as SeCspA, to produce transgenic wheat varieties exhibiting enhanced tolerance, reduced water loss, diminished wilting, and increased grain yield compared to non-transgenic wheat lines (Yu et al., 2017). Although there has been research on transgenesis to develop stress-tolerant transgenic plants, there is a notable lack of studies specifically addressing the miRNA-based transgenesis method. Moreover, in practice, the use of transgenic plants is often accompanied by crucial concerns that generally fall into two categories. The first is the impact of genetically modified substances on human health. For instance, transgenic plants may upregulate allergen–gene expression, which may cause allergies among consumers (Herman & Ladics, 2011). Furthermore, the population might be uncomfortable with ingesting genetic material from non-human sources such as viruses and bacteria, which carry the risk of being pathogenic. The second concern is the potential hindrance and alterations that crops produced using transgenesis may cause in the environment and the natural processes associated with it (Hansen Jesse & Obrycki, 2000; Losey et al., 1999). Hence, so far, the research conducted on crop development using transgenesis, especially targeting miRNAs, is limited, and its insufficiency makes it unsuitable for employing this approach for the development of climate-resistant crops.

14.2.1.2 Cisgenesis

Cisgenesis involves transferring miRNA genes between sexually crossable and compatible plant species to genetically modify recipient plants with naturally weather-resistant genes. This modification, performed using appropriate techniques with miRNA, aims to confer climate resistance to the plants. The introduced gene includes its introns and is flanked by its native promoter and terminator in the correct orientation. Cisgenic plants may carry one or more cis genes but do not contain transgenes. While both transgenesis and cisgenesis employ similar genetic modification techniques, cisgenesis exclusively utilizes genes from the plant itself or closely related species, which can also be transmitted through conventional breeding methods (Pavone & Martinelli, 2015). Cisgenesis does not modify the gene pool of the recipient species, introduce new traits, or result in fitness alterations beyond what can occur through conventional breeding or natural gene transfer. It presents no additional risks compared to traditional breeding regarding impacts on other organisms, potential toxicity or allergenicity of genetically modifiedfoods or animal feeds (Organisms (GMO) et al., 2022). This stands as the core contrast between cisgenesis and transgenesis. Thus, the deliberate introduction and commercialization of cisgenic plants are deemed equally secure as the introduction and commercialization of conventionally bred plants (van Hove & Gillund, 2017). In terms of safety considerations, regulatory bodies may regard cisgenic plants similarly to conventionally bred ones. Cisgenesis presents notable advantages over conventional breeding by circumventing the issue of linkage drag, commonly encountered in traditional crop cultivation methods. Linkage drag refers to the inadvertent transfer of undesirable traits closely associated with the target gene during breeding, potentially compromising the crop's natural characteristics. These transferred genes carry a heightened risk of encoding various allergens or toxins (Schnell et al., 2015). Moreover, while traditional breeding methods do not allow the preservation of original genetic traits in the progeny, plant varieties created using cisgenesis allow for the incorporation of desired traits while maintaining the overall genetic integrity of the crop. This can be particularly important for preserving important traits, such as environmental adaptability and disease resistance, that are already present within the species or its close relatives (Vasudevan et al., 2023). Despite several advantages offered by cisgenesis, there are certain limitations associated with the technique that need to be considered. One significant drawback of cisgenesis is the limitation in introducing traits from beyond the sexually compatible gene pool. Additionally, the creation of cisgenic crops requires exceptional expertise and time, meaning that the necessary genes or gene fragments may not be readily available and must instead be isolated from within the sexually compatible gene pool (Holme et al., 2013). Creating plants without markers often necessitates the development of novel methodologies, as existing ones are typically inadequate for the specific crop. Additionally, a significant proportion of transformed plants, ranging from 20% to 80%, carry the vector backbone sequence, necessitating the removal of numerous transgenic lines. Therefore, considerable effort must be expended to generate a large number of transformants, especially in crops with low transformation efficiency (Telem et al., 2013).

14.2.1.3 Intragenesis

Intragenesis involves transferring miRNA genes between different species, leading to GM in the recipient organism. This process combines specific gene fragments from donor organisms that are either the same species as the recipient or sexually compatible with it (Espinoza et al., 2013). These fragments can be organized in a specific sequence based on the alignment of the donor organism. Intragenic advancement entails incorporating a modified section of a gene's complete or partial coding region, frequently along with a different promoter and terminator sourced from a gene within the same species or one that can hybridize with it (Rijssen & Morris, 2018). In intragenesis, various plant genetic elements are combined in a laboratory setting to create gene expressions that are then inserted into plants belonging to the same sexually compatible gene pool (Rommens et al., 2007). However, unlike cis genes, intragenes are a combination of genes from different sources. For instance, an intragenic gene might contain genetic components from various genes with diverse promoter or terminator regions responsible for modulating gene expression. Thus, there exists the potential for creating novel gene combinations through in vitro reassembly of functional genetic elements. The characteristics resulting from intragenesis may not align with those of conventional breeding, as the expression levels of hybrid genes could vary from those naturally occurring (Holme et al., 2013). It's often contrasted with cisgenesis, as both methods offer alternatives within the realm of transgenic approaches. Both entail gene transfer between compatible species, omitting recombination among incompatible ones and sidestepping linkage drag. Consequently, intragenic or cisgenic crops generally enjoy greater public favorability compared to transgenic crops (Holme et al., 2013). However, there are distinctions within this field. Cisgenesis involves the genetic alteration wherein a gene native to one plant is transferred to another compatible plant, along with its promoter and terminator sequences (Moradpour & Abdullah, 2017). Intragenesis involves genetically modifying a recipient plant by introducing a gene along with its regulatory elements sourced from other compatible plants (Schouten & Jacobsen, 2008). In cisgenesis, the original genetic composition remains intact, unlike in intragenesis. These procedures effectively safeguard against the unwanted transfer of traits (Mujjassim et al., 2019).

14.2.2 miRNA Genome Editing

Conventional methods to bioengineer genetically modified crops resistant to climate change are relatively time-consuming and complex regarding their availability as food products in the market. Genome editing using miRNA can modify, silence, or insert target genes. Genome editing methods enable the addition of specific mutations with known precision, unlike mutations introduced through breeding methods, where the exact mutations may not be identified. Another advantage of genome editing over traditional approaches is the potential to regulate targeted gene expression by modifying regulatory sites (Gaba et al., 2021).

Genome editing operates by employing site-specific nucleases to alter specific regions of the genome, enhancing resistance to the effects of climate change (Gaba et al., 2021). These specialized enzymes induce double-stranded breaks at targeted locations. Subsequently, cellular repair mechanisms such as homologous recombination or non-homologous end joining mend these breaks (Shen & Li, 2022). These enzymes fall into categories such as zinc finger nucleases (ZFNs), transcription activator-like effector nucleases (TALENs), and the clustered regularly interspaced short palindromic repeats/CRISPR-associated protein system (Zenda et al., 2021).

14.2.2.1 Zinc Finger Nucleases

ZFNs represent genetically modified endonucleases incorporating a DNA-binding domain derived from ZNFs, which selectively attaches to a designated location, alongside a nuclease segment sourced from the FokI restriction enzyme, facilitating DNA cleavage (Zhang et al., 2018). A ZFN can bind to a specific base sequence due to the ZFP region. In this region, each of the Cys2-His2 fingers within the tandem array identifies approximately three base pairs of DNA (Wolfe et al., 2000). Recent studies have incorporated six such Cys2-His2 fingers to target longer and more discrete cleavage sites (Hajdu et al., 2023). Additionally, the Fok1 catalytic domain has two distinct characteristics to ensure specificity. The dimerization of Fok1 is essential for its cleavage ability, since this binding is weak (González Castro et al., 2021). Thus, DNA cleavage necessitates the independent and adjacent binding of the Fok1 domains, which must be accurately aligned and spaced to enable dimerization. As a result, this process ensures the specificity of the cleavage activity of the ZFN.

ZFNs have been employed to modify the genetic makeup of diverse plants such as figs, apples, rice, *Arabidopsis*, corn, and soybeans (Zhang et al., 2018). ZFNs have also been utilized for "trait stacking," allowing for the integration of several genetic alterations into one crop variety. This technique holds promise for enhancing crop resilience to climate change on multiple levels (Ainley et al., 2013). Additionally, ZFNs have been employed to identify specific locations within the rice genome suitable for the secure incorporation of genetic enhancements aimed at enhancing the crop (Cantos et al., 2014).

However, the primary challenges in the utilization of ZFNs are their expensive and time-consuming assembly, as well as their inability to target sites that are distantly located from each other and show lower guanine concentration (Scheben et al., 2016). Thus, the overall application of ZFNs for genomic editing is limited and needs further research in terms of miRNA-based applications.

14.2.2.2 Transcription Activator-Like Effector Nucleases

TALENs address the challenges posed by ZFNs by employing repeated variable di-residues to construct extended TAL repeat arrays that specifically target genomic sequences (Bhardwaj & Nain,

2021). Furthermore, they are also more adaptable in terms of targeting, as their only requirement is that the target site begins with a thymine residue (Scheben et al., 2016). As a result, TALENs can target a wider variety of crops compared to ZFNs.

TALENs are synthetic hybrid enzymes composed of a DNA-binding section known as a transcription activator-like effector (TALE) domain and a nuclease section. These TALENs identify and target specific DNA sequences, enabling the introduction of double-stranded breaks to facilitate genome editing (Godbey, 2022). Therefore, TALENs function as a type of adaptable restriction enzyme.

TALENs have enabled the editing of numerous crops, including sugar cane, thale cress, tomato, potato, barley, and flax (Zhang et al., 2018). To enhance wheat crop's resistance against powdery mildew, TALENs were utilized to deactivate three TaMLO homologs (Wang et al., 2014). However, challenges related to protein biosynthesis, engineering, and evaluation, as well as issues with TALE repeat creation, are some of the limitations to the widespread implementation of these nucleases (Yadav et al., 2021). Nevertheless, TALENs represent a significant advancement in GM and have the potential to create crop plants that are climate-adaptive by incorporating foreign genes.

14.2.2.3 CRISPR/Cas9 System

Over time, CRISPR/Cas9 has made significant strides and has been widely employed for genome editing in plant crops to enhance agricultural yields. It is recognized as a third-generation genome editing technology due to its remarkable adaptability (Yadav et al., 2021). In comparison to other genome editing methods such as ZFNs and TALENs, the CRISPR/Cas9 system stands out for its ability to target numerous genes simultaneously. Additionally, it offers advantages in terms of affordability, efficiency, and simplicity (Lopez-Obando et al., 2016). Currently, the predominant method for genome manipulation is the use of the *Streptococcus pyogenes* type II CRISPR system. Typically, plants are modified to enable the expression of Cas9 along with a chimeric guide RNA simultaneously (Jinek et al., 2012). This results in a double-stranded break at a position specified by a target sequence introduced into the guide RNA, which spans about 20 nucleotides. Additionally, there must be a protospacer adjacent motif of three nucleotides near the target site (Scheben et al., 2016). This method facilitates the effective modification of nearly any part of the genetic material, similar to TALENs and ZFNs. Unlike these other methods, which require extensive time and cost to produce specific proteins for each target sequence, CRISPR/Cas9 doesn't demand such resources. Additionally, the precision of genetic alterations achieved with CRISPR/Cas9 can be enhanced by utilizing Cas9 nickase (Shen et al., 2014).

Since its inception, the CRISPR/Cas9 system has been employed to edit the genomes of various crops such as soybean, maize, tomato, potato, wheat, and barley (Yadav et al., 2021). From enhancing the nutritional value of crops such as *Camelina sativa* to generating mutant rice varieties with increased grain quantity and improved grain size, this technique has shown significant potential for genome editing to improve agricultural crops (Jiang et al., 2017; Li et al., 2021). Similarly, the CRISPR/Cas9 technique has been utilized to strengthen the resilience of crops amid climate fluctuations. Previous research demonstrated enhancements in drought tolerance in thale cress (*Arabidopsis thaliana*) by deactivating the miRNA gene section known as MIR169a. Additionally, another study by Roca Paixao achieved similar results by activating the AREB1 promoter within the abscisic acid signaling pathway, a crucial factor in drought resilience (Roca Paixão et al., 2019; Zhao et al., 2016). Furthermore, to enhance rice's ability to withstand drought, researchers targeted and increased the expression of osmotic stress/ABA–activated protein kinase 2 (SAPK2), which is known respond to drought and salt stress (Lou et al., 2017). In a study conducted by Chung and colleagues in 2020, recombinant codon-optimized Cas9 (rCas9) was employed to target mature miRNA sequences or biogenesis sites in rice (*Oryza sativa*). The research led to the creation of 409 transgenic rice plants, focusing on 13 miRNAs associated with responses to drought tolerance, utilizing CRISPR/Cas9 technology (Chung et al., 2020).

Currently, there are no commercially available crops created through genome editing methods. However, these techniques have been applied to improve agronomic traits in crops sensitive to environmental factors, such as disease resistance. This study illustrates the potential of genome editing to enhance various characteristics of crops. The CRISPR/Cas9 system, known for its efficiency, precise targeting, and ability to edit multiple genes simultaneously, has broadened the scope of possibilities for both fundamental and practical research in plant biotechnology (Yadav et al., 2021).

14.2.3 miRNA Target Mimicry

To regulate the activity of miRNA, plants employ miRNA decoys, also called endogenous target mimics, consisting of short 200-nucleotide sequences that lack an ORF and exhibit varying expression in response to unfavorable conditions (Liu et al., 2019). By blocking miRNAs from pairing with the actual target transcripts through sequence complementarity, endogenous target mimics promote the accumulation of target mRNAs. The short tandem target mimicry strategy can be adopted to regulate the expression and accumulation of miRNA in plants and, in turn, control its associated processes within the crop in response to climatic conditions (Chaudhary et al., 2021). MiRNA target mimicry entails introducing into the genetically modified plant either custom-designed long non-coding RNA or circular RNA, commonly referred to as miRNA recognition elements, that exhibit significant sequence similarity to the target miRNA (Reichel et al., 2015). This engineered lncRNA or circular RNA contains several conserved binding sites for the specific miRNA target. However, the cleavage site shows minor sequence discrepancies, preventing the miRNA from being cleaved. Consequently, the miRNA remains hybridized but lacks physiological activity (Franco-Zorrilla et al., 2007).

A novel approach called short tandem miRNA target mimicry has been devised to analyze the role of miRNA in different crops. In a recent investigation, this method was utilized to examine how miR165/166 and miR160 interact and affect leaf development and response to drought stress in *Arabidopsis* plants (Yang et al., 2019). In another study, the use of a construct that mimicked the target of miR169n (MIM169n) effectively sequestered miR169n, resulting in heightened resilience to drought during seedling growth and improved tolerance to osmotic stress during the post-germination phase in rapeseed plants, also known as *Brassica napus* L. (Li et al., 2021). Furthermore, the short tandem target mimicry system was utilized to develop drought-tolerant rice crops by inactivating miR166 (Zhang et al., 2018).

Moreover, the imitation of miR319 targeting (MIM319) resulted in increased ethylene production in C_3 plants like rice and *Arabidopsis*. Additionally, it provided resistance against salt stress in these crops (Liu et al., 2019). Short tandem target mimicry represents a swift and effective approach to impede the function of specific small RNAs in plant systems. Upon the insertion of the target mimics into cells through transfection or transgenics, the target small RNAs/miRNAs in both plant systems show substantial decreases, indicating a mechanism of small RNA breakdown induced by the short tandem target mimics (Tang & Tang, 2013). Thus, they exhibit promising potential for combating stressors due to climate change in agronomically important crops.

14.2.4 Nanocarrier-Mediated Topical Delivery of miRNA

The use of nanomaterials in plant biotechnology is an emerging trend that holds significant promise. Nanotechnology is applied in agriculture to develop crops that are tolerant to stress and to boost overall growth. For example, modified nanomaterials have been shown to influence seed germination, plant growth, enhance resistance to diseases, and detect harmful agricultural chemicals (Nuruzzaman et al., 2016). Consequently, it may be beneficial to consider nanotechnology for improving plant health against various environmental stressors.

Nanoparticle engineering has enabled these tiny structures to navigate biological barriers due to their compactness and versatile physical and chemical properties. They serve as safe, non-viral vehicles for delivering various biomolecules, such as miRNA, into cells without causing harm. To ensure efficient delivery within cells, nanoparticle characteristics like size, shape, and biological activity have been fine-tuned. Numerous nanoparticle variants have been developed to respond to environmental cues such as temperature, pH, redox conditions, and enzymatic presence (Cunningham et al., 2018). Additionally, specific nanoparticles have been designed to achieve responsive cargo release for better site-specific management (Karimi et al., 2016).

The potential of miRNA delivery directly to target plants shows promise in enhancing crop yields through improved plant design and increased tolerance to abiotic stresses. Yet, the current challenge lies in the lack of a reliable delivery method for miRNAs due to the size limitation of nanoparticles, which must be less than 20 nm to penetrate the plant cell wall effectively (Yang et al., 2022). The low efficacy of gene silencing is also a result of miRNAs' resistance to the cell membrane's negative charge and their instability in vivo (Czauderna et al., 2003). To identify functional genes, these crucial delivery-related challenges to miRNAs must be addressed. Although the use of nanoparticles to transfer RNA has been thoroughly studied in diagnosis, therapeutic approaches, and pest management, further research is still required to fully understand its vast potential in agroecosystems. To overcome biological challenges, nano delivery systems utilize their control over morphology and surface functionalization, along with their ability to penetrate plant tissues effectively (Gelaw & Sanan-Mishra, 2022; Zhang et al., 2022). Advancements in nanomaterials offer promising prospects for the effective transport of biomolecules such as miRNA and CRISPR-Cas into plant cells (Mujtaba et al., 2021). It is anticipated that this will be a promising approach to overcoming the drawbacks of miRNA delivery. Yang and colleagues pioneered the development of the first star polycation nanocarrier delivery system, facilitating the transport of double-stranded precursor miRNA into the plant protoplasts of maize and *Arabidopsis*. This breakthrough allows for the delivery of specific miRNA targets, offering tailored solutions for enhancing resilience to diverse climatic conditions (Yang et al., 2022).

14.2.5 Cross-Kingdom miRNA Transfer

Some pathogenic bacteria, as well as those in symbiotic and mutualistic partnerships, can exhibit cross-kingdom miRNA transfer. Many studies have identified the presence of endogenous short RNAs, such as miRNA, in plants that are perfectly complementary to those found in animals or humans (Ivashuta et al., 2009). It was later discovered that plant-origin miRNA may also be absorbed and expressed by humans and pathogens, opening up a new form of genetic engineering in which a miRNA from one kingdom can be expressed in another to effect certain traits of interest to the host. For example, the process of miRNA cross-transfer presents an opportunity to reduce the removal of LDL from plasma by suppressing the low-density lipoprotein receptor adapter protein 1 in mammals through miR168a. This miRNA, derived from plants, can be absorbed and translated within the gastrointestinal tracts of various mammals (Zhang et al., 2012). Another approach involves employing the artificial miRNA strategy, which has been devised to generate particular miRNAs capable of efficiently suppressing targeted genes.

MiRNA horizontal transfer can occur in two ways: either passively during cellular consumption by the host organism or through interactions between the host and parasite. This transfer serves as a defense or counter-defense mechanism employed by either the host or the parasite (Basso et al., 2019). The presence of exosomes and specific RNA binding sites increases the stability of miRNA-mediated gene transfer. Both pre- and pri-miRNAs exhibit notable stability, enhancing the likelihood of effective delivery into cells. However, they are susceptible to digestion by the RNAi machinery within these organisms. It is thought to be achievable in some situations because these molecules are compatible with the host organism's RNAi machinery. Unfortunately, sensitive technologies, relevant methodologies, and efficient protocols for elucidating pri-miRNA or mature miRNA interkingdom cross-talk are still scarce.

Currently, this method finds extensive application in the cultivation of pathogen-resistant and disease-resistant crops. However, it also holds potential for enhancing plants' resilience to climate change. The cross-kingdom transfer is bidirectional; a miRNA can be transferred from the pathogen to the plant, usually resulting in disease-like conditions, while the pathogen can acquire miRNA from the plant, thus inhibiting its disease-causing tendencies. This was demonstrated by Ming Wang and his team, who showed that Botrytis, a type of fungal pathogen, alters host immunity via manipulation of small RNAs produced by Dicer-like protein 1 (Bc-DCL1) and Bc-DCL2 in Botrytis. By expressing sRNAs that inhibit Bc-DCL1 and Bc-DCL2 in plants, cross-transfer into the pathogenic fungi can occur, reducing its pathogenicity (Wang et al., 2016).

Zang et al. demonstrated another instance of miRNA transfer between the plant and fungal kingdoms. They found that miR159 and miR166 from the cotton variety *Gossypium hirsutum* provide resistance against *Verticillium dahliae*, a fungal pathogen (Zhang et al., 2016). These miRNAs were identified within fungal hyphae extracted from infected cotton tissues. They interact with virulence-related proteins, specifically HiC-15 (isotrichodermin C-15 hydroxylase) and Clp-1 (Ca2+-dependent cysteine protease). Consequently, this led to the expression of miRNA-resistant HiC-15 and Clp-1 in tobacco and *V. dahliae.*

14.2.6 Epigenetic Memory

Epigenetics refers to changes in gene expression patterns that occur without variations in DNA sequence. Epigenetic mechanisms involve covalent changes to DNA and histones, which influence chromatin expression levels and are heritable since chromatin modifications can be carried during mitotic and meiotic divisions. Epigenetic changes in plants can be brought about via two methods: one is methylation of DNA, and the second is histone modifications (Iwasaki & Paszkowski, 2014). Throughout their lifespans, plants confront a range of environmental and physiological fluctuations, such as droughts, excessive rainfall, low temperatures, and extreme heat. They strive to adapt to these challenges optimally. The changes in gene expression resulting from such adaptations are referred to as epigenetic memory. For example, due to seasonal or climatic changes, trees may have to survive in drought-like conditions, for which different species of trees have different adaptations, such as variations in tracheid size, reduced leaf area, embedded stomata, and rapid stomatal closure (Carvalho et al., 2015; Kijowska-Oberc et al., 2020; Sánchez-Salguero et al., 2015).

The involvement of epigenetics in abiotic stress tolerance can be demonstrated by comparing the stress-resistant and stress-sensitive variants of the same species. Methylation-sensitive amplified polymorphism analysis shows that salt-resistant and sensitive varieties exhibit many phenotypic variations. However, the tissue-specific loci are conserved but exhibit methylation polymorphisms (Wang et al., 2011). The study also concluded that demethylation is a cause of the epigenetic response against salt stress in rice.

Plants react to diverse non-living environmental cues, such as fluctuations in temperature, intense light, water scarcity, UV radiation, and osmotic pressure, through mechanisms that trigger the generation of ROS and RNS. This process induces hypermethylation of DNA, ultimately resulting in inheritable alterations in gene expression patterns (Guarino et al., 2022). Changes in genome methylation could be involved in the plant's adaptation to stress conditions. Additionally, abiotic stresses lead to changes in the expression of small RNAs. For instance, research has demonstrated that heat stress can trigger the release of small RNAs associated with pentatricopeptide repeat proteins, leading to the downregulation of genes essential for photosynthesis in chloroplasts (He et al., 2019).

Plants exhibit different epigenetic modifications to cope with climate change, which can be manipulated to develop higher-quality crops. Improved drought tolerance can be induced in cotton (*Gossypium herbaceum*) plants via the overexpression of AtHUB2 using the cauliflower mosaic virus (CaMV) 35S promoter, which monoubiquitylates H2B. The HUB2 gene encodes enzymes that govern various processes, such as salt tolerance and defense responses, and it also affects seed dormancy (Chen et al., 2019). Salt tolerance can be enhanced in plants by manipulating the genes responsible for ion transport. By downregulating the expression of TaHKT2;1 and TaHKT2;3, which are associated with potassium transporters in wheat (*Triticum aestivum* L.) via cysteine methylation, salt tolerance can be improved (Kumar et al., 2017).

Another strategy for using epigenetics to induce climate change resistance is to pre-stress the plants with minor abiotic challenges. These challenges prompt the plant to create an epigenetic response, which is then stored in the plant's DNA as epigenetic memories, allowing the plant to better adapt to climate change (do Amaral et al., 2020). A significant alteration in the genome of tolerant rice species was observed in a study after exposure to minor pre-stress conditions.

14.3 DEVELOPMENTAL CHALLENGES AND FUTURE PROSPECTS

Research has suggested that miRNA contributes to various aspects of plant growth and development, such as controlling shoot and flower development, regulating male fertility, and participating in molecular processes like anthocyanin synthesis (Sun et al., 2012). Many studies using miRNA as a tool to make plants more adaptable to climate change have yielded promising results. Many of these strategies have already been mentioned above. Approaches such as STTM and similar miRNA methods facilitate the examination of intricate processes involved in a plant's response to diverse biotic and abiotic stresses. Meanwhile, the cost-effectiveness of producing transgenic crops and precisely manipulating genes has increased with the advent of CRISPR/Cas gene-editing systems, in contrast to earlier methods like meganucleases, ZFNs, and TALENs (Chen et al., 2021). However, practical utilization of these technologies is hampered by various challenges associated with their application.

Each trait is controlled by several miRNAs, and to ensure effective results, multiple traits at various stages of development must be manipulated simultaneously. The exploration of miRNA's role in regulating plant growth mechanisms is of significant importance. Any modifications to miRNA or its target sites could potentially lead to adverse effects on plant development, triggering multifaceted changes (Djami-Tchatchou et al., 2017). Some miRNA expressions and their targets are highly specific, as miRNAs affect the targets of specific cells only under specific conditions. Hence, careful design of the methods is necessary to obtain the desired outcomes.

For efficient design of the miRNA technique, it is essential to first detect the miRNA of interest, deliver the miRNA modulators to upregulate or downregulate the specific miRNA, and take measures to minimize nonspecific interactions (Momin et al., 2021). Since individual miRNAs regulate multiple genes, it's essential to conduct an analysis when devising a method to modify a particular miRNA to ensure that there are no unintended consequences on other traits linked to the miRNA. Therefore, it's imperative to explore pri-miRNA-mediated regulatory networks in research (Chaudhary et al., 2021).

Research has demonstrated the significant role of miRNA in plant growth and development, gene expression regulation networks, and the maintenance of the shoot apical meristem. For instance, miRNA targets such as ARF10, ARF16, and ARF17 influence auxin response, thereby regulating leaf development (Dong et al., 2022). A changing global climate has led to increased drought, less rainfall, higher temperatures, saltwater intrusion, and other effects. During such abiotic stress, the plant can produce a regulatory response through the overexpression or downregulation of certain miRNAs. These reactions can be modified to create plants that are resistant to climate change. To achieve these objectives, we must first develop an efficient method for accurately and affordably sequencing the genetic information of the plant. There is also a need to develop protocols that can alter the genetic sequence or incorporate desired changes in the plant. Efforts are underway to enhance understanding of the diverse miRNAs present in plants and their roles in influencing plant responses to stress factors (Verma et al., 2023).

14.4 CONCLUSION

The complex regulatory functions of miRNAs in the growth and development of plants, especially in addressing environmental challenges triggered by climate fluctuations, highlight their promise in agricultural adaptation efforts. However, the effective deployment of miRNA-based strategies is impeded by the necessity for precise manipulation of multiple traits concurrently, as well as the need to mitigate unintended off-target effects. Furthermore, a comprehensive understanding of the complex regulatory networks orchestrated by miRNAs remains a critical area requiring further investigation. Advancements in high-throughput sequencing technologies, coupled with refined gene-editing tools like CRISPR/Cas systems, are pivotal in enabling precise genetic manipulations conducive to enhancing plant resilience against environmental adversities. Additionally, elucidating the intricate mechanisms governing miRNA-mediated regulatory pathways is imperative for optimizing the efficacy and specificity of miRNA-based interventions in agricultural contexts. Therefore, sustained research efforts aimed at unraveling the underlying molecular mechanisms driving miRNA-mediated responses in plants are essential for unlocking the full potential of these regulatory molecules in fostering agricultural sustainability amidst the challenges posed by climate change.

REFERENCES

Ainley, W. M., Sastry-Dent, L., Welter, M. E., Murray, M. G., Zeitler, B., Amora, R., Corbin, D. R., Miles, R. R., Arnold, N. L., Strange, T. L., Simpson, M. A., Cao, Z., Carroll, C., Pawelczak, K. S., Blue, R., West, K., Rowland, L. M., Perkins, D., Samuel, P., & Petolino, J. F. (2013). Trait stacking via targeted genome editing. *Plant Biotechnology Journal, 11*(9), 1126–1134. https://doi.org/10.1111/pbi.12107

Allen, R. S., Li, J., Stahle, M. I., Dubroué, A., Gubler, F., Millar, & A. A. (2007). Genetic analysis reveals functional redundancy and the major target genes of the Arabidopsis miR159 family. *Proceedings of the National Academy of Sciences of the United States of America, 104*(41), 16371–16376. https://doi.org/10.1073/pnas.0707653104

Anil, P., Dharmendra, P., Sachidanand, S., Yuriy, E., Snezhan, B., Dilfuza, J., & Chinmayi, J. (2023). Beneficial effects of biochar application on mitigating the drought and salinity stress implications on plants. *Plant Science Today, 10*(2), 188–193. http://doi.org/10.14719/pst.2591

Aukerman, M. J., & Sakai, H. (2003). Regulation of flowering time and floral organ identity by a microRNA and its *APETALA2*-like target genes. *The Plant Cell, 15*(11), 2730–2741. https://doi.org/10.1105/tpc.016238

Ayan, A., Meriç, S., Gümüş, T., Atak, Ç., Ayan, A., Meriç, S., Gümüş, T., & Atak, Ç. (2023). Transgenic plants in heat stress adaptation: present achievements and prospects. In M. Oliveira & A. Fernandes-Silva (Eds.), *Abiotic Stress in Plants—Adaptations to Climate Change*. IntechOpen, London. https://doi.org/10.5772/intechopen.111791

Basso, M. F., Ferreira, P. C. G., Kobayashi, A. K., Harmon, F. G., Nepomuceno, A. L., Molinari, H. B. C., & Grossi-de-Sa, M. F. (2019). microRNAs and new biotechnological tools for its modulation and improving stress tolerance in plants. *Plant Biotechnology Journal, 17*(8), 1482–1500. https://doi.org/10.1111/pbi.13116

Bhardwaj, A., & Nain, V. (2021). TALENs—an indispensable tool in the era of CRISPR: a mini review. *Journal of Genetic Engineering and Biotechnology, 19*(1), 125. https://doi.org/10.1186/s43141-021-00225-z

Cantos, C., Francisco, P., Trijatmiko, K. R., Slamet-Loedin, I., & Chadha-Mohanty, P. K. (2014). Identification of "safe harbor" loci in indica rice genome by harnessing the property of zinc-finger nucleases to induce DNA damage and repair. *Frontiers in Plant Science, 5*. https://www.frontiersin.org/journals/plant-science/articles/10.3389/fpls.2014.00302

Carlsbecker, A., Lee, J.-Y., Roberts, C. J., Dettmer, J., Lehesranta, S., Zhou, J., Lindgren, O., Moreno-Risueno, M. A., Vatén, A., Thitamadee, S., Campilho, A., Sebastian, J., Bowman, J. L., Helariutta, Y., & Benfey, P. N. (2010). Cell signalling by microRNA165/6 directs gene dose-dependent root cell fate. *Nature, 465*(7296), 316–321. https://doi.org/10.1038/nature08977

Carvalho, A., Nabais, C., Vieira, J., Rossi, S., & Campelo, F. (2015). Plastic response of tracheids in pinus pinaster in a water-limited environment: adjusting lumen size instead of wall thickness. *PLoS One, 10*(8), e0136305. https://doi.org/10.1371/journal.pone.0136305

Chaudhary, S., Grover, A., & Sharma, P. C. (2021). microRNAs: potential targets for developing stress-tolerant crops. *Life, 11*(4), 289. https://doi.org/10.3390/life11040289

Chen, H., Feng, H., Zhang, X., Zhang, C., Wang, T., & Dong, J. (2019). An *Arabidopsis* E3 ligase *HUB2* increases histone H2B monoubiquitination and enhances drought tolerance in transgenic cotton. *Plant Biotechnology Journal, 17*(3), 556–568. https://doi.org/10.1111/pbi.12998

Chen, J., Teotia, S., Lan, T., & Tang, G. (2021). microRNA techniques: valuable tools for agronomic trait analyses and breeding in rice. *Frontiers in Plant Science, 12*, 744357. https://www.frontiersin.org/journals/plant-science/articles/10.3389/fpls.2021.744357

Chung, P. J., Chung, H., Oh, N., Choi, J., Bang, S. W., Jung, S. E., Jung, H., Shim, J. S., & Kim, J.-K. (2020). Efficiency of recombinant CRISPR/rCas9-mediated miRNA gene editing in rice. *International Journal of Molecular Sciences, 21*(24), 9606. https://doi.org/10.3390/ijms21249606

Cunningham, F. J., Goh, N. S., Demirer, G. S., Matos, J. L., & Landry, M. P. (2018). Nanoparticle-mediated delivery towards advancing plant genetic engineering. *Trends in Biotechnology, 36*(9), 882–897. https://doi.org/10.1016/j.tibtech.2018.03.009

Czauderna, F., Fechtner, M., Dames, S., Aygün, H., Klippel, A., Pronk, G. J., Giese, K., & Kaufmann, J. (2003). Structural variations and stabilising modifications of synthetic siRNAs in mammalian cells. *Nucleic Acids Research, 31*(11), 2705–2716.

Djami-Tchatchou, A. T., Sanan-Mishra, N., Ntushelo, K., & Dubery, I. A. (2017). Functional roles of microRNAs in agronomically important plants—potential as targets for crop improvement and protection. *Frontiers in Plant Science, 8*, 378. https://doi.org/10.3389/fpls.2017.00378

do Amaral, M. N., Auler, P. A., Rossatto, T., Barros, P. M., Oliveira, M. M., & Braga, E. J. B. (2020). Long-term somatic memory of salinity unveiled from physiological, biochemical and epigenetic responses in two contrasting rice genotypes. *Physiologia Plantarum, 170*(2), 248–268. https://doi.org/10.1111/ppl.13149

Dong, Q., Hu, B., & Zhang, C. (2022). microRNAs and their roles in plant development. *Frontiers in Plant Science, 13*, 824240. https://www.frontiersin.org/journals/plant-science/articles/10.3389/fpls.2022.824240

Eckerstorfer, M. F., Engelhard, M., Heissenberger, A., Simon, S., & Teichmann, H. (2019). Plants developed by new genetic modification techniques—comparison of existing regulatory frameworks in the EU and non-EU countries. *Frontiers in Bioengineering and Biotechnology, 7*, 26. https://doi.org/10.3389/fbioe.2019.00026

Espinoza, C., Schlechter, R., Herrera, D., Torres, E., Serrano, A., Medina, C., & Arce-Johnson, P. (2013). Cisgenesis and intragenesis: new tools for improving crops. *Biological Research, 46*(4), 323–331. https://doi.org/10.4067/S0716-97602013000400003

Franco-Zorrilla, J. M., Valli, A., Todesco, M., Mateos, I., Puga, M. I., Rubio-Somoza, I., Leyva, A., Weigel, D., García, J. A., & Paz-Ares, J. (2007). Target mimicry provides a new mechanism for regulation of microRNA activity. *Nature Genetics, 39*(8), 8. https://doi.org/10.1038/ng2079

Gaba, Y., Pareek, A., & Singla-Pareek, S. L. (2021). Raising climate-resilient crops: journey from the conventional breeding to new breeding approaches. *Current Genomics, 22*(6), 450–467. https://doi.org/10.2174/1389202922666210928151247

Gelaw, T. A., & Sanan-Mishra, N. (2022). Nanomaterials coupled with microRNAs for alleviating plant stress: a new opening towards sustainable agriculture. *Physiology and Molecular Biology of Plants, 28*(4), 791–818. https://doi.org/10.1007/s12298-022-01163-x

Godbey, W. T. (2022). Genome editing. In W. T. Godbey (Ed.), *Biotechnology and Its Applications* (Second Edition) (pp. 339–355). Academic Press, New York. https://doi.org/10.1016/B978-0-12-817726-6.00014-9

González Castro, N., Bjelic, J., Malhotra, G., Huang, C., & Alsaffar, S. H. (2021). Comparison of the feasibility, efficiency, and safety of genome editing technologies. *International Journal of Molecular Sciences, 22*(19), 19. https://doi.org/10.3390/ijms221910355

Grover, A., Mittal, D., Negi, M., & Lavania, D. (2013). Generating high temperature tolerant transgenic plants: achievements and challenges. *Plant Science, 205–206*, 38–47. https://doi.org/10.1016/j.plantsci.2013.01.005

Guarino, F., Cicatelli, A., Castiglione, S., Agius, D. R., Orhun, G. E., Fragkostefanakis, S., Leclercq, J., Dobránszki, J., Kaiserli, E., Lieberman-Lazarovich, M., Sõmera, M., Sarmiento, C., Vettori, C., Paffetti, D., Poma, A. M. G., Moschou, P. N., Gašparović, M., Yousefi, S., Vergata, C., & Martinelli, F. (2022). An epigenetic alphabet of crop adaptation to climate change. *Frontiers in Genetics, 13*, 818727. https://doi.org/10.3389/fgene.2022.818727

Hajdu, B., Hunyadi-Gulyás, É., Kato, K., Kawaguchi, A., Nagata, K., & Gyurcsik, B. (2023). Zinc binding of a Cys2His2-type zinc finger protein is enhanced by the interaction with DNA. *Journal of Biological Inorganic Chemistry, 28*(3), 301–315. https://doi.org/10.1007/s00775-023-01988-1

Hammond, S. M. (2005). Dicing and slicing. *FEBS Letters, 579*(26), 5822–5829. https://doi.org/10.1016/j.febslet.2005.08.079

Hansen Jesse, L. C., & Obrycki, J. J. (2000). Field deposition of Bt transgenic corn pollen: lethal effects on the monarch butterfly. *Oecologia, 125*(2), 241–248. https://doi.org/10.1007/s004420000502

Hatfield, J. L., & Prueger, J. H. (2015). Temperature extremes: effect on plant growth and development. *Weather and Climate Extremes*, *10*, 4–10. https://doi.org/10.1016/j.wace.2015.08.001

He, J., Jiang, Z., Gao, L., You, C., Ma, X., Wang, X., Xu, X., Mo, B., Chen, X., & Liu, L. (2019). Genome-wide transcript and small RNA profiling reveals transcriptomic responses to heat stress1. *Plant Physiology*, *181*(2), 609–629. https://doi.org/10.1104/pp.19.00403

Health, N. R. C. (US) C. on I. and A. U. E. of G. E. F. on H. (2004). Methods and Mechanisms for Genetic Manipulation of Plants, Animals, and Microorganisms. In *Safety of Genetically Engineered Foods: Approaches to Assessing Unintended Health Effects*. National Academies Press (US). https://www.ncbi.nlm.nih.gov/books/NBK215771/

Herman, R. A., & Ladics, G. S. (2011). Endogenous allergen upregulation: transgenic vs. traditionally bred crops. *Food and Chemical Toxicology*, *49*(10), 2667–2669. https://doi.org/10.1016/j.fct.2011.07.018

Holme, I. B., Wendt, T., & Holm, P. B. (2013). Intragenesis and cisgenesis as alternatives to transgenic crop development. *Plant Biotechnology Journal*, *11*(4), 395–407. https://doi.org/10.1111/pbi.12055

Huntzinger, E., & Izaurralde, E. (2011). Gene silencing by microRNAs: contributions of translational repression and mRNA decay. *Nature Reviews Genetics*, *12*(2), 2. https://doi.org/10.1038/nrg2936

Hussain, H. A., Hussain, S., Khaliq, A., Ashraf, U., Anjum, S. A., Men, S., & Wang, L. (2018). Chilling and drought stresses in crop plants: implications, cross talk, and potential management opportunities. *Frontiers in Plant Science*, *9*, 393. https://www.frontiersin.org/journals/plant-science/articles/10.3389/fpls.2018.00393

Ipsaro, J. J., & Joshua-Tor, L. (2015). From guide to target: molecular insights into eukaryotic RNAi machinery. *Nature Structural & Molecular Biology*, *22*(1), 20–28. https://doi.org/10.1038/nsmb.2931

Ivashuta, S. I., Petrick, J. S., Heisel, S. E., Zhang, Y., Guo, L., Reynolds, T. L., Rice, J. F., Allen, E., & Roberts, J. K. (2009). Endogenous small RNAs in grain: semi-quantification and sequence homology to human and animal genes. *Food and Chemical Toxicology: An International Journal Published for the British Industrial Biological Research Association*, *47*(2), 353–360. https://doi.org/10.1016/j.fct.2008.11.025

Iwasaki, M., & Paszkowski, J. (2014). Epigenetic memory in plants. *The EMBO Journal*, *33*(18), 1987–1998. https://doi.org/10.15252/embj.201488883

Jiang, W. Z., Henry, I. M., Lynagh, P. G., Comai, L., Cahoon, E. B., & Weeks, D. P. (2017). Significant enhancement of fatty acid composition in seeds of the allohexaploid, *Camelina sativa*, using CRISPR/Cas9 gene editing. *Plant Biotechnology Journal*, *15*(5), 648–657. https://doi.org/10.1111/pbi.12663

Jinek, M., Chylinski, K., Fonfara, I., Hauer, M., Doudna, J. A., & Charpentier, E. (2012). A programmable dual RNA-guided DNA endonuclease in adaptive bacterial immunity. *Science*, *337*(6096), 816–821. https://doi.org/10.1126/science.1225829

Joshi, S., Patil, S., Shaikh, A., Jamla, M., & Kumar, V. (2024). Modern omics toolbox for producing combined and multifactorial abiotic stress tolerant plants. *Plant Stress*, *11*, 100301. https://doi.org/10.1016/j.stress.2023.100301

Karimi, M., Ghasemi, A., Zangabad, P. S., Rahighi, R., Basri, S. M. M., Mirshekari, H., Amiri, M., Pishabad, Z. S., Aslani, A., Bozorgomid, M., Ghosh, D., Beyzavi, A., Vaseghi, A., Aref, A. R., Haghani, L., Bahrami, S., & Hamblin, M. R. (2016). Smart micro/nanoparticles in stimulus-responsive drug/gene delivery systems. *Chemical Society Reviews*, *45*(5), 1457–1501. https://doi.org/10.1039/C5CS00798D

Khan, S., Anwar, S., Yu, S., Sun, M., Yang, Z., & Gao, Z. (2019). Development of drought-tolerant transgenic wheat: achievements and limitations. *International Journal of Molecular Sciences*, *20*(13), 3350. https://doi.org/10.3390/ijms20133350

Kijowska-Oberc, J., Staszak, A. M., Kamiński, J., & Ratajczak, E. (2020). Adaptation of forest trees to rapidly changing climate. *Forests*, *11*(2), 2. https://doi.org/10.3390/f11020123

Kobayashi, K., Wang, X., & Wang, W. (2023). Genetically modified rice is associated with hunger, health, and climate resilience. *Foods*, *12*(14), 2776. https://doi.org/10.3390/foods12142776

Kumar, S., Beena, A. S., Awana, M., & Singh, A. (2017). Salt-induced tissue-specific cytosine methylation downregulates expression of HKT genes in contrasting wheat (*Triticum aestivum* L.) genotypes. *DNA and Cell Biology*, *36*(4), 283–294. https://doi.org/10.1089/dna.2016.3505

Lamaoui, M., Jemo, M., Datla, R., & Bekkaoui, F. (2018). Heat and drought stresses in crops and approaches for their mitigation. *Frontiers in Chemistry*, *6*, 26. https://www.frontiersin.org/articles/10.3389/fchem.2018.00026

Li, J., Duan, Y., Sun, N., Wang, L., Feng, S., Fang, Y., & Wang, Y. (2021). The miR169n-*NF-YA8* regulation module involved in drought resistance in *Brassica napus* L. *Plant Science*, *313*, 111062. https://doi.org/10.1016/j.plantsci.2021.111062

Liu, Y., Li, D., Yan, J., Wang, K., Luo, H., & Zhang, W. (2019). MiR319 mediated salt tolerance by ethylene. *Plant Biotechnology Journal*, *17*(12), 2370–2383. https://doi.org/10.1111/pbi.13154

Lopez-Obando, M., Hoffmann, B., Géry, C., Guyon-Debast, A., Téoulé, E., Rameau, C., Bonhomme, S., & Nogué, F. (2016). Simple and efficient targeting of multiple genes through CRISPR-Cas9 in physcomitrella patens. *G3: Genes|Genomes|Genetics*, *6*(11), 3647–3653. https://doi.org/10.1534/g3.116.033266

Losey, J. E., Rayor, L. S., & Carter, M. E. (1999). Transgenic pollen harms monarch larvae. *Nature*, *399*(6733), 6733. https://doi.org/10.1038/20338

Lou, D., Wang, H., Liang, G., & Yu, D. (2017). OsSAPK2 confers abscisic acid sensitivity and tolerance to drought stress in rice. *Frontiers in Plant Science*, *8*, 993. https://www.frontiersin.org/journals/plant-science/articles/10.3389/fpls.2017.00993

Ma, C., Burd, S., & Lers, A. (2015). *miR408* is involved in abiotic stress responses in Arabidopsis. *The Plant Journal: For Cell and Molecular Biology*, *84*(1), 169–187. https://doi.org/10.1111/tpj.12999

Maheswari, M., Varalaxmi, Y., Vijayalakshmi, A., Yadav, S. K., Sharmila, P., Venkateswarlu, B., Vanaja, M., & Pardha Saradhi, P. (2010). Metabolic engineering using mtlD gene enhances tolerance to water deficit and salinity in sorghum. *Biologia Plantarum*, *54*(4), 647–652. https://doi.org/10.1007/s10535-010-0115-y

Meenakshi, S., Rahul, D., Sachidanand, S., Anita, G. M., & Rajpal, S. (2025). Plant microRNAs: an overview. In R. Dutta, C. Kapadia, S. Danish, & S.Singh (Eds.), *microRNA Advances and Application in Plant Biology Plant Biology, Sustainability and Climate Change*. Academic Press, New York.

Momin, M. Y., Gaddam, R. R., Kravitz, M., Gupta, A., & Vikram, A. (2021). The challenges and opportunities in the development of microRNA therapeutics: a multidisciplinary viewpoint. *Cells*, *10*(11), 3097. https://doi.org/10.3390/cells10113097

Moradpour, M., & Abdullah, S. (2017). *Cisgenesis and Intragenesis as New Strategies for Crop Improvemen.* Springer, Cham, pp. 191–216. https://doi.org/10.1007/978-3-319-65079-1_9

Mosquera-Rendón, J., Moreno-Herrera, C. X., Robledo, J., & Hurtado-Páez, U. (2023). Genome-Wide Association Studies (GWAS) Approaches for the Detection of Genetic Variants Associated with Antibiotic Resistance: A Systematic Review. Microorganisms, 11(12), 2866. https://doi.org/10.3390/microorganisms11122866

Mujjassim, N. E., Mallik, M., Rathod, N. K. K., & Nitesh, S. D. (2019). Cisgenesis and intragenesis a new tool for conventional plant breeding: a review. *Journal of Pharmacognosy and Phytochemistry*, *8*(1), 2485–2489.

Mujtaba, M., Wang, D., Carvalho, L. B., Oliveira, J. L., Espirito Santo Pereira, A. do, Sharif, R., Jogaiah, S., Paidi, M. K., Wang, L., Ali, Q., & Fraceto, L. F. (2021). Nanocarrier-Mediated Delivery of miRNA, RNAi, and CRISPR-Cas for Plant Protection: Current Trends and Future Directions. *ACS Agricultural Science & Technology*, *1*(5), 417–435. https://doi.org/10.1021/acsagscitech.1c00146

Mullins, E., Bresson, J.-L., Dalmay, T., Dewhurst, I. C., Epstein, M. M., Firbank, L. G., Guerche, P., Hejatko, J., Moreno, F. J., Naegeli, H., Nogué, F., Serrano, J. J. S., Savoini, G., Veromann, E., Veronesi, F., Casacuberta, J., Dumont, A. F., Gennaro, A., & Rostoks, N. (2022). Updated scientific opinion on plants developed through cisgenesis and intragenesis. *EFSA Journal*, *20*(10), e07621. https://doi.org/10.2903/j.efsa.2022.7621

Noopur, K., Pragati, K., Adarsh, K. S., Sachidanand, S., Tanvi, J., & Megha, B. (2025a). Role of non coding RNA in effector trigger immunity. In R. Dutta, C. Kapadia, S. Danish, & S.Singh (Eds.), *Non-Coding RNA in Plants Modulation and Stress Responses Plant Biology, Sustainability and Climate Change.* Academic Press, New York.

Noopur, K., Pragati, K., Megha, B., Arpan, M., Tanvi, J., Shrisendu, M., Adarsh, K. S., & Sachidanand, S. (2025b). Role of non-coding RNA in plant architecture. In R. Dutta, C. Kapadia, S. Danish, & S.Singh (Eds.), *Non-Coding RNA in Plants Modulation and Stress Responses Plant Biology, Sustainability and Climate Change*. Academic Press, New York.

Nuruzzaman, Md., Rahman, M. M., Liu, Y., & Naidu, R. (2016). Nanoencapsulation, nano-guard for pesticides: a new window for safe application. *Journal of Agricultural and Food Chemistry*, *64*(7), 1447–1483. https://doi.org/10.1021/acs.jafc.5b05214

Organisms (GMO), E. P. on G. M., Mullins, E., Bresson, J.-L., Dalmay, T., Dewhurst, I. C., Epstein, M. M., Firbank, L. G., Guerche, P., Hejatko, J., Moreno, F. J., Naegeli, H., Nogué, F., Serrano, J. J. S., Savoini, G., Veromann, E., Veronesi, F., Casacuberta, J., Dumont, A. F., Gennaro, A., & Rostoks, N. (2022). Updated scientific opinion on plants developed through cisgenesis and intragenesis. *EFSA Journal*, *20*(10), e07621. https://doi.org/10.2903/j.efsa.2022.7621

O'Brien, J., Hayder, H., Zayed, Y., & Peng, C. (2018). Overview of microRNA biogenesis, mechanisms of actions, and circulation. *Frontiers in Endocrinology*, *9*, 402. https://www.frontiersin.org/journals/endocrinology/articles/10.3389/fendo.2018.00402

Pavone, V., & Martinelli, L. (2015). Cisgenics as emerging bio-objects: bio-objectification and bio-identification in agrobiotech innovation. *New Genetics and Society*, *34*(1), 52–71. https://doi.org/10.1080/14636778.2014.998816

Pooja, G., Megha, B., Arpan, M., Chintan, K., Rahul, D., Subhan, D., & Sachidanand, S. (2025). Introduction to long noncoding RNA. In R. Dutta, C. Kapadia, S. Danish, & S.Singh (Eds.), *Non-Coding RNA in Plants Modulation and Stress Responses Plant Biology, Sustainability and Climate Change* (Vol. 1, pp. 1–12). Academic Press, New York.

Rai, G. K., Kumar, P., Choudhary, S. M., Kosser, R., Khanday, D. M., Choudhary, S., Kumar, B., Magotra, I., Kumar, R. R., Ram, C., Rouphael, Y., Corrado, G., & Behera, T. K. (2023). Biomimetic strategies for developing abiotic stress-tolerant tomato cultivars: an overview. *Plants*, *12*(1), 1. https://doi.org/10.3390/plants12010086

Reichel, M., Li, Y., Li, J., & Millar, A. A. (2015). Inhibiting plant micro RNA activity: molecular *SPONGE* s, target *MIMIC s* and STTM s all display variable efficacies against target micro RNA s. *Plant Biotechnology Journal*, *13*(7), 915–926. https://doi.org/10.1111/pbi.12327

Rijssen, W. J. V, & Morris, E. J. (2018). Chapter 13- Safety and risk assessment of food from genetically engineered crops and animals: the challenges. In A. M. Holban, & A. M. Grumezescu (Eds.), *Genetically Engineered Foods* (pp. 335–368). Academic Press, New York. https://doi.org/10.1016/B978-0-12-811519-0.00013-3

Riolo, G., Cantara, S., Marzocchi, C., & Ricci, C. (2020). miRNA targets: from prediction tools to experimental validation. *Methods and Protocols*, *4*(1), 1. https://doi.org/10.3390/mps4010001

Roca Paixão, J. F., Gillet, F.-X., Ribeiro, T. P., Bournaud, C., Lourenço-Tessutti, I. T., Noriega, D. D., Melo, B. P. de, de Almeida-Engler, J., & Grossi-de-Sa, M. F. (2019). Improved drought stress tolerance in arabidopsis by CRISPR/dCas9 fusion with a histone acetyltransferase. *Scientific Reports*, *9*(1), 8080.

Rommens, C. M., Haring, M. A., Swords, K., Davies, H. V., & Belknap, W. R. (2007). The intragenic approach as a new extension to traditional plant breeding. *Trends in Plant Science*, *12*(9), 397–403. https://doi.org/10.1016/j.tplants.2007.08.001

Sánchez-Salguero, R., Ortíz, C., Covelo, F., Ochoa, V., García-Ruíz, R., Seco, J. I., Carreira, J. A., Merino, J. Á., & Linares, J. C. (2015). Regulation of water use in the southernmost European Fir (*Abies pinsapo* Boiss.): drought avoidance matters. *Forests*, *6*(6), 6. https://doi.org/10.3390/f6062241

Scheben, A., Yuan, Y., & Edwards, D. (2016). Advances in genomics for adapting crops to climate change. *Current Plant Biology*, *6*, 2–10.

Schnell, J., Steele, M., Bean, J., Neuspiel, M., Girard, C., Dormann, N., Pearson, C., Savoie, A., Bourbonnière, L., & Macdonald, P. (2015). A comparative analysis of insertional effects in genetically engineered plants: considerations for pre-market assessments. *Transgenic Research*, *24*(1), 1–17. https://doi.org/10.1007/s11248-014-9843-7

Schouten, H. J., & Jacobsen, E. (2008). Cisgenesis and intragenesis, sisters in innovative plant breeding. *Trends in Plant Science*, *13*(6), 260–261. https://doi.org/10.1016/j.tplants.2008.04.005

Shen, B., Zhang, W., Zhang, J., Zhou, J., Wang, J., Chen, L., Wang, L., Hodgkins, A., Iyer, V., & Huang, X. (2014). Efficient genome modification by CRISPR-Cas9 nickase with minimal off-target effects. *Nature Methods*, *11*(4), 399–402.

Shen, H., & Li, Z. (2022). DNA double-strand break repairs and their application in plant DNA integration. *Genes*, *13*(2), 322. https://doi.org/10.3390/genes13020322

Singhal, P., Jan, A. T., Azam, M., & Haq, Q. M. R. (2016). Plant abiotic stress: a prospective strategy of exploiting promoters as alternative to overcome the escalating burden. *Frontiers in Life Science*, *9*(1), 52–63. https://doi.org/10.1080/21553769.2015.1077478

Sun, G., Stewart, C. N., Xiao, P., & Zhang, B. (2012). microRNA expression analysis in the cellulosic biofuel crop switchgrass (panicum virgatum) under abiotic stress. *PLoS One*, *7*(3), e32017. https://doi.org/10.1371/journal.pone.0032017

Tang, G., & Tang, X. (2013). Short tandem target mimic: a long journey to the engineered molecular landmine for selective destruction/blockage of microRNAs in plants and animals. *Journal of Genetics and Genomics*, *40*(6), 291–296.

Telem, R. S., Wani, S. H., Singh, N. B., Nandini, R., Sadhukhan, R., Bhattacharya, S., & Mandal, N. (2013). Cisgenics—A sustainable approach for crop improvement. *Current Genomics*, *14*(7), 468. https://doi.org/10.2174/13892029113146660013

van Hove, L., & Gillund, F. (2017). Is it only the regulatory status? Broadening the debate on cisgenic plants. *Environmental Sciences Europe*, *29*(1), 22. https://doi.org/10.1186/s12302-017-0120-2

Vasudevan, S. N., Pooja, S. K., Raju, T. J., & Damini, C. S. (2023). Cisgenics and intragenics: boon or bane for crop improvement. *Frontiers in Plant Science*, *14*, 1275145. https://www.frontiersin.org/journals/plant-science/articles/10.3389/fpls.2023.1275145

Verma, N., Sao, P., Srivastava, A., & Singh, S. (2021). Physiological and molecular responses to drought, submergence and excessive watering in plants. In A. Husen (Eds.), *Harsh Environment and Plant Resilience* (pp. 305–321). Springer International Publishing, Cham.

Wang, M., Weiberg, A., Lin, F.-M., Thomma, B. P. H. J., Huang, H.-D., & Jin, H. (2016). Bidirectional cross-kingdom RNAi and fungal uptake of external RNAs confer plant protection. *Nature Plants*, *2*(10), 10. https://doi.org/10.1038/nplants.2016.151

Wang, W., Zhao, X., Pan, Y., Zhu, L., Fu, B., & Li, Z. (2011). DNA methylation changes detected by methylation-sensitive amplified polymorphism in two contrasting rice genotypes under salt stress. *Journal of Genetics and Genomics = Yi Chuan Xue Bao*, *38*(9), 419–424. https://doi.org/10.1016/j.jgg.2011.07.006

Wang, Y., Cheng, X., Shan, Q., Zhang, Y., Liu, J., Gao, C., & Qiu, J.-L. (2014). Simultaneous editing of three homoeoalleles in hexaploid bread wheat confers heritable resistance to powdery mildew. *Nature Biotechnology*, *32*(9), 947–951.

Wolfe, S. A., Nekludova, L., & Pabo, C. O. (2000). DNA Recognition by Cys 2 His 2 Zinc Finger Proteins. *Annual Review of Biophysics and Biomolecular Structure*, *29*(1), 183–212. https://doi.org/10.1146/annurev.biophys.29.1.183

Yadav, R., Thankappan, R., & Kumar, A. (2021). Novel approaches for genome editing to develop climate smart crops. In S. A. Lone & A. Malik (Eds.), *Microbiomes and the Global Climate Change* (pp. 267–291). Springer, Singapore. https://doi.org/10.1007/978-981-33-4508-9_15

Yang, J., Yan, S., Xie, S., Yin, M., Shen, J., Li, Z., Zhou, Y., & Duan, L. (2022). Construction and application of star polycation nanocarrier-based microRNA delivery system in Arabidopsis and maize. *Journal of Nanobiotechnology*, *20*, 219. https://doi.org/10.1186/s12951-022-01443-4

Yang, T., Wang, Y., Teotia, S., Wang, Z., Shi, C., Sun, H., Gu, Y., Zhang, Z., & Tang, G. (2019). The interaction between miR160 and miR165/166 in the control of leaf development and drought tolerance in *Arabidopsis*. *Scientific Reports*, *9*(1), 2832.

Yu, T.-F., Xu, Z.-S., Guo, J.-K., Wang, Y.-X., Abernathy, B., Fu, J.-D., Chen, X., Zhou, Y.-B., Chen, M., Ye, X.-G., & Ma, Y.-Z. (2017). Improved drought tolerance in wheat plants overexpressing a synthetic bacterial cold shock protein gene *SeCspA*. *Scientific Reports*, *7*, 44050. https://doi.org/10.1038/srep44050

Zenda, T., Liu, S., Dong, A., & Duan, H. (2021). Advances in cereal crop genomics for resilience under climate change. *Life*, *11*(6), 502.

Zhang, H., Goh, N. S., Wang, J. W., Pinals, R. L., González-Grandío, E., Demirer, G. S., Butrus, S., Fakra, S. C., Flores, A. D. R., Zhai, R., Zhao, B., Park, S.-J., & Landry, M. P. (2022). Nanoparticle cellular internalization is not required for RNA delivery to mature plant leaves. *Nature Nanotechnology*, *17*(2), 197–205. https://doi.org/10.1038/s41565-021-01018-8

Zhang, J., Zhang, H., Srivastava, A. K., Pan, Y., Bai, J., Fang, J., Shi, H., & Zhu, J.-K. (2018). Knockdown of rice microRNA166 confers drought resistance by causing leaf rolling and altering stem xylem development. *Plant Physiology*, *176*(3), 2082–2094.

Zhang, L., Hou, D., Chen, X., Li, D., Zhu, L., Zhang, Y., Li, J., Bian, Z., Liang, X., Cai, X., Yin, Y., Wang, C., Zhang, T., Zhu, D., Zhang, D., Xu, J., Chen, Q., Ba, Y., Liu, J., & Zhang, C.-Y. (2012). Exogenous plant MIR168a specifically targets mammalian LDLRAP1: evidence of cross-kingdom regulation by microRNA. *Cell Research*, *22*(1), 1. https://doi.org/10.1038/cr.2011.158

Zhang, T., Zhao, Y.-L., Zhao, J.-H., Wang, S., Jin, Y., Chen, Z.-Q., Fang, Y.-Y., Hua, C.-L., Ding, S.-W., & Guo, H.-S. (2016). Cotton plants export microRNAs to inhibit virulence gene expression in a fungal pathogen. *Nature Plants*, *2*(10), 16153. https://doi.org/10.1038/nplants.2016.153

Zhang, Y., Massel, K., Godwin, I. D., & Gao, C. (2018). Applications and potential of genome editing in crop improvement. *Genome Biology*, *19*(1), 210. https://doi.org/10.1186/s13059-018-1586-y

Zhao, Y., Zhang, C., Liu, W., Gao, W., Liu, C., Song, G., Li, W.-X., Mao, L., Chen, B., & Xu, Y. (2016). An alternative strategy for targeted gene replacement in plants using a dual-sgRNA/Cas9 design. *Scientific Reports*, *6*(1), 23890.

Zuo, Z.-F., He, W., Li, J., Mo, B., & Liu, L. (2021). Small RNAs: the essential regulators in plant thermotolerance. *Frontiers in Plant Science*, *12*, 726762. https://www.frontiersin.org/journals/plant-science/articles/10.3389/fpls.2021.726762

15 Recent Approaches in Omics for Plant Resilience to Climate Change

Arpan Modi, Megha Barot, Raviraj Barot, Sachidanand Singh, and Dilfuza Jabborova*

15.1 INTRODUCTION

Climate change has severely affected agricultural output due to increasing temperature, diseases, plant physiology, and changes in microbial communities (Guo et al., 2010; Elad & Pertot, 2014). The productivity of wheat, rice, maize, and soybean yields has been shown to be reduced by an average of 6%, 3.2%, 7.4%, and 3.1% for every degree Celsius increase in the global mean temperature (Zhao et al., 2017). Approximately 230 statistical crop modeling studies reviewed by Hu et al. (2024) suggest that climate variability also affects yield predictions globally for the four major crops, namely maize, wheat, rice, and soybean. Omics technologies are advanced methods to improve the characteristics of agricultural crops by targeting genes, transcripts, proteins, and metabolites (Varshney et al., 2013). There is an urgent need to improve the quality and yield of agricultural crops to fulfill the increasing demand of the growing population. A multiomics approach has been used to decipher the genetic information pertaining to physiological and environmental stress components in important agricultural crops such as wheat, soybean, and millet (Singh et al., 2013; Deshmukh et al., 2014; Talukdar & Sinjushin, 2015; Muthamilarasan & Prasad, 2017; Shah et al., 2018; Yadav et al., 2018). The highly advanced technology in genomic sequencing is NGS (Next-Generation Technology), which can generate accurate and rapid data for genomes, transcriptomes, metabolomes. phenomes, epigenomes, and proteomes (Großkinsky et al., 2018). Proper application of this technology could provide genetic information to build crops resistant to disease, drought, and insects, benefiting the global scientific community. The most recently used techniques for crop improvement include next-generation sequencing, Sanger sequencing, microarrays, real-time PCR, 2D gel electrophoresis, and mass spectrometry (MS). These omics techniques have made it possible to compile a genomic information database to identify the genes and mutations associated with particular character. This chapter summarizes the updates in different omics techniques that have been used for crop improvement to withstand climate change.

Omics is a combination of four major techniques: genomics, transcriptomics, proteomics, and metabolomics. The first high-throughput technology developed was the DNA microarray (Schena et al., 1995). In accordance with the fundamental postulate, post-translational modifications, spatiotemporal proteome dynamics, and temporal transcriptome perturbations, as well as alternative splicing, have all been captured by omics technologies (Chakraborty et al., 2018). The omics technologies then expanded to include molecular interactions, disease associations, and epigenetic, epi-proteomic, and epitranscriptomic levels. The multiomics approach has facilitated our study of molecular markers and disease phenotypes, and single-cell sequencing is powerful enough to provide information at the single-cell level (Sinha et al., 2022). These rapidly developing omics technologies enable us to uncover the details of cellular-level changes, and the complexity of cells and their behavior will drive advancements in these techniques (Dai & Shen, 2022).

* *Corresponding Author*

DOI: 10.1201/9781003491378-15

15.2 NON-CONVENTIONAL CROP IMPROVEMENT STRATEGIES

Crop improvement strategies began with the domestication of cereal grains during ancient times in the region that covers Iraq, Iran, Turkey, and Syria, dating back around 10,000 years (Purugganan & Fuller, 2009). After the discovery of Mendel's laws of inheritance, the search for new genomic regions (genes) that govern particular characteristics or traits in plants began. Once the Green Revolution era commenced (after the 1960s), there was a remarkable increase in the yield potential of important field crops like rice and wheat. However, some major consequences of the revolution include the loss of several unique genes and excessive use of pesticides and fertilizers, which harmed the environment (Tilman et al., 2011). Additionally, anthropogenic activities contribute to climate change. As a result, crops are losing their biotic and abiotic stress tolerance. Moreover, the development of new varieties with enhanced resistance to climate stresses through conventional breeding strategies takes several years, highlighting the need for alternatives to such strategies (Razzaq et al., 2021). Considering time and resources, four strategies can be defined as alternatives to conventional breeding approaches: marker-assisted selection (MAS), speed breeding, genetic engineering, and genome editing. Two of these are subject to stringent regulatory compliance. The general procedure for the breeding strategy includes selecting parental lines containing gene(s) of interest (e.g., disease resistance, high yield, or wild-type); pollination under controlled conditions; selection of superior crosses; and testing, release, and commercialization of new cultivars. In this process, the selection of superior crosses is based on the phenotypic characteristics of the plants derived from the seeds generated. Such characteristics include plant growth parameters, yield, and survival percentage under stress conditions. After selection, crosses and back-crosses over a few years need to be carried out to develop the variety with specific traits of interest. Finally, before release and commercialization, multi-locational trials for the developed variety are carried out. There are numerous examples of the development of crop varieties suitable for changing climates using conventional breeding methodologies. Over time, breeders have discovered that the performance of different genotypes can vary in changing environments and management practices. Considering the combination of these three factors (genotype, environment, and management practices), Cooper et al. (2021) proposed a genotype X management setup for different environments. Using hybrid development in maize, they conducted trials in two different environments showing contrasting characteristics for evapotranspiration. Many factors can be targeted for the development of climate-resilient crops, one of which is root architecture. There are traits linked with root structure and performance. Ober et al. (2021) proposed that root phenotyping be considered as a selection criterion in wheat breeding programs. By focusing on tools and techniques used for computation, data analysis, phenotyping, and genotyping, conventional breeding techniques can be successfully employed for the development of climate-ready crops (Langridge et al., 2021). However, two major drawbacks are associated with conventional breeding techniques: they are time-consuming and have limited genetic diversity due to crosses of closely related species (Ahmar et al., 2020). Updates in conventional breeding methodologies include MAS, speed breeding, genetic engineering, and genome editing. According to crop biology, resources, and regulatory aspects, one of these strategies may be employed (Verma et al., 2023).

15.2.1 Marker-Assisted Selection

Classical breeding strategies are time-consuming processes where breeders must wait until the character of interest appears. However, there are molecular markers associated with each character that are transmitted over generations along with the trait of interest. In MAS, the selection of hybrids or superior genotypes is predicted based on the presence or absence of trait-linked markers (Gaba et al., 2021). These markers can be biochemical or DNA markers. Isozymes are biochemical markers whose presence or absence may be used for the characterization of different genotypes, although they are used less frequently in MAS. DNA markers can be divided into

PCR-based markers (where the presence or absence of a selected region of DNA is determined by amplification using thermostable polymerases) and PCR-free markers, such as restriction fragment length polymorphism, which cuts genomic DNA into several fragments; fingerprinting is performed based on the distribution of the recognition sites of restriction enzymes in the DNA. DNA markers are selected based on five fundamental considerations: reliability, DNA quality and quantity, methodology used, level of polymorphism, and cost (Mohler & Singrün, 2004). Few molecular markers or techniques fit these criteria and are widely used according to their application in MAS. These molecular markers in the genome are tightly linked to genes or agronomic traits governed by multiple genes. The most widely used DNA marker systems in plants are restriction fragment length polymorphism, randomly amplified polymorphic DNA, amplified fragment length polymorphism, simple sequence repeats (SSR), and single-nucleotide polymorphism (SNP). They differ in terms of genomic abundance, genomic coverage, inheritance, number of loci, level and type of polymorphism, reproducibility, genotyping throughput, marker index, time requirements, and cost (Hasan et al., 2021). The general flow in MAS involves selecting two parental lines (one susceptible and the other resistant) based on their phenotypes, crossing them, and then selfing the F1 populations to prepare the F2 population. Seeds from the F2 generation are raised and checked with the appropriate molecular marker to determine whether the trait of interest has been transmitted. Due to advancements in high-throughput screening and phenotyping tools, marker-assisted breeding has evolved into genomics-assisted breeding. The objective of MAS or genomics-assisted breeding is to identify and employ quantitative trait loci (QTL) associated with traits directly or indirectly involved in stress tolerance or adaptation. In wheat, several QTLs have been identified for yield reduction during abiotic stresses, as well as for adaptation under the same conditions (Kole et al., 2015). Xu et al. (2006) identified *Sub1* (*Submergence* 1) QTL in rice, and with just three back-crosses, they successfully developed a line tolerant to flooding stress. Similarly, in chickpea, databases for intra- and inter-specific genetic maps have been developed, allowing for the identification of genomic regions associated with biotic and abiotic stresses and agronomic traits (Kole et al., 2015).

15.2.2 Speed Breeding

MAS is known to save time compared to conventional methods of breeding for crop improvement. In parallel, the flowering process can be manipulated in a few ways that allow the breeder to skip the waiting phase between two generations. Physical factors like light, temperature, soil properties, and humidity can significantly affect plant growth. Manipulation of these conditions can boost plant growth, especially the flowering process. Thus, 3–4 breeding cycles can be attained in plant species that flower once a year. Controlled environments, high-density planting, and limited plant species are some of the major considerations for speed breeding. The concept of speed breeding began long ago when the use of carbon arc lamps showed acceleration in growth and the flowering cycle. Utah State University, in collaboration with NASA, developed dwarf wheat using perpetual lights, which demonstrated fast growth and development (Potts et al., 2023). In the 1990s, the use of LED lights made plant production under controlled environments more affordable, increasing productivity and introducing the term speed breeding (Houston et al., 2020). It has been observed that fluctuations in temperature, light intensities, and soil conditions can affect pollen viability and result in low yields. In a controlled environment, such fluctuations can be prevented. Thus, it is highly suitable for crop improvement strategies where harsh environmental conditions prevail (Wanga et al., 2021; Dupuis & Dumas, 1990). Speed breeding operates in a way that utilizes water more efficiently and enhances resistance to climate change, as crop production is facilitated in a controlled environment (Begna, 2022). A fully enclosed speed breeding environment resulted in six generations per year instead of 2–3 under glasshouse conditions for wheat, barley, pea, and chickpea. The results indicated that growth patterns and other phenotypic characteristics showed no deviation from conventional practices. The system demonstrated promising results for phenotyping, mutant studies, and transformation (Watson et al., 2018). The speed breeding method, in combination with *in vitro* embryo culture, showed a significantly higher success rate in the development of new lines in lentils. With this approach, normal growth patterns and four generations per year were achieved. It was proposed that with this technique, the development of recombinant inbred lines can be much faster than with conventional techniques (Bermejo et al., 2016). Although speed breeding methodologies have been successfully applied in many plant species (Sysoeva et al., 2010), they also showed severe injury in eggplant and tomato (Velez-Ramirez et al., 2011).

15.2.3 GM Crops

Both MAS and speed breeding methods have limitations when working within the genus to introduce the desired trait into the genotype of interest, considering intergenic crosses. Crosses of plant species within or between families are difficult due to genetic factors (Kumar et al., 2020). At the same time, stress-resistance genes are often available across plant species. Genetic engineering enables researchers to identify, isolate, and incorporate such genes into almost any plant species. However, these techniques also bring stringent regulatory monitoring, and the release of transgenic plants requires approval from various government regulatory bodies in almost every country (Choudhary et al. 2014). Development of genetically modified plants requires laboratory, greenhouse, and field conditions for the successful release of the transgenic plant. To transfer foreign genetic material into a new species, various transformation techniques are used. The most widely used methods are *Agrobacterium*-mediated transformation, where the soil bacterium (a natural genetic engineer) is modified to contain the gene of interest (GOI) in its extrachromosomal material, which is then transferred to the host plant under controlled conditions (mostly *in vitro*), and the particle-gun method, where gold or platinum nanoparticles containing plasmids carrying the GOI are bombarded directly onto the plant cells, resulting in random incorporation of the GOI at any site in the host genome. Development of genetically modified plants using *Agrobacterium*-mediated gene transformation is shown in Figure 15.1. The process starts with the isolation of the GOI from the source organism and its incorporation into a suitable vector. The vector is then transformed

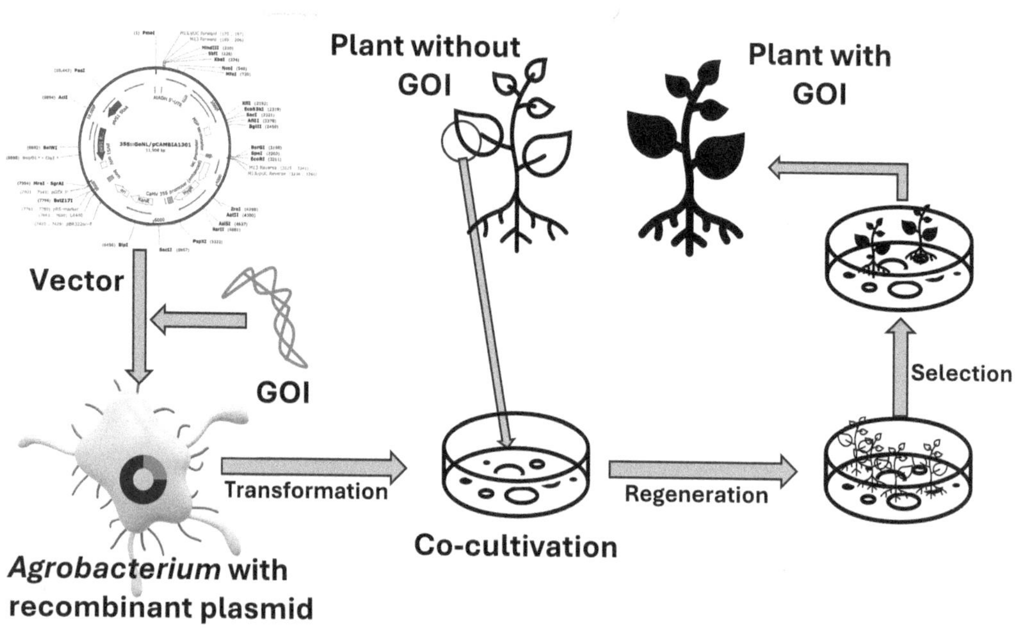

FIGURE 15.1 The process of *Agrobacterium*-mediated transformation in plants. (Created with SnapGene)

into *Agrobacterium*, and infection is initiated in the host plant tissue. After infection, 2–3 days are allowed for co-cultivation, followed by the removal of *Agrobacterium* by incubating the plant cells on antibiotic medium. As plant cells can grow into plantlets, the selection of transformants can be done using an antibiotic to which resistance has already been incorporated into the plant genome through the vector (Rahman et al., 2023).

Genetically modified crops developed specifically for herbicide tolerance and insect resistance showed a reduction in carbon emissions between 1996 and 2000. The estimation was carried out using two important parameters: fuel consumption and soil carbon sequestration. A total of 14,662 million liters of fuel could be saved, which accounts for a reduction in carbon dioxide emissions of up to 39,147 million kg where GM crops were practiced (Brookes, 2022; Brookes & Barfoot, 2020). Among the various targeted genes to be overexpressed for the development of GM crops, transcription factors (TFs) are a promising choice for researchers. These factors help the plant adapt to and/or resist changing environmental or stressful conditions. They are part of the stress-resistance network of gene regulation. Frequently, disease resistance genes are already present in the plants, but they are poorly expressed due to a lack of signals from gene regulatory networks. This makes the plant susceptible to stressful environments. TFs fill this gap. Widely overexpressed TFs include DREB (Dehydration Responsive Element-Binding factor), bZIP (basic leucine zipper), MYB, AP2/ERF (Ethylene Responsive Factor), zinc finger, and WRKY (Shahzad et al., 2021). Drought is a major consequence of climate change, and it severely reduces crop yield. Dong et al. (2024) developed a transgenic line of apple that overexpresses the *MdWRKY115* TF. Two lines were used (Gala and Golden Delicious) for the experiment. The gene was isolated from the Golden Delicious line and fused with a scorable marker (*GUS* gene, which shows a blue-colored product where the expression occurs). Osmotic stress was induced using mannitol treatment in transgenic *Arabidopsis*, and the plants showed enhanced resistance compared to the control (Dong et al., 2024).

15.2.4 Genome Editing (CRISPR/Cas9)

Development of genetically modified plants is based on the overexpression of resistant genes or genes of interest in the host. However, the same can be achieved through silencing or downregulating unwanted or susceptible genes. There are site-directed nucleases that cut genes at specific sites, resulting in permanent mutations. This leads to the loss of function of the undesired gene. In a similar manner, susceptible genes are silenced, which improves the plant's performance during adverse environmental conditions (Abdallah et al., 2021). Examples of site-directed nucleases include meganucleases, zinc finger nucleases, transcription activator-like effector nucleases, and clustered regularly interspaced short palindromic repeats (CRISPR)/CRISPR-associated protein 9 (Cas9). Among them, the CRISPR-Cas9 system is a highly versatile and relatively easy technique for genome editing in plants (Voytas & Gao, 2014; Ma et al., 2015; Jinek et al., 2012). It utilizes an RNA molecule to direct or guide the Cas9 enzyme, which cleaves the gene near the protospacer adjacent motif where the guide RNA binds. This stimulates the natural DNA repair mechanism, which is also error-prone, leading to the insertion or deletion of nucleotides. Usually, the guide RNA is designed from the functional part of the protein so that loss-of-function mutants can be expected. The process of genome editing in plants using CRISPR-Cas9 is shown in Figure 15.2. To date, many plant species have been edited using the CRISPR-Cas9 genome editing tool. However, their release as a variety for crop production is still limited as it must go through regulatory approvals (Ma & Liu, 2016).

Figure 15.2 describes how genome editing in plants using CRISPR/Cas9 works. Here, yellow petal color is the target. The gene responsible for this trait is identified, and a guide RNA construct is prepared from the selected region of the sequence. The construct is then transformed into the actively dividing plant cell, which gives rise to a whole new plant. If the editing works well, some of the transformants will lose the ability to show yellow color on the petals, and such transformants are screened based on molecular tools like PCR and sequencing, as well as morphological traits.

Plant genetic engineering and genome editing have several similarities, such as both requiring *in vitro* operations to develop the modified plant, methods of transformation and vectors, and

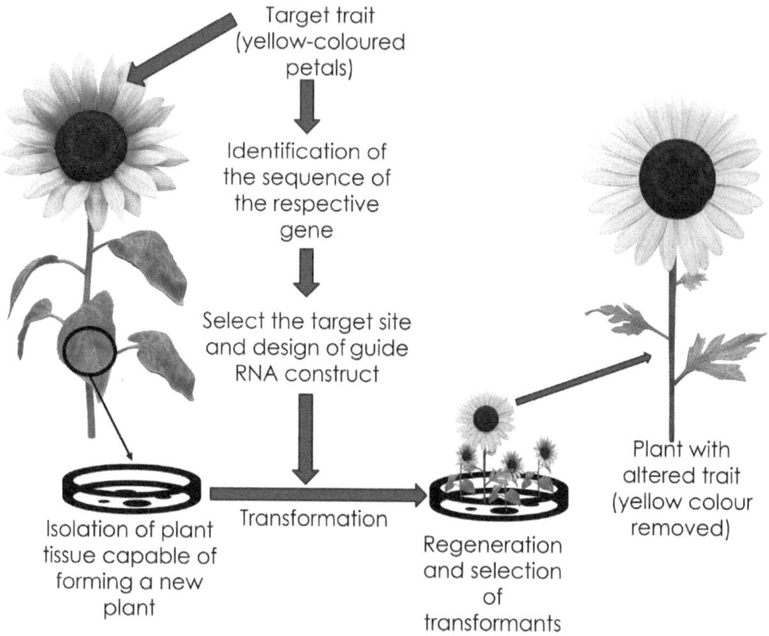

FIGURE 15.2 CRISPR/Cas9-mediated genome editing in plants targeting petal color.

sequence information of the targeted trait to some extent. However, they differ in terms of regulatory constraints, gene expression, and analyses of transformants. The outcomes of genome-edited plants can be expected to include biotic and abiotic stress tolerance. Various efforts are being made by researchers to create climate-resilient crops using CRISPR/Cas9 genome editing tools. Based on the observations made by Lu et al. (2015) and Mao et al. (2010) regarding higher tiller numbers and grain size improvement in the mutant lines of *OsPIN5b* and *GS3* genes, respectively, Zeng et al. (2020) targeted these two genes along with *OsMYB30* (responsible for cold stress susceptibility) using CRISPR/Cas9 technology. They introduced mutations at two different sites in each gene, with a minimum efficiency of 42% (in *OsPIN5b* site 2) and a maximum of 66% (in *GS3* site 1). The resultant mutants showed increased panicle numbers per plant, increased grain numbers per panicle, increased yield, and good plant vigor under cold stress conditions compared to wild-type control plants. Similarly, in tomato, *SlARF4* (auxin responsive factor 4) is directly involved in salinity stress. CRISPR lines of *ARF4* mutants showed upregulation of ABA biosynthesis genes and downregulation of ABA catabolism, which is a key regulator during salinity and osmotic stress. In the CRISPR lines of tomato, other growth parameters were unaffected due to mutations in the *ARF4* gene (Bouzroud et al., 2020).

The key players behind these genetic improvement strategies are various techniques developed in omics science. All regulatory processes start from DNA and extend through RNA, protein, metabolite, or phenotype levels. Omics is the collection of biological sciences required to assess and/or quantify various biomolecules. According to the biomolecules being targeted, the name of the omics science is designated, such as genomics (DNA), transcriptomics (RNA), proteomics (protein), metabolomics (metabolites), and phenomics (phenotypes).

15.3 TECHNIQUES IN OMICS

While developing plant resilience to climate change, researchers must establish measures of success to conserve time, effort, and resources. Omics-based approaches help develop the resources required before crop improvement strategies are designed, during experiments for monitoring, and after project completion to verify the variables (Yang et al., 2021). Based on the biomolecules to be studied, omics can be divided into five different techniques: next-generation sequencing, microarray, and real-time PCR for DNA and RNA; 2D-gel electrophoresis and MS for proteins; and MS for metabolites. Sequencing technology began with the Sanger sequencing method (Sanger et al., 1977), which was used to complete the human genome project (Sanger, 2004). To date, it holds the record for long read lengths (around 1,000–1,200 bp) and high per-base accuracy (99.999%). It is still widely used for single-product assays; however, it is not recommended for whole genomes and transcriptomes due to high costs and low throughput.

15.3.1 Next-Generation Sequencing

The basic difference between Sanger sequencing and NGS is the chemistry utilized to achieve the goal. In Sanger sequencing, chain termination technology is used, where the nucleotides passing through the capillary are monitored. Therefore, DNA fragments containing different alleles can be difficult to sequence, resulting in mixed peaks of nucleotides. In the case of NGS, the large DNA molecule is fragmented into random sizes, and each fragment is tagged. Using the known sequence of tagged DNA, sequencing by synthesis occurs. A significant portion of the work in NGS involves post-run analysis (Chen et al., 2021). From the raw data, genome and transcriptome assemblies are created and confirmed with known sequences in similar plant species. The development of NGS technologies began with the pyrosequencing created by Roche. It is based on a sequencing by synthesis approach, where the newly incorporated nucleotide releases pyrophosphate, which is then converted into ATP. The generated ATP is used by luciferase to produce light, which is detected in the channel. An advanced version of this technique was developed by Thermo Fisher (known as the Ion Torrent platform), which detects changes in pH to sequence the fragments (Reuter et al., 2015; Soon et al., 2013). With the help of pyrosequencing, it was possible to sequence the human genome within 2 months and at a cost of $100,000, which is 0.01% of the total cost of the human genome project completed by Sanger sequencing (Kircher & Kelso, 2010). Subsequently, SOLiD (Applied Biosystems), MiSeq, HiSeq, and NovaSeq 6000 (Illumina) platforms were developed, achieving read lengths of up to 300 bases per read, outputs of up to 50,000 Mb per day, and costs as low as $0.5 per Mb. The third-generation NGS platforms include single molecule real-time sequencing by Pacific Biosciences and Nanopore sequencing by Oxford. The advantage of third-generation NGS technologies is the detection of epigenetic markers without pre-processing techniques like chromatin immunoprecipitation (Akintunde et al., 2023). The sequence information available after sequencing is extremely valuable for mining new genes that can be incorporated into other plant species using genetic engineering or into susceptible plants of the same species using breeding strategies (Thottathil et al., 2016). Additionally, various molecular markers have been developed as a result of sequencing, such as SSR, EST-SSR, SNPs, and eQTL. Based on these markers, MAS is carried out (Salgotra et al., 2014; Edwards & Batley, 2010). From the sequencing data, copy number variations among different species can be calculated, which is used in heterosis breeding (Bolger et al., 2014). Plant genotyping is now also possible with NGS technologies termed genotyping by sequencing. The advantage of using NGS technology for genotyping is the elimination of ascertainment bias toward a particular population (Deschamps et al., 2012). Using the targeted exome-based genotyping by sequencing technique, Bhoite et al. (2023) identified 39 SNP locations, of which 8 produced polymorphic signals. Based on these results, high-throughput genotyping for MAS could be performed in wheat.

As compared to genomics, transcriptomics generates different types of data with distinct applications in crop improvement strategies. Transcriptomic datasets represent particular plant species under specific conditions, tissue types, plant ages, or treatments. Transcriptomics encompasses gene sets, patterns, functions, and the regulation of transcripts in a given cell type. Transcriptome data are more complex than genomics. The process includes the isolation of mRNA populations from plant tissue and the conversion to complementary DNA (cDNA) for sequencing. After sequencing, assemblies are prepared by comparing them with a reference genome, or *de novo* assemblies are reported (Wang et al., 2009). The transcriptomics approach is well suited for studying stress conditions. Abiotic and biotic stresses, climate change, and environmental degradation severely affect crop yield, and to mitigate these situations, researchers must identify the plasticity of gene expression in response to adverse environmental conditions. Such identification allows researchers to select regulatory elements for crop improvement strategies (Nejat et al., 2018). Seong et al. (2020) identified that the *OsC3H10* gene is responsible for the upregulation of certain stress-resistant genes, such as pathogenesis-related genes, late embryogenesis abundant proteins (leas), and germin-like proteins, when transcriptomic data were collected from transgenic rice. Overexpression of *OsC3H10* resulted in enhanced drought tolerance. Similarly, in wheat, Bi et al. (2021) identified 1,455 TF-encoding genes from full-length cDNA libraries under salt stress. They observed that *TabZIP15* was a key TF gene responsive to salt stress. Overexpression of this gene resulted in a larger root system, longer spike length, and more grains per spike compared to non-transgenic plants under salt stress conditions.

15.3.2 Microarray

The first DNA microarray was developed by Grunstein and Hogness (1975) using the colony hybridization approach (Fodor et al., 1991). In 1979, Gergen et al. (1979) reported a paper-based method for high-density colony replicas using this colony hybridization approach (Pease et al., 1994). Following these scientists, others further developed this technique. In 1980, Crampton et al. developed cDNA from targeted RNA and checked differential expressions in two samples; Schena et al. 1996 used two-color fluorescence hybridization on glass to analyze gene expression. Also in the same year, DeRisi et al. (1996) reported a method wherein they introduced cDNA microarrays using fluorescent probes on glass plates to observe gene patterns in tumors (Inaoka et al., 2015). Since then, there have been many developments in this technique, and it has been widely used in various fields.

Microarray technique utilizes a platform or solid surface that contains oligonucleotides or cDNAs representing an organism's complete or partial genome and are immobilized on a glass slide or other substrate, such as a nylon membrane. The platforms consist of labeled cDNAs from treated and control tissues in order to analyze gene expression. There are two different types of microarrays according to immobilized probes: (1) cDNA embedded on the surface, produced from PCR, which is the most commonly used technique (Viemann et al., 2005) and (2) oligonucleotide microarrays produced with oligos ranging from 10 to 120 mer, corresponding to known coding targets (Pervaiz et al., 2021). Plant growth, development, and responses to environmental stimuli are influenced by molecular pathways that can be understood through the simultaneous examination of hundreds of genes by high-throughput instruments like microarrays. Microarrays are widely utilized in agriculture to provide enhanced features like disease resistance, improved production, and stress tolerance through genotyping, gene expression profiling, and marker-assisted breeding. DNA microarray is a method that helps to understand the details of genes involved in regulatory networks and signal transduction associated with resistance to multiple abiotic stresses (Ghorbani et al., 2019; Raza et al., 2021). Ozturk et al. (2002) prepared cDNA libraries from drought- and salinity-stressed barley plants and screened them using microarray. They identified several proteins involved in jasmonate response, metallothionein-like proteins, LEA, and ABA-responsive genes that were upregulated under drought stress conditions. Targeting such genes in further crop improvement programs helps researchers save time and resources. To study stress-responsive loci and for engineering transgenic plants, microarray is one of the powerful techniques (Pervaiz et al., 2021). Transgenic wheat was created by expressing the TaPIMP1 gene, and this wheat showed exceptional resistance to the fungal pathogen *Bipolaris sorokiniana* and drought; TaPIMP1 hyper-expression was verified via microarray analysis (Liu et al., 2011). Twenty genes associated with cold and drought-inducible responses were identified by Seki et al. (2001). Of these, five were drought-specific, four were unique (FL5-2D23, FL5-3J4, FL2-56, and FL6-55), and two were cold-specific inducible (FL5-90). There have been reports of several members of stress-regulated gene families, including zinc finger proteins related to drought and cold, bZIP proteins related to drought, AP2 family proteins related to drought and cold, and MYB, NAC, and bHLH proteins related to drought, ABA, and salinity (Liu et al., 2013; Mittal et al., 2018; Pervaiz et al., 2017). Furthermore, a large number of modern, drought-sensitive genes have been identified in pine and poplar (Street et al., 2006; Lorenz et al., 2006).

15.3.3 Real-Time PCR

Sequencing technologies are front-line players in crop improvement programs, covering everything from the identification of novel genes to monitoring and confirmation. However, from the sequencing results, large and small gene sets can be verified using microarray and real-time PCR, respectively. In real-time PCR, the expression of a particular gene in terms of quantity can be assessed. The application of real-time PCR in crop improvement includes copy number detection in transgenic plants (Raj et al., 2015); expression patterns and relative or absolute values of gene expression during stress conditions, developmental stages (Modi et al., 2014a), and treatment effects (Modi et al., 2014b); SNP genotyping (Kalendar et al., 2022); and plant disease detection (Gachon et al., 2004; Alemu, 2014). Real-time PCR is an advanced version of end-point PCR, and detection of amplification is carried out with fluorescence emission during the cycling process. As the name suggests, detection is done in real time, thus saving time compared to conventional PCR, where the PCR product has to be run on gel electrophoresis (Alemu, 2014). Identification of differentially expressed genes can be performed through *in silico* analysis of available databases, and validation is done through qPCR. Magar et al. (2024) identified 22 TF genes belonging to the AP2/ERF family that showed differential gene expression in tolerant (W156) and susceptible (Brazil 32) genotypes of wheat. Based on the expression pattern of these TFs, they were categorized as 1 expressed in the tolerant genotype, 2 expressed in the susceptible genotype, 3 expressed in both genotypes, and 4 expressed in none. The results suggested that TaAP2/ERF TFs play an important role during heat stress at the grain filling stage. Targeting these genes for the development of heat-tolerant wheat lines would be a promising strategy.

15.3.4 2D-Gel Electrophoresis

By process, 2D-gel electrophoresis is a microarray for proteins. It applies to all the conditions where microarray (to study gene expression) is used, but at the proteomics level. The study of plant proteomics addresses the gap between genotype and phenotype. Transcriptome analyses do not reveal post-translational modifications, which are important aspects of gene regulation. Therefore, exploration of the final product (protein) is necessary during the characterization of plants. Characterization or genotyping of plants allows researchers to screen potential genotypes for the development of climate-resilient lines (Jorrin-Novo et al., 2019). Just like transcriptome studies, which target the complete set of mRNA populations, proteomic studies target all the proteins at a particular stage. First, these proteins are run and separated based on their isoelectric points on a strip containing polyacrylamide gel. Afterwards, these strips are placed on a second gel, and proteins are again separated based on their molecular weight. When two such strips (samples to be compared) are run simultaneously, the protein dots that appear larger at specific locations indicate upregulation of the protein under specific conditions (Yang et al., 2007).

15.3.5 Mass Spectrometry

The central dogma elaborates the process of gene expression and regulation up to the protein level. However, the process does not end when the final product is an enzyme that catalyzes metabolic reactions, leading to the development of products (metabolites). Such metabolites are important biochemicals involved in various biological processes and stress resistance. For crop improvement strategies (either through breeding or through biotechnological interventions), metabolites are usually monitored, and further approaches or strategies are designed. Additionally, the metabolome includes the identification of novel compounds or metabolites that have promising potential in crop improvement studies (Manickam et al., 2023). Detection of existing compounds and identification of novel compounds can be done using a single platform that combines chromatography with MS. Different compounds with varying polarities are separated in chromatography, and the separated compounds are ionized, producing different daughter compounds. In a mass spectrometer, the mass-to-charge ratio is determined and compared with the available database. Here, compounds appear as peaks, and the height of the peaks represents the concentration (Chong et al., 2022).

15.4 *IN SILICO* TOOLS FOR OMICS

Omics tools comprise both dry- and wet-lab work. Wet-lab work includes sample preparation, running experiments, monitoring, and data collection. Once the data acquisition process is completed, processing the large datasets requires online platforms or tools. Tools are available for all the omics platforms. For some plants, dedicated database repositories and analysis tools are also available (Tan et al., 2022). Some of the important tools used in the omics studies are mentioned in Table 15.1.

15.5 CONCLUSION AND FUTURE PROSPECTS

The 21st century has brought several alarming signals for agricultural productivity in terms of crop loss. Elevated temperatures, increased salinity, drought, and plant diseases are four major consequences of climate change. At the same time, the world population is increasing incessantly, and so is the demand for food. Therefore, it is extremely important to make field crops climate-ready. Several strategies (conventional breeding, advanced breeding, and biotechnology) are used for this purpose. In classical breeding, a stress-resistant genotype is screened and crossed with the

TABLE 15.1
Publicly Available Bioinformatics Platforms and Tools for the Analysis during Omics Approaches

Sr. No.	Name	Omics Covered	Features	Link
1	EnsemblPlants	Genomics	• Provides assemblies of 153 plant species • Additional tools like BLAST, HMMER, and Variant Effect Predictor	https://plants.ensembl.org/index.html
2	Metlin	Metabolomics	• Multiple metabolite searching • 240,000 metabolites • 72,000 high-resolution MS/MS spectra	https://metlin.scripps.edu
3	PlantCyc	Metabolomics	• Information for biochemical pathways and their catalytic enzymes and genes	https://plantcyc.org/
4	PLEXdb	Transcriptomics	• Database of gene expression datasets • Also includes structural genomics and phenotypic data	https://plantgdb.org/prj/PLEXdb
5	The Bio-Analytic Resources	Genomics, Transcriptomics	• Database for gene expression • Mapping of molecular markers	https://bar.utoronto.ca/
6	GelMap	Proteomics	• Analysis of 2D gels • Allows functional annotation of the protein based on the available database	https://www.gelmap.de/
7	Musite	Proteomics	• Site prediction of protein post-translation modifications	https://musite.net/

susceptible one. Screening is done based on phenotypic characteristics when the plant is actually subjected to stress conditions. Afterwards, further selection is made, which makes the process time- and resource-consuming. Advanced breeding consists of MAS, where molecular markers linked to the traits are key players; speed breeding, where controlled conditions are manipulated to attain the reproductive phase; and somatic hybridization, where tissue culture is the platform to develop the new genotype (a rarely used technique). Lastly, in biotechnology, a resistant gene is identified from the source plant or another living organism and overexpressed in the susceptible genotype using transformation techniques (termed genome modification), or the susceptible gene from the containing plant is edited to be non-functional (genome editing). The latter strategy is more promising for future trends. In all cases for the development of climate-ready crops, resources and monitoring of improvement are key factors for success. Omics tools play a vital role in fulfilling both factors. These tools help researchers select the GOI from the available database and map the path to the successful development of a new line. Out of the four major omics—genomics, transcriptomics, proteomics, and metabolomics—genomics provides a wealth of information for plant breeders and researchers. Next-generation sequencing and microarrays are the backbone of genomics. Both provide valuable information to start the experiment and are also useful in monitoring gene expression profiles, other genes of the regulatory network, loci identification, treatment effects, and genotyping. Other techniques like real-time PCR, 2-D gel electrophoresis, and MS are also important measures for scrutiny. The current version of omics relies mainly on all four omics mentioned here. However, related omics technologies are also emerging and occupy a significant position in plant improvement strategies. These include phenomics, mutagenomics, cytogenomics, epigenomics, glycomics, lipidomics, plant secretomics, signalomics, and transplastomics. Research in these omics has been carried out to fulfill specific needs. In the future, it is expected that by using a consortium of these omics technologies, tailor-made plants will be developed that respond precisely to the changing environment.

REFERENCES

Abdallah, N.A., Hamwieh, A., Radwan, K., Fouad, N. and Prakash, C., 2021. Genome editing techniques in plants: a comprehensive review and future prospects toward zero hunger. *GM Crops & Food, 12*(2), 601–615.

Ahmar, S., Gill, R.A., Jung, K.H., Faheem, A., Qasim, M.U., Mubeen, M. and Zhou, W., 2020. Conventional and molecular techniques from simple breeding to speed breeding in crop plants: recent advances and future outlook. *International Journal of Molecular Sciences, 21*(7), 2590.

Akintunde, O., Tucker, T. and Carabetta, V.J., 2023. The evolution of next-generation sequencing technologies. arXiv:2305.08724v1.

Alemu, K., 2014. Real-time PCR and its application in plant disease diagnostics. *Advances in Life Science and Technology, 27*, 39–49.

Begna, T., 2022. Speed breeding to accelerate crop improvement. *International Journal of Agricultural Science and Food Technology, 8*(2), 178–186.

Bermejo, C., Gatti, I. and Cointry, E., 2016. In vitro embryo culture to shorten the breeding cycle in lentil (*Lens culinaris* Medik). *Plant Cell, Tissue and Organ Culture (PCTOC), 127*, 585–590.

Bhoite, R., Smith, R., Bansal, U., Dowla, M., Bariana, H. and Sharma, D., 2023. Exome-based new allele-specific PCR markers and transferability for sodicity tolerance in bread wheat (*Triticum aestivum* L.). *Plant Direct, 7*(8), e520.

Bi, C., Yu, Y., Dong, C., Yang, Y., Zhai, Y., Du, F., Xia, C., Ni, Z., Kong, X. and Zhang, L., 2021. The bZIP transcription factor TabZIP15 improves salt stress tolerance in whcat. *Plant Biotechnology Journal, 19*(2), 209.

Bolger, M.E., Weisshaar, B., Scholz, U., Stein, N., Usadel, B. and Mayer, K.F., 2014. Plant genome sequencing—applications for crop improvement. *Current Opinion in Biotechnology, 26*, 31–37.

Bouzroud, S., Gasparini, K., Hu, G., Barbosa, M.A.M., Rosa, B.L., Fahr, M., Bendaou, N., Bouzayen, M., Zsögön, A., Smouni, A. and Zouine, M., 2020. Down regulation and loss of auxin response factor 4 function using CRISPR/Cas9 alters plant growth, stomatal function and improves tomato tolerance to salinity and osmotic stress. *Genes, 11*(3), 272.

Brookes, G. and Barfoot, P., 2020. Environmental impacts of genetically modified (GM) crop use 1996–2018: impacts on pesticide use and carbon emissions. *GM Crops & Food, 11*(4), 215–241.

Brookes, G., 2022. Genetically modified (GM) crop use 1996–2020: impacts on carbon emissions. *GM Crops & Food, 13*(1), 242–261.

Chakraborty, S., Hosen, M. I., Ahmed, M. and Shekhar, H.U., 2018. Onco-multi-OMICS approach: a new frontier in cancer research. *BioMed Research International, 2018*(1), 9836256.

Chen, X., Dong, Y., Huang, Y., Fan, J., Yang, M. and Zhang, J., 2021. Whole-genome resequencing using next-generation and nanopore sequencing for molecular characterization of T-DNA integration in transgenic poplar 741. *BMC Genomics, 22*(1), 329.

Chong, L., Hsu, C.C. and Zhu, Y., 2022. Advances in mass spectrometry-based phosphoproteomics for elucidating abscisic acid signaling and plant responses to abiotic stress. *Journal of Experimental Botany, 73*(19), 6547–6557.

Choudhary, B., Gheysen, G., Buysse, J., van der Meer, P. and Burssens, S., 2014. Regulatory options for genetically modified crops in India. *Plant Biotechnology Journal, 12*(2), 135–146.

Cooper, M., Voss-Fels, K.P., Messina, C.D., Tang, T. and Hammer, G.L., 2021. Tackling G× E× M interactions to close on-farm yield-gaps: creating novel pathways for crop improvement by predicting contributions of genetics and management to crop productivity. *Theoretical and Applied Genetics, 134*, 1625–1644.

Dai, X. and Shen, L., 2022. Advances and trends in omics technology development. *Frontiers in Medicine, 9*, 911861.

DeRisi, J., Penland, L., Bittner, M., Meltzer, P. S., Ray, M., Chen, Y. and Trent, J. M. (1996). Use of a cDNA microarray to analyse gene expression. *Nat. genet, 14*, 457–46.

Deschamps, S., Llaca, V. and May, G.D., 2012. Genotyping-by-sequencing in plants. *Biology, 1*(3), 460–483.

Deshmukh, R., Sonah, H., Patil, G., Chen, W., Prince, S., Mutava, R. and Nguyen, H.T., 2014. Integrating omic approaches for abiotic stress tolerance in soybean. *Frontiers in Plant Science*, 5, 244.

Dong, Q., Tian, Y., Zhang, X., Duan, D., Zhang, H., Yang, K., Jia, P., Luan, H., Guo, S., Qi, G. and Mao, K., 2024. Overexpression of the transcription factor MdWRKY115 improves drought and osmotic stress tolerance by directly binding to the MdRD22 promoter in apple. *Horticultural Plant Journal, 10*(3), 629–640.

Dupuis, I. and Dumas, C., 1990. Influence of temperature stress on in vitro fertilization and heat shock protein synthesis in maize (*Zea mays* L.) reproductive tissues. *Plant Physiology, 94*(2), 665–670.

Edwards, D. and Batley, J., 2010. Plant genome sequencing: applications for crop improvement. *Plant Biotechnology Journal, 8*(1), 2–9.

Elad, Y. and Pertot, I., 2014. Climate change impacts on plant pathogens and plant diseases. *Journal of Crop Improvement, 28*(1), 99–139.

Fodor, S.P., Read, J.L., Pirrung, M.C., Stryer, L., Lu, A.T. and Solas, D., 1991. Light-directed, spatially addressable parallel chemical synthesis. *Science, 251*(4995), 767–773.

Gaba, Y., Pareek, A. and Singla-Pareek, S.L., 2021. Raising climate-resilient crops: journey from the conventional breeding to new breeding approaches. *Current Genomics, 22*(6), 450.

Gachon, C., Mingam, A. and Charrier, B., 2004. Real-time PCR: what relevance to plant studies?. *Journal of Experimental Botany, 55*(402), 1445–1454.

Gergen, J.P., Stern, R.H. and Wensink, P.C., 1979. Filter replicas and permanent collections of recombinant DNA plasmids. *Nucleic Acids Research, 7*, 2115–2136. doi: 10.1093/nar/7.8.2115.

Ghorbani, R., Alemzadeh, A. and Razi, H., 2019. Microarray analysis of transcriptional responses to salt and drought stress in Arabidopsis thaliana. *Heliyon, 5*(11), e02614.

Großkinsky, D.K., Syaifullah, S.J. and Roitsch, T., 2018. Integration of multi-omics techniques and physiological phenotyping within a holistic phenomics approach to study senescence in model and crop plants. *Journal of Experimental Botany*, 69(4), 825–844.

Grunstein, M. and Hogness, D.S., 1975. Colony hybridization: a method for the isolation of cloned DNAs that contain a specific gene. *Proceedings of the National Academy of Sciences of the United States of America, 72*(10), 3961–3965.

Guo, R., Lin, Z., Mo, X. and Yang, C., 2010. Responses of crop yield and water use efficiency to climate change in the North China Plain. *Agricultural Water Management, 97*(8), 1185–1194.

Hasan, N., Choudhary, S., Naaz, N., Sharma, N. and Laskar, R.A., 2021. Recent advancements in molecular marker-assisted selection and applications in plant breeding programmes. *Journal of Genetic Engineering and Biotechnology, 19*(1), 128.

Houston, R.D., Bean, T.P., Macqueen, D.J., Gundappa, M.K., Jin, Y.H., Jenkins, T.L., Selly, S.L.C., Martin, S.A., Stevens, J.R., Santos, E.M. and Davie, A., 2020. Harnessing genomics to fast-track genetic improvement in aquaculture. *Nature Reviews Genetics*, *21*(7), 389–409.

Hu, T., Zhang, X., Khanal, S., Wilson, R., Leng, G., Toman, E.M., Wang, X., Li, Y. and Zhao, K., 2024. Climate change impacts on crop yields: a review of empirical findings, statistical crop models, and machine learning methods. *Environmental Modelling & Software*, *179*, 106119.

Inaoka, K., Inokawa, Y. and Nomoto, S., 2015. Genomic-wide analysis with microarrays in human oncology. *Microarrays*, *4*(4), 454–473.

Jinek, M., Chylinski, K., Fonfara, I., Hauer, M., Doudna, J.A. and Charpentier, E., 2012. A programmable dual-RNA–guided DNA endonuclease in adaptive bacterial immunity. *Science*, *337*(6096), 816–821.

Jorrin-Novo, J.V., Komatsu, S., Sanchez-Lucas, R. and de Francisco, L.E.R., 2019. Gel electrophoresis-based plant proteomics: past, present, and future. *Journal of Proteomics*, *198*, 1–10.

Kalendar, R., Baidyussen, A., Serikbay, D., Zotova, L., Khassanova, G., Kuzbakova, M., Jatayev, S., Hu, Y.G., Schramm, C., Anderson, P.A. and Jenkins, C.L., 2022. Modified "Allele-specific qPCR" method for SNP genotyping based on FRET. *Frontiers in Plant Science*, *12*, 747886.

Kircher, M. and Kelso, J., 2010. High-throughput DNA sequencing–concepts and limitations. *Bioessays*, *32*(6), 524–536.

Kole, C., Muthamilarasan, M., Henry, R., Edwards, D., Sharma, R., Abberton, M., Batley, J., Bentley, A., Blakeney, M., Bryant, J. and Cai, H., 2015. Application of genomics-assisted breeding for generation of climate resilient crops: progress and prospects. *Frontiers in Plant Science*, *6*, 563.

Kumar, K., Gambhir, G., Dass, A., Tripathi, A.K., Singh, A., Jha, A.K., Yadava, P., Choudhary, M. and Rakshit, S., 2020. Genetically modified crops: current status and future prospects. *Planta*, *251*(4), 91.

Langridge, P., Braun, H., Hulke, B., Ober, E. and Prasanna, B.M., 2021. Breeding crops for climate resilience. *Theoretical and Applied Genetics*, *134*(6), 1607–1611.

Liu, H., Zhou, X., Dong, N., Liu, X., Zhang, H. and Zhang, Z., 2011. Expression of a wheat MYB gene in transgenic tobacco enhances resistance to Ralstonia solanacearum, and to drought and salt stresses. *Functional & Integrative Genomics*, *11*, 431–443.

Liu, M., Shi, J. and Lu, C., 2013. Identification of stress-responsive genes in Ammopiptanthus mongolicus using ESTs generated from cold-and drought-stressed seedlings. *BMC Plant Biology*, *13*, 1–14.

Lorenz, W.W., Sun, F., Liang, C., Kolychev, D., Wang, H., Zhao, X., Cordonnier-Pratt, M.-M., Pratt, L.H. and Dean, J.F., 2006. Water stress-responsive genes in loblolly pine (*Pinus taeda*) roots identified by analyses of expressed sequence tag libraries. *Tree Physiology*, *26*, 1–16.

Lu, G., Coneva, V., Casaretto, J.A., Ying, S., Mahmood, K., Liu, F., Nambara, E., Bi, Y.M. and Rothstein, S.J., 2015. OsPIN5b modulates rice (*Oryza sativa*) plant architecture and yield by changing auxin homeostasis, transport and distribution. *The Plant Journal*, *83*(5), 913–925.

Ma, X. and Liu, Y.G., 2016. CRISPR/Cas9-based multiplex genome editing in monocot and dicot plants. *Current Protocols in Molecular Biology*, *115*(1), 1–36.

Ma, X., Zhang, Q., Zhu, Q., Liu, W., Chen, Y., Qiu, R., Wang, B., Yang, Z., Li, H., Lin, Y. and Xie, Y., 2015. A robust CRISPR/Cas9 system for convenient, high-efficiency multiplex genome editing in monocot and dicot plants. *Molecular Plant*, *8*(8), 1274–1284.

Magar, M.M., Liu, H. and Yan, G., 2024. Wheat TaAP2/ERF genes regulate heat tolerance through ethylene signaling at grain-filling stage. *Journal of Plant Growth Regulation*, *43*, 1–18.

Manickam, S., Rajagopalan, V.R., Kambale, R., Rajasekaran, R., Kanagarajan, S. and Muthurajan, R., 2023. Plant metabolomics: current initiatives and future prospects. *Current Issues in Molecular Biology*, *45*(11), 8894–8906.

Mao, H., Sun, S., Yao, J., Wang, C., Yu, S., Xu, C., Li, X. and Zhang, Q., 2010. Linking differential domain functions of the GS3 protein to natural variation of grain size in rice. *Proceedings of the National Academy of Sciences of the United States of America*, *107*(45), 19579–19584.

Mittal, S., Banduni, P., Mallikarjuna, M.G., Rao, A.R., Jain, P.A., Dash, P.K. and Thirunavukkarasu, N., 2018. Structural, functional, and evolutionary characterization of major drought transcription factors families in maize. *Frontiers in Chemistry*, *6*, 177.

Modi, A., Litoriya, N., Prajapati, V., Rafalia, R. and Narayanan, S., 2014a. Transcriptional profiling of genes involved in steviol glycoside biosynthesis in Stevia rebaudiana bertoni during plant hardening. *Developmental Dynamics*, *243*(9), 1067–1073.

Modi, A.R., Raj, S., Kanani, P., Patel, A. and Narayanan, S., 2014b. Analysis of differentially expressed genes involved in stevioside biosynthesis in cultures of *Stevia rebaudiana* Bertoni treated with steviol as an immediate precursor. *Journal of Plant Growth Regulation*, *33*, 481–488.

Mohler, V. and Singrün, C., 2004. General considerations: marker-assisted selection. In Lörz, H. and Wenzel, G. (eds) *Molecular Marker Systems in Plant Breeding and Crop Improvement* (pp. 305–317). Berlin, Heidelberg: Springer.

Muthamilarasan, M. and Prasad, M., 2017. Genetic determinants of drought stress tolerance in Setaria. In Doust, A. and Diao, X. (eds) *Genetics and Genomics of Setaria. Plant Genetics and Genomics: Crops and Models* (Vol. 19, pp. 267–289). Cham: Springer. doi: 10.1007/978-3-319-45105-3_16.

Nejat, N., Ramalingam, A. and Mantri, N., 2018. Advances in transcriptomics of plants. *Plant Genetics and Molecular Biology*, *164*, 161–185.

Ober, E.S., Alahmad, S., Cockram, J., Forestan, C., Hickey, L.T., Kant, J., Maccaferri, M., Marr, E., Milner, M., Pinto, F. and Rambla, C., 2021. Wheat root systems as a breeding target for climate resilience. *Theoretical and Applied Genetics*, *134*(6), 1645–1662.

Ozturk, Z.N., Talamé, V., Deyholos, M., Michalowski, C.B., Galbraith, D.W., Gozukirmizi, N., Tuberosa, R. and Bohnert, H.J., 2002. Monitoring large-scale changes in transcript abundance in drought-and salt-stressed barley. *Plant Molecular Biology*, *48*, 551–573.

Pease, A.C., Solas, D., Sullivan, E.J., Cronin, M.T., Holmes, C.P. and Fodor, S.P., 1994. Light-generated oligonucleotide arrays for rapid DNA sequence analysis. *Proceedings of the National Academy of Sciences of the United States of America*, *91*(11), 5022–5026.

Pervaiz, T., Amjid, M.W., El-Kereamy, A., Niu, S.H. and Wu, H.X., 2021. microRNA and cDNA-microarray as potential targets against abiotic stress response in plants: advances and prospects. *Agronomy*, 12(1), 11.

Pervaiz, T., Songtao, J., Faghihi, F., Haider, M.S. and Fang, J., 2017. Naturally occurring anthocyanin, structure, functions and biosynthetic pathway in fruit plants. *Journal of Plant Biochemistry & Physiology*, *5*(2), 1–9.

Potts, J., Jangra, S., Michael, V.N. and Wu, X., 2023. Speed breeding for crop improvement and food security. *Crops*, *3*(4), 276–291.

Purugganan, M.D. and Fuller, D.Q., 2009. The nature of selection during plant domestication. *Nature, 457*(7231), 843–848.

Rahman, S.U., Khan, M.O., Ullah, R., Ahmad, F. and Raza, G., 2023. Agrobacterium-mediated transformation for the development of transgenic crops; present and future prospects. *Molecular Biotechnology*, *66*, 1–17.

Raj, R.S., Singh, C., Modi, A. and Subhash, N., 2015. Genetic transformation of lowland rice variety GR11 for drought tolerance and its ratification for upland paddy cultivation. *Indian Journal of Genetics and Plant Breeding*, *75*(01), 30–40.

Raza, A., Su, W., Hussain, M.A., Mehmood, S.S., Zhang, X., Cheng, Y. and Lv, Y., 2021. Integrated analysis of metabolome and transcriptome reveals insights for cold tolerance in rapeseed (*Brassica napus* L.). *Frontiers in Plant Science*, *12*, 721681.

Razzaq, A., Kaur, P., Akhter, N., Wani, S.H. and Saleem, F., 2021. Next-generation breeding strategies for climate-ready crops. *Frontiers in Plant Science*, *12*, 620420.

Reuter, J.A., Spacek, D.V. and Snyder, M.P., 2015. High-throughput sequencing technologies. *Molecular Cell*, *58*(4), 586–597.

Salgotra, R.K., Gupta, B.B. and Stewart, C.N., 2014. From genomics to functional markers in the era of next-generation sequencing. *Biotechnology Letters*, *36*, 417–426.

Sanger, F., 2004. Determination of nucleotide sequences in DNA. *Bioscience Reports*, *24*(4), 237–253.

Sanger, F., Nicklen, S. and Coulson, A.R., 1977. DNA sequencing with chain-terminating inhibitors. *Proceedings of the National Academy of Sciences of the United States of America*, *74*(12), 5463–5467.

Schena, M., Shalon, D., Davis, R.W. and Brown, P.O., 1995. Quantitative monitoring of gene expression patterns with a complementary DNA microarray. *Science*, 270, 467–470. doi: 10.1126/science.270.5235.467

Seki, M., Narusaka, M., Abe, H., Kasuga, M., Yamaguchi-Shinozaki, K., Carninci, P. and Shinozaki, K., 2001. Monitoring the expression pattern of 1300 Arabidopsis genes under drought and cold stresses by using a full-length cDNA microarray. *The Plant Cell*, *13*(1), 61–72.

Seong, S.Y., Shim, J.S., Bang, S.W. and Kim, J.K., 2020. Overexpression of OsC3H10, a CCCH-zinc finger, improves drought tolerance in rice by regulating stress-related genes. *Plants*, *9*(10), 1298.

Shah, T., Xu, J., Zou, X., Cheng, Y., Nasir, M. and Zhang, X., 2018. Omics approaches for engineering wheat production under abiotic stresses. *International Journal of Molecular Sciences*, *19*, 2390–2405. doi: 10.3390/ijms19082390.

Shahzad, R., Jamil, S., Ahmad, S., Nisar, A., Amina, Z., Saleem, S., Iqbal, M.Z., Atif, R.M. and Wang, X., 2021. Harnessing the potential of plant transcription factors in developing climate resilient crops to improve global food security: current and future perspectives. *Saudi Journal of Biological Sciences*, *28*(4), 2323–2341.

Singh, U. M., Sareen, P., Sengar, R. S. and Kumar, A., 2013. Plant ionomics: a newer approach to study mineral transport and its regulation. *Acta Physiologiae Plantarum*, *35*, 2641–2653. doi: 10.1007/s11738-013-1316-8.

Sinha, A., John, J., Singh, S. and Johri, P., 2022. Microarray-based detection and identification of bacterial and viral plant pathogens. *Rhizosphere Microbes: Biotic Stress Management*, *40*, 47–69.

Soon, W.W., Hariharan, M. and Snyder, M.P., 2013. High-throughput sequencing for biology and medicine. *Molecular Systems Biology*, *9*(1), 640.

Street, N.R., Skogström, O., Sjödin, A., Tucker, J., Rodríguez-Acosta, M., Nilsson, P., Jansson, S. and Taylor, G., 2006. The genetics and genomics of the drought response in Populus. *Plant Journal*, *48*, 321–341.

Sysoeva, M.I., Markovskaya, E.F. and Shibaeva, T.G., 2010. Plants under continuous light: a review. *Plant Stress*, *4*(1), 5–17.

Talukdar, D. and Sinjushin, A., 2015. Cytogenomics and mutagenomics in plant functional biology and breeding. *PlantOmics: The Omics of Plant Science*, *5*, 113–156.

Tan, Y.C., Kumar, A.U., Wong, Y.P. and Ling, A.P.K., 2022. Bioinformatics approaches and applications in plant biotechnology. *Journal of Genetic Engineering and Biotechnology*, *20*(1), 106.

Thottathil, G.P., Jayasekaran, K. and Othman, A.S., 2016. Sequencing crop genomes: a gateway to improve tropical agriculture. *Tropical Life Sciences Research*, *27*(1), 93.

Tilman, D., Balzer, C., Hill, J. and Befort, B.L., 2011. Global food demand and the sustainable intensification of agriculture. *Proceedings of the National Academy of Sciences of the United States of America*, *108*(50), 20260–20264.

Varshney, R.K., Song, C., Saxena, R.K., Azam, S., Yu, S., Sharpe, A.G. and Cannon, S., 2013. Draft genome sequence of chickpea (*Cicer arietinum*) provides a resource for trait improvement. *Nature Biotechnology*, *31*(3), 240–248.

Velez-Ramirez, A.I., van Ieperen, W., Vreugdenhil, D. and Millenaar, F.F., 2011. Plants under continuous light. *Trends in Plant Science*, *16*(6), 310–318.

Verma, A., Kumar, R., Shukla, S., Singh, S., Johri, P. and Singh, M.S., 2023. Genome wide association studies: an outlook. *Journal of Innovation in Applied Research*, *6*, 56–64. doi: 10.51323/JIAR.2023/6/1/56-62

Viemann, D., Schulze-Osthoff, K. and Roth, J., 2005. Potentials and pitfalls of DNA array analysis of the endothelial stress response. *Biochimica et Biophysica Acta (BBA)-Molecular Cell Research*, *1746*(2), 73–84.

Voytas, D.F. and Gao, C., 2014. Precision genome engineering and agriculture: opportunities and regulatory challenges. *PLoS Biology*, *12*(6), e1001877.

Wang, Z., Gerstein, M. and Snyder, M., 2009. RNA-Seq: a revolutionary tool for transcriptomics. *Nature Reviews Genetics*, *10*(1), 57–63.

Wanga, M.A., Shimelis, H., Mashilo, J. and Laing, M.D., 2021. Opportunities and challenges of speed breeding: a review. *Plant Breeding*, *140*(2), 185–194.

Watson, A., Ghosh, S., Williams, M.J., Cuddy, W.S., Simmonds, J., Rey, M.D., Asyraf Md Hatta, M., Hinchliffe, A., Steed, A., Reynolds, D. and Adamski, N.M., 2018. Speed breeding is a powerful tool to accelerate crop research and breeding. *Nature Plants*, *4*(1), 23–29.

Xu, K., Xu, X., Fukao, T., Canlas, P., Maghirang-Rodriguez, R., Heuer, S., Ismail, A.M., Bailey-Serres, J., Ronald, P.C. and Mackill, D.J., 2006. Sub1A is an ethylene-response-factor-like gene that confers submergence tolerance to rice. *Nature*, *442*(7103), 705–708.

Yadav, C. B., Pandey, G., Muthamilarasan, M. and Prasad, M., 2018. Epigenetics and epigenomics of plants. In Varshney, R., Pandey, M., and Chitikineni, A. (eds) *Plant Genetics and Molecular Biology. Advances in Biochemical Engineering/Biotechnology* (Vol. 64, pp. 237–261). Cham: Springer. doi: 10.1007/10_2017_51.

Yang, Y., Saand, M.A., Huang, L., Abdelaal, W.B., Zhang, J., Wu, Y., Li, J., Sirohi, M.H. and Wang, F., 2021. Applications of multi-omics technologies for crop improvement. *Frontiers in Plant Science*, *12*, 563953.

Yang, Y., Thannhauser, T.W., Li, L. and Zhang, S., 2007. Development of an integrated approach for evaluation of 2-D gel image analysis: impact of multiple proteins in single spots on comparative proteomics in conventional 2-D gel/MALDI workflow. *Electrophoresis*, *28*(12), 2080–2094.

Zeng, Y., Wen, J., Zhao, W., Wang, Q. and Huang, W., 2020. Rational improvement of rice yield and cold tolerance by editing the three genes *OsPIN5b*, *GS3*, and *OsMYB30* with the CRISPR–Cas9 system. *Frontiers in Plant Science*, *10*, 1663.

Zhao, C., Liu, B., Piao, S., Wang, X., Lobell, D. B., Huang, Y. and Asseng, S., 2017. Temperature increase reduces global yields of major crops in four independent estimates. *Proceedings of the National Academy of Sciences of the United States of America*, *114*(35), 9326–9331.

16 Bioinformatics Approaches to Study Plant miRNA Implications in Plant Stress

*Samiksha Garse, Soham Sunil Gumaste, Pallavi Gurav, Kanchanlata Tungare, Sekar Vijayakumar, and Shine Devarajan**

16.1 INTRODUCTION

16.1.1 Plant microRNAs

microRNAs (miRNAs) are a broad category of non-coding RNAs consisting of 22 base pairs, believed to regulate the functioning of certain genes in metazoans (Noopur et al., 2025a,b; Pooja et al., 2025). Since plants also contain miRNAs, it is clear that eukaryotic evolution gave rise to this class of non-coding RNAs (Reinhart et al., 2002). The majority of biological processes in plants have now been connected to miRNAs. Although their discovery in plants dates back less than 20 years, research has demonstrated their importance in controlling growth factors such as leaf morphology, vegetative phase transition, blooming duration, and responsiveness to environmental stimuli. Furthermore, every plant species contains a large number of poorly conserved miRNAs, suggesting that the functional influence of these miRNAs permeates almost every aspect of plant biology (Millar, 2020).

16.1.2 Biogenesis

In *Arabidopsis*, the production of miRNAs is quite similar to that of mammals, as these miRNAs integrate into a protein complex called the RNA-induced silencing complex after being processed by RNAse III-like enzymes. Moreover, during plant miRNA biogenesis, ribose methylation occurs on the terminal nucleotide of the miRNAs. Plant miRNAs and their target mRNAs demonstrate higher sequence complementarity, which has facilitated the bioinformatic prediction of miRNA targets; many of these predictions have already been confirmed (Chen, 2005).

16.1.3 Function of miRNA

It has been hypothesized or demonstrated that plant miRNAs control several functions, including metabolism, stress responses, and development. Much of the miRNA target population consists of genes encoding transcription factors, which are essential for establishing plant morphology. Consequently, predicting miRNA target genes using bioinformatics has become more straightforward; many of these genes have since been confirmed through experimentation as genuine miRNA targets. It has also been hypothesized or demonstrated that plant miRNAs regulate genes encoding various types of proteins. A significant portion of miRNA target genes includes transcription factors and other proteins that modulate signal transduction pathways or plant growth signaling pathways. While miRNA targets encoding transcription factors have received the most attention thus far, those encoding proteins involved in metabolism, stress responses, and protein degradation have also been predicted and validated. The

* *Corresponding Author*

DOI: 10.1201/9781003491378-16

identification of targets has led to an understanding of the outcomes of *in vivo* cleavage. Therefore, it follows that miRNAs regulate a wide range of biological functions (Chen, 2005).

16.1.4 miRNA and Plant Stress

miRNAs have proven to be expert controllers of plant development and growth. The finding that miR398 target genes are involved in stress tolerance provides evidence of the role of miRNAs in regulating plant stress responses (Sunkar et al., 2012). By regulating the expression of specific miRNAs or synthesizing new ones, plants under abiotic stress can strengthen their resilience to environmental stressors, ultimately helping them endure and thrive in harsh conditions. The expression of specific miRNAs may be up- or downregulated depending on the circumstances. Various stresses, including high temperatures (HTs), chilling temperatures, droughts, high salinities, oxidation, and heavy metals, can also influence the functional regulation of particular miRNAs in plants (Jian et al., 2010). Research on

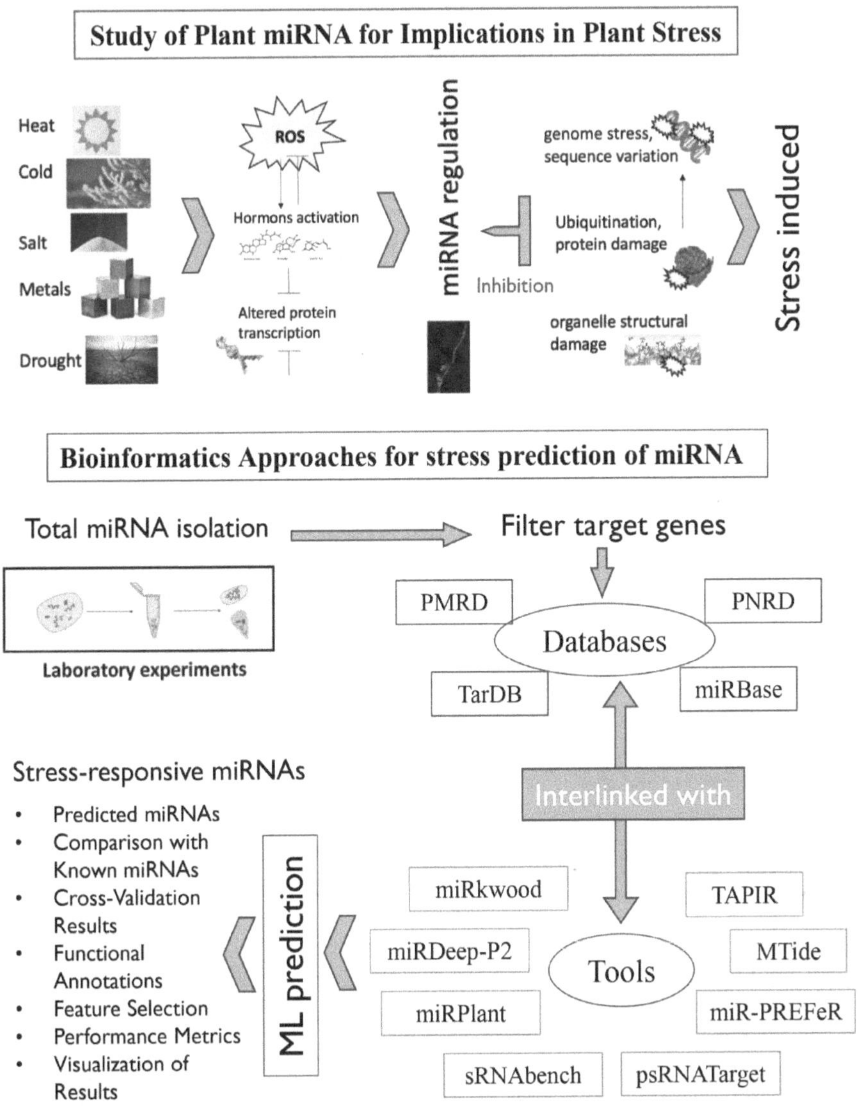

FIGURE 16.1 Summary of plant stress and target identification related to stress tolerance using bioinformatics approaches and machine learning models.

stress-responsive miRNAs can provide valuable insights for breeding stress-tolerant plants. Moreover, gene expression analysis has proven to be a powerful tool for uncovering previously unknown information about plant adaptive processes. Figure 16.1 illustrates a summary of various databases and tools that integrate plants' stress responses with the miRNA mechanism (Zhang et al., 2022).

16.2 FACTORS OF PLANT STRESS

16.2.1 Salinity and Drought Stress

Two of the most significant environmental issues limiting agricultural productivity globally are salinity and drought. Both stresses can alter morphology, physiology, biochemistry, and metabolism through various pathways, ultimately affecting a plant's ability to grow, develop, and produce. Plants respond to these stress conditions in a highly complex manner that depends on various parameters, including age, size, genotype, species, progression rate, and the duration and intensity of the stresses (Ma et al., 2020).

During a drought, stomata gradually close, correlating with a decrease in net photosynthetic activity and water-use efficiency. Drought tolerance is closely linked to changes in photosynthetic pigments, independent of other factors. To prevent irreversible damage to the photosynthetic machinery, self-protective responses against stress at the leaf level should therefore be swiftly induced (Ma et al., 2020).

Plants respond to salinity using a biphasic model, in which ion toxicity is induced over time, and the early phase resembles drought (osmotic stress) (Anil et al., 2023; Jabborova et al., 2023; Verma et al., 2021). Growth drastically slows down in the initial phase due to stomatal closure in response to the drop in water potential. In the second stage, the accumulation of ions, especially Na+, affects pigments and enzymes involved in photosynthetic processes and increases oxidative stress. In addition to acting as harmful by-products of stress, reactive oxygen species (ROS) also functions as a signaling molecule, activating numerous pathways that lead to defensive responses in the salt stress response, typically triggered by drought stress (Ma et al., 2020).

16.2.2 Cold Stress

One significant environmental issue that limits plant productivity in hilly areas is cold stress. To survive under stressful conditions, plants react and adjust at the molecular, cellular, physiological, and biochemical levels (Sanghera et al., 2011). Previous research has shown that low temperature limits plant growth and productivity in temperate climates. Plants' ability to withstand cold is a complex process. Plant tissues may become ice-covered when temperatures drop below freezing, leading to cell dehydration. Conversely, plants can defend themselves by preventing the formation of ice. The mechanism of cold stress is intricate and is influenced by ICE-CBF-COR genes, as well as other factors like hormones, light, and circadian rhythms. *ICE* is a C-repeat-binding factor (CBF) regulator; during cold stress, it induces and modulates the expression of the *COR* gene, also known as cold-regulated genes. This gene is crucial for plants' ability to endure low temperatures and adapt to them. Under normal conditions, the photoperiod and the circadian clock regulate *CBFs*. Nonetheless, *CBFs* activate several genes associated with cold stress to enhance plants' cold tolerance (Ritonga & Chen, 2020).

16.2.3 Multiple Abiotic Stress

Any unfavorable impact on a plant's growth and development in a particular environment is known as abiotic stress. Drought, salinity, and HTs are examples of stressors that frequently interact or work in tandem with one another. To recognize these environmental stresses and adapt their growth for survival and reproduction, plants have evolved mechanisms (Zhang et al., 2023). One of the key ways plants respond to different abiotic stimuli is through their abiotic stress response network, which is mediated by miRNA. Worldwide, crop losses are primarily caused by abiotic stressors. Most abiotic stimuli, such as excessive salinity, extreme temperatures, droughts, and heavy metals, result in oxidative stress. In both modeled and crop plants, abiotic and oxidative stressors limit plant growth and reduce production (Kerchev et al., 2020). Because of the enhanced fluidity of membrane lipids, the denaturation and aggregation of proteins caused by these abiotic stressors lead to direct injury in plants. Apart from the direct consequences, indirect damage includes inhibition of protein synthesis, disintegration of proteins, and loss of membranes in mitochondria and chloroplasts. During the reproductive stage, abiotic stress causes ovule abortion, floral drop, pollen sterility, pollen tube deformations, and decreased yield (Kumari et al., 2022).

16.2.4 Nutritional Stress

Enzymes like catalase (CAT), peroxidase (POD), and superoxide dismutase (SOD) are generally more concentrated in tissues containing N, K, Ca, and Mg. These enzymes lower the concentration of ROS and are referred to as antioxidant enzymes. By enhancing water absorption, nutrients like calcium and potassium aid in stomatal control and osmotic adjustments. These nutrients also help maintain a high tissue water potential under stress conditions. Micronutrients including zinc, boron, and iron aid in activating defense mechanisms, facilitating multiple physiological alterations in plants, and enhancing metabolic processes that enable plants to respond to various harmful stimuli. Certain combinations of nutrients can reduce stress more effectively when taken as a whole, and vice versa. There is still much to learn, even if using plant nutrients to manage abiotic stressors is a cost-effective and sustainable solution (Kumari et al., 2022).

16.2.5 Oxidative Stress

Plants are regularly susceptible to biotic and abiotic challenges because of climate change, which extensively jeopardizes plant productivity and growth. The rapid generation of ROS that causes oxidative stress is a crucial indicator of these molecular stressors (Hasanuzzaman & Fujita, 2022). Environmental stresses negatively impact agricultural output, plant growth, and development. These unfavorable conditions alter the metabolism of ROS and reactive nitrogen species. The disruption of redox homeostasis results from highly elevated concentrations of these reactive species, eventually surpassing the capacity of antioxidant defense enzymes. Moreover, this can adversely affect macromolecules such as proteins, nucleic acids, and membrane lipids, ultimately leading to nitro-oxidative stress and cell death in plants (Chaki et al., 2020).

16.2.6 UV-B Stress

UV-B stress is defined as intense, continuous full-wavelength UV-B that causes abnormal plant growth and development and harms plants. Heritable variation results from the effect of UV-B stress on pyrimidine dimer formation, which alters DNA synthesis and replication (Britt, 1995). In addition to directly causing DNA damage, UV-B generates ROS that lead to oxidative stress as well as the oxidation of proteins and lipids. Plants perceive UV-B as an environmental cue and a potential source of abiotic stress that may interfere with growth and adaptation. UV-B inhibits hypocotyl elongation, cotyledon growth, and flavonoid accumulation. However, high-intensity UV-B can also damage DNA, increase the amount of ROS in the environment, and hinder photosynthesis. Under UV-B exposure, plants have evolved "sunscreen" flavonoids that accumulate to prevent or minimize harm. A UV-B receptor, UV-resistance locus 8 (UVR8), is essential for enhancing resistance to UV-B stress by promoting flavonoid synthesis. Recent research has clarified numerous UVR8-mediated and UVR8-independent mechanisms that control UV-B stress tolerance (Shi & Liu, 2021).

16.3 IMPACT OF RAPIDLY CHANGING ENVIRONMENT ON PLANT

Terrain plants are always exposed to a variety of unfavorable, sometimes harmful, environmental circumstances. Abiotic stresses that negatively impact plant development and productivity include salinity, drought, extreme temperatures, heavy metals, ozone, UV radiation, and nutrient deficiencies.

These stresses are becoming more significant due to the immediate and secondary consequences of changing climatic conditions. Abiotic stressors cause plants to respond in various ways, affecting physiology, gene expression, primary and secondary metabolism, and plant structure. The complex changes in plants allow them to endure and/or adapt to a wide range of adverse situations. The length and severity of the stress, the genetic makeup of the plant, the combination of multiple stresses, the type of susceptible tissues and cells, and the growth stage at which the plant senses the stress can all influence the complexity of the plant's response. Thus, it is critical to understand more about the impact of stressed conditions perceived by plants and their adaptive responses (in both natural and man-made contexts) (Mareri et al., 2022). The following are some environmental factors (Figure 16.1) that cause various plant stresses.

16.3.1 Temperature Stress

Globally, HT stress is a major environmental stressor that hinders plant growth, metabolic processes, and agricultural potential. A multitude of temperature-sensitive metabolic pathways impacts the growth and development of plants. The type of plant and the extent and duration of HT influence how different plants respond to it. Crop yields are currently greatly affected by HT, making it crucial to develop methods for maintaining high crop yields during HT stress. To cope with these HT conditions, plants have a variety of acclimation, avoidance, and adaptive strategies. To counteract stress-induced biochemical and physiological changes, significant tolerance processes involving proteins, ion carriers, osmoprotectants, antioxidants, and other components related to signaling cascades and transcriptional regulation are also activated (Hasanuzzaman et al., 2013).

Plants that are exposed to HT collect many metabolites, including antioxidants, osmoprotectants, heat shock proteins, and others, triggering their metabolic pathways and activities. These modifications highlight the importance of conducting physiological and molecular research to understand the mechanisms behind stress responses. Developing stress-tolerant plants would also benefit from an understanding of the specific genes produced in response to HT and the mechanisms of the signaling cascades. Molecular techniques that identify responses and tolerance mechanisms may pave the way for engineering plants that can withstand heat and serve as the foundation for creating crop varieties capable of generating profitable yields under heat stress (Rodríguez et al., 2005; Ahmad & Prasad, 2012).

16.3.2 Drought Stress

Major staple crops are currently the subject of significant research to determine the mechanisms that respond to drought in order to maximize grain yield and quality; however, future research should focus on how drought and HTs intersect to affect the reproductive phases of primary grain crops (Seleiman & Kheir, 2018). Plant growth and development heavily depend on an optimal level of water availability; deviations from this level can impact grain quality and production. Conversely, an inadequate supply of water in the rhizosphere hinders plant growth, preventing the plant from absorbing nutrients (Elemike et al., 2019). Reduced or negligible rainfall diminishes the soil's moisture content and limits water's capacity to reach plant components above soil level, such as leaves and stems, which is the root cause of water scarcity outbreaks (Ristvey et al., 2019). This occurs in arid environments when the rate at which water moves through roots is less than the rate at which it evaporates through transpiration from leaves (Goche et al., 2020). When there is water scarcity, plants strive to take in more water by spreading their roots in hopes of reducing stomatal water loss (Martínez-Vilalta & Garcia-Forner, 2016). Plants experiencing drought stress often exhibit leaf rolling, dwarfism, yellowing, blistering, and persistent wilting (Corso et al., 2020). Additionally, the frequency and intensity of previous drought stress events, as well as the presence of other stresses, significantly impact how plants respond to a specific water deficit (Thomason & Battaglia, 2020).

16.3.3 Salt and Osmotic Stress

Salinity constitutes one of the most harmful environmental stressors that drastically lowers agricultural productivity and quality globally. Over 20% of the agricultural land in the world is susceptible to salt stress, and as a result of both natural and human activity, the percentage of these salt-prone areas is constantly increasing (Arora, 2019). However, because of the rising demand for water for irrigation over the past 20 years, this difficulty has worsened significantly in dry and semi-arid countries (Acosta-Motos et al., 2017). Salinity has been shown to negatively impact photosynthesis, as magnesium resides in the center of chlorophyll, resulting in limitations in plant and leaf growth as well as a decrease in chlorophyll levels. Moreover, intracellular osmotic pressure rises in response to salt stress, which can lead to hazardous sodium accumulation (Zhao et al., 2021).

Salt stress has numerous deleterious effects on plant growth and reproduction, similar to other abiotic stresses. It causes ion toxicity, oxidative and osmotic stress, nutritional and hormonal abnormalities, and heightened plant sensitivity to disease. Furthermore, salt stress can harm or even kill plants. Excessive concentrations of salt in the soil alter the soil's hydraulic conductivity and porosity, which lowers the soil's water potential and creates water stress. Eventually, this leads to physiological drought conditions, disruption of the cellular membrane, and protein degradation from the toxicity of various ions (primarily Na+). In addition to the impacts listed above, salt stress causes a range of physiological and metabolic changes in plants, as well as modifications to seed germination behavior, photosynthesis, and other biosynthetic processes that restrict growth (Hasanuzzaman & Fujita, 2022).

16.3.4 UV and Light Stresses

Sunlight is an essential environmental cue that plants use to regulate their growth and development, in addition to serving as an energy source. Light signals are employed by higher plants to regulate several growth processes and modifications, including adaptation, stomatal development, physical shifts in response to light, shadow avoidance, germination, tracking time and season through photoreceptors, and blooming. One example of this is the model plant *Arabidopsis thaliana*, commonly known as *Arabidopsis*. In addition to controlling growth and development, or what is now referred to as "normal" photomorphogenesis, intense light, particularly that which has an inherent ultraviolet (UV) component, can induce stress reactions in plants. Through photosynthesis, sunlight provides plants with the energy they need to thrive. However, excessive light, especially that which contains UV radiation, can stress cells and significantly damage proteins, DNA, and other biological components. Plants exhibit what is often called "plant memory," or their ability to adapt to environmental challenges. There is mounting evidence that plants use cellular-level epigenetic processes to remember exposure to biotic or abiotic stressors. The fluence rate, duration of exposure, and presence or absence of prior light acclimation all influence the type of stress response that light elicits. UV-B receptor UV-RESISTANCE LOCUS 8 (UVR8) studies have confirmed that UV signaling occurs rapidly, wherein UV target genes, including chalcone synthase, respond immediately to UV treatment. In light of these findings, a UV memory may seem unnecessary. However, some research avenues suggest that plants may establish an epigenetic memory of exposure to light and UV radiation (Müller-Xing et al., 2014).

16.3.5 Nutrient Stress

Lack of vital nutrients can result in stunted growth, necrosis, and chlorosis, ultimately leading to significant crop losses and impacting human nutrition. When nutrients are scarce, plants adapt morphologically, biochemically, and through rhizosphere-mediated processes to enhance nutrient uptake. Changes in morphology occur at the (plant's apex roots) as well as in the aerial parts(shoots and leaves). In nutrient-deficient plants, shoot branching is restricted, and leaf senescence is encouraged. Plants can alter the architecture of their roots by inhibiting the growth of primary roots while promoting the growth of secondary roots and root hairs in an effort to search for the limited available nutrients. Plants and bacteria can collaborate symbiotically to improve nutrient intake. Additionally, plants increase the regulation of nutrient transporters, enhance the synthesis of acid phosphatase, promote proton signaling processes, and produce organic acids to ensure effective nutrient absorption. Plants cultivated in nutrient-deficient soil also show signs of anthocyanin accumulation and ROS emission (Francis et al., 2023). Once stress-related genes are activated, antioxidant enzymes, osmoprotectants, and heat shock proteins are biosynthesized; ROS are detoxified; proteins are structurally or functionally protected; DNA is repaired; and the stability of the cellular membrane is maintained. Moreover, other plant mechanisms are supported by plant nutrients in mitigating abiotic stress (Kumari et al., 2022).

16.3.6 Heavy Metal Stress

Exposure to heavy metals impairs the structural morphology, biochemical processes, and molecular functioning of plants. The levels of lead (Pb) and cadmium (Cd) in the environment have a significant impact on plant growth and productivity. Plants, on the other hand, absorb zinc from the soil for essential purposes (Mareri et al., 2022). Heavy metals, including nickel, copper, cobalt, chromium, and mercury, are major environmental pollutants that negatively affect plant productivity and pose a serious risk to agro-ecosystems. They alter the physiology of plants and induce stress (Sethy & Ghosh, 2013). Zinc facilitates the accumulation and movement of lead and copper through the aerial parts of maize plants. Furthermore, Zn, Pb, and Cd interact to impede the uptake and translocation of other divalent metals. This work examines whether the methylation of DNA and histone acetylation affects metal stress tolerance using Zn transporters. It also warns against excessive use of Zn fertilizer on soils contaminated with metals (Shafiq et al., 2020).

With the exception of cadmium, which does not seem to have a specific transporter protein, certain transporter proteins carry heavy metals into roots through the plasma membrane's barrier, and then coplanar or plastid extracellular signaling networks transport them to the xylem or phloem. A variety of concerning anomalies, such as decreased seed germination, delayed growth, leaf discoloration, chlorophyll disruption, darkening of tissues due to cellular death, and stunted roots, are exhibited by stressed plants. Although plants have varying levels of tolerance to different heavy metals, these anomalies are triggered by the effects of excessive heavy metal exposure on plant growth, development, and reproduction. Heavy metals have antagonistic effects on plant physiological processes. They alter protein structures, cause denaturation and inactivation of intracellular enzymes in plants, and disrupt membrane integrity. These effects subsequently impact normal plant metabolic processes such as transpiration, respiration, and photosynthesis. Furthermore, the introduction of toxic heavy metals into plant tissues causes an increase in ROS, which disrupts the redox balance (Feng et al., 2023).

16.4 BIOINFORMATICS OF PLANT miRNA

Bioinformatics approaches provide multifaceted components that enable researchers and botany experts to explore plant miRNA through computational databases and predict its structure and functional targets, as well as their gene-regulating profiles, using machine learning algorithms (Verma et al., 2023). A timeline of various databases and target identification tools is provided in Table 16.1.

16.4.1 miRNA Databases

16.4.1.1 PMRD

The PMRD (Plant microRNA Database) is a database that incorporates data from internal sources, the latest publications, and publicly available information on plant miRNA. Together with the primary sequences of miRNA, it includes a genome browser, which allows users to browse various genes of interest, identify their targets, and study their secondary structures along with their gene-regulating profiles. The database focuses on crucial crops that can be modeled, such as

TABLE 16.1
Various Databases and Target Identification Tools

Tool	Published Year	Significance	Website	References
		Databases		
miRBase	2008	The miRNA Database: Comprehensive database of animal and plant microRNAs.	https://www.mirbase.org/	Griffiths-Jones et al. (2007)
PMRD	2010	Plant miRNA Database: Stores experimentally validated plant miRNA–target interactions.	http://bioinformatics.cau.edu.cn/PMRD/	Zhang et al. (2009)
PNRD	2015	Plant miRNA and Novel Disease Relationship Database: Links plant miRNAs to stress response.	http://structuralbiology.cau.edu.cn/PNRD/	Yi et al. (2014)
TarDB	2021	Target RNA Database: Curated database of experimentally validated plant miRNA–target interactions.	http://www.biosequencing.cn/TarDB/	Liu et al. (2021)
		Databases		
TAPIR	2010	Target Prediction Algorithm for miRNA and piRNA: Employs sequence and structure features for miRNA target prediction.	http://bioinformatics.psb.ugent.be/webtools/tapir/	Bonnet et al. (2010)
miR-PREFeR	2014	miRNA Prediction and Regulation for Efficient Forward Engineering: Predicts plant miRNA targets using sequence complementarity.	https://github.com/hangelwen/miR-PREFeR	Lei and Sun (2014)
miRPlant	2014	A Probabilistic Model for Detecting Plant miRNA Targets: Employs probabilistic modeling to predict plant miRNA targets.	https://sourceforge.net/projects/mirplant/	An et al. (2014)
MTide	2014	microRNA Target Identification and Evaluation System: Integrates various prediction algorithms for miRNA target identification.	https://bis.zju.edu.cn/MTide/	Zhang et al. (2014)
psRNATarget	2018	Plant Small RNA (sRNA) Target Analysis Server: Offers a range of tools for miRNA target prediction, including sequence and structure analysis.	https://www.zhaolab.org/psRNATarget/	Dai and Zhao (2011)
miRDeep-P2	2019	Deep learning algorithms: To learn complex patterns from large datasets, can better distinguish between bona fide miRNAs and other sRNA species, reducing the chances of false positives for highly accurate miRNA target prediction.	https://sourceforge.net/projects/mirdp2/	Kuang et al. (2018)
miRkwood	2019	A Plant microRNA Target Prediction Tool Using Support Vector Machine: Utilizes machine learning (SVM) for plant miRNA target prediction.	https://bioinfo.univ-lille.fr/mirkwood/	Guigon et al. (2019)
sRNAbench	2022	Perform differential expression analysis: Compare miRNA expression between different samples to predict stress response.	https://arn.ugr.es/srnatoolbox/	Aparicio-Puerta et al. (2022)

Arabidopsis, *Oryza* (rice), *Triticum* (wheat), *Glycine max* (soybean), *Zea mays* (maize), *Sorghum* (millets), and *Hordeum vulgare* (barley). Around 8,000 samples of miRNAs are extracted from over a 221 species across the globe (Zhang et al., 2009).

The PMRD facilitates the study of plant miRNAs in relation to plant stress by providing miRNA expression profiles for various plant species, including data related to stress conditions. For example, the database includes miRNA expression data related to oxidative stress in rice, cold stress-responsive miRNAs in poplar, stress-regulated miRNAs in *Arabidopsis*, and miRNAs responding to stress in maize. Additionally, the database integrates information about target genes and predicted interaction sites between miRNAs and mRNA, which can provide an understanding of the structural stability and functional roles of miRNAs and their impact on stressed plants. PMRD also offers access to supplementary data, such as miRNA sequences, and is designed to be updated regularly as new plant miRNA data becomes available (Cui et al., 2012; Wang et al., 2019; Zhang et al., 2009).

Table 16.1: Timeline for tools and databases issued. Various databases and Target Identification Tools is to be followed.

16.4.1.2 TarDB

TarDB is a database freely available on the internet that has compiled three classes of plant miRNA targets with the highest probability of expression levels, also known as likely high-confidence miRNA targets. These classes are (1) conserved miRNA targets across species; (2) miRNA targets sequenced by the degradome/parallel analysis of RNA ends approach; and (3) miRNA-induced phasiRNA (phased, secondary siRNAs) loci. With its intuitive interface, users can quickly search for, explore, and extract miRNA targets and a wide range of miRNA-induced phasiRNAs across various plant species (Liu et al., 2021).

The database includes a collection of trustworthy plant miRNA targets from well-researched model species, including miRNA-triggered phasiRNAs and previously unreported miRNA targets. By identifying the miRNA targets associated with species- or lineage-specific stress-related pathways, this database enables researchers to gain functional insights into how miRNAs trigger plant responses to stress. TarDB provides high-confidence miRNA target sites and details on miRNA-triggered phasiRNAs. It is publicly accessible. Researchers can learn more about regulatory networks involving miRNAs and their targets in the context of plant stress by using the data in TarDB (Liu et al., 2021; Sihag et al., 2024).

16.4.1.3 PNRD

The Plant Non-coding RNA Database (PNRD) is an open resource for analyzing non-coding RNA (ncRNA) in plants. It was built on a conventional LAMP stack, which includes Linux, Apache, MySQL, and PHP systems, offering a sophisticated interface with a broad search range and authentic content for users. PNRD also incorporates two prevalent tools for ncRNA analysis, miR analyzer and CPC toolkits, to help researchers predict genome coding potential or efficiency, discover new miRNAs, and utilize basic tools for functional annotation within the database. PNRD supports 11 different types of known ncRNAs extracted from various sources, including internal data, published references, and public data repositories. It includes information on miRNA targets and extends to other non-coding RNAs, such as endogenous target mimics. This integrated data can provide a comprehensive understanding of the regulatory role of plant miRNAs in stress responses. PNRD offers sequence information for 25,739 ncRNA entries supporting 150 plant species. Each entry includes detailed information on targets, RNA secondary structure, expression profile, gene structure, and function-related publications. Future developments of the database are anticipated to include frequent updates to the text-mining service and full utilization of high-throughput technologies for identifying small RNA (sRNA) cleavage sites and discovering new miRNA targets that mediate stress responses, such as the parallel analysis of RNA ends (degradome sequencing method) for miRNA-mediated gene regulation. In an effort to serve as an invaluable resource and cohesive framework for plant ncRNA research, PNRD offers researchers a one-stop shop (Yi et al., 2014).

16.4.1.4 miRBase

miRBase preserves a broad library of miRNA sequences across various taxa of plants, animals, and viruses. Plant-specific miRNA sequences can be retrieved and analyzed in research on the impact of plant stress on the expression and regulation of miRNAs. miRBase maps miRNAs involved in stress response pathways to their locations in the genome and highlights clusters of miRNAs. It has developed tools to integrate deep-sequencing data, which is particularly valuable for studying plant miRNA expression under stress conditions. This allows for tracking the regulatory patterns of miRNAs in response to stress and identifying novel stress-responsive miRNAs. By incorporating short RNA deep-sequencing data as evidence for miRNA annotation, further developments could enhance the understanding of the implications of plant miRNAs in stress by providing additional insights into miRNA expression under stress conditions (Griffiths-Jones et al., 2007; Kozomara & Griffiths-Jones, 2010; Kozomara et al., 2018).

16.4.1.5 miRDB

Target prediction information across the transcriptome is available in the miRDB database, with approximately 3.5 million targets predicted for their functional properties influenced by 7,000 miRNAs across distinct species. This comprehensive data can be utilized to identify potential miRNA targets involved in plant stress responses. miRDB now features the investigation of cell-specific miRNA targets to analyze miRNA regulation within certain cell models. This enables the identification of miRNA targets relevant to stress response pathways in specific plant cell types. Moreover, target prediction and gene ontology data can be integrated and analyzed by miRDB, which can help anticipate miRNA-regulated biological processes associated with plant stress. miRDB provides a valuable resource for investigating the implications of plant miRNAs in stress by facilitating the identification of miRNA targets and their functional annotations in the context of plant stress responses (Chen & Wang, 2019; Liu & Wang, 2019).

16.4.2 Tools for Identification of miRNA and Their Target

16.4.2.1 miR-PREFeR

miR-PREFeR is a plant miRNA prediction tool that utilizes short RNA-Seq data to accurately predict plant miRNAs. This tool follows standard annotation for plant miRNAs and utilizes the expression patterns of miRNAs. The procedure involves inputting one or more aligned files of short RNA-Seq reads in SAM format corresponding to the same species, along with the whole genome file for that species in FASTA format.

With its probabilistic model for identifying miRNA precursors and mature miRNAs, miR-PREFeR can detect stress-responsive miRNAs and investigate their regulatory functions and characteristics under stress conditions. By comparing predicted miRNA expression patterns to known stress-responsive miRNAs and evaluating their expression patterns in response to various stress situations, miR-PREFeR can uncover miRNAs that may be involved in the plant stress response. Additionally, miR-PREFeR provides accurate and sensitive predictions, along with a low false-positive rate and fast processing, making it a valuable tool for studying the role of stress-related miRNA expression profiles (Lei & Sun, 2014)

The output of the miR-PREFeR tool is an HTML table that lists all the sequences of pre-miRNAs with their secondary structures; it establishes the anticipated length distribution of mature miRNA sequences along with the number of predictions and the expression profiles of completely aligned reads against the pre-miRNA sequences. A few separate files are also provided for simple downstream analysis. miR-PREFeR has a small tracking system that serves as a memory footprint for previous analyses by the user, making it quick, handy, and easy to use. Its limited dependence on other applications and checkpoint functionality makes it simple to resume an interrupted task from where it was left off. The workflow is now being integrated into the genome annotation engine MAKER-P (Lei & Sun, 2014).

16.4.2.2 miRPlant

An extension of miRDeep to a wider range of plant species, miRPlant is a plant miRNA prediction tool that uses appropriate techniques to find hairpin excision sites and screen hairpin structures in plants. In contrast to previous tools, miRPlant doesn't need any additional tools for its functioning, such as mapping or prediction tools for the secondary structure of RNA. It efficiently maps miRNA hairpin structures with small reads to discover novel miRNAs, allowing for the observation of novel pre-miRNA structures and the positioning of short RNA reads specific to the hairpin. Furthermore, it may identify and describe the characteristics of miRNAs associated with plant stress responses by analyzing sRNA datasets directly linked to plant stress, thereby enhancing our understanding of the molecular mechanisms underlying plant stress tolerance (Kumar et al., 2017).

Using a customizable mechanism, the prediction approach outputs precursor candidates from the genomic region encompassing RNAseq readings. To determine whether mapped reads form hairpin structures, it expands the surrounding genomic areas by 200 bp. Next, it searches for short plant miRNAs in 100 bp regions, with promising identification of genuine miRNAs that might go unnoticed if the larger (200 bp) precursor candidate length were the only one employed. miRPlant also allows for parameter customization, as various plant species may have distinct miRNA biogenesis. It generates output files similar to miRDeep and uses a minimum score of four for prediction. The tool has been tested on multiple plant species, achieving an increased accuracy of at least 10% compared to other popular plant miRNA prediction tools, thanks to its machine learning approach and scoring system. miRPlant can predict stress-responsive miRNAs, assess their expression levels, and further predict miRNA precursors and mature miRNAs associated with responses generated against plant stress (An et al., 2014).

16.4.2.3 MTide

MTide is a tool designed for identifying miRNA–target interactions in plants, integrating the algorithms of miRDeep2 and CleaveLand4, scripted in Perl and incorporating modules from the MTide Package. It includes features such as differential expression analysis of known miRNAs and support for multiple threads. The tool is available for use on Linux platforms.

MTide operates on different modules to analyze the networks of target mRNAs interacting with plant miRNAs, identifying target mRNAs for known miRNAs as well as uncovering new miRNA genes. MTide provides insights into the regulatory networks linked with plant stress responses by exploring the regulatory patterns of established miRNA genes and discovering novel ones. To better understand the functional significance of miRNAs in plant stress, an additional component that predicts miRNA targets can rank the predicted targets according to gene ontology similarity. Furthermore, MTide offers a differential expression analysis option for known miRNAs using DESeq, allowing for the comparison of miRNA expression patterns under varied stress conditions and providing valuable information about the specific roles of miRNAs in plant stress responses (Zhang et al., 2014).

16.4.2.4 miRDeep-P2

miRDeep-P2 works by employing several new strategies to enhance the analysis of the miRNA transcriptome in plants. The program normalizes read copies, which keeps the reads within a limit of 19–24 base pairs specific to sRNA datasets generated by the next-generation sequencing approach, and excises the precursors using user-defined normalized copy counts. Additionally, conserved reads from mature miRNAs of plants cataloged in the miRBase database undergo separate processing. Therefore, it can be used to identify and measure the expression of miRNA in response to stress conditions by examining short RNA sequencing data. Furthermore, to minimize the occurrence of false-positive results in miRNA candidates, miRDP2 has modified its annotation criteria for plant miRNAs and developed specific guidelines for miRNAs with 23 or 24 nucleotides. The new scoring system introduced in miRDP2 significantly improves the sensitivity of the tool, enabling the precise identification of miRNA candidates with longer precursors. These strategies have been shown to reduce run time, improve accuracy and sensitivity, and make the tool faster and more reliable for analyzing the miRNA transcriptome in plants. These features make the tool suitable for detecting changes in miRNA expression under stress conditions. By using miRDeep-P2, researchers can now analyze sRNA libraries for a wide variety of plant species to study the implications of miRNA as a response to plant stress. The tool's efficacy in evaluating miRNA expression under various conditions has been demonstrated by testing with short RNA libraries from several plant species, despite the increase in genome sizes (Kuang et al., 2018; Wang et al., 2020).

16.4.2.5 TAPIR

The TAPIR web server predicts plant miRNA targets using two different algorithms: a fast algorithm and a precise algorithm. Compared to the fast algorithm, which enables speedy searches, the precise algorithm is slower and ensures the discovery of less completely paired miRNA–target duplexes. This approach also allows the prediction of mimicry targets that appear as large loop-like structures specific to the miRNA–target duplex, making them hard to detect by classic methods. These target mimics play an important role in modulating the miRNA activity of plants in response to stress. Additionally, it is known that plant miRNAs play a role in various stress responses. TAPIR can locate putative miRNA targets connected to pathways associated with stress. By using the TAPIR web server, researchers can input the miRNA sequences of interest and search for target mRNA sequences within plant genomes. This makes it possible to identify putative miRNA targets involved in stress response pathways, offering important insights into the regulatory networks governing plant stress responses (Bonnet et al., 2010).

16.4.2.6 sRNAbench

sRNAbench is a tool for analyzing sRNA data generated through high-throughput sequencing, which is especially useful for multi-species studies. It is built on the miRanalyzer program and includes features that allow the investigation of multiple species and libraries to support whole genome mapping against the reference sequence, enhancing novel miRNA prediction. In addition, this tool supports isomiR variations, a fuzzy classification technique to identify variants in read sequence or read length. sRNAbench consists of various output components for isomiR in the form of plots (box, volcano, etc.) and heat maps, with a visualization tool linked to packages in R programming.

It accepts multiple input formats, automatically detects adapter sequences, and can filter out some classes of sRNA prior to expression profiling. The method involves general mapping strategies, read length distribution, differential expression analysis, isomiR statistics, quality checks, and non-redundant profiles regulating plants under various stress conditions. It also includes specific steps for clustering reads and extracting read clusters for further analysis. sRNAbench is available as both downloadable software and a web server, making it user-friendly for those with less bioinformatics background (Aparicio-Puerta et al., 2022; Barturen et al., 2014).

16.4.2.7 miRkwood

miRkwood is a technique for accurately identifying miRNAs within plants by sequencing the genomes of sRNAs. It provides a simple interface for navigating the data and offers export options for additional analysis on a local computer. It includes a comprehensive approach that incorporates data consisting of conserved regions of miRNA precursors along with secondary structure information. The procedure begins with aligning short RNA sequence reads against the reference genome and filtering them. Next, miRNA precursor candidates are identified, which can be used to find stress-triggered miRNAs and explore their regulatory roles in plant stress responses. This tool uses a k-mer-based approach for identifying miRNA precursors and mature miRNAs. Researchers can use miRkwood to analyze sRNA sequencing data from stressed plant samples and identify stress-associated miRNAs. It can predict both canonical miRNAs and long miRNAs and assigns a quality score to each prediction, enabling users to evaluate them. Since it is specifically designed for plant miRNAs, it allows users to manage the diversity of plant pre-miRNAs while leveraging their unique features. This tool can be used to assess the function of miRNAs in plant stress responses, characterize stress-triggered miRNAs with their target loci, and study the regulatory networks involved in plant stress pathways (Guigon et al., 2019).

16.4.2.8 psRNATarget

The psRNATarget method predicts miRNA targets in plants by considering various factors such as seed region, conservation, hybridization, function, and accessibility. It computes the energy needed to open the secondary structure surrounding the target site, which may be modulated under stress, to determine the target site's specificity using programs like RNAup and RNAduplex. The method accounts for the secondary structures around the target site to predict and regulate the miRNA–mRNA complex. Additionally, users can provide their own mRNA and/or miRNA sequences to psRNATarget, which will then provide predictions based on the inputs. This may shed light on the potential role of miRNAs in controlling pathways associated with plant responses to stress.

The psRNATarget server employs a scoring scheme to assess the complementarity level between the sRNA and the targeted gene to understand the associated protein functionality. Like the tools discussed so far, this server also evaluates target-site specificity by estimating the free energy needed to "open" the secondary structure surrounding the short RNA's target site within the mRNA. This service combines current findings in plant miRNA target identification, discriminates between the inhibition processes of translational and post-transcriptional phases, and provides the number of pairs of sRNA and target sites that may influence the binding activity of sRNA to the target transcript. The server features an effective distributed computing network handling a back-end pipeline operating on a Linux cluster, designed for processing high-throughput next-generation sequencing data. The graphical user interface offers three user-friendly options for inputting an sRNA sequence with its transcript or choosing an example upload. The server provides a comprehensive list of interacting complexes of miRNAs with their target mRNAs. This can help in identifying specific genes regulated by miRNAs under stress conditions. It has two important features for analysis functions: The Reverse Complementary Matching function allows for matching sRNAs with target transcripts using a scoring schema. This helps in identifying potential target genes for specific miRNAs involved in plant stress responses. psRNATarget calculates the free energy required to "open" the secondary structure surrounding an sRNA's target site on mRNA. This helps to decode how specific target sites reveal the potential regulatory activity of miRNAs under stress conditions, eventually leading to the identification of potential stress-related target genes. By analyzing the predicted target genes and their functions, researchers can gain insights into the regulatory roles of plant miRNAs in response to stress (Dai & Zhao, 2011).

16.5 CASE STUDY

Meher et al. (2022) developed a computational model using machine learning algorithms (Figure 16.2) to predict abiotic stress-responsive miRNAs, using attributes obtained from miRNA sequences. This technique is designed to support evidence prior to wet lab experiments and sequencing technologies for predicting miRNAs in abiotic environments.

Pseudo-K-tuple nucleotide compositional (PseKNC) characteristics convert miRNA sequences into feature vectors with numeric properties. These features provide a comprehensive representation of sequence characteristics by capturing the sequence order influenced by correlation factors. This enables machine learning algorithms to accurately identify trends and their connections within the data. In the context of miRNA sequences, producing PseKNC characteristics can result in a large feature space, where machine learning methods like support vector machine-recursive feature elimination (SVM-RFE) assist in determining the most important features while reducing dimensionality.

Each miRNA and pre-miRNA sequencing yielded 1,372 numeric characteristics. To address the issue of highly correlated data, the SVM-RFE feature selection approach can be used to identify essential characteristics for more accurately categorizing stress-responsive miRNAs. It is possible

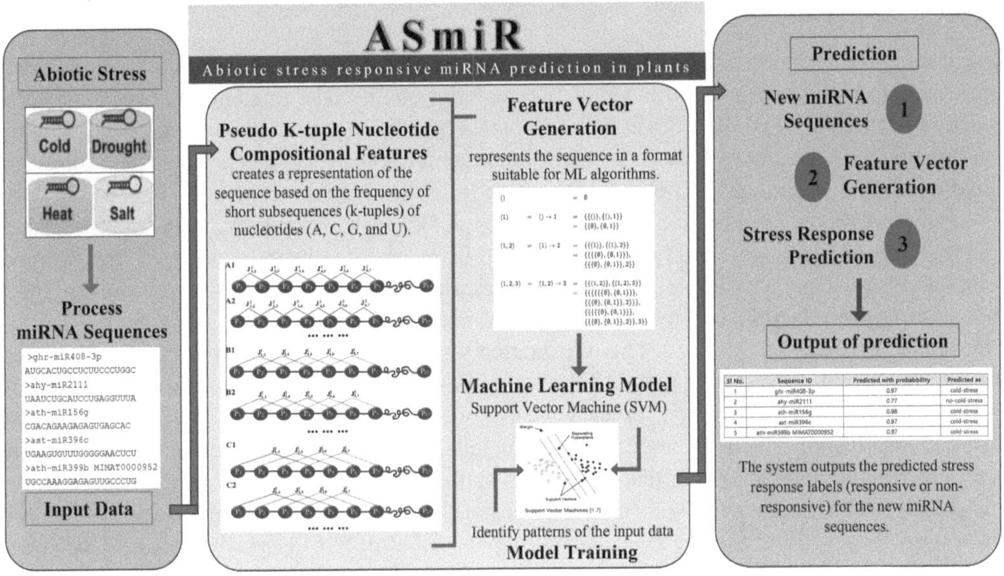

FIGURE 16.2 Machine learning model to predict miRNAs for abiotic stress response.

to generate feature sets with 200 miRNA sequences, 250 pre-miRNA sequences, and 500 combined feature datasets in order to get the greatest classification with the best accuracies. Machine learning methods can better capture the sequence-specific information that is required to differentiate between stress-responsive and non-stress-responsive miRNAs by using PseKNC. The feature selection procedure ensures that the model concentrates on the most discriminative characteristics, resulting in greater classification accuracy and prediction performance.

Each miRNA and pre-miRNA sequencing yielded 1,372 numeric characteristics. To address the issue of highly correlated data, the SVM-RFE feature selection approach can be used to identify essential characteristics for more accurately categorizing stress-responsive miRNAs. It is possible to generate feature sets with 200 miRNA sequences, 250 pre-miRNA sequences, and 500 combined feature datasets to achieve the greatest classification with the best accuracies. Machine learning methods can better capture the sequence-specific information required to differentiate between stress-responsive and non-stress-responsive miRNAs by using PseKNC. The feature selection procedure ensures that the model focuses on the most discriminative characteristics, resulting in greater classification accuracy and prediction performance.

Some important implications of using machine learning include the following:

1. A deeper understanding of plant stress responses through the identification and characterization of abiotic stress-responsive miRNAs, which allows scientists to better comprehend the molecular mechanisms underlying these responses. This knowledge can be used to design strategies for improving crop stress tolerance.
2. Precision breeding for stress resistance: This can aid precision breeding efforts to create agricultural varieties with greater stress tolerance. By targeting specific miRNAs involved in stress responses, breeders can introduce genetic alterations that enhance crops' stress tolerance while being environmentally friendly.
3. Machine learning techniques expedite crop improvement by identifying crucial regulatory molecules involved in stress responses.
4. By understanding the role of miRNAs in plant stress responses, tailored strategies can be developed to mitigate the negative effects of abiotic stressors on agricultural productivity. Hence, altering the production of stress-responsive miRNAs may enhance crop resilience to environmental stresses.

16.6 CONCLUSION

Researching the role of small miRNAs in plant development and growth is crucial for selecting crop types that are high-yield, stress-tolerant, and of superior quality. The ongoing discovery and accumulation of miRNAs and their target genes in plants have been predicted and validated through a combination of molecular biology and bioinformatics approaches. miRNAs promote post-transcriptional regulation; specifically, when a transcription factor has a direct connection with a conserved cis-regulatory promoter, a broader effect is observed. However, since most of these target genes are transcription factors, the mechanism of miRNA involvement in plant response to stress is more complex. Because plant miRNAs can regulate complex agronomic variables that positively influence improved production, quality, and stress tolerance, they emerge as intriguing targets for crop enhancement (Zhang et al., 2019). The information discussed in the preceding sections highlights the significance of miRNAs for plants. Target gene prediction associated with miRNAs, using the aforementioned tools and servers, serves as an eye opener for researchers to identify various transcriptional activation functions based on their subunits. These functions are crucial for controlling plant growth and adaptation. On the other hand, some target genes code for proteins or enzymes that induce or catalyze the regulation of metabolic pathways in plants. Plants may exhibit pleiotropic phenotypes as a result of the overexpression or repression of miRNAs, which can cause the downstream target genes to be upregulated or downregulated. Simultaneous studies on miRNAs and their target mRNAs are necessary for an accurate understanding (Zhang et al., 2022).

Despite this, there are still numerous obstacles in the way of plant miRNA research due to challenges related to conserved and non-conserved miRNAs involved in plant growth and development, which promote tolerance and survival against abiotic stress, respectively. Nevertheless, a significant number of miRNAs have been shown to be substantially conserved in plants. Prior research has demonstrated that miRNAs exhibit specificity in different species, influenced by time and spatial precision associated with the regulation of particular gene expression, thereby acting as extensive regulators of overall plant growth and development, including aerial and apical parts of plants, against both biotic and abiotic stress (Tiwari et al., 2014; Zhao et al., 2021). Furthermore, many miRNAs and the genes they target, associated with agronomic qualities, remain unidentified, and a single miRNA may participate in several regulatory pathways connected to plant development and abiotic stress. Thus, it is essential to investigate novel miRNAs to identify their targets and explore the spread of regulatory networks mediated by miRNAs. Plant adaptive behavior is driven by implementing miRNA-based strategies to strengthen the regulation of metabolic networks and their associated signaling pathways (Zhang et al., 2022). Utilizing miRNA databases and prediction algorithms allows for prior investigation of miRNA behavior under stressful conditions. Meher et al. (2022) emphasize the implementation of machine learning integrated with PseKNC characteristics to accurately predict stress-responsive miRNAs in plants. Hence, this approach serves as an advanced platform for breeding plants with high tolerance against environmental stress through precision breeding and accelerating crop improvement.

REFERENCES

Acosta-Motos, J., Ortuño, M., Bernal-Vicente, A., Diaz-Vivancos, P., Sanchez-Blanco, M., & Hernandez, J. (2017). Plant responses to salt stress: adaptive mechanisms. *Agronomy*, *7*(1), 18. https://doi.org/10.3390/agronomy7010018.

Ahmad, P., & Prasad, M. (2012). *Environmental Adaptations and Stress Tolerance of Plants in the Era of Climate Change*. Springer, Cham. https://doi.org/10.1007/978-1-4614-0815-4.

Anil, P., Dharmendra, P., Sachidanand, S., Yuriy, E., Snezhan, B., Dilfuza, J., & Chinmayi, J. (2023). Beneficial effects of biochar application on mitigating the drought and salinity stress implications on plants. *Plant Science Today*, *10*(2), 188–193. https://doi.org/10.14719/pst.2591.

An, J., Lai, J., Sajjanhar, A., Lehman, M. L., & Nelson, C. C. (2014). miRPlant: an integrated tool for identification of plant miRNA from RNA sequencing data. *BMC Bioinformatics*, *15*(1), 275. https://doi.org/10.1186/1471-2105-15-275.

Aparicio-Puerta, E., Gómez-Martín, C., Giannoukakos, S., Medina, J. M., Scheepbouwer, C., García-Moreno, A., Carmona-Saez, P., Fromm, B., Pegtel, M., Keller, A., Marchal, J. A., & Hackenberg, M. (2022). sRNAbench and sRNAtoolbox 2022 update: accurate miRNA and sncRNA profiling for model and non-model organisms. *Nucleic Acids Research*, *50*(W1), W710–W717. https://doi.org/10.1093/nar/gkac363.

Arora, N. K. (2019). Impact of climate change on agriculture production and its sustainable solutions. *Environmental Sustainability*, *2*(2), 95–96. https://doi.org/10.1007/s42398-019-00078-w.

Barturen, G., Rueda, A., Hamberg, M., Alganza, A., Lebron, R., Kotsyfakis, M., Shi, B., Koppers-Lalic, D., & Hackenberg, M. (2014). sRNAbench: profiling of small RNAs and its sequence variants in single or multi-species high-throughput experiments. *Methods in Next Generation Sequencing*, *1*(1). https://doi.org/10.2478/mngs-2014-0001.

Bonnet, E., He, Y., Billiau, K., & Van De Peer, Y. (2010). TAPIR, a web server for the prediction of plant microRNA targets, including target mimics. *Bioinformatics*, *26*(12), 1566–1568. https://doi.org/10.1093/bioinformatics/btq233.

Britt, A. B. (1995). Repair of DNA damage induced by ultraviolet radiation. *Plant Physiology*, *108*(3), 891–896. https://doi.org/10.1104/pp.108.3.891.

Chaki, M., Begara-Morales, J. C., & Barroso, J. B. (2020). Oxidative stress in plants. *Antioxidants*, *9*(6), 481. https://doi.org/10.3390/antiox9060481.

Chen, X. (2005). microRNA biogenesis and function in plants. *FEBS Letters*, *579*(26), 5923–5931. https://doi.org/10.1016/j.febslet.2005.07.071.

Chen, Y., & Wang, X. (2019). miRDB: an online database for prediction of functional microRNA targets. *Nucleic Acids Research*, *48*(D1), D127–D131. https://doi.org/10.1093/nar/gkz757.

Corso, D., Delzon, S., Lamarque, L., Cochard, H., Ruiz, J. M. T., King, A., & Brodribb, T. (2020). Neither xylem collapse, cavitation, or changing leaf conductance drive stomatal closure in wheat. *Plant Cell & Environment*, *43*(4), 854–865. https://doi.org/10.1111/pce.13722.

Cui, X., Wang, Q., Yin, W., Xu, H., Wilson, Z. A., Wei, C., Pan, S., & Zhang, D. (2012). PMRD: a curated database for genes and mutants involved in plant male reproduction. *BMC Plant Biology*, *12*(1). https://doi.org/10.1186/1471-2229-12-215.

Wang, K., Peng, X., Ji, Y., Yang, P., Zhu, Y., & Li, S. (2013). Gene, protein, and network of male sterility in rice. *Frontiers in Plant Science*, *4*, 92. https://doi.org/10.3389/fpls.2013.00092.

Dai, X., & Zhao, P. X. (2011). psRNATarget: a plant small RNA target analysis server. *Nucleic Acids Research*, *39*(suppl_2), W155–W159. https://doi.org/10.1093/nar/gkr319.

Elemike, E., Uzoh, I., Onwudiwe, D., & Babalola, O. (2019). The role of nanotechnology in the fortification of plant nutrients and improvement of crop production. *Applied Sciences*, *9*(3), 499. https://doi.org/10.3390/app9030499.

Feng, D., Wang, R., Sun, X., Liu, L., Liu, P., Tang, J., Zhang, C., & Liu, H. (2023). Heavy metal stress in plants: ways to alleviate with exogenous substances. *The Science of the Total Environment*, *897*, 165397. https://doi.org/10.1016/j.scitotenv.2023.165397.

Francis, B., Aravindakumar, C., Brewer, P. B., & Simon, S. (2023). Plant nutrient stress adaptation: a prospect for fertilizer limited agriculture. *Environmental and Experimental Botany*, *213*, 105431. https://doi.org/10.1016/j.envexpbot.2023.105431.

Goche, T., Shargie, N. G., Cummins, I., Brown, A. P., Chivasa, S., & Ngara, R. (2020). Comparative physiological and root proteome analyses of two sorghum varieties responding to water limitation. *Scientific Reports*, *10*(1), 3. https://doi.org/10.1038/s41598-020-68735-3.

Griffiths-Jones, S., Saini, H. K., Van Dongen, S., & Enright, A. J. (2007). miRBase: tools for microRNA genomics. *Nucleic Acids Research*, *36*(Database), D154–D158. https://doi.org/10.1093/nar/gkm952.

Guigon, I., Legrand, S., Berthelot, J., Bini, S., Lanselle, D., Benmounah, M., & Touzet, H. (2019). miRkwood: a tool for the reliable identification of microRNAs in plant genomes. *BMC Genomics*, *20*(1), 9. https://doi.org/10.1186/s12864-019-5913-9.

Hasanuzzaman, M., & Fujita, M. (2022). Plant oxidative stress: biology, physiology and mitigation. *Plants*, *11*(9), 1185. https://doi.org/10.3390/plants11091185.

Hasanuzzaman, M., Nahar, K., Alam, M., Roychowdhury, R., & Fujita, M. (2013). Physiological, biochemical, and molecular mechanisms of heat stress tolerance in plants. *International Journal of Molecular Sciences*, *14*(5), 9643–9684. https://doi.org/10.3390/ijms14059643.

Jabborova, D., Nigora, Z., Yuriy, E., Abdujalil, N., Anand, D., Khurshid, S., Zafarjon, J., Sachidanand, S., & Rahul, D. (2023). Growth of spinach as influenced by biochar and *Bacillus endophyticus* IGPEB 33 in drought condition. *Pakistan Journal of Botany*, *55*, 6. http://dx.doi.org/10.30848/PJB2023-SI(6).

Jian, X., Zhang, L., Li, G., Zhang, L., Wang, X., Cao, X., Fang, X., & Chen, F. (2010). Identification of novel stress-regulated microRNAs from *Oryza sativa* L. *Genomics*, *95*(1), 47–55. https://doi.org/10.1016/j.ygeno.2009.08.017.

Kerchev, P., Van Der Meer, T., Sujeeth, N., Verlee, A., Stevens, C. V., Van Breusegem, F., & Gechev, T. (2020). Molecular priming as an approach to induce tolerance against abiotic and oxidative stresses in crop plants. *Biotechnology Advances*, *40*, 107503. https://doi.org/10.1016/j.biotechadv.2019.107503.

Kozomara, A., & Griffiths-Jones, S. (2010). miRBase: integrating microRNA annotation and deep-sequencing data. *Nucleic Acids Research*, *39*(Database), D152–D157. https://doi.org/10.1093/nar/gkq1027.

Kozomara, A., Birgaoanu, M., & Griffiths-Jones, S. (2018). miRBase: from microRNA sequences to function. *Nucleic Acids Research*, *47*(D1), D155–D162. https://doi.org/10.1093/nar/gky1141.

Kuang, Z., Wang, Y., Li, L., & Yang, X. (2018). miRDeep-P2: accurate and fast analysis of the microRNA transcriptome in plants. *Bioinformatics*, *35*(14), 2521–2522. https://doi.org/10.1093/bioinformatics/bty972.

Kumar, K. V. V., Srikakulam, N., Padbhanabhan, P., & Pandi, G. (2017). Deciphering microRNAs and their associated hairpin precursors in a non-model plant, abelmoschus esculentus. *Non-Coding RNA*, *3*(2), 19. https://doi.org/10.3390/ncrna3020019.

Kumari, V. V., Banerjee, P., Verma, V. C., Sukumaran, S., Chandran, M. S., Gopinath, K. A., Venkatesh, G., Yadav, S. K., Singh, V. K., & Awasthi, N. K. (2022). Plant nutrition: an effective way to alleviate abiotic stress in agricultural crops. *International Journal of Molecular Sciences*, *23*(15), 8519. https://doi.org/10.3390/ijms23158519.

Lei, J., & Sun, Y. (2014). miR-PREFeR: an accurate, fast and easy-to-use plant miRNA prediction tool using small RNA-Seq data. *Bioinformatics*, *30*(19), 2837–2839. https://doi.org/10.1093/bioinformatics/btu380.

Liu, J., Liu, X., Zhang, S., Liang, S., Luan, W., & Ma, X. (2021). TarDB: an online database for plant miRNA targets and miRNA-triggered phased siRNAs. *BMC Genomics*, *22*(1), 5. https://doi.org/10.1186/s12864-021-07680-5.

Liu, W., & Wang, X. (2019). Prediction of functional microRNA targets by integrative modeling of microRNA binding and target expression data. *Genome Biology*, *20*(1), 1629. https://doi.org/10.1186/s13059-019-1629-z.

Ma, Y., Dias, M. C., & Freitas, H. (2020). Drought and salinity stress responses and microbe-induced tolerance in plants. *Frontiers in Plant Science*, *11*, 591911. https://doi.org/10.3389/fpls.2020.591911.

Mareri, L., Parrotta, L., & Cai, G. (2022). Environmental stress and plants. *International Journal of Molecular Sciences*, *23*(10), 5416. https://doi.org/10.3390/ijms23105416.

Martínez-Vilalta, J., & Garcia-Forner, N. (2016). Water potential regulation, stomatal behaviour and hydraulic transport under drought: deconstructing the iso/anisohydric concept. *Plant Cell & Environment*, *40*(6), 962–976. https://doi.org/10.1111/pce.12846.

Meher, P. K., Begam, S., Sahu, T. K., Gupta, A., Kumar, A., Kumar, U., Rao, A. R., Singh, K. P., & Dhankher, O. P. (2022). ASRmiRNA: abiotic stress-responsive miRNA prediction in plants by using machine learning algorithms with pseudo K-tuple nucleotide compositional features. *International Journal of Molecular Sciences*, *23*(3), 1612. https://doi.org/10.3390/ijms23031612.

Millar, A. A. (2020). The function of miRNAs in plants. *Plants*, *9*(2), 198. https://doi.org/10.3390/plants9020198.

Müller-Xing, R., Xing, Q., & Goodrich, J. (2014). Footprints of the sun: memory of UV and light stress in plants. *Frontiers in Plant Science*, 5, 474. https://doi.org/10.3389/fpls.2014.00474.

Noopur, K., Pragati, K., Adarsh, K. S., Sachidanand, S., Tanvi, J., & Megha, B. (2025a). Role of non coding RNA in effector trigger immunity. In Dutta, R., Kapadia, C., Danish, S., & Singh, S. (Eds), *Non-Coding RNA in Plants Modulation and Stress Responses Plant Biology, Sustainability and Climate Change*. Academic Press, New York.

Noopur, K., Pragati, K., Megha, B., Arpan, M., Tanvi, J., Shrisendu, M., Adarsh, K. S., & Sachidanand, S. (2025b). Role of non-coding RNA in plant architecture. In Dutta, R., Kapadia, C., Danish, S., & Singh, S. (Eds), *Non-Coding RNA in Plants Modulation and Stress Responses Plant Biology, Sustainability and Climate Change*. Academic Press, New York.

Pooja, G., Megha, B., Arpan, M., Chintan, K., Rahul, D., Subhan, D., & Sachidanand, S. (2025). Introduction to long noncoding RNA. In Dutta, R., Kapadia, C., Danish, S., & Singh, S. (Eds), *Non-Coding RNA in Plants Modulation and Stress Responses Plant Biology, Sustainability and Climate Change*. Vol. 1, pp. 1–12. Academic Press, New York.

Reinhart, B. J., Weinstein, E. G., Rhoades, M. W., Bartel, B., & Bartel, D. P. (2002). microRNAs in plants. *Genes & Development*, *16*(13), 1616–1626. https://doi.org/10.1101/gad.1004402.

Ristvey, A., Belayneh, B., & Lea-Cox, J. (2019). A comparison of Irrigation-Water containment methods and management strategies between two ornamental production systems to minimize water security threats. *Water*, *11*(12), 2558. https://doi.org/10.3390/w11122558.

Ritonga, F. N., & Chen, S. (2020). Physiological and molecular mechanism involved in cold stress tolerance in plants. *Plants*, *9*(5), 560. https://doi.org/10.3390/plants9050560.

Rodríguez, M., Canales, E., & Borrás-Hidalgo, O. (2005). Molecular aspects of abiotic stress in plants. *Biotecnología Aplicada*, *22*(1), 1–10.

Sanghera, G. S., Wani, S. H., Hussain, W., & Singh, N. B. (2011). Engineering cold stress tolerance in crop plants. *Current Genomics*, *12*(1), 30–43. https://doi.org/10.2174/138920211794520178.

Seleiman, M. F., & Kheir, A. M. (2018). Saline soil properties, quality and productivity of wheat grown with bagasse ash and thiourea in different climatic zones. *Chemosphere*, *193*, 538–546. https://doi.org/10.1016/j.chemosphere.2017.11.053.

Sethy, S. K., & Ghosh, S. (2013). Effect of heavy metals on germination of seeds. *Journal of Natural Science Biology and Medicine*, *4*(2), 272. https://jnsbm.org/effect-of-heavy-metals-on-germination-of-seeds.

Shafiq, S., Ali, A., Sajjad, Y., Zeb, Q., Shahzad, M., Khan, A. R., Nazir, R., & Widemann, E. (2020). The interplay between toxic and essential metals for their uptake and translocation is likely governed by DNA methylation and histone deacetylation in maize. *International Journal of Molecular Sciences*, *21*(18), 6959. https://doi.org/10.3390/ijms21186959.

Shi, C., & Liu, H. (2021). How plants protect themselves from ultraviolet-B radiation stress. *Plant Physiology*, *187*(3), 1096–1103. https://doi.org/10.1093/plphys/kiab245.

Sihag, N., Singh, T., Sheoran, S., Singh, O., Malik, R., Kumar, L., & Singh, J. (2024). Role of RNA interference in drought stress management: physiological, biochemical and molecular approach. *Crop and Pasture Science*, *75*(1), 183. https://doi.org/10.1071/cp23183.

Sunkar, R., Li, Y., & Jagadeeswaran, G. (2012). Functions of microRNAs in plant stress responses. *Trends in Plant Science*, *17*(4), 196–203. https://doi.org/10.1016/j.tplants.2012.01.010.

Thomason, W., & Battaglia, M. (2020). Early defoliation effects on corn plant stands and grain yield. *Agronomy Journal*, *112*(6), 5024–5032. https://doi.org/10.1002/agj2.20402.

Tiwari, M., Sharma, D., & Trivedi, P. K. (2014). Artificial microRNA mediated gene silencing in plants: progress and perspectives. *Plant Molecular Biology*, *86*(1–2), 1–18. https://doi.org/10.1007/s11103-014-0224-7.

Verma, N., Prachi, S., Aashna, S., & Sachidanand, S. (2021). Physiological and molecular responses to drought, submergence and excessive watering in plants. In Husen, A. (Ed), *Harsh Environment and Plant Resilience*. Springer International Publishing, Cham.

Verma, A., Kumar, R., Shukla, S., Singh, S., Johri, P., & Singh, M.S. (2023). Genome wide association studies: an outlook. *Journal of Innovation in Applied Research*, *6*(1), 56–64. https://jiar.in/admin/public/uploads/187/120_pdf.pdf.

Wang, Y., Kuang, Z., Li, L., & Yang, X. (2020). A bioinformatics pipeline to accurately and efficiently analyze the microRNA transcriptomes in plants. *Journal of Visualized Experiments*, *155*, 59864. https://doi.org/10.3791/59864.

Yi, X., Zhang, Z., Ling, Y., Xu, W., & Su, Z. (2014). PNRD: a plant non-coding RNA database. *Nucleic Acids Research*, *43*(D1), D982–D989. https://doi.org/10.1093/nar/gku1162.

Zhang, F., Yang, J., Zhang, N., Wu, J., & Si, H. (2022). Roles of microRNAs in abiotic stress response and characteristics regulation of plant. *Frontiers in Plant Science*, *13*, 919243. https://doi.org/10.3389/fpls.2022.919243.

Zhang, M., An, P., Li, H., Wang, X., Zhou, J., Dong, P., Zhao, Y., Wang, Q., & Li, C. (2019). The MIRNA-mediated post-transcriptional regulation of maize in response to high temperature. *International Journal of Molecular Sciences*, *20*(7), 1754. https://doi.org/10.3390/ijms20071754.

Zhang, Y., Xu, J., Li, R., Ge, Y., Li, Y., & Li, R. (2023). Plants' response to abiotic stress: mechanisms and strategies. *International Journal of Molecular Sciences*, *24*(13), 10915. https://doi.org/10.3390/ijms241310915.

Zhang, Z., Jiang, L., Wang, J., Gu, P., & Chen, M. (2014). MTide: an integrated tool for the identification of miRNA–target interaction in plants. *Bioinformatics*, *31*(2), 290–291. https://doi.org/10.1093/bioinformatics/btu633.

Zhang, Z., Yu, J., Li, D., Zhang, Z., Liu, F., Zhou, X., Wang, T., Ling, Y., & Su, Z. (2009). PMRD: plant microRNA database. *Nucleic Acids Research*, 38(suppl_1), D806–D813. https://doi.org/10.1093/nar/gkp818.

Zhao, S., Zhang, Q., Liu, M., Zhou, H., Ma, C., & Wang, P. (2021). Regulation of plant responses to salt stress. *International Journal of Molecular Sciences*, *22*(9), 4609. https://doi.org/10.3390/ijms22094609.

Zhao, Y., Kuang, Z., Wang, Y., Li, L., & Yang, X. (2021). microRNA annotation in plants: current status and challenges. *Briefings in Bioinformatics*, *22*(5), 75. https://doi.org/10.1093/bib/bbab075.

17 Databases for Plant miRNA Analysis

Deepa Dhatwalia, R Shubhra Maithreyi, Manisha Negi, and Murali Krishna Kormutla*

17.1 INTRODUCTION

17.1.1 Background and Driving Forces

In the fascinating world of plant genomics, microRNAs (miRNAs) have emerged as key players in regulating gene expression, influencing a wide range of biological processes such as growth, stress responses, and disease resistance. As research in this area has flourished, the need for organized and accessible databases to manage the ever-growing volume of miRNA data has become essential. These databases are invaluable to scientists, providing crucial resources for identifying, annotating, and analyzing miRNAs across different plant species.

These specialized databases have been created in response to the rapid accumulation of miRNA data and the need to make this information easily accessible and usable. Plant miRNA databases can be broadly categorized into four types: comprehensive databases, miRNA–target interaction (MTI) databases, species-specific databases, and specialized databases. Each type serves a unique purpose, helping researchers navigate the complex landscape of miRNA research and uncover new insights into plant biology.

This chapter provides an in-depth look at the various categories of plant miRNA databases. We explore their unique features, data scope, and importance in advancing our understanding of miRNAs in plants. By offering a clear categorization and overview of these resources, we aim to guide researchers in selecting the most suitable database for their specific needs, thereby supporting and accelerating progress in plant miRNA research.

17.1.2 Overview of miRNAs in Plant Biology

17.1.2.1 Definition and Discovery of miRNAs

Plant microRNAs (miRNAs) are single-stranded, predominantly 21-nt RNAs (with a specialised 24-nt sub-class) that down-regulate genes by guiding AGO-RISC to cleave target mRNAs, block their translation, or, in some cases, trigger RNA-directed DNA methylation. Since their discovery in Arabidopsis two decades ago, miRNAs have been shown to orchestrate key developmental programmes—leaf architecture, flowering time, shoot apical meristem maintenance—and to fine-tune responses to abiotic stresses such as drought, salinity, heat, cold, heavy-metal toxicity and nutrient limitation (Mallory & Vaucheret 2006,Basso et al. 2019; Millar 2020; Dong et al. 2022; Šečić et al. 2021). Deciphering these regulatory circuits offers actionable targets for breeding and genome editing aimed at enhanced yield and stress resilience.

* *Corresponding Author*

DOI: 10.1201/9781003491378-17

17.1.2.2 Biological Significance and Functions in Plants

miRNAs are small non-coding RNA (ncRNA) molecules that play crucial roles in regulating gene expression in plants (Noopur et al., 2025a,b; Pooja et al., 2025; Rahul et al., 2024; Subhan et al., 2025; Tanvi et al., 2025). The discovery of miRNAs in plants dates back about 25 years, with these molecules being key players in gene expression regulation during various developmental processes (Dong et al., 2022; Millar, 2020). Advances in genomic research and the development of bioinformatic tools have led to the identification of numerous miRNAs in plants, including novel ones, through high-throughput sequencing and in silico analysis (Khemka et al., 2021; Djami et al., 2017; Li et al., 2021). These miRNAs have been found to be involved in diverse processes such as DNA damage repair, metabolic pathways, RNA degradation, and responses to abiotic stresses like salinity, highlighting their significance in plant biology and stress adaptation (Fang et al., 2015; Feng et al., 2017; Jones et al., 2006). Further research and technological advancements continue to enhance our understanding of miRNA functions and their potential for genetic improvement in plants (Jones et al., 2006; Samynathan et al., 2023).

17.1.2.3 Historical Development and Milestones in miRNA Research

The discovery of plant miRNAs has evolved significantly over time, marked by key milestones. Initially, miRNAs were identified through experimental techniques like forward genetic screening and microarray analysis, which were costly and time-consuming (Jones et al., 2006). The advent of computational tools in the last decade revolutionized miRNA research, leading to the discovery of numerous novel plant miRNAs through bioinformatics approaches (Akter et al., 2014; Bonnet et al., 2004). However, traditional tools for miRNA discovery have shown limited effectiveness, especially in plants where pre-miRNAs are complex (Chu et al., 2016). Recent advancements include the development of deep learning systems like miWords, which significantly outperformed existing tools in accurately identifying pre-miRNAs in plants, achieving accuracy rates above 94% and even up to 98% (Azahar et al., 2020). Additionally, studies have highlighted the importance of understanding miRNA biogenesis processes, such as transcription, processing, stability, and export, in regulating miRNA levels in plants (Cuperus et al., 2011). High-throughput sequencing methods coupled with computational analyses have also expanded the identification of miRNA targets on a genome-wide scale, revealing the significance of relaxed criteria for target prediction (Wang et al., 2015; Xu et al., 2011).

17.2 IMPORTANCE OF miRNA DATABASES

17.2.1 Role in Advancing Plant Genomics

Plant miRNA databases offer essential resources for researchers by providing information on miRNA targets, expression profiles, and regulatory mechanisms (Li et al., 2021). These databases, such as TarDB (Lu et al., 2021), mirEX (Wang et al., 2010), Plant miRNA Encyclopedia (PmiREN) (Fan et al., 2018), and MepmiRDB (Zhou et al., 2011), encompass various key features. They include high-confidence miRNA targets such as cross-species conserved targets, degradome-supported targets, and miRNA-triggered phased small interfering RNA (phasiRNA) loci (Lu et al., 2021). Additionally, databases like mirEX focus on exploring and visualizing plant primary miRNA (pri-miRNA) expression profiles, aiding in the understanding of miRNA biogenesis regulation (Wang et al., 2010). PmiREN stands out for its comprehensive annotation of miRNA loci in numerous plant species, providing information on precursor sequences, expression patterns, and potential targets supported by PARE sequencing (Fan et al., 2018). Furthermore, MepmiRDB caters specifically to medicinal plants, offering data on miRNA sequences, expression patterns, and validated MTIs, thereby enhancing research on medicinal plant miRNAs (Zhou et al., 2011). Overall, these databases play a crucial role in advancing plant miRNA research by consolidating valuable information for data-driven studies.

17.2.2 Benefits for Researchers and Plant Breeders

Plant miRNA databases play a crucial role in agricultural advancements by providing valuable resources for research and crop improvement (Fan et al., 2018). These databases, such as PmiREN2.0 (Lu et al., 2021) and PmiREN (Fan et al., 2018), offer extensive collections of miRNA entries from various plant species, along with regulatory information, predicted targets, and design tools, facilitating experimental practices and accelerating miRNA research in plants (Fan et al., 2018; Lu et al., 2021). Additionally, the development of plant-specific miRNA target predictors, like P-TarPmiR, using databases such as TarDB (Li et al., 2021), enhances the accuracy of miRNA target prediction in plant species, aiding in the identification of potential gene targets for crop improvement and stress resilience (Li et al., 2021). Overall, these databases serve as essential tools for data-driven miRNA research, thereby enabling a deeper understanding of miRNA functions and regulatory relationships in plants, ultimately contributing to advancements in agriculture.

17.2.3 Challenges and Opportunities in miRNA Data Management

Plant miRNA databases play a crucial role in comparative genomics by providing comprehensive resources for researchers to explore and analyze miRNA sequences, targets, and functions across various plant species (Li et al., 2021). These databases, such as the PmiREN (Fan et al., 2018), sRNAanno (Li et al., 2018), and TarDB (Lu et al., 2021), offer valuable information on miRNA loci, precursor sequences, expression patterns, potential targets, and evolutionary relationships (Fan et al., 2018; Li et al., 2018; Lu et al., 2021). By integrating data from multiple sources and applying stringent annotation criteria, these databases facilitate the identification of conserved miRNA targets, phasiRNAs, and other small RNA (sRNA) classes in plants (Li et al., 2018). Researchers can leverage these databases to study miRNA evolution, regulatory networks, and genetic improvements in crops, enhancing our understanding of plant development, stress responses, and genome stability (Llave et al., 2002; Lu et al., 2021).

17.3 PLANT miRNA DATABASES

Based on the content of miRNA databases, they can be divided into four categories: comprehensive databases, MTI databases, species-specific databases, and specialized databases. Each category serves a unique purpose, helping researchers navigate the complex landscape of miRNA research and uncover new insights into plant biology (Table 17.1).

17.3.1 Comprehensive Plant miRNA Databases

17.3.1.1 miRBase: The Cornerstone of Plant miRNA Databases

miRBase serves as the primary public repository and online resource for miRNA sequences and annotations. Established in 2002 as the miRNA Registry, it has played a critical role in standardizing miRNA nomenclature and has been instrumental in documenting new miRNA discoveries across various organisms, including plants (Griffiths-Jones et al., 2008; Kozomara et al., 2019).

17.3.1.1.1 Database Content and Structure

The latest release of miRBase (version 22, http://www.mirbase.org/) contains 38,589 entries representing hairpin precursor miRNAs from 271 organisms, producing a total of 48,860 different mature miRNA sequences. This includes entries from well-annotated genomes of both vertebrates and plants. For example, the *Arabidopsis thaliana* genome includes 326 hairpin precursors and 428 mature miRNA sequences.

TABLE 17.1
Plant miRNA Database

S. No.	Database	Category	Brief Introduction	Organism Types	Publication Span	Latest Update	Availability	References	Latest Version	Link
1	miRBase	Comprehensive	Central repository for miRNA sequences and annotations	Multiple plants and other organisms	2002-present	Oct-18	Free, public	Kozomara et al. (2019) Nucleic Acids Res	Release 22.1	http://www.mirbase.org/
1	PmiREN	Comprehensive	Plant miRNA encyclopedia	170+ plant species	2019-present	Dec-23	Free, public	Guo et al. (2020) Nucleic Acids Res	Version 2.0	http://www.pmiren.com/
2	sRNAanno	Comprehensive	Repository for various miRNA data types	Multiple species	2020-present	Mar-24	Free, public	Zhou et al. (2020) Nucleic Acids Res	Version 1.5	http://srnnanno.com/
3	PNRD	Comprehensive	Plant non-coding RNA database	Multiple plant species	2014-present	Mar-24	Free, public	Yi et al. (2015) Nucleic Acids Res	Version 3.0	http://structuralbiology.cau.edu.cn/PNRD/
4	TarBase	miRNA–target interaction	Experimentally validated miRNA–target interactions	Multiple species including plants	2006-present	Jan-24	Free, public	Karagkouni et al. (2018) Nucleic Acids Res	Version 8.0	http://carolina.imis.athena-innovation.gr/diana_tools/web/index.php?r=tarbasev8
5	miRTarBase	miRNA–target interaction	miRNA–target interactions with experimental validation	Multiple species including plants	2011-present	Nov-23	Free, public	Huang et al. (2020) Nucleic Acids Res	Version 9.0	http://mirtarbase.cuhk.edu.cn/
6	starBase	miRNA–target interaction	Integrates data on miRNA and target gene regulation	Multiple species including plants	2011-present	Feb-24	Free, public	Li et al. (2014) Nucleic Acids Res	Version 3.0	http://starbase.sysu.edu.cn/
7	P-TarPmiR	miRNA–target interaction	Plant miRNA target prediction and functional analysis	Multiple plant species	2022-present	Apr-24	Free, public	Jabbar et al. (2022) Plant Physiol	Version 1.5	https://ptarpmir.cu-bic.ca/
8	ASRP	Species-specific	*Arabidopsis* Small RNA Project	*Arabidopsis thaliana*	2005-2012	Aug-12	Free, public	Backman et al. (2008) Nucleic Acids Res	Version 4.0	https://asrp.danforthcenter.org/
9	PmiRKB	Species-specific	Plant miRNA knowledge base	*Arabidopsis* and rice	2011-present	Jun-23	Free, public	Meng et al. (2011) Nucleic Acids Res	Version 2.5	http://bis.zju.edu.cn/pmirkb/

(Continued)

TABLE 17.1 (*Continued*)
Plant miRNA Database

S. No.	Database	Category	Brief Introduction	Organism Types	Publication Span	Latest Update	Availability	References	Latest Version	Link
10	MepmiRDB	Species-specific	Medicinal plant miRNA database	Multiple medicinal plants	2019-present	Nov-23	Free, public	Zhou et al. (2019) BMC Plant Biol	Version 2.0	http://mepmirdb.cn/mepmirdb/index.html
11	miRVIT	Species-specific	Grapevine miRNA database	*Vitis vinifera*	2018-present	Sep-23	Free, public	Chitarra et al. (2018) Front Plant Sci	Version 1.5	http://mirvit.ipsp.cnr.it/
12	mirEX	Species-specific	miRNA expression database	Multiple plant species	2012-present	Dec-23	Free, public	Zielezinski et al. (2015) Nucleic Acids Res	Version 2.5	http://www.comgen.pl/mirex2/
13	miRFANs	Species-specific	miRNA functional association networks	Multiple plant species	2016-present	Oct-23	Free, public	Liu et al. (2018) Nucleic Acids Res	Version 2.0	http://www.mirfans.org/
14	MirPub	Specialized	Collection of miRNA-related literature	Multiple species including plants	2013-present	Jan-24	Free, public	Manchanda et al. (2015) Database	Version 2.0	http://www.mirpub.org/
15	mirtronDB	Specialized	Collection of mirtrons	Multiple species including plants	2017-present	Jul-23	Free, public	Wen et al. (2018) Nucleic Acids Res	Version 1.5	http://mirtrondb.cp.utfpr.edu.br/
16	PmiRExAt	Specialized	Plant miRNA expression atlas	Multiple plant species	2013-present	Jan-24	Free, public	Gurjar et al. (2016) Database	Version 3.0	http://pmirexat.nabi.res.in/
17	PASmiR	Specialized	miRNA regulation in plant abiotic stress responses	Multiple plant species	2013-present	Dec-23	Free, public	Zhang et al. (2013) BMC Plant Biol	Version 2.0	http://pcsb.ahau.edu.cn:8080/PASmiR/
18	miSolRNA	Specialized	Solar stress-responsive miRNAs	*Solanum lycopersicum*	2015-present	Feb-24	Free, registration required	Bresso et al. (2018) BMC Plant Biol	Version 3.0	http://www.misolrna.org/
19	TransmiR	Specialized	Transcription factor-miRNA regulations	Multiple species including plants	2010-present	Mar-24	Free, public	Tong et al. (2019) Nucleic Acids Res	Version 2.5	http://www.cuilab.cn/transmir
20	PMTED	Specialized	Plant miRNA target expression database	Multiple plant species	2015-present	Oct-23	Free, public	Sun et al. (2015) BMC Bioinformatics	Version 2.0	http://pmted.agrinome.org/
21	PmiRNASSRdb	Specialized	Plant miRNA-SSR database	Multiple plant species	2021-present	May-24	Free, public	Biswas et al. (2021) Agronomy	Version 1.2	http://www.genomicsres.x10.mx/plantmirssr/

17.3.1.1.2 Data Submission and Quality Control

miRBase relies primarily on author submissions for sequence data, with most miRNAs discovered through sRNA deep sequencing techniques. To maintain precision, miRBase has implemented protocols for retrospective analysis of published miRNA sequences, ensuring data reliability.

17.3.1.1.3 Functional Annotations and Tools

miRBase provides extensive functional annotations, including miRNA sequences, biogenesis precursors, genome coordinates, and context, as well as literature references and deep sequencing expression data. It links to resources that include predicted and experimentally validated targets of miRNAs. To enhance the availability of miRNA functional information, miRBase disseminates Gene Ontology (GO) terms annotated against miRBase sequences and uses a text-mining approach to extract functional information from scientific literature.

miRBase offers extensive coverage of miRNA sequences from a wide range of species, including both plants and animals, making it a comprehensive resource for miRNA research:

Functional Annotations: miRBase integrates diverse data types, making it an essential tool for researchers studying miRNA functions and their roles in various biological processes.

Quality Control: miRBase implements measures to improve miRNA annotations by collecting sRNA deep sequencing datasets and mapping reads to miRNA sequences, ensuring the accuracy of miRNA annotations.

miRBase stands out as a cornerstone resource for miRNA research, particularly in plants. Its extensive collection of miRNA sequences, combined with advanced functional annotations and quality control measures, makes it an invaluable platform for researchers. miRBase is publicly and freely accessible at (http://https://rfam.org/; http://mirbase.org).

17.3.1.2 Rfam: An Expansive Database for Plant miRNA Families

Rfam is a comprehensive database that houses a collection of RNA families, each represented by a multiple sequence alignment of known RNA sequences and a covariance model used to identify additional members. This database is integral to the study of ncRNAs across various species, including plants. As of Rfam 14, the database includes 3,444 RNA families, reflecting a 28% increase from its previous version (Kalvari et al., 2018). Rfam is publicly and freely accessible at (https://rfam.org/).

17.3.1.2.1 Database Content and Structure

RNA Family Representation: Each RNA family in Rfam is represented by a multiple sequence alignment (the "seed"), a consensus secondary structure, and a covariance model, enabling the annotation of ncRNAs in nucleotide datasets using the internal software.

Expanding Coverage: Recent developments in Rfam have focused on broadening coverage, particularly for miRNAs, viral RNAs, and bacterial RNAs. Collaboration with miRBase has resulted in the creation of 356 new Rfam families and updates to 40 existing ones, significantly enhancing the database's utility for miRNA research.

Plant miRNAs: Rfam 14 has improved the coverage of plant miRNAs, providing comprehensive annotations for model plants like *Arabidopsis thaliana*, rice (*Oryza sativa*), and maize (*Zea mays*).

Metagenomic and Viral Families: Rfam 14 has increased the coverage of bacterial and metagenome-based RNA families, adding 253 new families from the Zasha Weinberg Database. A new workflow for annotating viral RNA families has also been established.

17.3.1.2.2 Functional Annotations and Tools

miRNA Synchronization: Key updates in Rfam 14 include the synchronization of miRNA families with miRBase, standardizing their classification and annotation.

Advanced Search and Visualization Tools: New features include an improved sequence similarity search powered by RNAcentral and the ability to for and visualize families with pseudoknots.

Community Contributions: The Rfam Cloud family curation system allows expert users to build and submit new RNA families, ensuring Rfam remains dynamic and up to date.

Species Coverage: Rfam offers extensive coverage of ncRNA families across a broad range of species. Compared to databases like miRNEST and miRBase, Rfam encompasses a wider variety of RNA types.

Functional Annotations: Rfam provides detailed functional annotations that are essential for understanding the roles of ncRNAs in various biological processes and for the functional characterization of newly discovered RNAs.

Integration with Other Resources: The synchronization of miRNA families between Rfam and miRBase enhances the utility of both databases, providing a comprehensive view of RNA functions and interactions.

Rfam stands out as a pivotal resource for studying plant miRNAs and other ncRNAs. Its extensive coverage, rigorous annotation processes, and integration with other RNA databases make it an invaluable tool for researchers (Kalvari et al., 2021, Rfam).

17.3.1.3 PMRD: Plant microRNA Database

The Plant microRNA Database (PMRD) serves as a comprehensive tool for investigating plant miRNAs. Established by scholars at China Agricultural University, PMRD functions as a crucial archive of miRNA sequences, expression patterns, and relationships with target genes across diverse plant species. PMRD is publicly and freely accessible at (Zhang et al., 2010; http://bioinformatics.cau.edu.cn/PMRD/).

17.3.1.3.1 Database Content and miRNA Collection

PMRD integrates a vast array of miRNA data sourced from miRBase, Rfam, and recent literature. It contains 8,433 non-redundant miRNA sequences from 121 plant species. For example, the database holds 1,427 Arabidopsis miRNAs, 2,540 rice miRNAs, and 2,780 poplar miRNAs, significantly expanding the entries available in miRBase for these species.

17.3.1.3.2 miRNA Expression Profiling

One critical feature of PMRD is its inclusion of miRNA expression profiles from several public datasets, covering various stress responses and developmental stages. For instance, PMRD provides detailed fold-change data for miRNAs like ptc-miR474b in poplar under cold stress, highlighting significant expression changes across seven time points.

17.3.1.3.3 Database Architecture and Access

PMRD is built using HTML, Perl CGI, and the MySQL database engine, running on a Linux server. The architecture is divided into three sections: miRNAs, target genes, and miRNA expression profiles, allowing users to browse and search for specific miRNAs, view sequences, secondary structures, chromosomal locations, and explore expression patterns and target interactions.

17.3.1.3.4 Future Updates and Enhancements

PMRD continuously updates its repository with newly identified plant miRNAs, their respective target genes, and expression patterns, ensuring it remains a relevant and essential tool for plant scientists globally. PMRD is recognized for offering thoroughly curated information on miRNA sequences, expression patterns, and interactions with targets, facilitating a wide array of research endeavors from fundamental exploration to functional genomics investigations.

17.3.1.4 miRNEST: A Comprehensive Plant miRNA Database

miRNEST is an integrative database designed to address the need for a comprehensive repository of miRNAs from various species, including plants, animals, and viruses. Developed by the Laboratory of Bioinformatics at Adam Mickiewicz University in Poznań, Poland, it serves as a significant resource for identifying and annotating miRNAs across a wide range of species. The database is accessible online (Szcześniak et al., 2012; http://mirnest.amu.edu.pl).

17.3.1.4.1 Features and Functionality

Data Integration and Prediction: miRNEST integrates miRNA data from multiple sources, including Expressed Sequence Tags (ESTs), sRNA sequencing data, and existing miRNA databases such as miRBase and PMRD. This integration allows for a comprehensive dataset covering species that were previously underrepresented in miRNA research.

Prediction Pipeline: The miRNA prediction process involves sequence similarity searches, filtering low-complexity regions, secondary structure prediction, and removing sequences with high similarity to known proteins or ncRNAs. This ensures high specificity and sensitivity, minimizing false positives and providing a high-quality set of miRNA candidates.

Degradome Analysis and HuntMi Predictions: miRNEST incorporates degradome analysis to predict miRNA targets, particularly in plant species. The database also uses HuntMi, a machine learning (ML) tool, to discriminate between true and false pre-miRNAs.

Splice Sites Prediction: miRNEST provides information on miRNA splice sites, adding another layer of functional annotation.

17.3.1.4.2 User Interface and Accessibility

miRNEST offers a user-friendly web interface that allows easy browsing, searching, and downloading of miRNA data. Researchers can access detailed information on miRNA loci, including genomic coordinates, sequence alignments, and expression data. The database supports data submissions from the research community, promoting collaboration and data sharing.

Species Coverage: miRNEST covers a broader range of species compared to other miRNA databases like miRBase. It includes data from 544 species, integrating miRNA sequences from multiple sources to provide a comprehensive view of miRNA diversity across the tree of life.

Functional Annotations: The database offers extensive functional annotations, including target predictions, expression data, polymorphisms, and links to external miRNA resources. This integration of diverse data types makes miRNEST a valuable tool for researchers studying miRNA functions and their roles in various biological processes.

Deep Sequencing Data: miRNEST incorporates sRNA deep sequencing data from 192 libraries across 29 species. This feature allows researchers to explore isomiRs and changes in sRNA counts under different conditions, providing insights into the dynamic nature of miRNA expression.

miRNEST stands out as a comprehensive and integrative resource for miRNA research, particularly in plants. Its extensive species coverage, rigorous prediction pipeline, and integration of diverse data types make it an invaluable tool for researchers. By providing a unified platform for miRNA data, miRNEST streamlines miRNA research and facilitates the discovery of new miRNAs and their functional roles in various species.

17.3.1.5 PmiREN: A Comprehensive Encyclopedia of Plant miRNAs

PmiREN, the Plant microRNA Encyclopedia, is a robust database designed to support the research community by providing extensive data on plant miRNAs. The latest version, PmiREN2.0, represents a significant upgrade from its predecessor, incorporating a wealth of new data and features aimed at advancing the study of miRNA functions and regulatory networks in plants (Guo et al., 2022; https://www.pmiren.com/).

17.3.1.5.1 Data Content and Updates

PmiREN2.0 has expanded its repository to include 39,751 miRNA entries from 179 species/subspecies, almost doubling the data content from the previous version. This expansion is made possible by the incorporation of sRNA sequencing (sRNA-seq) datasets from 66 plant species, leading to the identification of 17,798 high-confidence miRNA loci. The database also includes 1,565 unique miRNA loci from lycophytes and ferns, enhancing the evolutionary understanding of miRNA functions across diverse plant lineages.

17.3.1.5.2 Functional Exploration Tools

To aid in the functional exploration of miRNAs, PmiREN2.0 includes several new tools and data collections. Notably, the database provides predictions of cis-acting regulatory elements using PlantCare and PlantRegMap, annotations of miRNA target genes via InterProScan and KEGG, and phylogenetic analyses of conserved miRNA families. These features enable researchers to investigate the regulatory roles of miRNAs and their evolutionary trajectories in various plant species.

17.3.1.5.3 Regulatory Network Construction

One of the standout features of PmiREN2.0 is the construction of genome-wide regulatory networks centered on miRNAs. For instance, a regulatory network has been established for *Arabidopsis*, incorporating MTIs and TF–miRNA interactions. This network provides valuable insights into the complex regulatory mechanisms mediated by miRNAs and facilitates the identification of key regulatory modules involved in plant development and stress responses.

17.3.1.5.4 Community and Experimental Support

PmiREN2.0 also fosters community engagement and experimental practice through its newly added forum, “PmiREN Community.” This platform allows researchers to discuss findings, share resources, and collaborate on new discoveries. Additionally, the database includes built-in tools for experimental design, such as OE-miRNA for overexpression constructs, AmiRNA for artificial miRNA design, and STTM for target mimicry constructs, all of which are crucial for functional studies of miRNAs.

PmiREN2.0 represents a significant advancement in plant miRNA research, providing a comprehensive and user-friendly platform for the annotation, functional exploration, and regulatory network analysis of miRNAs. By continuously updating its data and integrating new analytical tools, PmiREN2.0 supports the growing needs of the research community, fostering a deeper understanding of miRNA functions and their impact on plant biology.

17.3.1.6 sRNAanno: A Comprehensive Database for Plant sRNAs

sRNAanno is a specialized database dedicated to the comprehensive annotation of sRNAs in plants. This database, hosted by the XIALAB at the College of Horticulture, South China Agricultural University, provides an extensive collection of miRNA, phasiRNA, and heterochromatic-siRNA (hc-siRNA)-generating loci annotated from approximately 140 plant species using consistent and high-confidence criteria. It serves as a valuable resource for researchers focused on the roles and functions of sRNAs in plant biology (Cui et al., 2022; http://www.plantsrna.com/).

17.3.1.6.1 Features and Functionality

Comprehensive sRNA Collection: sRNAanno integrates data on various types of sRNAs, including miRNAs, phasiRNAs, and hc-siRNAs. This extensive collection allows researchers to explore the diversity and functional roles of different sRNA species in plant biology. The database ensures high-confidence data through rigorous annotation criteria.

High-Confidence Annotations: The annotations in sRNAanno are based on a combination of sequence similarity, structural features, and expression patterns. This method ensures that the data is accurate and reliable, providing a high-quality resource for the study of sRNAs in plants.

User Interface and Accessibility: sRNAanno offers a user-friendly web interface that facilitates easy browsing, searching, and downloading of sRNA data. Researchers can access detailed information on sRNA loci, including genomic coordinates, sequence alignments, and expression data. The database supports data submissions from the research community, promoting collaboration and data sharing.

Integration with Other Databases: sRNAanno integrates data from various RNA-related databases and resources, providing a comprehensive platform for sRNA research. This integration facilitates access to a wide array of sRNA information, enhancing the utility of sRNAanno as a central hub for plant sRNA research.

17.3.1.6.2 Comparative Analysis with miRNEST and miRBase

Species Coverage: sRNAanno covers a broad range of plant species, offering a more specialized focus on plant sRNAs compared to miRNEST and miRBase. While miRNEST and miRBase include data from both plants and animals, sRNAanno specializes exclusively in plant sRNAs, providing a detailed resource for plant RNA research.

Functional Annotations: Similar to miRNEST and miRBase, sRNAanno offers extensive functional annotations for sRNA sequences. This includes information on RNA biogenesis, processing, and regulatory functions. However, sRNAanno places a stronger emphasis on integrating various types of sRNAs, including miRNAs, phasiRNAs, and hc-siRNAs.

Tools and Resources: sRNAanno provides a suite of bioinformatics tools tailored to the analysis of sRNAs. These tools complement those available in miRNEST and miRBase, offering additional capabilities for secondary structure prediction and functional characterization of sRNAs. The database also includes features for comparative genomics and evolutionary studies, facilitating a comprehensive understanding of RNA functions across different plant species.

sRNAanno stands out as a specialized and comprehensive resource for plant sRNA research. Its extensive collection of sRNA sequences, combined with advanced bioinformatics tools and a user-friendly interface, makes it an invaluable platform for researchers. By focusing exclusively on plant sRNAs, sRNAanno complements the broader scope of miRNEST and miRBase, providing detailed insights into the roles and functions of sRNAs in plant biology.

17.3.2 miRNA–Target Interaction Databases

These databases focus on the interactions between miRNAs and their target genes, collecting experimental evidence of these interactions.

17.3.2.1 TarBase: A Comprehensive miRNA–Target Interaction Database

In the landscape of miRNA research, TarBase has established itself as a cornerstone resource, providing an extensive repository of experimentally supported MTIs. Initially launched to consolidate MTIs validated through various methodologies, TarBase has evolved significantly. The latest iteration (TarBase v9.0, https://dianalab.e-ce.uth.gr/tarbasev9) extends experimentally supported miRNA–gene interactions to cell types and virally encoded miRNAs, solidifying its position as a vital tool for both basic and applied RNA research (Karagkouni et al., 2018).

TarBase v9.0 boasts approximately 6 million entries, encompassing around 2 million unique miRNA–target pairs, making it the largest collection of experimentally validated MTI s to date. The database integrates data from 37 experimental protocols and covers interactions in 172 tissues and cell types. The inclusion of interactions involving virally encoded miRNAs expands its utility to virology and infectious disease research. Additionally, the latest version documents interactions leading to target-directed miRNA degradation events, adding a new layer of functional insight into miRNA biology.

17.3.2.1.1 Data Annotation and Accessibility

Each interaction in TarBase is richly annotated with metadata, including information on genes, transcripts, miRNAs, experimental conditions, and associated publications. The database also provides millions of miRNA-binding locations with cell-type resolution, allowing researchers to explore the spatial dynamics of miRNA interactions in detail.

17.3.2.1.2 User Interface and Functionality

The user interface of TarBase v9.0 has been completely redesigned to enhance usability and flexibility. The new interface supports sophisticated query designs, enabling users to filter interactions based on various criteria such as cell lines, experimental conditions, species, and tissues. This allows for highly customized searches tailored to specific research needs. Additionally, the platform supports fine-tuning of search results based on miRNA confidence levels and expression data.

17.3.2.1.3 Integration with Other Resources

TarBase is seamlessly integrated with other prominent biomedical databases. Interactions with exact binding sites can be visualized in the Ensembl genome browser. Gene and miRNA information can be accessed through Ensembl and miRBase, respectively. TarBase is also included in RNACentral, ensuring users have access to a broad network of related data.

17.3.2.1.4 Advanced Features

Among the advanced features of TarBase v9.0 is a visualization tool that allows users to assess the combinatorial effects of multiple miRNAs on single or multiple genes. This network graph feature enables researchers to visualize complex interaction networks, facilitating a better understanding of miRNA regulatory mechanisms.

TarBase v9.0 represents a significant leap forward in the cataloging and analysis of MTIs. Its vast and meticulously curated dataset, combined with an intuitive and powerful user interface, makes it an indispensable resource for miRNA researchers.

17.3.2.2 miRTarBase: A Cornerstone for Experimentally Validated miRNA–Target Interactions

miRNAs are pivotal in regulating gene expression, impacting various physiological and pathological processes. As research into miRNAs has advanced, the demand for comprehensive, experimentally validated MTI databases has grown. miRTarBase stands out as one of the most significant resources in this domain, providing extensive data on MTIs supported by robust experimental evidence (Huang et al., 2020; https://mirtarbase.cuhk.edu.cn/~miRTarBase/miRTarBase_2022/php/index.php).

17.3.2.2.1 Evolution and Scope of miRTarBase

Established in 2011, miRTarBase has undergone numerous updates, with each version enhancing its data collection and user interface. The latest version, miRTarBase 9.0, includes over 2.2 million validated MTIs sourced from more than 13,000 research articles and CLIP-seq datasets. This extensive collection covers interactions in 37 species and includes data from high-throughput techniques such as reporter assays, western blots, and next-generation sequencing (NGS).

17.3.2.2.2 Key Features and Enhancements

1. **Comprehensive Data Collection**: The database includes a vast number of MTIs, integrating data from 440 CLIP-seq datasets and providing detailed interaction information.
2. **User-Friendly Interface**: The updated web interface offers an intuitive way for researchers to access data, browse MTIs by miRNA, target gene, or disease association, and visualize MTIs in CLIP-seq datasets.
3. **Disease Associations and Biomarker Discovery**: miRTarBase links miRNAs to specific diseases, aiding in the identification of potential biomarkers for diagnosis and therapy, including miRNAs involved in the COVID-19 response.
4. **Integration with Other Databases**: miRTarBase collaborates with various biological databases, such as TransmiR, miRSponge, and HMDD, to provide a comprehensive view of miRNA functions and their regulatory networks.

17.3.2.2.3 Plant miRNA Data in miRTarBase

While miRTarBase predominantly covers human MTIs, it also includes valuable data on plant miRNAs. This collection encompasses interactions validated in various plant species, providing crucial insights into plant miRNA functions and their regulatory roles.

17.3.2.2.4 Applications and Impact

The extensive and meticulously curated data in miRTarBase make it a critical resource for scientists investigating miRNA functions and their implications in diseases. Researchers can use miRTarBase to validate their experimental findings, explore new MTIs, and understand the broader regulatory networks involving miRNAs.

miRTarBase has established itself as a cornerstone resource in the field of miRNA research. Its comprehensive and experimentally validated data, coupled with a user-friendly interface, make it indispensable for researchers aiming to unravel the complexities of MTIs.

17.3.2.3 starBase: A Comprehensive Resource for miRNA Interaction Networks

In the expansive field of RNA research, starBase has emerged as a pivotal database for understanding the intricate interactions between miRNAs and other RNA molecules, including long ncRNAs (lncRNAs), pseudogenes, circular RNAs (circRNAs), and mRNAs. Initially developed to map miRNA–mRNA interactions, starBase has significantly expanded to include a wide array of RNA–RNA and protein–RNA interactions (Li et al., 2014).

17.3.2.3.1 Evolution and Scope of starBase

starBase was first introduced to provide a systematic exploration of MTIs using data from high-throughput sequencing methods such as CLIP-Seq. The latest version, starBase v2.0, integrates data from 108 CLIP-Seq datasets across 37 studies, covering a diverse range of species and tissues. This dataset includes approximately 9,000 miRNA–circRNA, 16,000 miRNA-pseudogene, and 285,000 protein–RNA interactions.

17.3.2.3.2 Key Features and Enhancements

1. **Comprehensive Interaction Networks**: The database includes detailed interaction networks among miRNAs, lncRNAs, pseudogenes, circRNAs, and mRNAs, which are crucial for understanding miRNA regulatory roles.
2. **User-Friendly Interface**: starBase provides an intuitive web interface for exploring interaction data, supporting interactive visualizations for in-depth exploration of complex regulatory networks.
3. **Functional Annotations and Predictive Tools**: Tools such as miRFunction and ceRNAFunction predict the functions of miRNAs and other ncRNAs based on their interactions, utilizing 13 functional genomic annotations.

4. **Integration with Other Databases**: starBase incorporates data from other key databases, enhancing its utility for comprehensive analysis of miRNA functions and regulatory networks.

17.3.2.3.3 Plant miRNA Data in starBase

starBase's inclusion of plant miRNA data significantly broadens its applicability. It provides detailed annotations of miRNA interactions in various plant species, facilitating research into plant miRNA functions and their roles in development, stress responses, and disease resistance.

17.3.2.3.4 Applications and Impact

The extensive and meticulously curated data in starBase make it an essential resource for scientists investigating miRNA functions and their broader implications. Researchers can utilize starBase to validate experimental findings, discover new MTIs, and understand the complex regulatory networks involving miRNAs.

starBase stands out as a premier resource in the field of miRNA research, offering an unmatched depth of data and advanced tools for analyzing miRNA interaction networks. Its user-friendly interface, comprehensive datasets, and integration with other databases make it an invaluable asset for researchers.

17.3.2.4 P-TarPmiR: A Plant-Specific miRNA Target Prediction Tool

In the realm of plant genomics, accurately predicting miRNA targets is crucial for understanding gene regulation mechanisms. P-TarPmiR, a cutting-edge plant-specific miRNA target predictor, addresses the unique challenges associated with miRNA–mRNA interactions in plants. Developed by leveraging a newly created plant miRNA–mRNA interaction database, P-TarPmiR outperforms traditional animal-based predictors by providing highly accurate predictions tailored for plant species (Lu et al., 2021).

17.3.2.4.1 Evolution and Scope of P-TarPmiR

P-TarPmiR was developed to improve the prediction accuracy of miRNA targets in plants, which often exhibit different binding characteristics compared to animals. Traditional miRNA target predictors, primarily trained on animal data, tend to perform poorly when applied to plants due to differences in miRNA biogenesis, binding specificity, and target site characteristics. P-TarPmiR overcomes these limitations by using plant-specific training data sourced from the TarDB database.

17.3.2.4.2 Key Features and Enhancements

1. **Plant-Specific Training Data**: P-TarPmiR is trained exclusively on plant miRNA–mRNA interaction data, ensuring high accuracy for plant species. This approach eliminates the noise introduced by animal data, which often detracts from the performance of miRNA target predictors in plants.
2. **Advanced Prediction Algorithm**: P-TarPmiR employs a random forest-based ML model that integrates conventional features such as seed match accessibility and AU content with new features like m/e motifs and stem conservation. This comprehensive feature set enhances the model's ability to accurately predict miRNA target sites in plants.
3. **User-Friendly Web Interface**: The P-TarPmiR web server provides a convenient platform for researchers to submit miRNA and mRNA sequences for target prediction. The interface allows for both file uploads and direct text input, and results are returned with detailed information about the predicted target sites and their confidence levels.
4. **Application to Soybean Genome**: P-TarPmiR has been applied to the entire soybean genome, providing a comprehensive list of high-confidence miRNA–mRNA interactions. These predictions are accompanied by GO term enrichment analyses, offering insights into the potential biological functions of each miRNA. The results are shared with the research community, facilitating further studies on soybean miRNA functions.

17.3.2.4.3 Plant miRNA Data in P-TarPmiR

P-TarPmiR's focus on plant-specific interactions makes it a valuable tool for plant scientists. The database includes detailed annotations of miRNA interactions in various plant species, enabling researchers to explore the regulatory roles of miRNAs in plant development, stress responses, and adaptation to environmental changes. This specificity is particularly important for studying economically important crops like soybean, where miRNA-mediated gene regulation can impact traits such as yield and disease resistance.

17.3.2.4.4 Applications and Impact

P-TarPmiR's extensive and accurately curated data make it an essential resource for scientists investigating plant miRNA functions. Researchers can use the tool to validate experimental findings, discover new MTIs, and understand the broader regulatory networks involving miRNAs. By providing a high-confidence prediction tool tailored to plant species, P-TarPmiR supports significant advancements in plant biology and agricultural biotechnology.

P-TarPmiR stands out as a premier resource in the field of plant miRNA research, offering unparalleled accuracy and specificity in miRNA target predictions. Its user-friendly interface, comprehensive datasets, and focus on plant-specific interactions make it an invaluable asset for researchers. As P-TarPmiR continues to evolve, it will undoubtedly remain at the forefront of plant miRNA research, enhancing our understanding of gene regulation and supporting the development of innovative agricultural practices.

17.3.3 Species-Specific Databases

17.3.3.1 *Arabidopsis* Small RNA Project: A Cornerstone for Plant miRNA Research

The *Arabidopsis* Small RNA Project (ASRP) database has been instrumental in advancing our understanding of sRNAs in plants, particularly miRNAs. Established to support the burgeoning field of sRNA research, ASRP provides an extensive repository of sRNA sequences and related data from the model plant *Arabidopsis thaliana* (Gustafson et al., 2005; https://asrp.danforthcenter.org/).

17.3.3.1.1 Scope and Evolution of ASRP

The ASRP database was developed to manage and analyze the vast amounts of data generated by high-throughput sequencing technologies. It includes over 218,585 unique sRNA sequences derived from various tissues and developmental stages of *Arabidopsis*, as well as from mutants with defects in sRNA silencing pathways. This comprehensive collection allows researchers to study miRNAs alongside other sRNA classes like siRNAs, providing a holistic view of RNA-mediated gene regulation in plants.

17.3.3.1.2 Key Features

1. **Extensive Data Collection**: ASRP catalogs sRNAs from wild-type and mutant *Arabidopsis* plants. The database encompasses data from multiple developmental stages and various sRNA biogenesis mutants, facilitating detailed studies of miRNA functions and regulatory mechanisms.
2. **Advanced Search and Visualization Tools**: The ASRP web interface offers robust tools for searching and visualizing sRNA data. Users can explore sRNA classes, view their genomic positions, and analyze expression profiles using an integrated genome browser. These tools enhance the accessibility and utility of the data for researchers.
3. **Integration with Other Databases**: ASRP integrates data from significant sources, such as the David Bartel lab, enhancing its comprehensiveness. This integration allows users to cross-reference and validate their findings with external datasets, supporting more reliable research outcomes.

4. **Support for Data Mining and Analysis**: The database employs efficient algorithms for data handling and analysis, ensuring rapid processing and retrieval of information. This capability is critical for managing the large volumes of data generated by high-throughput sequencing and for conducting in-depth analyses of sRNA populations.

ASRP has significantly impacted plant miRNA research by providing a centralized, accessible repository of sRNA data. It supports the identification and functional analysis of miRNAs in *Arabidopsis*, a model organism whose findings often extend to other plant species. The detailed annotation and extensive data provided by ASRP facilitate the exploration of miRNA-mediated regulatory networks, contributing to our understanding of plant development, stress responses, and gene regulation mechanisms.

The ASRP database is a vital resource for plant biologists, offering extensive data and advanced tools for studying sRNAs, particularly miRNAs. Its comprehensive collection, user-friendly interface, and integration with other databases make it an indispensable tool for advancing miRNA research in plants. By leveraging the capabilities of ASRP, researchers can gain deeper insights into the complex regulatory roles of miRNAs in plant biology.

17.3.3.2 PmiRKB: A Comprehensive Plant miRNA Knowledge Base

In the evolving landscape of plant genomics, the Plant microRNA Knowledge Base (PmiRKB) stands out as a crucial resource for researchers focusing on the functional roles of miRNAs in plants. Developed by Zhejiang University, PmiRKB offers an extensive, meticulously curated database dedicated to miRNAs in the model plants *Arabidopsis thaliana* and *Oryza sativa*. This database is designed to facilitate the understanding of miRNA-mediated gene regulation through its comprehensive data and specialized tools (You et al., 2017; https://bis.zju.edu.cn/).

17.3.3.2.1 Key Features and Functional Modules

PmiRKB is structured into four major functional modules, each designed to provide detailed insights into various aspects of miRNA biology:

1. **SNP Module**: This module collects single-nucleotide polymorphisms (SNPs) within pre-miRNAs and miRNA–target RNA duplexes across multiple accessions of Arabidopsis and rice subspecies. SNPs are analyzed for their potential to affect the secondary structures of pre-miRNAs and their interactions with target RNAs, providing insights into the genetic variations that influence miRNA function and stability.
2. **Pri-miR Module**: This module investigates the tissue-specific transcriptional contexts of pre- and pri-miRNAs using massively parallel signature sequencing data. It is the first large-scale attempt to elucidate the transcriptional ranges of pri-miRNAs in plants, offering valuable information for understanding miRNA transcription and processing.
3. **MiR-Tar Module**: Designed to validate MTIs, this module uses parallel analysis of RNA ends data to identify cleavage signals within predicted miRNA–target sites. This high-throughput approach enables large-scale validation of miRNA–target pairs, facilitating studies on miRNA-mediated gene regulation.
4. **Self-reg Module**: This module explores miRNA precursor metabolism, including precursor processing and the effects of miRNA-mediated self-regulation on their host precursors. By analyzing PARE data, it provides insights into the processing dynamics and regulatory mechanisms of miRNA genes.

17.3.3.2.2 Impact and Applications

PmiRKB is an invaluable resource for plant scientists, offering detailed annotations and functional data on plant miRNAs. The database supports various research applications, including the study of miRNA functions in development, stress responses, and adaptation. By providing a comprehensive and user-friendly platform, PmiRKB enhances researchers' ability to explore and validate MTIs, contributing to a broader understanding of gene regulation in plants.

17.3.3.3 MepmiRDB: A Medicinal Plant microRNA Database

MepmiRDB is a pioneering database dedicated to the study of miRNAs in medicinal plants. Recognizing the unique regulatory roles that miRNAs play in secondary metabolism and potential cross-kingdom gene regulation, MepmiRDB provides a comprehensive resource for researchers investigating these sRNA molecules in medicinal plant species (Zhang et al., 2019; http://mepmirdb.cn/mepmirdb).

MepmiRDB encompasses miRNA data from 29 medicinal plant species, many of which are listed in the Pharmacopoeia of the People's Republic of China. This database includes thousands of miRNA candidates, offering detailed information on their sequences, expression patterns, and regulatory networks. The data is organized into several functional modules:

1. **Sequence Module**: This module provides sequences of mature miRNAs and their precursors, including structural information and processing signals. It also identifies conserved and non-conserved miRNAs, enhancing the understanding of miRNA evolution and functionality in medicinal plants.
2. **Expression Module**: This section details organ-specific and condition-specific expression patterns of miRNAs, helping researchers understand how miRNA expression varies across different tissues and environmental conditions. Such insights are crucial for elucidating the roles of miRNAs in plant development and stress responses.
3. **Interaction Module**: Utilizing degradome-seq data, this module offers validated miRNA–target pairs for several species. It provides visual networks that depict the regulatory relationships between miRNAs and their targets, facilitating the study of miRNA-mediated gene regulation.
4. **Search and Download Modules**: These modules enable users to perform precise searches by species, miRNA families, or sequences and to download the entire dataset for further analysis. This accessibility ensures that researchers can efficiently obtain and utilize the data for their specific research needs.

MepmiRDB significantly advances the field of plant miRNA research by providing a specialized platform focused on medicinal plants. The database supports the exploration of miRNA roles in secondary metabolism, which is particularly relevant for medicinal plants that produce bioactive compounds used in traditional and modern medicine. Moreover, the potential for plant miRNAs to engage in cross-kingdom gene regulation underscores their importance not only in plant biology but also in therapeutic applications.

MepmiRDB is an invaluable resource for researchers studying medicinal plant miRNAs. Its comprehensive data and user-friendly interface facilitate detailed analyses of miRNA functions, contributing to our understanding of gene regulation in medicinal plants. By enabling the discovery and validation of MTIs, MepmiRDB supports significant advancements in plant biology and pharmacology.

17.3.3.4 miRVIT: A Comprehensive Database for Grapevine miRNAs

miRVIT is a specialized database dedicated to the miRNAs of grapevine (*Vitis vinifera*), a key crop with significant economic and cultural importance. Developed to enhance the understanding of miRNA-mediated regulatory mechanisms in grapevine, miRVIT provides a comprehensive repository of miRNA sequences, expression profiles, and target interactions specific to this plant species (Pantaleo et al., 2010; https://mirvit.ipsp.cnr.it/).

miRVIT includes a wide array of data specifically curated for grapevine miRNAs. The database encompasses miRNA sequences derived from high-throughput sequencing, as well as computational predictions and experimental validations. This extensive collection allows researchers to explore the roles of miRNAs in grapevine biology, particularly in response to biotic and abiotic stresses, development, and fruit quality.

1. **miRNA Sequences**: miRVIT provides detailed information on the sequences of both mature miRNAs and their precursors in grapevine. This includes annotations of conserved and novel miRNAs, helping researchers identify key regulatory RNAs involved in various physiological processes.
2. **Expression Profiles**: The database offers organ-specific and condition-specific expression data for grapevine miRNAs. By detailing how miRNA expression varies across different tissues and under various environmental conditions, miRVIT aids in understanding the dynamic regulation of gene expression.
3. **miRNA–Target Interactions**: miRVIT includes validated and predicted MTIs, providing insights into the regulatory networks controlled by miRNAs. This information is crucial for identifying target genes and understanding how miRNAs modulate gene expression to influence grapevine growth, development, and stress responses.
4. **Search and Analysis Tools**: The database features user-friendly tools for searching and analyzing miRNA data. Researchers can perform searches based on miRNA names, sequences, or target genes and visualize interaction networks to better understand miRNA functions.

miRVIT significantly advances the study of plant miRNAs by focusing on a single, economically important species. The database supports grapevine research by providing high-quality, specific data that facilitates the exploration of miRNA roles in viticulture. This targeted approach helps uncover the molecular mechanisms underlying important traits such as disease resistance, fruit development, and environmental adaptation.

miRVIT is an invaluable resource for researchers studying grapevine miRNAs. Its comprehensive data collection and specialized focus on *Vitis vinifera* enhance the ability to investigate and understand miRNA-mediated gene regulation in this key crop. By offering detailed sequence information, expression profiles, and interaction data, miRVIT supports significant advancements in grapevine biology and viticulture.

17.3.3.5 mirEX: An Integrated Environment for Plant miRNA Expression Profiling

mirEX/mirEX 2 is a specialized database designed to facilitate the exploration of miRNA expression profiles in plants. Developed to address the need for comprehensive and accessible miRNA expression data, mirEX provides a valuable resource for researchers investigating the roles of miRNAs in plant development and stress responses (Fujita and Matsui 2014; http://www.combio.pl/mirex2).

mirEX includes expression data for miRNAs from several plant species, including the model plants *Arabidopsis thaliana*, *Hordeum vulgare* (barley), and *Pellia endiviifolia* (liverwort). The database encompasses data from quantitative real-time PCR, Northern blot, and NGS experiments, offering a detailed view of miRNA expression across different developmental stages and tissues.

1. **Comprehensive miRNA Expression Data**: mirEX provides expression profiles for 461 miRNAs representing 268 families. The database includes data on both pri-miRNAs and mature miRNAs, enabling researchers to comprehensively study miRNA biogenesis and function.
2. **User-Friendly Interface**: The mirEX web portal offers an intuitive interface for exploring miRNA expression data. Users can query the database using a graphical query builder, allowing them to select specific species, developmental stages, and tissues. The results are presented in modular, thematic windows with advanced visualization tools, such as line graphs, bar charts, and heat maps.

3. **Multi-Condition Analysis**: mirEX supports the comparison of miRNA expression levels across multiple conditions, including different plant species, developmental stages, and tissues. This feature is particularly useful for identifying miRNAs involved in specific biological processes or stress responses.
4. **Data Accessibility**: All data in mirEX are downloadable, facilitating further analysis and integration with other datasets. The database also includes extensive documentation and tutorials to assist users in navigating and utilizing the available resources.

mirEX significantly advances plant miRNA research by providing a centralized platform for miRNA expression data. Its comprehensive coverage and advanced analytical tools enable researchers to explore the dynamic roles of miRNAs in plant biology, including their regulatory functions in development and responses to environmental stresses. By integrating data from multiple experimental platforms, mirEX supports robust, multi-faceted analyses of miRNA expression.

mirEX is an indispensable resource for plant scientists, offering detailed and accessible miRNA expression profiles across a variety of plant species. Its comprehensive dataset, user-friendly interface, and advanced visualization tools make it a powerful platform for studying miRNA functions and regulatory mechanisms in plants. By facilitating the exploration and comparison of miRNA expression data, mirEX significantly enhances our understanding of plant biology and the complex roles that miRNAs play in development and stress responses.

17.3.3.6 miRFANs: An Integrated Database for Plant miRNA Functional Annotations

miRFANs is a comprehensive database designed to annotate and explore the functions of miRNAs in *Arabidopsis thaliana*. Given the crucial roles miRNAs play in developmental control, hormone regulation, and responses to environmental stresses, miRFANs provides an invaluable resource for researchers studying plant gene regulation (Li et al., 2012).

miRFANs integrates a wide range of datasets, including MTIs, TF networks, expression profiles, genomic annotations, and pathway information. This holistic approach allows for a detailed analysis of miRNA functions in plants.

1. **miRNA–Target Interactions**: miRFANs consolidates data from multiple prediction tools (psRNATarget, TargetAlign, and UEA target-finder) and experimentally validated sources (TarBase and miRTarBase). This results in a comprehensive list of MTIs, facilitating the study of miRNA-mediated gene regulation.
2. **Expression Profiles**: The database includes expression data for miRNAs and their target genes under various conditions, allowing researchers to investigate co-expression patterns and regulatory dynamics. This is particularly useful for understanding the spatial and temporal regulation of miRNAs.
3. **Functional Annotations and Pathways**: miRFANs provides functional annotations for miRNA targets, including GO terms and KEGG pathways. This helps in deciphering the biological roles and pathways influenced by miRNAs.
4. **User-Friendly Interface**: The web interface of miRFANs supports diverse queries, including searches by miRNA, gene, or pathway. It also features tools for differential expression analysis, clustering, and network visualization, making it easy to analyze complex datasets.

miRFANs significantly enhances the study of plant miRNAs by providing a centralized platform that integrates multiple types of data. Its comprehensive coverage and advanced analytical tools allow researchers to explore the regulatory networks controlled by miRNAs in *Arabidopsis thaliana*. This resource supports the discovery of new miRNA functions and their roles in plant development and stress responses.

miRFANs is an essential resource for plant scientists, offering detailed functional annotations and expression profiles of miRNAs in *Arabidopsis thaliana*. By integrating diverse datasets and providing robust analytical tools, miRFANs facilitates a deeper understanding of miRNA-mediated gene regulation in plants.

17.3.4 Specialized Databases

These databases focus on specific aspects or types of miRNAs, providing specialized information relevant to certain research areas.

17.3.4.1 PVsiRNAdb: A Database for Plant-Exclusive Virus-Derived Small Interfering RNAs

The PVsiRNAdb is an essential resource for researchers focused on plant–virus interactions, particularly those studying the role of virus-derived small interfering RNAs (vsiRNAs) in plant antiviral defense mechanisms. Developed by the Bioinformatics Laboratory at the National Institute of Plant Genome Research, this database serves as a comprehensive repository for vsiRNAs identified in various virus-infected plants (Kumar et al., 2015; http://14.139.61.8/PVsiRNAdb/index.php).

PVsiRNAdb integrates extensive miRNA expression data, which is crucial for understanding the regulatory networks in plants during viral infections. The database houses detailed information on 322,214 entries and 282,549 unique vsiRNA sequences derived from 20 different viral strains infecting 12 plant species. By offering a platform that combines vsiRNA sequences with corresponding plant and virus data, PVsiRNAdb enables researchers to study the expression patterns and regulatory roles of these sRNAs across various plant tissues and developmental stages.

17.3.4.1.1 Exploring miRNA–Target Interactions in Plants

The database provides tools for exploring MTIs, which are fundamental for elucidating the mechanisms by which vsiRNAs mediate antiviral responses. PVsiRNAdb includes functionalities such as BLAST, Smith–Waterman align, and mapping tools, allowing users to align their sequences with vsiRNAs and investigate potential target sites on viral genomes or host plant transcripts. This facilitates a deeper understanding of how vsiRNAs contribute to gene silencing and antiviral defense by targeting viral RNA for degradation or inhibiting its translation.

17.3.4.1.2 Applications in Understanding Plant miRNA Biology

PVsiRNAdb is instrumental in advancing the understanding of plant miRNA biology. The detailed annotations and comprehensive search options allow researchers to retrieve specific vsiRNA information, including sequence length, tissue of origin, and predicted secondary structures. This data is invaluable for studying the biogenesis of vsiRNAs, their incorporation into RNA-induced silencing complexes, and their role in transcriptional and post-transcriptional gene silencing. Additionally, the database supports the development of transgenic plants with enhanced viral resistance, thereby contributing to agricultural biotechnology and crop improvement strategies.

By providing a specialized and detailed repository for plant-exclusive vsiRNAs, PVsiRNAdb significantly enhances researchers' ability to investigate and manipulate sRNA-mediated defense mechanisms in plants, paving the way for innovative approaches to plant protection and genetic research.

17.3.4.2 miSolRNA: A Specialized Database for Tomato miRNA Function

miSolRNA is a specialized database designed to provide comprehensive information on miRNAs in tomato (*Solanum lycopersicum*). Recognizing the importance of miRNAs in regulating gene expression, miSolRNA offers valuable insights into their roles in tomato development and stress responses. It focuses on miRNAs identified in tomato, providing detailed annotations and expression profiles. The database integrates data from high-throughput sequencing and bioinformatics predictions, ensuring a robust and comprehensive resource for researchers (Panchy et al., 2016; http://www.misolrna.org/).

1. **miRNA Sequences**: miSolRNA includes sequences of mature miRNAs and their precursors in tomato. It provides information on conserved and novel miRNAs, highlighting their evolutionary significance and functional diversity.
2. **Expression Profiles**: The database offers expression data for miRNAs across various tissues and developmental stages of tomato, enabling researchers to study the temporal and spatial regulation of miRNAs and their involvement in different physiological processes.
3. **miRNA–Target Interactions**: miSolRNA provides validated and predicted MTIs, offering insights into the regulatory networks mediated by miRNAs in tomato. This information is crucial for understanding how miRNAs influence gene expression and contribute to traits such as fruit development and stress tolerance.
4. **Functional Annotations**: The database includes functional annotations of miRNA targets, integrating GO terms and pathway information. This helps researchers link miRNAs to specific biological functions and metabolic pathways.

miSolRNA significantly advances the study of plant miRNAs by focusing on a single, economically important species. The database supports research into the roles of miRNAs in tomato development, stress responses, and fruit quality. By providing detailed expression profiles and interaction data, miSolRNA enables researchers to explore the regulatory mechanisms controlled by miRNAs in tomato, facilitating the development of improved crop varieties.

miSolRNA is an invaluable resource for scientists studying tomato miRNAs. Its comprehensive data and specialized focus on *Solanum lycopersicum* enhance the investigation of miRNA functions and regulatory networks in this key crop. By offering detailed sequence information, expression profiles, and functional annotations, miSolRNA supports significant advancements in tomato biology and agriculture.

17.3.4.3 PNRD: A Comprehensive Plant Non-Coding RNA Database

The Plant Non-Coding RNA Database (PNRD) is a critical resource designed to facilitate the study of ncRNAs in plants. Expanding upon its predecessor, PMRD, PNRD integrates extensive data on various types of ncRNAs, including miRNAs, providing a comprehensive platform for plant RNA research (Yi et al., 2015; http://bioinformatics.cau.edu.cn/PMRD/).

PNRD includes data on 25,739 ncRNA entries from 150 plant species, encompassing 15,041 miRNAs and their predicted targets. This extensive collection allows researchers to explore miRNA sequences, expression profiles, and functional annotations in detail.

1. **miRNA Sequences and Targets**: PNRD offers detailed information on miRNA sequences and their predicted targets. The database includes target information for 178,138 pairs across 46 species, incorporating data from both public repositories and published literature.
2. **Expression Profiles**: The database includes expression profiles for miRNAs derived from NGS and microarray studies. This data covers various tissues and developmental stages, enabling researchers to understand the spatial and temporal dynamics of miRNA expression.
3. **Functional Annotations**: PNRD provides functional annotations for miRNA targets, integrating GO terms and pathway information. This feature helps researchers link miRNAs to specific biological processes and regulatory networks.

4. **User-Friendly Interface and Tools**: PNRD features an intuitive web interface with advanced search and visualization tools. Researchers can browse ncRNAs by species or category, perform keyword searches, and explore miRNA relationships with histone modifications using integrated genome browsers.

PNRD significantly enhances the study of plant miRNAs by offering a centralized, comprehensive repository of ncRNA data. Its detailed annotations and extensive datasets support the identification and functional analysis of miRNAs, facilitating research on gene regulation, development, and stress responses in plants.

PNRD is an essential resource for plant scientists, providing a one-stop platform for ncRNA research. Its comprehensive data collection, user-friendly tools, and integration with various analytical resources make it a valuable asset for exploring the complex roles of ncRNAs in plant biology.

17.3.4.4 PMTED: Plant miRNA Target Expression Database

The Plant miRNA Target Expression Database (PMTED) is a comprehensive resource designed to provide detailed expression profiles of miRNA targets in plants. By integrating data from various high-throughput sequencing technologies, PMTED supports researchers in deciphering the complex regulatory networks mediated by miRNAs (Yang et al., 2019; http://pmted.agrinome.org/).

PMTED focuses on collating and analyzing expression data for miRNA targets across a range of plant species, developmental stages, and environmental conditions. This database is pivotal for understanding the dynamic nature of miRNA-mediated gene regulation in plants.

1. **miRNA Target Data**: PMTED includes both experimentally validated and computationally predicted miRNA targets. The database compiles data from numerous studies, providing a robust dataset that enhances the reliability of miRNA target predictions.
2. **Expression Profiles**: The database offers extensive expression profiles for miRNA targets obtained from RNA-Seq, microarray, and other high-throughput technologies. These profiles are essential for investigating the spatial and temporal patterns of miRNA target gene expression under various physiological and stress conditions.
3. **Functional Annotations**: PMTED provides functional annotations of miRNA targets, integrating GO terms and KEGG pathways. This feature helps researchers link miRNAs to specific biological processes and regulatory pathways, offering insights into their functional roles.
4. **User Interface and Tools**: PMTED features a user-friendly web interface that allows researchers to search, visualize, and analyze miRNA target expression data easily. Advanced tools for differential expression analysis and network visualization support detailed studies of miRNA functions and their regulatory mechanisms.

PMTED significantly advances the field of plant miRNA research by providing a centralized repository of miRNA target expression data. Its comprehensive coverage and detailed annotations facilitate the identification and functional analysis of miRNA targets, enhancing our understanding of gene regulation, plant development, and responses to environmental stresses.

PMTED is an invaluable resource for plant scientists, offering detailed and accessible data on miRNA target expression. Its extensive datasets, user-friendly tools, and integration with functional annotations make it a powerful platform for studying the regulatory roles of miRNAs in plants.

17.3.4.5 PASmiR: A Literature-Curated Database for Plant miRNA Regulation in Response to Abiotic Stress

The Plant Abiotic Stress miRNA (PASmiR) database is a specialized repository developed to consolidate and standardize data on miRNA regulation under various abiotic stress conditions in plants.

By curating data from approximately 200 published studies, PASmiR provides a comprehensive platform for researchers studying the molecular mechanisms of miRNA-mediated stress responses (Wang et al., 2018; http://pmted.agrinome.org/PASmiR).

PASmiR focuses on miRNA regulatory relationships in response to abiotic stresses across 33 plant species. The database includes detailed information on 1,038 regulatory interactions between 682 miRNAs and 35 different abiotic stresses.

1. **miRNA–Stress Regulatory Data**: PASmiR encompasses miRNA expression patterns, detection methods, and target genes, sourced from literature and databases like miRBase. Each entry is annotated with species name, miRNA identifier, stress type, and experimental details, ensuring comprehensive coverage of miRNA–stress interactions.
2. **Expression Profiles**: The database provides detailed expression patterns of miRNAs under different stress conditions, such as drought, salinity, and temperature extremes. These profiles help elucidate the roles of miRNAs in plant stress physiology.
3. **Functional Annotations**: PASmiR includes functional annotations for miRNA targets, integrating GO terms and KEGG pathways. This feature aids in linking miRNAs to specific biological processes and pathways involved in stress responses.
4. **User Interface and Tools**: The PASmiR web interface allows for keyword-based searches and offers fuzzy search capabilities to facilitate data retrieval. Users can search by plant species, abiotic stress, or miRNA, and can contribute new regulatory entries through a submission page.

PASmiR significantly advances the study of plant miRNAs by providing a centralized, curated repository of miRNA–stress regulatory data. Its detailed annotations and extensive datasets support the identification and functional analysis of miRNAs, enhancing our understanding of plant responses to abiotic stresses.

PASmiR is an invaluable resource for plant scientists, offering detailed and accessible data on miRNA regulation in response to abiotic stress. Its comprehensive datasets, user-friendly tools, and integration with functional annotations make it a powerful platform for studying the regulatory roles of miRNAs in plants.

17.3.4.6 mirtronDB: A Comprehensive Mirtron Knowledge Base

mirtronDB is a specialized database designed to centralize and organize data on mirtrons, a unique class of miRNAs that originate from short introns. These non-canonical miRNAs use the splicing mechanism as the first step in their biogenesis, distinguishing them from traditional miRNAs. mirtronDB provides an invaluable resource for researchers by compiling data from multiple species and offering tools for detailed analysis (Chung et al., 2019; http://mirtrondb.cp.utfpr.edu.br).

mirtronDB contains data on 1,407 mirtron precursors and 2,426 mature mirtrons across 18 species, including chordates, invertebrates, and plants. This extensive dataset is curated from literature spanning over a decade, ensuring comprehensive coverage and high-quality information.

1. **Mirtron Sequences and Annotations**: mirtronDB includes detailed sequences of both precursor and mature mirtrons. The database provides functional annotations and categorizes mirtrons based on their organism and type, facilitating easy retrieval and comparison.
2. **Expression Profiles and Target Predictions**: The database offers predicted target genes for mirtrons in plants, using advanced bioinformatics tools like psRNATarget. It also includes ceRNA predictions, highlighting potential regulatory interactions between mirtrons and lncRNAs.
3. **Similarity Analysis**: mirtronDB performs similarity analyses among mirtron sequences across different species, helping researchers understand evolutionary conservation and

functional similarities. This feature is particularly useful for identifying conserved regulatory mechanisms across taxa.

4. **User Interface and Tools**: The mirtronDB web portal provides a user-friendly interface that supports searches by organism, group, type, and name. It also includes tools for visualizing mirtron networks and browsing sequence alignments, enhancing the usability and accessibility of the data.

mirtronDB significantly contributes to the field of plant miRNA research by offering a centralized repository of mirtron data. Its detailed annotations and comprehensive datasets support the study of miRNA biogenesis, function, and regulatory roles in plants. By facilitating the exploration of mirtron–target interactions and their potential impact on gene regulation, mirtronDB advances our understanding of plant biology and enhances the development of innovative agricultural strategies.

mirtronDB is an essential resource for researchers studying mirtrons and their roles in gene regulation. Its extensive datasets, user-friendly tools, and advanced analytical capabilities make it a valuable platform for exploring the unique aspects of mirtron biology.

17.3.4.7 PmiRExAt: Plant miRNA Expression Atlas

PmiRExAt, the Plant miRNA Expression Atlas, is an advanced database designed to provide a comprehensive overview of miRNA expression across multiple plant species, tissues, and developmental stages. This resource is crucial for researchers aiming to understand the regulatory roles of miRNAs in plant biology and utilize this knowledge for crop improvement (Bertolini et al., 2018; http://pmirexat.nabi.res.in).

PmiRExAt includes miRNA expression profiles from high-throughput sRNA sequencing (sRNA-Seq) datasets of three major crops: wheat, rice, and maize. The database encompasses a total of 1,859 wheat miRNAs, 2,330 rice miRNAs, and 283 maize miRNAs, offering a rich resource for comparative expression analysis.

1. **miRNA Expression Data**: The database contains detailed expression profiles of miRNAs in various tissues and developmental stages. This data is sourced from public repositories and includes 21 datasets for wheat, 53 for rice, and 43 for maize.
2. **Functional Annotations**: PmiRExAt provides functional annotations for miRNA targets, integrating information from GO terms and KEGG pathways. This assists researchers in linking miRNAs to specific biological processes and regulatory networks.
3. **Data Normalization and Visualization**: Expression data is normalized using transcripts per million values, and heatmaps are generated for visualization. This facilitates easy identification of differentially expressed miRNAs and their tissue-specific expression patterns.
4. **User Interface and Tools**: The PmiRExAt web portal offers an intuitive interface with advanced search and visualization tools. Users can conduct custom expression analyses, compare miRNA expression across different species, and generate high-resolution heatmaps for a detailed examination of expression data.

PmiRExAt significantly advances plant miRNA research by providing a centralized platform for miRNA expression data. Its comprehensive coverage and detailed annotations support the identification and functional analysis of miRNAs, facilitating research on gene regulation, development, and stress responses in plants. This database is particularly valuable for identifying miRNAs that are differentially expressed under various conditions, aiding in the development of biotechnological approaches for crop improvement.

PmiRExAt is an essential resource for plant scientists, offering a detailed and accessible platform for miRNA expression profiling. Its extensive datasets, user-friendly tools, and integration with functional annotations make it a powerful asset for studying the regulatory roles of miRNAs in plants.

17.3.4.8 PlantMiR-SSR: A Plant microRNA Simple Sequence Repeat Marker Database

The Plant microRNA Simple Sequence Repeat Marker Database (PlantMiR-SSR) is an innovative and comprehensive resource designed to facilitate genetic improvements in crops through the utilization of miRNA-associated simple sequence repeats (SSRs). By integrating extensive data on miRNA-SSRs, PlantMiR-SSR helps researchers understand the evolutionary dynamics of pre-miRNAs and develop functional molecular markers for plant breeding programs (Xie et al., 2018; http://plantmirssr.enset-project.org/ or http://genomicsres.org/plantmirssr/).

PlantMiR-SSR includes data on 16,892 pre-miRNA sequences from 292 plant species across six taxonomic groups, ranging from algae to dicots. This extensive dataset shows that 51% of pre-miRNA sequences contain SSRs, highlighting their prevalence and potential utility in genetic studies.

1. **miRNA-SSR Data**: The database provides detailed information on the identification, distribution, and density of SSRs within pre-miRNAs. It catalogs 9,498 microsatellite loci with potential as pre-miRNA-SSR markers, of which 3,573 are non-redundant, and 2,341 primer pairs can be transferred across different plant taxonomic groups.
2. **Functional Annotations**: PlantMiR-SSR offers functional annotations for miRNA targets, integrating GO terms and pathway information. This helps researchers link miRNAs and their SSRs to specific biological processes and regulatory networks.
3. **User-Friendly Interface**: The database features an intuitive web interface that allows users to search for miRNA-SSR markers by species, miRNA identifier, or SSR type. It also provides tools for visualizing SSR distributions and designing primers, facilitating the practical application of this data in genetic research.

PlantMiR-SSR significantly advances the field of plant genetics by providing a comprehensive repository of miRNA-SSR data. Its detailed annotations and extensive datasets support the development of molecular markers for crop improvement, enhancing our understanding of miRNA evolution and function. This database is particularly valuable for researchers working in plant breeding, as it provides the tools necessary to develop SSR markers that can accelerate genetic advancements in a wide range of crops.

PlantMiR-SSR is an essential resource for plant scientists, offering detailed and accessible data on miRNA-SSR markers. Its comprehensive datasets, user-friendly tools, and integration with functional annotations make it a powerful platform for studying the regulatory roles of miRNAs in plants and developing new genetic markers for crop improvement.

17.3.4.9 Plant Small RNA Genes Database

The Plant Small RNA Genes (PSRG) database is a valuable resource that provides comprehensive annotations and analyses of sRNA-producing loci from 47 diverse plant species. This database includes both miRNAs and small interfering RNAs (siRNAs), offering a wide range of data that is crucial for understanding the roles of these sRNAs in plant gene regulation (Johnson et al., 2016; https://plantsmallrnagenes.science.psu.edu/index.php).

PSRG encompasses data from 1,333 sRNA sequencing (sRNA-seq) libraries containing over 20 billion reads. This extensive dataset allows for detailed annotations of more than 2.7 million sRNA-producing loci across various plant species, including model plants and economically important crops.

1. **miRNA and siRNA Data**: PSRG provides comprehensive information on miRNA and siRNA loci, including their sequences, expression profiles, and genomic locations. The database classifies sRNA loci based on their size and origin, offering insights into their biogenesis and functional roles.

2. **Expression Profiles**: The database includes expression data for sRNAs from various tissues and developmental stages, enabling researchers to explore the spatial and temporal patterns of sRNA expression. This feature is particularly useful for identifying differentially expressed sRNAs under various physiological conditions.
3. **Functional Annotations**: PSRG integrates functional annotations for sRNA targets, including GO terms and KEGG pathways. This helps link sRNAs to specific biological processes and pathways, providing a deeper understanding of their regulatory roles.
4. **User Interface and Tools**: The PSRG web portal offers an intuitive interface with advanced search and visualization tools. Users can browse genome browsers, perform sequence searches, and visualize RNA secondary structures, facilitating the analysis and interpretation of sRNA data.

PSRG significantly advances the field of plant miRNA research by providing a centralized and accessible repository of sRNA data. Its comprehensive coverage and detailed annotations support the identification and functional analysis of miRNAs and siRNAs, enhancing our understanding of gene regulation, plant development, and stress responses. This database is particularly valuable for researchers in plant genomics and molecular biology, as it provides the necessary tools to explore the complex regulatory networks mediated by sRNAs.

PSRG is an essential resource for plant scientists, offering detailed and accessible data on sRNAs in plants. Its extensive datasets, user-friendly tools, and integration with functional annotations make it a powerful platform for studying the regulatory roles of miRNAs and siRNAs in plants.

17.4 PRACTICAL APPLICATIONS AND FUTURE DIRECTIONS

17.4.1 Applications in Crop Improvement

The advent of comprehensive miRNA databases has revolutionized our approach to crop improvement, offering unprecedented opportunities for genetic engineering and breeding programs (Zhang et al., 2006). These databases serve as invaluable resources for researchers and plant breeders, providing a wealth of information on miRNA sequences, expression patterns, and predicted targets across various plant species. This knowledge forms the foundation for innovative strategies in crop enhancement, addressing challenges such as yield improvement, stress tolerance, and nutritional quality (Jones et al., 2006).

For instance, the manipulation of miR156, a highly conserved miRNA family in plants, has shown remarkable potential in improving various crop traits. Studies documented in miRNA databases have revealed that overexpression of miR156 in switchgrass (*Panicum virgatum*) led to increased biomass production, delayed flowering, and enhanced drought tolerance (Fu et al., 2012). This discovery, facilitated by the wealth of data available in plant miRNA databases, has significant implications for bioenergy crop improvement and sustainable agriculture.

Another successful application stemming from miRNA database research is the engineering of disease resistance in crops. The miR399 family, well characterized and extensively annotated in miRNA databases, has been implicated in regulating phosphate homeostasis in plants (Bari et al., 2006). Researchers utilized this information to develop transgenic tomato plants overexpressing miR399, resulting in enhanced resistance to *Phytophthora infestans*, the causative agent of late blight disease (Gu et al., 2010).

miRNA databases have also played a crucial role in improving nutritional quality in crops. The miR156-SPL (squamosa promoter binding protein-like) regulatory module, extensively documented in these databases, has been manipulated to enhance carotenoid content in tomato fruits (Zhang et al., 2018). By suppressing specific SPL TFs through miR156 overexpression, researchers achieved a significant increase in lycopene and β-carotene levels, thereby improving the nutritional value of this important crop.

In the realm of plant breeding, miRNA databases offer valuable insights for marker-assisted selection (MAS) programs. By identifying miRNAs and their target genes associated with desirable traits, breeders can develop molecular markers for the efficient selection of superior genotypes (Kantar et al., 2011). For example, variations in miR172 binding sites within the Q gene of wheat, as revealed through miRNA target prediction tools linked to these databases, have been associated with differences in spike morphology and threshability (Debernardi et al., 2017). This discovery has led to the development of molecular markers that aid in the selection of wheat varieties with improved grain yield and quality.

17.4.2 Stress Response and Adaptation

Understanding plant responses to biotic and abiotic stresses is crucial for developing resilient crop varieties capable of thriving in challenging environments. miRNA databases have emerged as powerful tools in elucidating the complex regulatory networks underlying stress responses in plants, offering insights that are instrumental in breeding stress-tolerant crops (Sunkar et al., 2012).

Abiotic stress tolerance, including resistance to drought, salinity, and extreme temperatures, is a major focus of crop improvement efforts (Anil et al., 2023; Jabborova et al., 2023; Verma et al., 2021). miRNA databases provide comprehensive information on stress-responsive miRNAs and their targets, enabling researchers to unravel the molecular mechanisms of plant adaptation to adverse conditions (Zhang, 2015). For instance, the miR398 family, well characterized in various plant species and documented in miRNA databases, plays a crucial role in regulating plant responses to oxidative stress, drought, and heat (Zhu et al., 2011).

Utilizing this knowledge, researchers have successfully engineered enhanced drought tolerance in *Arabidopsis* by overexpressing miR398 (Guan et al., 2013). This approach led to improved antioxidant capacity and reduced water loss under water-deficit conditions. Similar strategies have been applied to crop species, with promising results. In rice, overexpression of miR319, another well-documented miRNA in databases, resulted in enhanced tolerance to cold and salt stress, demonstrating the practical application of miRNA database information in developing climate-resilient crops (Yang et al., 2013).

Biotic stress resistance, particularly against pathogens and pests, is another area where miRNA databases have made significant contributions. These repositories contain valuable data on miRNAs involved in plant immune responses and their regulation during pathogen interactions (Katiyar-Agarwal and Jin, 2010). For example, the miR393 family, extensively annotated in miRNA databases, is implicated in basal defense against bacterial pathogens in *Arabidopsis* (Navarro et al., 2006). This information has been translated to crop species, with transgenic tomato plants overexpressing miR393 showing enhanced resistance to *Pseudomonas syringae* (Jiang et al., 2013).

miRNA databases also facilitate comparative analyses of stress-responsive miRNAs across different plant species, enabling the identification of conserved regulatory modules that can be targeted for broad-spectrum stress tolerance (Zhang et al., 2018). This approach is particularly valuable for developing crops with multiple stress tolerance traits, which are necessary in the face of climate change and evolving pathogen pressures.

The integration of miRNA data with other omics information, such as transcriptomics and metabolomics, provides a holistic view of plant stress responses (Zhu and Helliwell, 2011). Many miRNA databases now offer tools for integrative analysis, allowing researchers to correlate miRNA expression patterns with global changes in gene expression and metabolite profiles under stress conditions. This systems biology approach, powered by comprehensive databases, is paving the way for more sophisticated strategies in engineering stress-tolerant crops (Mallory and Vaucheret, 2010).

17.4.3 Future Trends in miRNA Research

The field of plant miRNA research is rapidly evolving, driven by technological advancements and the growing recognition of miRNAs' importance in plant biology. As we look to the future, several emerging trends and potential developments are poised to shape the landscape of miRNA research and databases.

One of the most promising emerging technologies in miRNA research is single-cell sequencing (Svensson et al., 2018). This approach allows for the profiling of miRNA expression at unprecedented resolution, revealing cell-specific regulatory networks and heterogeneity within plant tissues. As this technology matures, we can expect miRNA databases to incorporate single-cell miRNA expression data, providing researchers with a detailed view of miRNA dynamics across different cell types and developmental stages (Liu et al., 2021).

CRISPR-Cas9 technology is another game-changer in miRNA research (Wolter and Puchta, 2018). While it is already utilized for studying miRNA function, future applications of CRISPR are likely to include high-throughput screening of miRNA targets and the fine-tuning of miRNA expression in crops. miRNA databases will need to evolve to capture and represent the wealth of functional data generated through CRISPR-based studies, potentially incorporating modules for designing and predicting the outcomes of CRISPR experiments targeting miRNAs or their regulatory elements (Zhang et al., 2019).

Artificial intelligence (AI) and ML algorithms are set to revolutionize miRNA target prediction and functional annotation (Zhao et al., 2019). As these technologies advance, we can expect more accurate and context-specific predictions of MTIs. Future miRNA databases will likely incorporate AI-driven tools for predicting miRNA functions, expression patterns, and potential applications in crop improvement, making them increasingly valuable resources for researchers and breeders (Budak and Akpinar, 2015).

The integration of miRNA data with other omics datasets is another area poised for significant development (Subramanian et al., 2020). Future miRNA databases are likely to offer seamless integration with transcriptomics, proteomics, metabolomics, and epigenomics data, providing a comprehensive view of gene regulatory networks in plants. This multi-omics approach will enable more sophisticated modeling of plant biological processes and aid in the design of targeted crop improvement strategies (Sun et al., 2022).

Nanopore sequencing technology holds great promise for miRNA research, offering the potential for real-time, long-read sequencing of miRNA precursors and their genomic contexts (Wang et al., 2019). As this technology matures, miRNA databases may need to adapt to accommodate longer sequencing reads and provide tools for analyzing the structural features of miRNA precursors in greater detail.

The application of synthetic biology principles to miRNA research is an exciting frontier (de Lange et al., 2018). Researchers are beginning to design artificial miRNAs and regulatory circuits, opening new possibilities for precise control of gene expression in plants. Future miRNA databases may include modules for designing synthetic miRNAs and predicting their efficacy, supporting this emerging field of miRNA engineering (Liu and Stewart, 2015).

As climate change continues to pose challenges to agriculture, there will be an increased focus on studying miRNAs involved in plant adaptation to extreme environments (Shriram et al., 2016). miRNA databases are likely to expand their coverage of stress-responsive miRNAs and offer specialized tools for analyzing miRNA-mediated stress adaptation mechanisms across diverse plant species.

The growing interest in plant–microbe interactions and the plant microbiome may lead to the development of specialized databases focusing on miRNAs involved in these interactions (Couzigou et al., 2016). These resources could provide insights into how plants regulate their microbiome through miRNA-mediated mechanisms and how beneficial microbes might influence plant miRNA expression.

Lastly, we can anticipate a trend toward more user-friendly, interactive interfaces for miRNA databases (Kozomara et al., 2019). Future databases may incorporate virtual reality or augmented reality features for data visualization, making complex miRNA regulatory networks more intuitive to explore and analyze (Johri et al., 2022).

In conclusion, the future of plant miRNA research and databases is bright with possibilities. As our understanding of miRNA biology deepens and technologies advance, these databases will continue to evolve, offering ever more powerful tools for unlocking the potential of miRNAs in plant science and agriculture (Zhang et al., 2006; Verma et al., 2023). The integration of cutting-edge technologies, from AI to synthetic biology, promises to transform miRNA databases into dynamic platforms that not only store data but also drive discovery and innovation in crop improvement and plant biotechnology. As we face global challenges such as food security and climate change, these enhanced miRNA resources will play a crucial role in developing sustainable solutions and shaping the future of agriculture (Jones et al., 2006).

REFERENCES

Akter, J., Khan, M. G., Hoque, T. S., Islam, S. M., and Haque, M. M. "Genome-wide identification of drought-responsive microRNAs in wheat." *BioMed Research International* 2014 (2014), art. 965164, doi: 10.1155/2014/965164

Anil, P., Dharmendra, P., Sachidanand, S., Yuriy, E., Snezhan, B., Dilfuza, J., and Chinmayi, J. "Beneficial effects of biochar application on mitigating the drought and salinity stress implications on plants." *Plant Science Today* 10, no. 2 (2023): 188–193. http://doi.org/10.14719/pst.2591.

Azahar, A., Sobahan, M. A., and Islam, M. M. "A comprehensive overview of plant microRNA databases and computational tools." *International Journal of Molecular Sciences* 21, no. 23 (2020), art. 8392, doi: 10.3390/ijms21238392

Backman, T. W., Sullivan, C. M., and Brown, R. H. "The Arabidopsis Small RNA Project (ASRP): a database for over 120 000 small RNAs and associated genes." *Nucleic Acids Research* 36, no. S1 (2008): D982–D989, doi: 10.1093/nar/gkm905

Basso, M. F., Ferreira, P. C. G., Kobayashi, A. K., and Grossi-de-Sá, M. F. "MicroRNAs and new biotechnological tools for their modulation and the improvement of stress tolerance in plants." *Plant Biotechnology Journal* 17, no. 8 (2019): 1777–1791, doi:10.1111/pbi.13116

Bari, R., et al. "PHO2, microRNA399, and PHR1 define a phosphate-signaling pathway in plants." *Plant Physiology* 141, no. 3 (2006): 988–999.

Bertolini, E., et al. "PmiRExAt: the plant miRNA expression atlas." *Nucleic Acids Research* 46, no. D1 (2018): D1186–D1193.

Biswas, M. K., Islam, M. S., Adhikari, B. and Bairagi, S. D. "PmiRNASSRdb: a database of simple sequence repeats in plant microRNAs." *Agronomy* 11, no. 7 (2021): 1291, doi: 10.3390/agronomy11071291

Bonnet, E., Wuyts, J., Rouzé, P. and Van de Peer, Y. "Detection of 91 potential conserved plant microRNAs in Arabidopsis thaliana and Oryza sativa identifies important target genes." *Proceedings of the National Academy of Sciences USA* 101, no. 31 (2004): 11511–11516, doi: 10.1073/pnas.0404025101

Bresso, E. G., Chitarra, W., Sicard, A., Gambino, G. and Grimplet, J. "miSolRNA: a comprehensive resource of stress-responsive microRNAs in tomato." *BMC Plant Biology* 18 (2018): 167, doi: 10.1186/s12870-018-1379-2

Budak, H., and Akpinar, B. A. "Plant miRNAs: biogenesis, organization and origins." *Functional & Integrative Genomics* 15, no. 5 (2015): 523–531.

Chitarra, W., Perrone, I., Boccacci, P., Pagliarani, C., Riccio, A. and Gambino, G. "miRVIT: a grapevine microRNA expression atlas and functional database." *Frontiers in Plant Science* 9 (2018): 992, doi: 10.3389/fpls.2018.00992

Chu, Z., et al. "Identification and comparative analysis of microRNA in wheat (*Triticum aestivum* L.) callus derived from mature and immature embryos during in vitro culture." *Frontiers in Plant Science* 7 (2016): 1302. https://doi.org/10.3389/fpls.2016.01302.

Chung, W. J., et al. "mirtronDB: a comprehensive database for mirtron-derived microRNAs." *Nucleic Acids Research* 47, no. D1 (2019): D175–D181.

Couzigou, J. M., et al. "microRNAs: small regulators with a big impact on symbiotic associations." *Molecular-Microbial Ecology of the Rhizosphere* 16, no. 4 (2016): 1021–1037.

Cuperus, J. T., Fahlgren, N., and Carrington, J. C. "Evolution and functional diversification of miRNA genes." *The Plant Cell* 23, no. 2 (2011): 431–442. https://doi.org/10.1105/tpc.110.082784.

Cui, J., et al. "sRNAanno—a database repository of uniformly annotated small RNAs in plants." *Horticulture Research* 9 (2022): uhab043.

Dai, X., Zhang, Z., and Zhao, P. X. "PmiREN: a comprehensive encyclopedia of plant miRNAs." *Nucleic Acids Research* 46, no. D1 (2018): D1181–D1187.

Debernardi, J. M., et al. "microRNA172 plays a crucial role in wheat spike morphogenesis and grain threshability." *Nature Plants* 3, no. 10 (2017): 17024.

de Lange, O., et al. "Programmable DNA-binding proteins from Burkholderia provide a fresh perspective on the TALE-like repeat domain." *Nucleic Acids Research* 46, no. 10 (2018): 5041–5050.

Djami-Tchatchou, A.T., Sanan-Mishra, N., Ntushelo, K. and Dubery, I. A. "Functional roles of microRNAs in agronomically important plants—Potential as targets for crop improvement and protection." *Frontiers in Plant Science* 8 (2017): 378, doi:10.3389/fpls.2017.00378

Dong, Q., Hu, B., and Zhang, C. "microRNAs and their roles in plant development." *Frontiers in Plant Science* 13 (2022), art. 824240, doi:10.3389/fpls.2022.824240

Fan, C., Chen, C., Fang, J., Xu, C., Yin, X. and Liu, M. "PmiREN: a comprehensive encyclopedia of plant microRNAs." *Nucleic Acids Research* 48 (2020), suppl. 1: D1114–D1121, doi: 10.1093/nar/gkz806

Fang, X., Cui, Y., Li, Y. and Qi, Y. "Transcription and processing of primary microRNAs are coupled by the Elongator complex in *Arabidopsis*." *Nature Plants* 1 (2015): 15075, doi:10.1038/nplants.2015.75

Fu, C., et al. "Overexpression of miR156 in switchgrass (*Panicum virgatum* L.) results in various morphological alterations and leads to improved biomass production." *Plant Biotechnology Journal* 10, no. 4 (2012): 443–452.

Fu, R., Zhang, M., Zhao, Y., He, X., Ding, C., Wang, S., Feng, Y., Song, X., Li, P. and Wang B. "Identification of salt-tolerance-related microRNAs and their targets in maize (Zea mays L.) using high-throughput sequencing and degradome analysis." *Frontiers in Plant Science* 8 (2017): 864, doi:10.3389/fpls.2017.00864

Fujita, M., and Matsui, M. "mirEX 2.0: an integrated expression atlas for plant microRNAs." *Plant and Cell Physiology* 55, no. 1 (2014): e2.

Griffiths-Jones, S., et al. "miRBase: tools for microRNA genomics." *Nucleic Acids Research* 36, no. suppl_1 (2008): D154–D158.

Gu, M., et al. "Expression analysis suggests potential roles of microRNAs for phosphate and arbuscular mycorrhizal signaling in *Solanum lycopersicum*." *Physiologia Plantarum* 138, no. 2 (2010): 226–237.

Guan, Q., Lu, X., Zeng, H., Zhang, Y. and Zhu, J. "Heat-inducible expression of MIR398 enhances antioxidant capacity and drought tolerance in Arabidopsis thaliana." *Plant Physiology* 162, no. 3 (2013): 1259 – 1273, doi: 10.1104/pp.113.217299

Guo, Z., et al. "PmiREN2.0: plant microRNA encyclopedia version 2.0." *Nucleic Acids Research* 50, no. D1 (2022): D1388–D1393.

Guo, Z., Kuang, Z., Wang, Y., Zhao, Y., Tao, Y., Cheng, J., Liu, M. and Liu, X. "PmiREN: a comprehensive encyclopedia of plant microRNAs." *Nucleic Acids Research* 48, no. S1 (2020): D1114–D1121, doi: 10.1093/nar/gkz806

Gurjar, S., Rane, T., Pokle, A., Sharma, R. and Sharma, A. K. "PmiRExAt: plant microRNA expression atlas." *Database (Oxford)* 2016 (2016): baw030, doi: 10.1093/database/baw030

Gustafson, A. M., et al. "ASRP: the arabidopsis small RNA project database." *Nucleic Acids Research* 33, no. suppl_1 (2005): D637–D640.

Huang, H.-Y., et al. "miRTarBase 2020: updates to the experimentally validated microRNA–target interaction database." *Nucleic Acids Research* 48, no. D1 (2020): D148–D154.

Jabbar, A., You, Q., Xu, S., Liu, Y., Li, L., Pan, Y., Luo, M., Jiang, L. and Zheng, Y. "P-TarPmiR: a plant microRNA target-prediction tool based on cross-species validated interactions." *Plant Physiology* 190, no. 2 (2022): 1125–1139, doi: 10.1093/plphys/kiac123

Jabborova, D., Nigora, Z., Yuriy, E., Abdujalil, N., Anand, D., Khurshid, S., Zafarjon, J., Sachidanand, S., and Rahul, D. Growth of spinach as influenced by biochar and *Bacillus endophyticus* IGPEB 33 in drought condition. *Pakistan Journal of Botany* 55, no. SI (2023): 1–7. https://doi.org/10.30848/PJB2023-SI(6).

Jiang, N., et al. "Overexpression of microRNA169 confers enhanced drought tolerance to tomato." *Biotechnology Letters* 35, no. 8 (2013): 1231–1238.

Johnson, N. R., et al. "PSRG: the plant small RNA genes database." *Nucleic Acids Research* 44, no. D1 (2016): D1114–D1119.

Johri, P., Rajput, M. S., Trivedi, M., and Singh, A. Artificial intelligence and deep learning: current impact and perspectives in life sciences and medicine. *Advances in Bioresearch* 13, no. 1 (2022): 184196. https://doi.org/10.15515/abr.0976-4585.13.1.184196.

Jones, R. M. W., Bartel, D. P., and Bartel, B. "microRNAs and their regulatory roles in plants." *Annual Review of Plant Biology* 57 (2006): 19–53. https://doi.org/10.1146/annurev.arplant.57.032905.105218.

Kalvari, I., et al. "Rfam 13.0: shifting to a genome-centric resource for non-coding RNA families." *Nucleic Acids Research* 46, no. D1 (2018a): D335–D342.

Kalvari, I., Argasinska, J., Quinones-Olvera, N., Nawrocki, E. P., Rivas, E., Eddy, S. R., Bateman, A., Finn, R. D. and Petrov, A. I. "Rfam 14: expanded coverage of metagenomic, viral and microRNA families." *Nucleic Acids Research* 49, suppl. 1 (2021): D192–D200, doi: 10.1093/nar/gkaa1047

Kalvari, I., Nawrocki, E. P., Argasinska, J., Quinones-Olvera, N., Finn, R. D., Bateman, A., and Petrov, A. I. "Non-coding RNA analysis using the Rfam database." *Current Protocols in Bioinformatics* 62, no. 1 (2018b): e51.

Kantar, M., et al. "Subgenomic analysis of microRNAs in polyploid wheat." *Functional & Integrative Genomics* 11, no. 4 (2011): 559–568.

Karagkouni, D., et al. "TarBase v8: a decade-long collection of experimentally supported miRNA–gene interactions." *Nucleic Acids Research* 46, no. D1 (2018): D239–D245.

Katiyar-Agarwal, S., and Jin, H. "Role of small RNAs in host-microbe interactions." *Annual Review of Phytopathology* 48 (2010): 225–246.

Khemka, N., Rajkumar, M., Garg, R., and Jain, M. "Genome-wide profiling of miRNAs during seed development reveals their functional relevance in seed size/weight determination in chickpea." *Plant Direct* 5, no. 3 (2021): e299.

Kozomara, A., Birgaoanu, M., and Griffiths-Jones, S. "miRBase: from microRNA sequences to function." *Nucleic Acids Research* 47, no. D1 (2019): D55–D62.

Kruszka, K., Pacak, A., Swida-Barteczka, A., Nuc, P., Alaba, S. and Karlowski, W. M. "mirEX: a platform for comparative exploration of plant microRNA expression data." *BMC Plant Biology* 10 (2010), art. 144, doi: 10.1186/1471-2229-10-144

Kumar, V., et al. "PVsiRNAdb: a comprehensive database of plant virus-derived small interfering RNAs." *BMC Genomics* 16, no. 1 (2015): 165.

Li, J.-H., et al. "starBase v2.0: decoding miRNA-ceRNA, miRNA-ncRNA and protein-RNA interaction networks from large-scale CLIP-Seq data." *Nucleic Acids Research* 42, no. D1 (2014): D92–D97.

Li, M. and Yu, B. "Recent advances in the regulation of plant miRNA biogenesis." *RNA Biology* 18 no. 12 (2021): 2087–2096, doi:10.1080/15476286.2021.1899491

Li, R., Li, D., Li, J. and Yang, X. "Plant microRNA databases and bioinformatic tools: advancements and perspectives." *Frontiers in Genetics* 12 (2021), art. 798587, doi: 10.3389/fgene.2021.798587

Li, Y., et al. "miRFANs: an integrated database for *Arabidopsis thaliana* microRNA function annotations." *Nucleic Acids Research* 40, no. D1 (2012): D195–D200.

Liu, S., et al. "Single-cell analysis of long non-coding RNAs in the developing human neocortex." *Genome Biology* 22, no. 1 (2021): 1–19.

Liu, W., and Stewart Jr, C. N. "Plant synthetic biology." *Trends in Plant Science* 20, no. 5 (2015): 309–317.

Llave, C., Xie, Z., Kasschau, K. D. and Carrington, J. C. "Cleavage of Scarecrow-like mRNA targets directed by a class of Arabidopsis microRNA." *Science* 297, no. 5589 (2002): 2053–2056, doi: 10.1126/science.1076305

Lu, S., Sun, Y., Yuan, J., Wang, X. and Liu, Z. "TarDB: a database for plant microRNA targets with cross-species evidence." *Nucleic Acids Research* 49, suppl. 1 (2021): D1588–D1595, doi: 10.1093/nar/gkaa1037

Manchanda, A., Li, Y., Ding, G., Ghosh, D., MacDiarmid, R. and Ipsaro, J. "MirPub: a database of microRNA–publication relationships." *Database (Oxford)* 2015 (2015): bav085, doi: 10.1093/database/bav085

Mallory, A. C., and Vaucheret, H. "Form, function, and regulation of ARGONAUTE proteins." *The Plant Cell* 22, no. 12 (2010): 3879–3889.

Mallory, A. C., and Vaucheret, H. "Functions of microRNAs and related small RNAs in plants." *Nature Genetics* 38, no. 6 (2006): S31–S36.

Meng, Y., Shao, C., Wang, H., Chen, M. and Yang, B. "PmiRKB: a plant microRNA knowledge base." *Nucleic Acids Research* 39, no. S1 (011): D1128–D1133, doi: 10.1093/nar/gkq917

Millar, A. A. "The function of miRNAs in plants." *Plants* 9, art. 198, doi:10.3390/plants9020198

Navarro, L., et al. "A plant miRNA contributes to antibacterial resistance by repressing auxin signaling." *Science* 312, no. 5772 (2006): 436–439.

Noopur, K., Pragati, K., Adarsh, K. S., Sachidanand, S., Tanvi, J., and Megha, B. (2025a). Role of non coding RNA in effector trigger immunity. In Dutta, R., Kapadia, C., Danish, S., and Singh, S. (Eds), *Non-Coding RNA in Plants Modulation and Stress Responses Plant Biology, Sustainability and Climate Change.* Academic Press, New York.

Noopur, K., Pragati, K., Megha, B., Arpan, M., Tanvi, J., Shrisendu, M., Adarsh, K. S., and Sachidanand, S. (2025b). Role of non-coding RNA in plant architecture. In Dutta, R., Kapadia, C., Danish, S., and Singh, S. (Eds), *Non-Coding RNA in Plants Modulation and Stress Responses Plant Biology, Sustainability and Climate Change*. Academic Press, New York.

Pantaleo, V., et al. "miRVIT: a microRNA database for grapevine (*Vitis vinifera*)." *BMC Genomics* 11, no. 1 (2010): 631.

Panchy, N., et al. "miSolRNA: a comprehensive database for microRNAs in *Solanum lycopersicum*." *Plant Physiology* 172, no. 1 (2016): 206–218.

Pooja, G., Megha, B., Arpan, M., Chintan, K., Rahul, D., Subhan, D., and Sachidanand, S. (2025). Introduction to long noncoding RNA. In Dutta, R., Kapadia, C., Danish, S., and Singh, S. (Eds), *Non-Coding RNA in Plants Modulation and Stress Responses Plant Biology, Sustainability and Climate Change*. Academic Press, New York.

Rahul, D., Chintan, K., Subhan, D., and Sachidanand, S. (2024). *Non-Coding RNA in Plants: Modulation and Stress Responses*. Academic Press, New York.

Samynathan, P., Venkidasamy, B., Shanmugam, S., Ramalingam, S. and Thiruvengadam M. "Functional role of microRNA in the regulation of biotic and abiotic stress in agronomic plants." *Frontiers in Genetics* 14 (2023): 1272446, doi:10.3389/fgene.2023.1272446

Šečić, E., Kogel, K.-H., and Ladera-Carmona, M. J. "Biotic stress-associated microRNA families in plants." *Journal of Plant Physiology* 263 (2021), art. 153451, doi:10.1016/j.jplph.2021.153451.

Sun, W., Li, J., Xie, F., Lei, Y., Tang, M. and Liu, J. "PMTED: a plant microRNA target expression database." *BMC Bioinformatics* 16 (2015): 321, doi: 10.1186/s12859-015-0735-3

Sun, Y., Fan, C., Liu, M. and Yang, J. "Integrating multi-omics datasets to decode plant microRNA regulatory networks." *Trends in Plant Science* 27, no. 11(2022): 1157–1173, doi: 10.1016/j.tplants.2022.06.006

Subhan, D., Uzma, Y., Niaz, A., Khurram, S., Rahul, D., and Sachidanand, S. (2025). Fertilizer-responsive non-coding RNA and their significance. In Dutta, R., Kapadia, C., Danish, S., and Singh, S. (Eds), *Non-Coding RNA in Plants Modulation and Stress Responses Plant Biology, Sustainability and Climate Change*. Academic Press, New York.

Sunkar, R., et al. "Functions of microRNAs in plant stress responses." *Trends in Plant Science* 17, no. 4 (2012): 196–203.

Shriram, V., et al. "microRNAs as potential targets for abiotic stress tolerance in plants." *Frontiers in Plant Science* 7 (2016): 817.

Subramanian, I., et al. "Multi-omics data integration, interpretation, and its application." *Bioinformatics and Biology Insights* 14 (2020): 1177932219899051.

Svensson, V., et al. "Exponential scaling of single-cell RNA-seq in the past decade." *Nature Protocols* 13, no. 4 (2018): 599–604.

Szcześniak, M. W., et al. "miRNEST database: an integrative approach in microRNA search and annotation." *Nucleic Acids Research* 40, no. D1 (2012): D198–D204.

Tanvi, J., Arpita, S., Praneta, Y., Palak, S., Sachidanand, S. (2025). Epigenetic modulation of plant developmental pathways by non-coding RNA. In Dutta, R., Kapadia, C., Danish, S., and Singh, S. (Eds), *Non-Coding RNA in Plants Modulation and Stress Responses Plant Biology, Sustainability and Climate Change*. Academic Press, New York.

Tong, Z., Cui, Q., Wang, J. and Zhou, Y. "TransmiR v2.0: an updated transcription factor–microRNA regulation database." *Nucleic Acids Research* 47, no. D1 (2019): D253–D258, doi: 10.1093/nar/gky1023

Verma, N., Prachi, S., Aashna, S., and Sachidanand, S. (2021). Physiological and molecular responses to drought, submergence and excessive watering in plants. In Husen, A. (Eds), *Harsh Environment and Plant Resilience*. Springer International Publishing, Cham.

Verma, A., Kumar, R., Shukla, S., Singh, S., Johri, P., and Singh, M.S. Genome wide association studies: an outlook. *Journal of Innovation in Applied Research* 6, no. 1 (2023): 56–64. https://doi.org/10.51323/JIAR.2023/6/1/56-62.

Wang, J., et al. "PASmiR: a literature-curated database for miRNA molecular regulation in plant response to abiotic stress." *BMC Plant Biology* 18, no. 1 (2018): 210.

Wang, Y., et al. "Nanopore sequencing technology, bioinformatics and applications." *Nature Biotechnology* 37, no. 11 (2019): 1155–1171.

Wang, H., Zhang, P., Chen, S. and Tong, Z. "High-throughput sequencing reveals novel microRNAs associated with heat stress in maize." *Plant Physiology and Biochemistry* 98 (2015): 19–29, doi: 10.1016/j.plaphy.2015.10.006

Wen, M., Shen, Y., Shi, S. and Tang, T. "mirtronDB: a comprehensive database of mirtron-derived small RNAs in plants and animals." *Nucleic Acids Research* 46, no. D1 (2018): D165–D170, doi: 10.1093/nar/gkx1047

Wolter, F., and Puchta, H. "The CRISPR/Cas revolution reaches the RNA world: Cas13, a new Swiss Army knife for plant biologists." *The Plant Journal* 94, no. 5 (2018): 767–775.

Xie, F., et al. "PlantMiR-SSR: a database for microRNA-associated simple sequence repeats in plants." *Nucleic Acids Research* 46, no. D1 (2018): D1190–D1195.

Xu, M. Y., et al. "High-throughput sequencing identifies small regulatory RNAs in *Medicago truncatula* involved in nitrogen fixation." *PLoS One* 6, no. 11 (2011): e27494. https://doi.org/10.1371/journal.pone.0027494.

Yang, C., et al. "Overexpression of microRNA319 impacts leaf morphogenesis and leads to enhanced cold tolerance in rice (*Oryza sativa* L.)." *Plant, Cell & Environment* 36, no. 12 (2013): 2207–2218.

Yang, J., et al. "PMTED: a plant miRNA target expression database." *BMC Genomics* 20, no. 1 (2019): 210.

You, B., Zheng, Y., Chen, G. and Yang, W. "PmiRKB: a plant microRNA knowledge base." *Database (Oxford)* 2017 (2017), art. baw101, doi: 10.1093/database/baw101

Yi, X., et al. "PNRD: a plant non-coding RNA database." *Nucleic Acids Research* 43, no. D1 (2015): D982–D989.

Zhang, B. "microRNA: a new target for improving plant tolerance to abiotic stress." *Journal of Experimental Botany* 66, no. 7 (2015): 1749–1761.

Zhang, B., et al. "Plant microRNA: a small regulatory molecule with big impact." *Developmental Biology* 289, no. 1 (2006): 3–16.

Zhang, F., et al. "Precision editing of the plant genome made easy: progress in CRISPR/Cas9 technology." *National Science Review* 6, no. 3 (2019): 421–437.

Zhang, X., et al. "Suppression of endogenous gene silencing by bidirectional cytoplasmic RNA decay in Arabidopsis." *Science* 360, no. 6389 (2018): 701–705.

Zhang, Z., et al. "PMRD: plant microRNA database." *Nucleic Acids Research* 38, no. suppl_1 (2010): D806–D813.

Zhang, P., Zhou, X., Fu, Y., Huang, Q., Liu, L., Cao, Y., Lin, W., Hu, G. and Wang, P. "PASmiR: a literature-curated database for plant abiotic stress-responsive microRNAs." *BMC Plant Biology* 13 (2013): 33, doi: 10.1186/1471-2229-13-33

Zhou, T., Li, S., Liu, M., Li, F. and Liu, D. "MepmiRDB: a medicinal-plant microRNA database." *Database (Oxford)* 2021 (2021), art. baab009, doi: 10.1093/database/baab009

Zhou, Y., Jin, J., Li, J., Xiao, X., Zhu, L. and Guo, L. "sRNAanno: an integrated database for small-RNA annotation in plants." *Nucleic Acids Research* 48, no. S1 (2020): D128–D136, doi: 10.1093/nar/gkz1046

Zhu, C., et al. "The interaction between miR398 and its targets is important for drought tolerance in rice." *The Plant Journal* 68, no. 3 (2011): 538–548.

Zhu, Q. H., and Helliwell, C. A. "Regulation of flowering time and floral patterning by miR172." *Journal of Experimental Botany* 62, no. 2 (2011): 487–495.

Zhao, Y., et al. "A review of new features for improving miRNA-target prediction." *Briefings in Bioinformatics* 20, no. 4 (2019): 1565–1583.

Zielezinski, A., Karlowski, W. M., Kozlowski, P. and Makalowska, I. "mirEX 2.0—an integrated environment for plant microRNA expression profiling." *Nucleic Acids Research* 43, no. D1 (2015): D135–D140, doi: 10.1093/nar/gku1106

Index

FSC
www.fsc.org
MIX
Papier | Fördert gute Waldnutzung
FSC® C083411

Kolibri 360 GmbH
Ferdinand-Jühlke-Straße 7
99095 Erfurt, Deutschland
produktsicherheit@kolibri360.de